CHRONOLOGY
OF SCIENCE

CHRONOLOGY OF SCIENCE

From Stonehenge to the Human Genome Project

Consulting Editor
Lisa Rosner

A B C 🜚 C L I O

Santa Barbara, California   Denver, Colorado   Oxford, England

Library of Congress Cataloging-in-Publication Data is available.

06 05 04 03 02 10 9 8 7 6 5 4 3 2 1

Published in the United Kingdom by
Helicon Publishing Ltd.
Clarendon House
Shoe Lane
OX1 2DP
England

Published in the United States of America by
ABC-CLIO, Inc.
130 Cremona Drive, P.O. Box 1911
Santa Barbara, California 93116-1911

This book is printed on acid-free paper.

Printed and bound in Spain

EDITORS

Editorial Director
Hilary McGlynn

Managing Editor
Elena Softley

Project Editors
Michael Lacewing
Heather Slade

Editorial Assistant
Ruth Collier

Text Editors
Gerard Delaney
Michael Elliot
Catherine Thompson

Technical Editor
Stuart Brown

Picture Editor
Sophie Evans

Production Director
John Normansell

Production Controller
Stacey Penny

Page Design
Ken Wilson

Illustrations
Lorraine Hodghton

Contents

Contributors

CONSULTING EDITOR
Lisa Rosner, Richard Stockton College of New Jersey

CONTRIBUTORS
David Applin, Board of Continuing Education, University of Cambridge
Joe Cain, University College London
Margaret W Carruthers, freelance science writer specializing in earth sciences
John Cartwright, Chester College of Higher Education affiliated to the University of Liverpool
John O E Clarke, freelance science writer specializing in mathematics
Celia Deane-Drummond, freelance writer on science and religion
Tim Furniss, space correspondent of *Flight International*
Robert Goldsmith, St Mary's College of Maryland
Peter Higgins, University of Essex
Bruce J Hunt, University of Texas
Keith B Hutton, University of Oxford
Gren Ireson, Loughborough University
Peter Lafferty, freelance science writer
M Susan Lindee, University of Pennsylvania
Paul Lucier, Dibner Institute for the History of Science and Technology, MIT
John O'Connor, University of St Andrews
Julian Rowe, freelance science writer
F W Taylor, University of Oxford
Phill Watts, University of Liverpool
Paul Wymer, freelance science writer

Acknowledgements

Some of the entries in this work were published in shorter form in the four-volume *Hutchinson Chronology of World History* (3rd edition, 1999). The publishers gratefully acknowledge the contribution to the present work of the authors and editors of the *Chronology of World History*: Vol I, *The Ancient and Medieval World*, H E L Mellersh; Vol II, *The Expanding World*, Neville Williams; Vol III, *The Changing World*, Neville Williams; Vol IV, *The Modern World*, Neville Williams and Philip Waller. The Editorial Board of *The Chronology of World History* comprises T C W Blanning, David Feldman, Roy Porter, Bruce Schulman, Alison Scott, and John Sutherland.

Preface

Welcome to *The Chronology of Science*, your detailed and comprehensive guide to the progress of science internationally, from earliest discoveries to the present day.

Material within the *Chronology* is organized into chapters covering five broad time periods, including Earliest Discoveries, The Medieval World, The Scientific Revolution, The Industrializing World, and The 20th Century. Each chapter has been subdivided into smaller time periods, and within those time periods by subject area. This allows you to identify quickly the subject you are interested in at any particular time in history. Each chapter has an introductory essay to give a concise overview of developments within each subject, whilst photographs and illustrations throughout help to illuminate the past.

As well as chronological events, this book contains topical essays, discussing notable ideas, experiments, trends, and issues in science, along with the people who helped to shape those developments.

Following the final chapter are supplementary sections, containing biographical information on selected scientists, suggestions for further reading, useful Web sites, and a glossary of terms. In addition, there are appendices with information on Nobel prizewinners, Fields Medal prizewinners, and the discoveries of chemical elements.

Introduction

According to tradition, the Greek word 'Eureka', meaning 'I have found it' was first used by the mathematician Archimedes, on discovering a method to determine the proportion of gold in the crown of the King of Syracuse. Ever since, the exclamation 'Eureka!' has been associated with new and exciting discoveries, and the purpose of *The Chronology of Science* is to document the many 'Eureka!' moments in the history of science from earliest discoveries until the present day.

Discoveries can occur in different ways. Some arise as solutions to practical problems, like the invention of the alloy bronze or of sewage pipes. Some come from deliberate innovation, like Galileo's decision to turn the telescope, used in shipping, into a precision instrument for observing the heavens. Some develop from scientists pursuing exciting new lines of research, like the many studies linking specific bacteria to individual diseases following the acceptance of the germ theory in the 1880s and 1890s. And some are pure accident, like Röntgen's discovery of X-rays. Discoveries can arise out of voyages of discovery, whether to the other side of the world or the other side of the Solar System, or they can occur close to home, by reinterpreting something so familiar other people take it for granted.

These chronologies are sub-divided into scientific disciplines: astronomy, biology, chemistry, earth sciences, ecology and environment, mathematics, and physics. The practice of dividing science and scientists into disciplines is relatively recent. The word 'science' only came to have the specific meaning 'the study of the natural world', from the 1830s. Before that period, science was more commonly known as 'natural philosophy', and scientists as 'natural philosophers', and it was not at all unusual for scientists to make discoveries in several fields. Even more recent work, such as research on vitamins or the structure of DNA, can straddle disciplines. In the interests of clarity, these chronologies divide up science by the discovery, not the discoverer: for example, Robert Boyle's experiments will be found under both chemistry and physics. And wherever possible, research which clearly leads up to later, highly significant research – a crescendo of 'Eureka!' – has been kept together under the same discipline.

Historical categories are also recent and subject to interpretation. Early scholars of the natural world did not think of themselves as ancient but as innovators. Natural philosophers during the Scientific Revolution drew on the work of their Greek and Roman predecessors at the same time as they transformed it. Discoveries during the period of the late 18th through the 19th centuries both reflected and contributed to the process of global industrialization. And the scientific disciplines that dominated fields of inquiry at the start of the 20th century continued to develop, interact, split into subdisciplines, and recombine to yield new insights at the start of the 21st.

That first 'Eureka!' has found a multitude of echoes. *The Chronology of Science* is their history.

Lisa Rosner
Richard Stockton College of New Jersey
March 2001

Earliest Discoveries
30,000 BC–AD 499

Activities such as healing, star-watching, and engineering have been practised in many societies since ancient times. Our knowledge of megalithic societies comes from burial mounds, archaeological sites, and stone circles. Later civilizations, such as the cultures around the Mediterranean Sea from Egypt (*c.* 3000 BC onwards) through the Roman Empire (6th century BC–AD 476), in China from the Shang dynasty (*c.* 1500–1066 BC) through early Imperial China (*c.* 200 BC–AD 200), and in Mesoamerica from the Olmec (1200–400 BC) and Maya (*c.* 300 BC–AD 925) through the Mexica (Aztec; *c.* 1110–1500) and Inca (flourished 1400–1500s), left more extensive records of their understanding of the natural world.

Astronomy

Although evidence suggests that the grouping of stars into constellations was done before a fairly accurate calendar was drawn up, the calendar was probably the first systematic and practical application of astronomy. In order to predict when the Nile would flood and surrounding lands would be fertile enough for crops to be planted, the Egyptians made observations of Sirius, the brightest star in the night sky. They discovered that the date when Sirius could first be seen in the dawn sky (the heliacal rising) enabled the date of the flooding to be calculated. This also enabled the length of the year to be calculated quite accurately; so by 2780 BC the Egyptians knew that the time between successive heliacal risings was about 365 days. More precise observations enabled them to show that the year was about 365¼ days long.

The prediction of phenomena was also being carried out by other ancient civilizations, such as the Chinese; the existence in Europe and elsewhere of megalithic sites such as Stonehenge in England (some dating back to almost 3000 BC), suggests that early civilizations were concerned with the cyclical nature of astronomical events.

As far as we can tell from existing evidence, the Greeks first deduced the Earth to be a sphere and attempted to measure its size. Star catalogues were drawn up, the most celebrated being that of Hipparchus (*c.* 190–*c.* 120 BC). The *Almagest*, by Ptolemy of Alexandria (*c.* AD 100–*c.* AD 170), provided a synthesis of Greek astronomical theory and centuries of celestial observation. It became the basic astronomical text for the Mediterranean and European worlds for the next 1,500 years. The Greeks regarded the Earth as the centre of the universe, although they discussed the hypothesis, attributed to Aristarchus of Samos (*c.* 320–240 BC), that the Earth moves around the Sun.

1

Biology

The anatomical knowledge of the prehistoric world was very limited. Cave paintings have shown that prehistoric peoples were aware of the heart and its location. Such anatomical observations may have been made when preparing animal carcasses.

The Greek philosopher Aristotle (384–322 BC) wrote several works that laid the foundations for biological study until the 1500s. These included treatises on comparative anatomy, taxonomy (classification), and embryology. Aristotle's approach to anatomy was functional: he believed that questions about structure and function always go together and that each biological part has its own special uses. Aristotle believed that animals could be placed on a vertical, hierarchical scale ('scale of being'), extending from humans down through quadrupeds, birds, snakes, and fishes to insects, molluscs, and sponges. One of Aristotle's pupils, Theophrastus (c. 372–287 BC), extended to botany many of the ideas that Aristotle had applied in zoology. In medicine, the writings attributed to Hippocrates of Cos (c. 460 BC) became the standard corpus of medical knowledge, studied and commented on until the 18th century.

After the classical period of Greek thought, the most important biomedical thinker was Greek physician Galen (c. 129–c. 199), who was a shrewd anatomist and the most brilliant experimental physiologist of antiquity.

Chemistry

Chemistry seems to have originated in Egypt and Mesopotamia over 5,000 years ago. Certainly by about 3000 BC the Egyptians had produced the copper–tin alloy bronze by heating the ores of copper and tin together, and this new material was soon common enough to be made into tools, ornaments, armour, and weapons.

By 600 BC the Greeks began to turn their attention to the nature of the universe and to the structure of its materials. Aristotle proposed that there were four elements – earth, air, fire, and water – and that everything was a combination of these four. The idea of the four elements persisted for 2,000 years. The Greek philospher Thales (c. 624–c. 547 BC) promoted the idea of water as the primary ingredient of material nature. His follower, Anaximander (c. 610–c. 546 BC), believed all things came from the Apeiron, the indefinite and finite, which generated the worlds that were composed of heath and cold. Certain Greeks also believed that matter ultimately consisted of small indivisible particles, *atomos* – the origin of our word 'atom'.

From the Egyptians and the Greeks comes *khemeia*, alchemy and eventually chemistry as we know it today. In antiquity and the medieval world, one of its greatest aims was to transform base metals such as lead and copper into silver or gold. From the four-element theory, it seemed that it should be possible to perform any such change, if only the proper technique could be found.

Earth Sciences

An understanding of the Earth grew out of traditions of thought that took shape in the Middle East and the Eastern Mediterranean. Early civilizations needed to adapt to the seasons, to topography and geography, volcanoes, floods, and earthquakes. Yet inhabitants of Mesopotamia, the Nile Valley, and the Mediterranean littoral had experience of only a fraction of the Earth. Antiquity advanced a human-centred view of the Earth, designed purposely and specifically as a habitat for humans, a view held throughout much of human history.

Like many other Greek philosophers, Empedocles (*c.* 500–*c.* 430 BC) was concerned with change and stability, order and disorder, unity and plurality. He believed that in the beginning, the Earth had brought forth living structures more or less at random. Some had died out. The survivors became the progenitors of modern species.

Aristotle considered the world to be eternal and drew attention to natural processes continually changing its surface features. Earthquakes and volcanoes were due to the wind coursing about in underground caves. Rivers took their origin from rain. Fossils indicated that parts of the Earth had once been covered by water.

In the 2nd century AD, Ptolemy composed a geography that summed up the Ancients' learning. Ptolemy believed the equatorial zone was too torrid to support life, but he postulated an unknown land mass to the south, the *terra australis incognita*.

Ecology and Environment

The notion of ecology as a science was probably not developed by any of the great ancient civilizations, but knowledge of the interconnectedness of living systems was passed down from generation to generation.

Neolithic people probably began the change from nomadic, hunter-gatherer populations, to settled agricultural ones in the 7th century BC.

By the time of the Greeks, intensive agriculture was the norm, and Plato (*c.* 427–347 BC) is the first documented to carefully observe and decry the damage this was doing to the environment. Other Greek philosophers, notably Theophrastus, observed the relationships between plants and animals in *Historia Plantarum*, although this was written from a human-centred point of view.

Mathematics

Prehistoric humans probably learned to count. The ancient Egyptians (3rd millennium BC), Sumerians (2000–1500 BC), and Chinese (1500 BC) had systems for writing down numbers and could perform calculations using various types of abacus. They used some fractions. Mathematicians in ancient Egypt could solve simple problems, which involved finding a quantity that satisfied a given linear relationship. Sumerian mathematicians knew how to solve problems that involved quadratic equations. What is now known as Pythagoras' theorem was known in various forms in these cultures and also in Vedic India (1500 BC).

The first theoretical mathematician is held to be Thales who is reputed to have proposed the first theorems in plane geometry. His disciple Pythagoras (lived *c.* 530 BC) established geometry as a recognized science among the Greeks. Pythagoras began to insist that mathematical statements must be proved using a logical chain of reasoning starting from acceptable assumptions. The use of logical reasoning, the methods of which were summarized by Aristotle, enabled Greek mathematicians to make general statements instead of merely solving individual problems.

One of the most lasting achievements of Greek mathematics was the *Elements* by Euclid (*c.* 330–*c.* 260 BC). This is a complete treatise on geometry in which the

3

entire subject is logically deduced from a handful of simple assumptions. The ancient Greeks lacked a convenient notation for numbers and nearly always relied on expressing problems geometrically.

Physics

One of the earliest discoveries in physics, apart from observations of effects such as magnetism, was the relation between musical notes and the lengths of vibrating strings. The Greek mathematician Pythagoras found that harmonious sounds were given by strings whose lengths were in simple numerical ratios, such as 2:1, 3:2, and 4:3. From this discovery the belief grew that all explanations could be found in terms of numbers. This was developed by Plato into a conviction that the cause underlying any effect could be expressed in mathematical form.

Aristotle's writings on physics, like those on biology, provided the conceptual framework and much of the concrete information for the study of the nature of force and motion for the next 1,500 years. Also significant were the achievements of the Greek mathematician Archimedes (*c.* 287–212 BC), who deduced the principles of the lever and of buoyancy.

Earliest Discoveries
30,000 BC–AD 499

30,000 BC–1001 BC

Astronomy

c. 2800 BC The Neolithic monument Stonehenge is built in England near Salisbury, Wiltshire, comprising a circular earthwork 97.5 m/320 ft in diameter with 56 small pits around the circumference (later known as the Aubrey holes). The position of the 'heel stone' outside the circle suggests a connection with Sun worship and observation. It is probably an astronomical observatory with religious functions; the motions of the Sun and Moon are followed with the aid of carefully aligned rocks.

c. 2700 BC A lunar calendar is developed in Mesopotamia in which new months begin at each new Moon. A year is 354 days long and the calendar is used primarily for administrative purposes.

c. 2100 BC An intercalated month is added to the Sumerian calendar to bring the lunar calendar in line with the solar year.

Debate is ongoing as to the true purpose of Stonehenge, but one thing is certain – it could never have been accomplished without the invention of the lever, a simple tool that made seemingly superhuman feats possible.
Corel

5

During a total solar eclipse the Sun's corona or atmosphere becomes visible. *National Aeronautical Space Agency*

1500 BC Chinese astronomers make the earliest record of an appearance of a comet.

They describe comets as 'hairy stars' or 'broom stars' and consider them to be consistently evil omens. There is also some evidence that Egyptian astronomers record the apparition of a comet at around the same time, in the reign of Thutmosis III (1504–1450 BC). This may well be the same as that recorded by the Chinese, and could be the bright, long-period comet we now know as Halley, although the dates from this period are not known with sufficient precision to confirm either possibility.

1361 BC Chinese astronomers make the first recording of an eclipse of the Moon.

Their records show that they believe the Moon is being devoured by a dragon; in early Chinese the word for 'eclipse' is the same as the word for 'eat'. Their oracles interpret eclipses as omens, both good and bad.

1302 BC The first recording of an eclipse of the Sun is made by Chinese astronomers.

The records, inscribed on tortoiseshell, are used by US scientists more than 3,000 years later to obtain the length of the day at the earlier time. The inscription says that on the '52nd day, fog until next dawn. Three flames ate the Sun, and big stars were seen', apparently a description of a total eclipse of the Sun. The calculated dates and paths of total solar and lunar eclipses visible in Shang China fix the date as 5 June 1302 BC.

***c.* 1300 BC** The Egyptians have identified 43 constellations and are familiar with those planets visible to the naked eye: Mercury, Venus, Mars, Jupiter, and Saturn.

The Egyptians also begin to devise a sophisticated calendar of 365 days, with twelve months of 30 days each and five additional days at the end of the year.

1300 BC The Shang dynasty in China establishes the solar year at 365¼ days. The calendar consists of 12 months of 30 days each, with intercalary months added to adjust the lunar year to the solar.

Biology

c. **15,000 BC** Cave paintings made over the next 5,000 years at Lascaux in France give an impression of our ancestors' observations of the living world. The images include paintings of wild ox, deer, and other animals.

c. **2650 BC** Egyptian healer Imhotep is believed to have studied causes of diseases from natural phenomena.

c. **2000 BC** Papyri documents discovered in the Nile valley contain information on the treatment of wounds and diseases.

c. **1750 BC** Hammurabi, king of Babylon, has laws relating to medical practice inscribed on a stone pillar. They detail regulations for fees, and contain harsh penalties for malpractice, such as the amputation of hands for killing a patient.

c. **1500 BC** The Chinese breed silkworms for the production of fine cloth. Farmers hang bags of ants on citrus trees to protect the fruit from insects that destroy the crop – the first recorded use of biological control.

Chemistry

c. **4000 BC** The first metal to be recovered from an ore is probably copper. Around this time stone is the material used for tools and weapons, as metals are rarely found in their pure state.

    The discovery that metals could be recovered from certain rocks called ores is probably an accident caused by building a hot fire on ground containing copper ore. The reaction between the carbon in the wood and the ore at high temperature could produce pure metallic copper. This discovery begins the science of metallurgy, the recovery of metals from their ores.

c. **3400 BC** The first important metallic alloy, bronze, is discovered. Alloying copper with tin makes bronze.

    Pure copper is too soft a metal to replace stone as the principal material for weapons and tools. However, bronze is not only a much harder material than copper, but is also much easier to melt and can therefore be cast more readily into the shapes required for tools. Around this time, bronze is in widespread use in Sumeria; however, the process for making bronze has not been perfected and it will be some time before it totally replaces stone.

c. **3000 BC** The widespread use of bronze to replace stone tools and weapons heralds the beginning of the Bronze Age in Egypt and western Asia.

    Advances in the understanding of the composition of bronze result in alloys that have significantly improved properties compared with early bronze. The increased hardness of the alloy allows its use in the fabrication of tools, farming implements, and weapons of war. By the time of *The Iliad* by Homer (8th century BC), the Greeks are waging war in bronze chariots, wearing bronze armour, and wielding bronze weapons.

c. **1800 BC** Fermentation – the process by which sugar and starch are converted to alcohol and carbon dioxide by yeast – is perfected in Egypt.

    It is known that fruit juice becomes alcoholic if left standing for a period of time and that moistened dough sometimes rises during baking to produce leavened bread. These

7

are results of fermentation caused by contamination by airborne yeast. The Egyptians are the first to discover the link between yeast and fermentation and are the first to devise a controlled process for the production of leavened bread and beer.

c. 1500 BC The Hittites of Asia Minor are the first to discover how to obtain iron from iron ore. Although iron is the second most abundant metal in the Earth's crust, metallic iron is only known from the rare discovery of iron-rich meteorites.

The use of wood fires to separate metals from their ores did not provide enough energy to separate iron. The Hittites discover that burning charcoal will provide a higher temperature than wood and that this is sufficient to convert iron ore to metallic iron.

c. 1200 BC The first permanent natural red-purple dye, Tyrian purple, is developed in the city of Tyre in Phoenicia. Unlike other natural dyes of the period, Tyrian purple does not bleach in sunlight or fade after washing.

The dye is extracted from snails belonging to the genera *Murox* and *Purpura*, which live in the eastern Mediterranean. Since the amount of dye obtained from each snail is minute, the extraction process is very time consuming. The dye is so expensive that only the rich can afford to buy it and the wearing of clothes of this colour becomes associated with high status, such as the Roman emperors.

Earth Sciences

c. 3000 BC Flint mines are in operation in Belgium and England.

c. 2500 BC Copper mines are in operation in Yugoslavia.

c. 1800 BC Chinese historical records refer to earthquakes.

The astronomical office of the Chinese court keeps records of earthquakes so that they can be interpreted (like the stars) for court horoscopes.

c. 1500 BC Egyptian paintings show metal foundries.

c. 1150 BC The Turin Papyrus, the oldest map still in existence, is created. The map shows a mountain, now identified as Wadi Hammamat in Egypt, where gold is being mined.

Ecology

c. 7000 BC The Neolithic or New Stone Age begins in southern Europe, Asia, North Africa, and South America. It is characterized by polished stone tools, settlement in permanent villages, a more complex social structure, and the domestication of plants and animals.

People begin to exploit a few species of plants that they sow and harvest. Species of animals such as sheep and goats that do not directly compete with humans for food are herded and semi-domesticated and used to provide food for humans. As a result of over-killing, wild gazelle become scarce. The hunting of food now only supplies a small part of the diet compared to the harvesting of crops and the killing of domesticated animals. The Neolithic Revolution is beginning.

Mathematics

c. 30,000 BC Palaeolithic peoples record tallies on bone in central Europe and France; one wolf bone has 55 cuts arranged in groups of five – the earliest counting system.

c. **3400 BC** The first symbols for numbers, simple straight lines, are used in Egypt.

c. **3000 BC** A decimal number system is in use in Egypt.

c. **3000 BC** The abacus, which uses rods and beads for making calculations, is developed in the Middle East and adopted throughout the Mediterranean. A form of the abacus is also used in China at this time.

c. **3000 BC** The Sumerians of Babylon develop a sexagesimal (based on 60) numbering system. Used for recording financial transactions, the order of the numbers determines their relative, or unit value (place-value), although no zero value is used. It continues to be used for mathematics and astronomy until the 17th century AD.

This number system is developed before the introduction of writing. The sexagesimal system has come down to us today in the use of seconds and minutes to measure time and angles (60 seconds = 1 minute, 60 minutes = 1 hour or one degree of arc).

c. **1900 BC** The Golenishev (or Moscow) papyrus is written. It documents Egyptian knowledge of geometry.

Multiplication tables appear for the first time in Mesopotamia.

c. **1750 BC** The Babylonians under Hammurabi use the sexagesimal system to solve linear and quadratic algebraic equations, compile tables of square and cube roots, and extend

Arabic	Egyptian	Ionic (Greek)	Hebrew	Chinese	Mayan	Babylonian	Roman
1	I	α	א	一	•	𒁹	I
2	II	β	ב	二	••	𒐖	II
3	III	γ	ג	三	•••	𒐗	III
4	II II	δ	ד	四	••••	𒐘	IV
5	III II	ε	ה	五	—	𒐙	V
6	III III	Γ	ו	六	÷	𒐚	VI
7	IIII III	ζ	ז	七	••	𒑆	VII
8	IIII IIII	η	ח	八	•••	𒑇	VIII
9	III III III	θ	ט	九	••••	𒑊	IX
10	∩	I	'	十	=	‹	X

The numbers one to ten as represented by ancient civilizations.

The Origins of Mathematics

by Peter Higgins

FIRST MATHEMATICS

It is often said that people first encountered mathematics when they began to count their livestock or tried to measure the size of a field. The two sides of the mathematical coin, discrete mathematics based on counting, and continuous mathematics that arises through measurement, still form the basis of modern mathematics.

INVENTING NAMES FOR NUMBERS

Counting and arithmetic did not come about easily for a variety of reasons. To count we just need tally marks; but in order to talk about counting, we need a name for each number we use. Every language combined names of small numbers as a way of expressing larger ones, as in the French word for 80, *quatre-vingt* (four twenties). The ancient Greeks used letters to represent numbers, so that α was 1 while κ stood for 20, and in that way would write κα for 21. They could equally have written ακ to convey the same meaning, one-and-twenty.

Roman numerals were based on ten with the basic symbols being I, X, C, and M for 1, 10, 100, and 1,000 respectively, although they also introduced V to stand for 5, L for 50, and D for 500. The symbols were generally written in descending order, so that

$$1{,}944 = \text{MDCCCCXXXXIIII}$$

Sometimes they made use of *position*: a smaller unit placed before a larger one indicated subtraction of the smaller from the larger – for instance 9 was written as IX instead of VIIII. So 1,944 = MCMXLIV. But this representation is not always as easy to understand or employ in arithmetic, which may be why the Romans did not always make use of it.

POSITIONAL SYSTEMS AND '0'

In our number system, unlike the Greek system, order matters. Take 21. Swapping the places of the numerals 2 and 1 gives 12, a different number, for the 1 now represents 1 ten, while the 2 means two units.

No ancient European society devised a complete positional numbering system in which the meaning of a numeral depends on its position within the number and full use is made of a zero symbol.

There were nonetheless many practical and sophisticated counting systems in the ancient world. Commercial and trading societies often constructed good systems of arithmetic and the peoples of ancient Mesopotamia did have a sexagesimal positional system, one based on 60, over 4,000 years ago. Numbers exceeding 60 were written according to the positional principle, while combinations of the symbols from one to ten were used to make the basic numbers of their system, which ran from 1 to 59. For instance, the ancient clay tablets reveal examples like:

$$524{,}551 = 2 \times 60^3 + 25 \times 60^2 + 42 \times 60 + 31$$

The number system we use in the West today was developed by Indian mathematicians in the 7th century. The invention of zero as a place marker and the use of a positional system based on multiples of powers of ten was passed to Europe via medieval Arabic scholars seven centuries later.

EARLY COMPUTING

Throughout Asia and Europe, arithmetic was carried out on the calculator of the ancient world, the abacus (Greek 'sand tray'). The main obstacle to written arithmetic was lack of cheap writing materials. The first example of long division is by Calandri in 1491. The decimal system of fractions did not firmly take root until after the French Revolution of 1789.

A typical abacus consisted of a wooden rectangular frame in which a series of parallel rods were housed. Along each were a number of identical beads. Cutting across the rods was a counting bar. One rod represented the units column and the beads on rods to its left each represented multiples of 10, 100, 1,000, and so forth. Beads above the counting bar counted for five while those below represented one. Addition using an abacus is easy, as we need only count the number of each bead type and carry over to higher units as the need arises, although to carry out subtractions may require borrowing from the next highest rod.

A great advantage of the pen and paper methods that emerged in the Renaissance is one of communication, for they allow working to be shown and checked. The scribes of the ancient Babylonian tablets left descriptions of numerous problems and their answers, but we would need to see the clerks of the ancient world in action on their counting frames to appreciate exactly how they did their sums.

GEOMETRY AND PARADOX

An early use of geometry and measurement arose in Egypt where the ancient Greek historian Herodotus tells us that the Nile's annual flood regularly washed away boundaries and landmarks, so that a system of accurate surveying was needed in order to reaffirm who owned what: indeed the Greek word 'geometry' means 'Earth measure'. The founder of geometry was the Greek philosopher and scientist Thales of Miletus, who is said to have impressed the Egyptians by measuring the height of the Great Pyramid of Cheops through use of shadows.

His successor was the Greek mathematician and philosopher Pythagoras of Samos, best known for his theorem that says that the square on the longest side (hypotenuse) of a right-angled triangle has an area equal to the sum of the area of the squares of the other two sides. Pythagoras's followers are also said to have discovered irrational numbers. If a right-angled triangle has two sides which are both 1 in length, the hypotenuse will be the square root of 2. The Pythagoreans proved this number was irrational, that is to say that it cannot be represented by any ordinary fraction a/b. Up to this point, it was taken as self-evident that, *in principle*, any constructed line could be measured *exactly* using a standard ruler, provided that we marked the ruler with a sufficiently fine scale. The Pythagoreans had proved this to be false.

This and some other paradoxes in classical mathematics were eventually resolved in the 4th century BC by the Greek mathematician and astronomer Eudoxus with his *Theory of Proportions*, an account of which is to be found in the *Elements*, the classical texts written by the Greek mathematician Euclid of Alexandria. Eudoxus introduced a theory that applied equally well to all lengths by making subtle use of inequalities to deal with equalities.

LATER ACHIEVEMENTS

Euclid's Alexandrian School remained the leading centre of thought during the later

classical period. Its greatest genius was the Greek mathematician Archimedes, who took both geometry and mechanics to new heights. He died in 212 BC, probably killed by the Roman invaders of his home city of Syracuse, which he brilliantly defended through the use of devastating war machines that capsized the vessels of the invaders. In his tract *The Method* Archimedes allied mathematics and physics by insisting on rigorous standards of proof while emphasizing the importance of physical intuition as a guide to the truth.

The Greek mathematician Apollonius of Perga, a younger contemporary of Archimedes, gave the definitive description of curves arising from cones that proved to be a major ingredient of the theory devised by English physicist and mathematician Isaac Newton nearly 2,000 years later to explain planetary orbits. The Greek mathematician and engineer Hero of Alexandria (lived AD 62) invented the first working steam engine, made a primitive thermometer, and proved the formula for the area of a triangle in terms of its three sides. Other outstanding figures were Greek geographer and mathematician Eratosthenes, who calculated the diameter of the Earth in 230 BC through the difference in the Sun's elevation at Syrene and Alexandria at the summer solstice, and Greek mathematician Diophantus, whose treatise on number theory inspired French mathematician Pierre de Fermat's last theorem in the 17th century, while *The Collection* of Greek mathematician, astronomer, and geographer Pappus was the final great intellectual work of classical times.

DECLINE AND FALL

Despite its success and staggering sophistication, Greek mathematics remained wedded to the geometric style of Euclid, in which even common algebraic facts about numbers were demonstrated in what appears to us a strange and unnatural fashion through areas of geometrical figures. Even the ancient Babylonians, one thousand years before the birth of Pythagoras, seemed more at home with algebra. Although they did not use symbols to stand for numbers as we would, they demonstrated how to solve quadratic equations (in which the unknown quantity x appears through its square, x^2) and compiled astonishingly extensive tables of number triples such as (3, 4, 5) and (4961, 6480, 8161) where the sum of the squares of the two smaller numbers equals the square of the larger. The precursors of

The Origins of Mathematics *continued*

modern algebraic methods are to be found in societies such as Mesopotamia, India, and China.

After the 4th century AD classical Greek mathematics entered terminal decline.

Mathematics was not revived until the dawn of the modern age when the need to solve problems in navigation and the physical world ushered in a fresh and progressive epoch.

The use of an abacus of beads or pebbles on a board or on strings has lasted from prehistoric times to the present day in countries such as Russia and China. The English word 'exchequer' refers to a checkered tablecloth on which accounts were kept with counters – a form of abacus – until the time of George II. *Oxford Museum of the History of Science*

Sumerian astronomical knowledge. They are also aware of the Pythagorean property of the right-angled triangle.

***c.* 1700 BC** The Rhind (or Ahmes) papyrus is written. It shows the position of Egyptian mathematics at the time; multiplication is done by repeated duplication, and fractions by successive halvings.

***c.* 1360 BC** A decimal number system (a system using numbers to the base 10) is introduced in China, but it has no zero.

1000 BC–701 BC

Astronomy

15 June 763 BC Assyrian archivists record an eclipse of the Sun. This is probably the event described in the Bible, in Amos 8:9.

Biology

***c.* 802 BC** Rose trees from Asia are introduced to Europe and cultivated there for the first time.

Chemistry

***c.* 1000 BC** Developments in the smelting of iron enable it to replace bronze as the principal metal used in Europe – this is the start of the Iron Age in Europe.

Some time in the first millennium BC it was discovered that iron could be made more durable by adding carbon to it to make steel, which was stronger and took a better cutting edge than bronze. *Corbis*

Although iron ore is significantly more abundant than copper ore, iron does not immediately replace bronze, as it is softer than the best quality bronze. It is discovered that under the correct conditions, charcoal will combine with iron to form the carbon–iron alloy, steel. This material is significantly stronger than iron. Steel weapons can cut through those made of bronze.

Earth Sciences

c. **1000 BC** The Chinese record finding 'dragon' teeth, now thought to be dinosaur fossils.

c. **800 BC** Greek poet Homer describes the Earth as a convex disc surrounded by ocean.

700 BC–401 BC

Astronomy

28 May 585 BC The first accurate prediction of an eclipse of the Sun is made by the Greek philosopher Thales of Miletus.

432 BC Athenian astronomer Meton discovers that the dates of the phases of the Moon are repeated exactly after a period of 19 years.

The mathematical reason is that 19 tropical (seasonal) years contain 6939.60 days while 235 synodic months contain 6939.69 days. This is almost equal to 20 eclipse years, 6932.4 days, with the result that a series of four or five eclipses can occur on the same dates 19 years apart. This is called the Metonic cycle, and is used to determine how months should be inserted into a lunar calendar so that the calendar year and the tropical year are kept in step.

Biology

c. 570 BC Greek philosopher Anaximander proposes that the first animals developed in water, then metamorphosed into land creatures.

c. 500 BC Heraclitos of Ephesos (modern Turkey) suggests that tension between opposing forces is essential to life. He believes fire to be the principal element.

During the 6th century the concept that earth, fire, air, and water are the basic components of all matter develops. Later the four humours of medical theory are derived from this concept.

c. 460 BC Over the next 90 years, the Greek physician Hippocrates lives and teaches on the Greek island of Cos.

Medicine increasingly adopts an empirical approach to diagnosis, treatment, and prognosis. Alcmaeon of Croton establishes the connection between the sense organs and the brain through dissection. Medical writings by Hippocrates almost certainly include the work of others centred at Cnidos on the Greek mainland. The medical school at Cnidos was established earlier than the school of Cos but its written records have not survived. The Hippocratic oath, long considered to define the ideal relationship between physician and patient, is taken by doctors today when they graduate from many medical schools.

Chemistry

c. 580 BC Greek philosopher Thales postulates that all matter is fundamentally one element, water.

Thales is concerned with what the universe is made from and he looks for an answer that does not depend on a supernatural explanation, such as the intervention of gods. He proposes that all matter is formed from water and that anything that does not appear to be water is either modified water or has originated in water. The concept that matter is composed of fundamental forms of matter is correct, even though Thales was wrong to think that it was water.

c. 445 BC Greek philosopher and scientist Empedocles distinguishes the 'four elements' – earth, fire, water, and air – which he claims all substances are made of, and which also explain the development of the universe by the forces of attraction and repulsion. The doctrine is embodied in Aristotle's works and influences Western thought until the 17th century AD.

Greek philosophers have been discussing what matter is composed of since Thales proposed that water is the primary element. Anaximenes proposes air and Heraclitus suggests fire. Empedocles is the first to suggest that matter is composed of more than one element and that it is the proportion of each of the four elements that defines what a substance is.

c. 435 BC Greek philosopher Leucippus is the first to state categorically that all events have a rational explanation that is not connected in any way to supernatural interventions. He may have been the first to suggest that when a substance is continually divided, a point will be reached when it can no longer be subdivided and will become indivisible.

This view expands upon the rational approach to the world proposed by Thales and is the first time that the theory of atomism has been stated. It is developed later by Leucippus's pupil Democritus.

***c*. 420 BC** Greek philosopher Democritus of Abdera develops Leucippus's atomic theory and states that space is a vacuum and that all things consist of eternal, invisible, and indivisible atoms. He also posits necessary laws by which they interact.

The word atom comes from the Greek word *atomos* meaning 'indivisible'. The view that matter is composed of atoms is a fruitful hypothesis that influences scholars for the next 2,000 years.

Earth Sciences

***c*. 560 BC** Greek philosopher Xenophanes suggests that fossil seashells are the result of a great flood that buried them in the mud.

***c*. 530 BC** Pythagoras of Samos proposes that the Earth might be spherical because it casts an arcuate shadow on the Moon during eclipses.

***c*. 515 BC** Greek philosopher Parmenides of Elea, Italy, claims the Earth is a sphere, and promulgates the idea of immutability of elements. Large fragments of his philosophical poem survive.

***c*. 450 BC** Greek historian Herodotus suggests that the Nile delta is caused by the deposition of mud carried by the Nile. He also states that earthquakes might cause large valleys and gorges on the Earth's surface and suggests that they may shape the landscape. He declares the Caspian Sea to be an inland sea and not part of the northern ocean, as most philosophers of the time believe.

His writings suggest that Herodotus believes the Earth to be very old.

Mathematics

575 BC Greek philosopher Thales tries to use Babylonian mathematical knowledge to solve practical problems such as the determination of the distance of ships from the shore.

Thales is also credited with introducing geometry to Greece from Egypt.

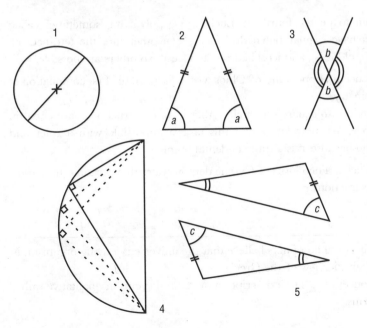

Some of the basic rules of geometry were first laid down by Thales: (1) a circle is bisected by its diameter; (2) in an isoceles triangle, the two angles opposite the equal sides are themselves equal; (3) when straight lines cross, opposite angles are equal; (4) the angle in a semicircle is a right angle; and (5) two triangles are congruent if they have two angles and one side identical.

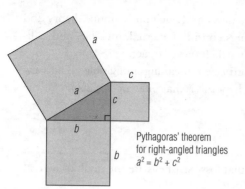

Pythagoras' theorem
for right-angled triangles
$a^2 = b^2 + c^2$

Pythagoras's theorem states that the area of a square drawn on the longest side of a right-angled triangle (the hypotenuse), will be equal to the sum of the areas of the squares drawn on the other two sides. The theorem is likely to have been known long before the time of Pythagoras. It was probably used by the ancient Egyptians to lay out the pyramids.

530 BC Pythagoras of Samos, scientist and philosopher, moves to Croton and starts researching and teaching theories of mathematics, geometry, music, and reincarnation. A mystic as well as a mathematician, he argues that the key to the universe lies in numbers, while preaching immortality and the transmigration of souls. He founds a brotherhood in Croton which is to remain influential for several generations. His work leads to a number of important results, including Pythagoras's theorem of right-angled triangles.

There is geometry in the humming of the strings.
There is music in the spacings of the spheres.

Pythagoras, quoted in Aristotle *Metaphysics*

c. 500 BC Astronomers in Mesopotamia, Greece, Alexandria, and India use the Babylonian sexagesimal mathematical system to predict the positions of the Sun, Moon, and planets.

c. 465 BC Pythagorean mathematician Hippasus mentions a 'sphere of 12 pentagons', which is taken to mean a dodecahedron (a solid figure with 12 faces that are all regular pentagons).

This will turn out to be the fourth of Euclid's five polyhedra (sometimes called Platonic solids) that have regular polyhedral faces. The others are the tetrahedron, cube, octahedron (eight faces), and icosahedron (20 faces). No others are possible.

c. 440 BC Greek mathematician Hippocrates of Chios writes *Elements*, the first compilation of the elements of geometry.

He fails in his attempt to square the circle – that is, to construct a square equal in area to a given circle. We now know that this task is impossible with a ruler and compass construction because π is a transcendental number.

c. 425 BC Greek mathematician Theodorus of Cyrene demonstrates that certain square roots cannot be written as fractions.

Physics

c. 500 BC In China, Confucian philosophers believe that the universe is a living organism, in contrast to the mechanistic view of the Greeks.

The Confucian concept leads to a description of natural events in qualitative rather than quantitative terms.

Astronomy

***c.* 366 BC** Greek mathematician and astronomer Eudoxus of Cnidus observes the heavens and constructs a model of 27 nested spheres to give the first systematic explanation of the motion of the Sun, Moon, and planets around the Earth.

Eudoxus is inspired by Plato's concept that uniform circular motion is fundamental to the cosmos, to propose a model where each planet is on a rotating sphere concentric with the Earth with its poles set in the adjacent sphere. By adjusting the positions of the axes and the rotations of the spheres, Eudoxus was able to mimic the main motions and retrogressions of the planets. The model predicts the motions of Jupiter and Saturn fairly well, but fails on the planets Mars and Venus. The model also fails to account for the observed differences in the lengths of the seasons, or for variations in the observed diameter of the Moon and in the brightness of planets, although Eudoxus makes it clear that he knows that this is because their distances are changing.

352 BC Chinese astronomers make the earliest known record of a 'visitor star', probably a supernova.

***c.* 350 BC** Aristotle defends the doctrine that the Earth is a sphere in *De caelo / Concerning the Heavens* and estimates its circumference to be about 400,000 stadia (one stadium varied from 154 m/505 ft to 215 m/705 ft). It is the first quantitative estimate of the circumference of the Earth.

Aristotle's argument is basically that only a spherical shape can result when pieces of matter have an innate tendency to move from every direction towards the same point, saying: 'the true statement that this takes place because it is the nature of whatever has weight to move towards the centre'. Observations of the eclipses of the Moon, where the Earth casts a round shadow, of the gradual disappearance of ships over the horizon, and of the different constellations visible from various latitudes, provide further evidence of the roundness of Earth.

***c.* 350 BC** Greek astronomer Heracleides is the first to suggest that the Earth rotates on its axis.

Heracleides also suggests that Mercury and Venus revolve around the Sun. Unfortunately, most of his writings are lost and only fragments remain, so the depth of his understanding is not known.

***c.* 300 BC** Babylonian astronomer Berosus invents the hemispherical sundial. It consists of a block of stone or wood with a hemispherical opening with arcs inscribed on the inner surface. Time is reckoned by the position of the shadow of a pointer, which is attached to the outer part of the hemisphere, as it crosses the arcs.

Eleven lines divide the day into twelve equal parts, and the pointer is positioned so that no shadow is cast at noon. Because the length of the day varies with season, the 'hours' which elapse between markings also vary and are known as temporal hours.

***c.* 280 BC** Greek astronomer Aristarchus of Samos writes *On the Sizes and Distances of the Sun and the Moon*. He uses a geometric method to calculate the distance of the Sun and Moon from the Earth.

Aristarchus was a student of Strato of Lampsacus, who was head of the Lyceum at Athens. He is credited, in histories written by Aristotle and Plutarch, with being the first to maintain that the Earth rotates and revolves around the Sun, although his book makes no mention of this theory. He estimates that the Sun must be at least 19 times further away from the Earth than the Moon. Modern calculations estimate the Sun to be 400 times further away. Aristarchus also invented a sundial in the shape of a hemispherical bowl with a pointer to cast shadows placed in the middle of the bowl.

240 BC Chinese astronomers make the first recorded observation that can definitely be associated with Halley's comet.

c. 200 BC Chinese astronomers are the first to recognize a relationship between tides and the phases of the Moon.

They begin to observe and record sunspots about this time, and keep continuous records from 28 BC to AD 1638.

c. 200 BC The Greeks invent the astrolabe. It is used for observing the altitudes of stars and performing astronomical calculations.

c. 150 BC Greek astronomer Hipparchus of Bithynia builds an observatory on the island of Rhodes, where he creates the first known star catalogue. It gives the latitude, longitude, and brightness of nearly 850 stars and is later referred to by Ptolemy.

127 BC Greek astronomer Hipparchus discovers the precession of the equinoxes and calculates the year to an accuracy of within 6.5 minutes of the modern value. He also makes an early formulation of trigonometry.

The astrolabe was an astronomical computer. Projections of the celestial sphere were engraved onto brass discs that were used to solve many astronomical problems. First invented by the Greeks, it remained more or less unchanged for centuries. This example is from the Middle East. *Oxford Museum of the History of Science*

Hipparchus wrote three books of which only one is known to have survived; regrettably this contains no mathematical astronomy and his work is only popularized by later commentaries. In his second and third books, Hipparchus gives an account of the rising and setting of the constellations. Towards the end of the third book he gives a list of bright stars always visible, and describes how the time at night can be accurately determined by observing them.

Biology

c. 320 BC Greek philosopher Theophrastus begins the science of botany with his books *De causis plantarum / The Causes of Plants* and *De historia plantarum / The History of Plants.*

In them he classifies over 500 plants, develops a scientific terminology for describing biological structures, distinguishes between the internal organs and external tissues of plants, and gives the first clear account of plant sexual reproduction, including how to pollinate the date palm by hand. He also describes the medicinal uses of plants.

c. 290 BC Aristotle's Lyceum inspires the organization of the great museum at Alexandria, Egypt. Collections of species of plants and animals are made for the museum, and a library is associated with it.

It is the largest and, despite its destruction in 414 AD, remains perhaps the most famous library in the world. Working here, Herophilus makes advances in anatomy by describing the digestive, nervous, and blood systems. His student Erasistratus continues the work in anatomy and investigates different aspects of physiology.

Chemistry

c. 400 BC Greek philosopher Plato first uses the term *stoicheia*, to describe the fundamental forms in which matter exists.

Although the concept of a fundamental form has been known for some time, Plato is the first assign the name stoicheia to describe the group of which the forms are part.

c. 350 BC Greek philosopher Aristotle summarizes and expands upon earlier ideas to propose that the world is made up from the four elements fire, air, earth, and water and that a fifth element, aether, composes the heavens.

Aristotle suggests that the Earth is formed from consecutive shells of the elements. The centre is composed of earth surrounded by a ball of water, itself surrounded by one of air and then one of fire. The heavens are separate from the Earth and composed of the incorruptible element aether. This theory is to dominate scientific thinking for the next 2,000 years.

Earth Sciences

c. 400 BC The first reference to a rain gauge is made in *The Science of Politics* by Chanakaya, the minister of the ruler of India from 321 to 296 BC.

19

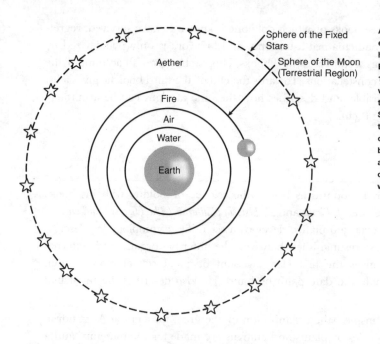

Sphere of the Fixed Stars

Sphere of the Moon (Terrestrial Region)

Aether

Fire

Air

Water

Earth

Aristotle's philosophy was embraced whole-heartedly by Christian Europe in the 12th and 13th centuries when his works were first translated into Latin. The Scientific Revolution of the 16th and 17th centuries, spearheaded by Copernicus, Galileo, and later Newton overthrew the Aristotelian world view.

c. 314 BC Greek philosopher Theophrastus writes *Peri lithon/On Stones*, a classification of 70 different minerals. It is the oldest known work on rocks and minerals and is the best treatise on the subject for nearly 2,000 years.

Peri lithon is essentially a catalogue of materials traded by Athenians in the 4th century BC. It includes descriptions of technology and manufacturing, as well as mineral names. Some of the mineral names (for example, alabaster, cinnabar, and amethyst) still survive today. Others he describes are now not considered to be minerals.

c. 300 BC Greek philosopher Theophrastus writes *On Weather Signs*, an early book on weather forecasting. He discusses 80 signs of rain, 50 signs of storms, 45 signs of wind, and 34 signs of fair weather.

c. 270 BC Cilician didactic poet Aratus of Soli writes *Phaenomena*, a poem about meteorological observations. It serves as the basis of ancient Greek knowledge of meteorology.

c. 250 BC Greek philosopher Aristotle writes of the weather and climate in *Meteorologica/Meteorology*.

Meteorologica consists of discussions of 'all phenomena that may be regarded as common to air and water, and the kinds and parts of the Earth,' including abundant observations of wind and weather, and a suggestion that earthquakes are produced by underground winds that are caused by the two fundamental forces of Hot and Cold. Its title becomes the root for the modern word 'meteorology'.

c. 250 BC In *Lü-shi Chun Qui/The Spring and Autumn Annals of Mr Lu*, Chinese scholars explicitly describe the hydrologic cycle. The idea has been known in China for a century.

c. 235 BC Greek scholar and philosopher Eratosthenes estimates the Earth's circumference to be 46,250 km/28,790 mi – about 15% greater than the modern value.

Eratosthenes makes his estimate by observing the difference in the angle of the Sun at midday in Alexandria, comparing it to that of Syene (Aswan), and then estimating the distance between the two places. Eratosthenes also makes a map coordinate system based on Sun angles.

Plato (depicted here on the left, with Aristotle) was one of the first to recognize the consequences of human activity on the natural world, observing the effect deforestation and overgrazing by sheep and goats had on the erosion of topsoil and the subsequent decline in the land's ability to sustain agriculture. *Corbis*

Ecology

c. **340 BC** Greek philosopher Plato laments the effects of soil erosion and deforestation on his native Greece in his work *Critias*.

Plato observes the effects of overgrazing and the felling of forests for human use. Much of Greece is mountainous with only a few areas suitable for agriculture. As population grows, so the trees are felled to provide wood for fuel, houses and ships, and to provide more agricultural land. The slopes are terraced to avoid soil erosion, but the topsoil is gradually washed away. Overgrazing by sheep and goats means that young trees and shrubs are unable to regenerate. What is left is a mixture of bare rock and hardy shrubs largely inedible to grazing animals.

Mathematics

c. **400 BC** Babylonian mathematicians use two wedge-shaped symbols to indicate an empty place. It is an early example of the use of zero, but is only used as a place marker.

c. **360 BC** Greek mathematician and astronomer Eudoxus of Cnidus develops the theory of proportion (dealing with irrational numbers), and the method of exhaustion (for calculating the area bounded by a curve) in mathematics.

Eudoxus' method also provides a way of finding the volumes of cones, pyramids, and spheres.

c. **300 BC** Alexandrian mathematician Euclid sets out the laws of geometry in his *Stoicheia / Elements*; it remains a standard text for 2,000 years. He also sets out the laws of reflection in *Catoptrics*.

There is no royal road to geometry.

Euclid, Greek mathematician, to Ptolemy I,
quoted in Proclus *Commentary on Euclid*, Prologue

Elements consists of 13 books. It deals with two-dimensional geometry, the theory of numbers, three-dimensional geometry, and the five regular polyhedra (tetrahedron, cube, octahedron, dodecahedron, and icosahedron). Euclid himself did not devise these laws; the book is a consolidation of previous mathematicians' work. From five postulates (unproven basic assumptions) Euclid logically expounds the entirety of contemporary geometrical knowledge. The fifth postulate – that given a line and a point not on the line, then it is possible to draw one and only one line through the point parallel to the line – is later questioned in the 19th century, leading to the development of non-Euclidean geometry.

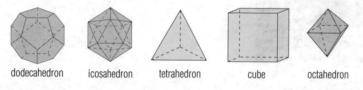

dodecahedron   icosahedron   tetrahedron   cube   octahedron

The five regular polyhedra or Platonic solids.

***c.* 250 BC** Greek mathematician and inventor Archimedes, in his *On the Sphere and the Cylinder*, provides the formulae for finding the volume of a sphere and a cylinder; in *Measurement of the Circle* he arrives at an approximation of the value of π; in *The Sand Reckoner* he creates a place-value system of notation for Greek mathematics; and in *Floating Bodies*, the first known work on hydrostatics, he discovers the principle that bears his name – that submerged bodies are acted upon by an upward or buoyant force equal to the weight of the fluid displaced.

Eureka! I have found it!

Archimedes, Greek mathematician,
Remark, quoted in Vitruvius Pollio *De Architectura* IX

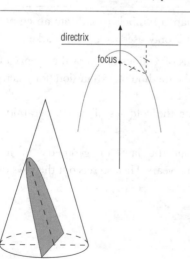

directrix

focus

The parabola is a curve produced when a cone is cut by a plane. It is one of the family of curves called conic sections that also includes the circle, ellipse, and hyperbola, developed by Apollonius of Perga.

c. **230 BC** Alexandrian mathematician Apollonius of Perga writes *Conics*, a systematic treatise on the principles of conics in which he introduces the terms 'parabola', 'ellipse', and 'hyperbola'.

These curves, the conic sections, can be obtained by cutting a right circular cone with a plane parallel to one side for a parabola, obliquely for an ellipse, and perpendicularly for a hyperbola.

c. **230 BC** Greek scholar Eratosthenes of Cyrene develops a method of finding all prime numbers. Known as the sieve of Eratosthenes it involves striking out the number 1 and every *n*th number following the number *n*. Only prime numbers then remain.

Eratosthenes is chief librarian at Alexandria, and the prefaces of some of Archimedes' texts are addressed to him.

Physics

c. **306 BC** Greek philosopher Epicurus supports the atomic theory of Democritus and Leucippus.

Democritus holds the view that all things are composed of indivisible particles called atoms, with heavier ones making the Earth and lighter ones the air. The work of Leucippus, sometimes called the founder of atomism, is incorporated into the writing of Democritus. The philosophical world view of Epicurus and Democritus is best known through the work of the Roman poet Lucretius in his work *De rerum natura / On the Nature of Things*.

100 BC–AD 199

Astronomy

c. **100 BC** Greek philosopher Poseidonius, member of the Stoic school, correlates tides with the lunar cycle.

80 BC Greek philosopher Poseidonius makes an estimate of the circumference of the Earth by measuring the distance from Rhodes to Alexandria and the angular height of the star Canopus above the horizon at each location. His estimate is only 11% larger than the modern value.

Poseidonius also makes estimates of the distance and magnitude of the Sun. The titles and subject matter of more than 20 books by Poseidonius on philosophy and observation of the natural world are known, but the works themselves are lost.

AD 128 Greek mathematician and astronomer Theon of Smyrna suggests a model of the universe in which Venus and Mercury orbit the Sun, which in turn orbits the Earth.

150 Egyptian astronomer Claudius Ptolemy publishes the work he calls 'The Mathematical Collection' but is later known as the *Almagest / The Greatest*. It is an astronomical encyclopedia in 13 volumes that is to be highly influential for over a millennium.

Ptolemy is building on the work of Hipparchus, cataloguing many additional stars (1,022 in all) and charting the motions of the Sun, Moon, and planets. Most importantly, Ptolemy uses years of observation to outline a theory of the universe in which the Earth is not only at the centre but immovable. Known as the Ptolemaic system, this model survived until the time of Copernicus in the 16th century AD.

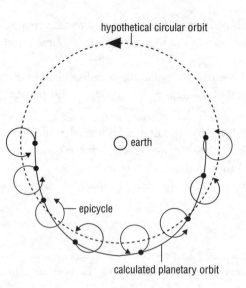

hypothetical circular orbit

○ earth

epicycle

calculated planetary orbit

Ptolemy developed the idea of epicycles to explain the observed departure of planetary orbits from perfect circles.

Biology

c. **100 BC** Greek physician Asclepiades of Bithynia is influenced by the work of Erasistratus and denies the doctrine of the four humours of Hippocrates. He teaches that nature cures disease, including mental illness.

95 BC Roman poet Lucretius writes that life on Earth is progressing and that the world began with the chance assembly of atoms.

Lucretius' work includes a theory of evolution that incorporates the idea of natural selection: some chance combinations of atoms are better suited to survive than other combinations, which are eliminated.

AD 129 Claudius Galen, celebrated Greek physician, is born in Pergamum, Asia Minor (now Burgama, Turkey); dies, possibly in Rome (aged around 70).

Galen becomes the greatest writer on medicine and the greatest experimentalist of his time. His work remains a major influence until the Renaissance. He dissects human cadavers (human dissection was not usually undertaken in his time), but much of his understanding of anatomy and physiology comes from dissecting animals. He also gains experience of surgery as chief physician to the gladiators of his home town, Pergamum. Later, in Rome, Galen is physician to a succession of four emperors. His humoral theory of health and disease derives from the earliest Greek writers and is the basis of a classification of all personalities: sanguine (cheerful), choleric (hot-tempered), phlegmatic (slow), and melancholic (depressed).

c. **150** *Physiologus/Naturalist*, a Greek work by an anonymous author, is written in Alexandria. All the medieval 'bestiaries' evolve from this work, which is an encyclopedia of real and imagined natural history.

Chemistry

c. **100 BC** Glassblowing is developed in Syria.

The discovery of how to make glass has been known for many centuries, the earliest examples being dated to around 4000 BC in Egypt. However, the process of making glass is such a difficult process that the applications of glass are restricted to expensive objects usually used for ceremonial purposes. It is discovered that if glass is attached to a hollow tube, it can be blown into a mould to shape it. This allows glass vessels to be produced both quicker and cheaper than before. Glass objects for everyday use become widespread throughout the Mediterranean.

Earth Sciences

c. **100 BC** The Greeks are the first known to measure and record wind direction. They install a wind vane on the Acropolis.

AD 62 Roman philosopher, playwright, and politician Lucius Annaeus Seneca writes *Quaestiones Naturales/Natural Questions*. Among other things, the book discusses volcanoes, earthquakes, and fish living in rocks. With this book, Seneca is given credit for transmitting the ideas of Aristotle through western Europe.

77 Pliny the Elder (Gaius Plinius Secundus) completes his *Historia naturalis/Natural History*. Divided into 37 books, it covers such topics as cosmology, astronomy, zoology, botany, agricultural techniques, medicine, drugs, minerals, and metals. The book brings together the scattered material of earlier writers and also makes original contributions, combining keen observation with hearsay.

Natural History is published in Venice in 1469 and translated into English in 1601. It is read throughout Europe in the 16th, 17th, and 18th centuries.

132 Chinese engineer and philosopher Zhang Heng develops the first known instrument for recording earthquakes.

It consists of a series of balls suspended in the mouths of eight carved dragons arranged around a central point. The balls that are dislodged during an earthquake indicate the direction of the earthquake's epicentre. The instrument is said to have been capable of registering earthquakes that cannot be felt by humans.

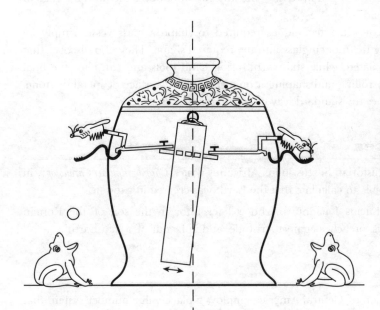

The invention of the seismograph for detecting earth tremors is credited to the Chinese. This ingenious example uses a series of balls delicately balanced in the mouths of eight carved dragons. At the slightest vibration a ball drops into the mouth of a frog poised to catch it below. The position of the dislodged ball indicates the direction of the earthquake.

Mathematics

c. 100 BC Chinese records show that their mathematicians are the first to begin using negative numbers in mathematics.

AD c. 1 Chinese mathematician Liu Hsin becomes the first to use decimal fractions in mathematics.

c. 60 Greek mathematician and engineer Hero of Alexandria writes *Metrica/Measurements*, containing many formulae for working out areas and volumes.

He also describes various hydraulic mechanisms and a simple reaction motor powered by steam.

100–150 The classical Chinese mathematics text *Jiuzhang Suanshu/Nine Chapters on the Mathematical Art* is assembled.

It contains a statement implying that the circumference of a circle is three times its diameter, equivalent to making π equal to 3. It also discusses algorithms for common fractions, proportional distribution, and applications of Pythagoras' theorem. It is arguably the most influential Chinese mathematical book and remains in use for many centuries.

200–499

Astronomy

6 June 346 During a total eclipse of the Sun, Chinese astronomers record that the stars become visible in daylight – the first time this phenomenon has been placed on record.

Chemistry

c. 350 Free glassblowing is developed in Syria – the innovation removes the need for glass to be blown into a mould.

Glassblowers discover the techniques required to shape a glass vessel simply by blowing and twisting the molten glass into the required shape. They also discover that glass can be manipulated, while still warm, using hand tools and can alter the final form of the glass by rolling and shaping it on a smooth flat surface of metal or stone. This method becomes the standard way of making glassware.

Earth Sciences

c. 221 Roman Christian historian Sextus Julius Africanus writes *Chronographiai/Chronology*, in which he uses the Bible to calculate that the Earth was created in 5499 BC.

300 Christian apologist Lucius Lactantius, who is later tutor to the son of the Roman emperor Constantius, preaches against Aristotle, and in favour of a 'flat Earth'.

Mathematics

c. 250 The Maya civilization of Central America employs a place-value number system that includes a version of zero as a place holder; it uses the base 20.

c. 250 Greek mathematician Diophantus of Alexandria writes *Arithmetica*, a study of problems in which only rational numbers are allowed as solutions.

Like Euclid's *Elements* it consists of 13 books and introduces a Greek form of algebra, involving linear and quadratic equations, in solving number theory problems.

263 By measuring polygons with up to 3,072 sides (which would approximate to a circle), the Chinese mathematician Liu Hui calculates the value of π as 3.14159, in agreement with the correct value to five decimal places.

340 Greek mathematician Pappus of Alexandria writes *Synagoge / Collections*, an invaluable guide to ancient mathematics and astronomy.

c. 400 Greek mathematician Hypatia writes commentaries on Diophantus and Apollonius. She is the first recorded female mathematician, although Pappus had implied that there were female mathematicians 50 years earlier.

450 Proclus, a mathematician and Neo-Platonist, is one of the last notable philosophers at Plato's Academy at Athens.

He is best known for his commentary on Book I of Euclid's *Elements*.

c. 460 Chinese mathematician and astronomer Tsu Ch'ung Chi gives the approximation 355/113 to π (the ratio of the circumference of a circle to its diameter), which is correct to 6 decimal places.

499 Indian astronomer and mathematician Aryabhata I produces his *Aryabhatiya*, a treatise on quadratic equations, the value of π, and other scientific problems, in which he adds tilted epicycles to the orbits of the planets to explain their movement.

Aryabhata calculates π as 3.1416, using a place-value decimal number system.

Physics

c. 400 Greek astronomer and mathematician Hypatia distinguishes herself as one of the first women scientists, becoming head of the Neo-Platonist school at Alexandria, and a widely consulted authority on matters of physics and mathematics.

Hypatia's activities lead to some resentment and it is believed that her death at the hands of a Christian riot is due to the Archbishop of Alexandria.

The Medieval World
500–1449

After the fall of the Roman Empire in the west in the 5th century, the tradition of ancient science was preserved in the Byzantine Empire, centred on Constantinople and the eastern Mediterranean. Following the rise of Islamic power throughout northern Africa and Spain in the 9th century, Muslim scholars began serious study and analysis of the science of classical antiquity. From about 1150, this combination of ancient Greco-Roman and medieval Arabic scholarship was transmitted to the countries of western Europe, becoming the basis for the university curriculum until about 1500.

Astronomy

The Arabs' mathematical skill and ingenuity with instruments enabled them to refine the observations and theories of the Greeks and produce better star maps, which were becoming increasingly useful for navigation, one of the spurs to astronomical research for many centuries to come.

Biology

The works of Aristotle and Galen remained the basis of biology and medicine. Muslim scholars like Ibn Sina (979–1037; known in the West as Avicenna) and Ibn Rushd (1126–1198; known in the West as Averroës) produced extensive commentaries analysing and in some cases extending the knowledge of the ancients.

Chemistry

From 7th century AD for the next five centuries *al-kimiya*, or alchemy, was developed by the Arabs. They drew on many ideas from the Greeks but they were also in contact with the Chinese – for example, the idea that gold possessed healing powers came from China. The Arabs believed that 'medicine' could be added to base metals to produce gold. Among later European writers, this 'medicine' became known as the 'philosopher's stone'. The idea that not only could the philosopher's stone heal 'sick' or base metals, but that it could also act as the elixir of life, was also originally Chinese. The Arab alchemists discovered new classes of chemicals such as the caustic alkalis (from the Arabic *al-qaliý*) and they improved technical procedures such as distillation. Western Europe had its first contact with the Islamic world as a result of the Crusades. Gradually the works of the Arabs were translated into Latin and made available to European scholars in the 12th and 13th centuries.

Earth Sciences

The centuries from Antiquity to the Renaissance accumulated knowledge of minerals, gems, fossils, metals, crystals, useful chemicals and medicaments, expounded in encyclopedic natural histories by Pliny (AD 23–79) and Isidore of Seville (560–636). At the same time, comprehensive philosophies of the Earth were being elaborated, influenced by the Christian revelation of Creation as set out in Genesis. In Christian eyes, time was directional, not cyclical. God had made the Earth perfect but, in response to Original Sin, He had been forced to send Noah's Flood to punish people by depositing them in a harsh environment, characterized by the niggardliness of Nature. This physical decline would continue until God had completed his purposes with humans.

Ecology and Environment

Ecology continued to be a marginal science, although some unhealthy effects of coal smoke were reported, as coal-burning became widespread.

Mathematics

When the Hellenic civilization declined, Greek mathematics (and the rest of Greek science) was kept alive by the Arabs, especially in the scientific academy at the court of the caliphs of Baghdad. The Arabs also learned of the considerable scientific achievements of the Indians, including the invention of a system of numerals (now called 'Arabic' numerals) which could be used to write down calculations instead of having to resort to an abacus. One mathematician can be singled out as a bridge between the ancient and medieval worlds: Muhammad ibn-Mūsā al-Khwārizmī (c. 780–c. 850) summarized Greek and Indian methods for solving equations and wrote the first treatise on the Indian numerals and calculating with them. Al-Khwārizmī's books and other Arabic works were translated into Latin and interest in mathematics in Western Europe began to increase in the 12th century. It was the demands of commerce that gave the major impetus to mathematical development and north Italy, the centre of trade at the time, produced a succession of important mathematicians beginning with Italian mathematician Leonardo Fibonacci (c. 1180–c. 1250) who introduced Arabic numerals.

Physics

Aristotle's works continued to stimulate commentary and some controversy. Medieval scholars in the Islamic empires and in European universities analysed types of motion (such as uniform motion and accelerated motion). They also developed the highly fruitful 'impetus theory', which stated that a body, when pushed, receives an impressed force, or impetus, from whatever pushed it; the body's motion continues until the impetus is worn out. These ideas were the starting point for Galileo's and Newton's investigations into force and motion.

The Medieval World
500–1449

500–599
Astronomy

545 Indian mathematician Varahamihira writes *The Complete System of Natural Astrology*. It is based on Greek astronomical texts from the period before Ptolemy, and provides a valuable record of these lost works.

Mathematics

594 Decimal notation is used for numbers in India. Using a multiplicative system to base ten, and employing the number zero, this becomes the basis of the modern western system.

598–665 Indian mathematician and astronomer Brahmagupta uses negative numbers in mathematics, and introduces a method of approximation for calculating the sines of small angles.

600–699
Astronomy

628 Indian mathematician Brahmagupta writes *Brahmasphutasiddanta/The Opening of the Universe*, a work which takes the calculation of planetary motions to a new degree of accuracy.

657 Pacal, the Lord of Pallenque, a Mayan religious centre in southern Mexico, founds an observatory and builds astronomical monuments.

Pacal ascends the throne at the age of only 12 and reigns until his death at the age of 80 in AD 683. His sarcophagus in the Temple of the Inscriptions is aligned to the winter solstice.

Chemistry

671 Syrian-born Byzantine scholar Kallinikos of Heliopolis invents 'Greek fire', a highly inflammable mixture, probably of bitumen, petroleum, resin, and an unknown essential ingredient.

The liquid is projected through tubes and can be directed at troops and enemy ships. It is used during the battle of Cyzicus by the Byzantines to repel a Saracen fleet that is attacking near Constantinople. It is particularly effective since it bursts into flames when in contact with water. The exact composition of Greek fire is kept a state secret that only the Byzantine emperor and the family of Kallinikos know. This knowledge has since been lost.

700–799

Astronomy

725 Chinese astronomer and monk I Hsing, using instruments made by engineer Liang Ling-Tsan, measures the deviation of star coordinates from their expected values, taking precession of the Earth's orbit into account.

765 Mayan scientists hold a meeting to discuss astronomy and to adjust their calendar.

772 Muslim astronomer Al-Fazāri translates the Indian astronomical compendium *Mahāsid-dhānta/Treatise on Astronomy*.

This begins the establishment of a golden age in Arabic astronomy lasting several centuries. Prominent astronomers include Muhammad ibn-Mūsā al-Khwārizmī, who writes treatises on astronomy and astronomical tables, which are translated into Latin in the early 12th century by Adelard of Bath and Gerard of Cremona, and Thabit ibn Qurrah. They produce some of the earliest criticisms of the Ptolemaic view of astronomy. Al-Battani, among other achievements, determines the length of the solar year as 365 days, 5 hours, 46 minutes and 24 seconds, a result very close to the modern value.

Chemistry

750 The most famous Arabian alchemist of the period, Jābir ibn-Hayyān of Kufa, known as Geber in Europe, distils vinegar to produce concentrated acetic acid.

Until now, the only agent known to produce chemical change is heat. Acids can also produce chemical changes, which are different from those produced by heat alone, but only if they are strong enough. By discovering the concentrated form of vinegar, Geber shows that acids can chemically alter a substance. He is also reputed to have discovered nitric, hydrochloric, citric, and tartaric acids.

Mathematics

703 English monk and historian Bede, probably the first English mathematician, compiles *De temporibus/On Times*, a brief treatise on chronology and the calculation of Easter.

729 Chinese astronomer and monk I Hsing, director of the Imperial Astronomical Bureau, introduces a new calendar, the *Dayan Li* into China, correcting many errors in the earlier version.

780–850 Arab mathematician Muhammad ibn-Mūsā al-Khwārizmī writes *Al-jam' w'al-tafriq ib hisab al-hind/Addition and Subtraction in Indian Arithmetic*, which introduces the Indian system of numbers to the West. His other book, *Hisab al-jabr w'almuqabala/Calculation by Restoration and Reduction*, gives us the word 'algebra', from *al-jabr*.

31

Some Thoughts About Nought

John O E Clark

The ancient Romans were great engineers. They built roads, bridges, and tall aqueducts, and their planners had the need to use measures and numbers. Roman numerals were based on 10 and employed comparatively few symbols: I (standing for 1), V (5), X (10), L (50), C (100), D (500), and M (1,000). However, simple arithmetic was not easy using these numerals. Addition and subtraction were possible but awkward – try adding XLIV, XXX, and XXVI (the answer is C, or 100). Multiplication and division were virtually impossible in Roman numerals. Multiplying was achieved by consecutive addition; for example, XXII times V is XXII + XXII + XXII + XXII + XXII = CX. Division was done by successive subtraction. The method is very slow, but it works and is still employed by today's electronic calculators, for which speed is no problem.

A Roman centurion was in charge of 100 soldiers. This fact is reflected in the title: *centum* is Latin for 'hundred' (and accounts for C as the symbol for 100 in Roman numerals). When the centurion handed out batches of swords to a *century* of soldiers, the tally – to give the number of swords left – might be C − XLVI − XXX − XXIV. The result, if you are a Latin speaker, is *nihil* ('nothing'). The answer has to be expressed as a word because the Romans had no symbol for zero.

The idea of a zero symbol was first used by the Mesopotamians and Babylonians, and was employed in the way that we do to distinguish between 74 and 704. The full potential of the system was not embraced, however, as the 0 was seldom used in the final place, the way we show the difference between 74 and 740.

The first complete positional system came into use in India around the 7th century with the invention of the symbol that has become known to us as '0', called variously nil, nothing, nought, cipher, or zero. The symbol for 0 was called sunya, the Hindu word for 'empty'. It represents the absence of a quantity, just right for the remainder of the centurion's swords. Zero did not come into general use in Europe until the late 15th century.

However, zero is more than merely another number: it is a place marker. In the decimal system the number following 9 is 10, which stands for one ten and no units. The number 100 is one hundred, no tens, and no units. In this way the same symbol, 1, takes on a different meaning when zeros are added to indicate its place. Thus 1,011 is one thousand, no hundreds, one ten, and one unit – or one thousand and eleven. With the introduction of zero, multiplication became straightforward, as did long division (after Fillip Calendar introduced the algorithm for doing it in 1491).

PROPERTIES OF ZERO

Zero is taken to be an even number and it has some properties not possessed by other numbers. Adding zero to or subtracting zero from another number is no problem – the number remains unchanged. Multiplication by zero is a bit different. Consider 0×53. The answer is zero, which may be easier to see if the multiplication is written the other way round: $53 \times 0 = 0$. After all, 53 lots of nothing are still nothing.

Any number divided into zero results in zero. For example, $0 \div 53 = 0$: divide nothing into 53 parts and each part still equals nothing. But what happens when one divides another number by zero? What is $53 \div 0$? To get a clue, let us see what happens when 53 is divided by numbers that get smaller and smaller: $53 \div 0.1 = 530$; $53 \div 0.01 = 5,300$; $53 \div 0.001 = 53,000$; $53 \div 0.0001 = 530,000$; $53 \div 0.00001 = 5,300,000$. The answers get bigger and bigger, and the numbers themselves – large and small – are a good demonstration of the role of zero as a place marker. In fact, any number divided by zero gives the answer infinity (∞) or negative infinity ($-\infty$), if one approaches zero from the negative side. Mathematicians have to be careful when manipulating formulae and equations in algebra so that they do not inadvertently divide an expression by a term that equals zero. In fact, one of the rules of algebra states that dividing by zero is not allowed. Thus multiplying by zero gives the smallest possible number (zero); and dividing by zero gives the largest possible number (infinity). It has to be said, though, that many mathematicians prefer to say that division by zero is undefined.

POSITIVE OR NEGATIVE?

The ordinary counting series of numbers are positive. Adding two positive numbers – even if one of them is zero – results in another positive number. Thus $2 + 3 = 5$ and $1,234,567 + 7,654,321 = 8,888,888$; they are all positive. But what happens when one subtracts one positive number from another that is smaller than it? An example is $5 - 8$. The answer is a negative number: $5 - 8 = -3$. It is as if zero acts as a watershed between positive and negative. To work out $5 - 8$, start at $+5$ and count back 8 places. The answer, as predicted, is -3. On this scale, as elsewhere, zero itself is neither positive nor negative. Any positive number added to its negative results in zero.

POWER OF ZERO

In arithmetic, a power (also called an index) is a shorthand way of writing certain multiplications. For example, 2^2 (2 to the power 2, or 2 squared) represents $2 \times 2 = 4$. Similarly, $2^3 = 2 \times 2 \times 2 = 8$, $2^4 = 2 \times 2 \times 2 \times 2 = 16$, and so on. Powers are also used to express large numbers without having to write out long strings of zeros. The distance between two planets might be $1,000,000,000,000$ km. This number can be expressed as 10^{12} km, which is a good way of saving zeros. The distance from Earth to the Sun, about $150,000,000$ km, can be written more compactly as 1.5×10^8 km.

BINARY NUMBERS

The decimal system, that is numbers with the base 10, requires ten digits in all (1 to 9 and 0). Numbers with the base 2, called the binary system, require only two digits, 1 and 0. Binary is still a place number system, and the significance of 1 or 0 depends on its position. The decimal numbers 0 to 11 are written in binary as 0, 1, 10, 11, 100, 101, 110, 111, 1000, 1001, 1010, 1011, respectively.

The binary numbers get larger (in terms of numbers of digits) very much more quickly, but they still need only two numbers. The binary system came into its own with the development of digital computers. Tiny voltages or on–off pulses of current represent numbers within a computer's circuits or memory devices. The presence of a signal (a voltage) represents a 1; the absence of a signal represents a 0. Any digitized information can thus take the form of a binary stream of on–off electric signals. The data in a computer, the conversation on a mobile telephone, and the music on a compact disc are all represented by a procession of the 1s and 0s of the binary code. The humble but versatile zero has come a long way.

DUCKS AND CHICKENS

When a batsman at cricket fails to score any runs before being out, the zero that appears on the scoreboard is referred to as a duck's egg. We say that the batsman has scored a duck. A score of zero during a game of tennis is referred to as 'love'. This derives from the French *l'oeuf* meaning 'the egg', because of the resemblance of the shape of a chicken's egg to the symbol for zero.

There are many other names for zero. The word itself comes from the Latin *zephirum*, which in turn derives from the Arabic *sifr* meaning 'empty' or 'null', the source also of the word cipher. Null owes its origins to the Latin *nullus*, meaning 'none' or 'not any', and nil is a contraction of the Latin *nihil* meaning 'nothing'. Nought, the starting point of these thoughts, is a variation of naught from the Old English *nowiht*, meaning 'no thing'. The Old English term is not totally dead. There is a phrase from the north of England: 'You can't get owt for nowt'. There is no direct word for zero in today's version from North America: 'There's no such thing as a free lunch' – not even a duck's egg!

800–899

Astronomy

800 An astronomical altar known as 'the hitching post of the Sun' is built in the Peruvian city of Machu Picchu, and is used to measure solar and lunar movements with great accuracy.

815 Muslim scholar Māshā'allah writes on astrology, the astrolobe, and meteorology.

Biology

800 Muslim scholar Al-Batrīq produces Arabic translations of major works by the Greek physicians Galen and Hippocrates; they will have a lasting effect on Arab medicine.

Earth Sciences

813 The brothers Ben Shaku, Hebrew astronomers, attempt to measure the length of a degree of meridian on the Earth's surface, and so determine the size of the world, at the order of Caliph al-Ma'mun.

Mathematics

816 The Council of Chelsea, a church council convened at King Offa's palace, introduces the Anno Domini system of dating into England.

900–999

Astronomy

c. **970** Muslim astronomer Abu al-Wafa' invents the mural quadrant for the accurate measurement of the declination of stars.

987 Toltec conquerors of the Central American Mayan city of Chichén Itzá construct monuments with ritual astronomical alignments to the rising and setting of the Sun and the sacred planet Venus.

Mathematics

c. **970** Muslim astronomer Abu al-Wafa' writes translations and commentaries (subsequently lost) on Euclid, among others. He also makes contributions to the development of trigonometry and geometry. In particular Abu al-Wafa' introduces theorems (with proofs) on spherical triangles and their trigonometry.

976 Arabs introduce the decimal number system into Spain. The *Codex Vigilanus* is the first evidence of the use of decimal numbers in Europe.

986 French monk Abbo of Fleury writes *De numero mensura et pondere / On Weights and Measures*, a mathematical commentary.

The Mayans of Central America devised complex calendars based on use of observatories, such as this one at Chichén Itzá, which were used in religious rituals and the drawing up of farmers' almanacs. The structure itself is aligned to the rising and setting of astronomical bodies.
Corel

c. **990** Muslim mathematician al-Karaji writes *Al-Fakhri*. He is one of the first mathematicians to study algebra without using geometry.

Al-Karaji gives the binomial theorem using Pascal's triangle.

c. **990** French churchman and scholar Gerbert of Aurillac (who will later become Pope Sylvester II) reintroduces the abacus into Europe, and uses Arabic numerals, although he appears not to have known about zero.

1000–1099

Astronomy

1048 Muslim astronomer al-Bīrunī ('the master') writes his *Description of India*, which contains much astronomical data. He attacks the idea of astrology as an influence over historical events.

4 July 1054 A bright new star, visible in daylight, appears in the constellation Taurus. The supernova is observed in China and Korea, and is recorded in rock paintings in southwestern America.

It is the closest and most spectacular supernova event ever recorded; the Chinese imperial astronomers who observe it note that it is visible even in daylight for a period of three weeks. The remnant, now known as the Crab Nebula, is about 6,500 light years away, so the explosion actually occurred around 5446 BC. At the centre of the mass of gas and dust making up the Crab Nebula is a pulsar, a very dense star some 20 km/12 mi across composed almost entirely of neutrons. The cloud glows with synchrotron radiation, produced by very fast moving electrons travelling in the strong magnetic fields that exist in the nebula.

1066 The comet later known as Halley's comet appears in the sky, and is taken as an omen by both the Norman and English sides before the Battle of Hastings. The victorious Normans record its appearance in the Bayeux Tapestry.

1074 The Seljuk Sultan Malik Shah of Baghdad builds a new observatory (at Isfahan, Persia), where the mathematician, astronomer, and poet 'Umar al-Khayyām (Omar Khayyám) is appointed to reform the old Persian calendar.

15 March 1079 An improved Persian calendar is presented by the astronomer and poet 'Umar al-Khayyām (Omar Khayyám), in his *Astronomical Calendar*. It fixes New Year (Naw Ruz) to the vernal equinox and has years of 365 days, except when the Sun fails to enter Aries before noon on the last day of the year, when a 366th day has to be added. Khayyám proposes a regular leap year cycle of 33 years, containing 8 leap years, in which the Persian year has a mean length of $365^8/_{33} = 365.2424$ days, more accurate than the Gregorian year.

1087 Al-Zarqellu and other Muslim and Jewish astronomers compile the *Toledan Tables*, named after Toledo in Spain, where they are produced.

The tables (one of the first almanacs – an Arabic word) describe the positions of the planets, and the use of instruments, especially the astrolabe, for making astronomical measurements. Since the tables are trigonometric, Al-Zarqellu writes an introduction explaining the tables, which he calls 'Canones sive regulae tabularum astronomiae'. The tables rapidly become noted for their accuracy, and are translated into Latin by Gherardo Cremonese, becoming more widely consulted than any other Arab astronomical tables. They remain popular for more than 300 years, with Nicolaus Copernicus, Tycho Brahe, and Johannes Kepler among those who use them in later years.

1091 Benedictine abbot and monastic reformer Wilhelm of Hirsau writes his astronomical treatise *De astronomia/On Astronomy*.

As well as writing about them, Wilhelm constructs various astronomical instruments, including a sundial that shows the solstices, equinoxes, and other astronomical phenomena.

1091 French-born Prior Walcher of Malvern Abbey, England, records his observations in Italy of an eclipse of the Moon. This is one of the earliest accurate western European observations of the phenomenon.

Biology

1059 Chinese scholar Fu Kung writes a treatise on crabs entitled *Hsieh-p'u*.

1070 Chinese scholar and horticulturist Wang Kuan writes *Yang-chou shao-yo-p'u* about the peony. The treatise describes 39 different varieties of the plant.

Chemistry

1044 The Chinese text *Wu Ching Tsung Yao* is written, including a recipe for black powder, which uses a mix of saltpetre (potassium nitrate), charcoal, and sulphur to produce the earliest form of gunpowder.

The Chinese use gunpowder in bamboo tubes and in rockets in their conflicts with the Mongols; however, the weapons are not very effective and may have been designed simply to scare horses.

Mathematics

12 May 1003 Pope Sylvester II, formerly known as Gerbert of Aurillac, French head of the Roman Catholic Church 999–1003, dies in Rome. A pioneering scientist, his works include treatises on the abacus, the astrolabe, and Spanish-Arabic numerals.

c. 1022 Persian scientist ibn Sina (latinized to Avicenna) completes his encyclopedic work *Kitab al-Shifa'/The Book of Healing*. As well as geometry and arithmetic, it also counts astronomy and music as sections of mathematics.

1040 Arab mathematician Ahmad al-Nasawi writes *al-Muqniʾfi al-Hisāb al-Hindi/The Convincer on Hindu Calculation* on fractions, square and cubic roots, and other mathematical phenomena using Hindu (or Arabic) numerals.

Physics

1038 Muslim mathematician and physicist Abu Alī al-Hassan ibn al-Haytham (latinized to Alhazen) is active. His major work on optics, translated into Latin in 1270, provides a synthesis of what is known about the science of vision, from the behaviour of light to the physiology of the eye.

1100–1199
Astronomy

1108 French-born Prior Walcher of Malvern Abbey, England, compiles tables mapping the movements of the Moon in the period 1036–1112, the earliest European attempt at such a difficult astronomical feat.

1110 Spanish astronomer Pedro Alfonsi visits the court of King Henry I of England, bringing Islamic astronomical knowledge to Christian scholars, including Prior Walcher of Malvern Abbey.

1145 A carving near a sacred lake at the Toltec sacred city of Chichén Itzá records a transit of Venus across the face of the Sun, considered to be an event of great religious significance.

1150 Hebrew scholar Solomon Jarchus produces his almanac, a table for the calculation of celestial movements, and a calendar.

Biology

1100 French poet and physician Odo of Meung writes *De viribus herbarum/On the Power of Herbs*, a herbal derived from classical sources. Written in verse as an *aide mémoire*, it becomes extremely popular.

1169 Muslim physicist and philosopher Ibn Rashd (latinized as Averroës) is active. His series of commentaries on Aristotle and on Galenic medicine are highly influential.

His series are translated into Hebrew and Latin and later became a standard part of the university curriculum in Western Europe.

Earth Sciences

1136 Silver-bearing ore is discovered at Freiburg in Saxony, southern Germany, triggering a 'silver rush' in which Freiburg becomes a centre for mining and metallurgy.

1154 Egyptian Muslim scholar al-Tīfāshī writes his pioneering work on mineralogy *Flowers of Knowledge of Precious Stones*.

Mathematics

1100 English scholar Robert of Chester translates the 9th-century *Astronomical Tables*, by Arab astronomer Muhammad ibn Musa al-Khwârizmi, into Latin.

1130–1185 Indian mathematician Bhaskara produces *Lilavat/The Beautiful* on arithmetic and geometry and *Bijaganita/Seed Arithmetic* on algebra.

1130 English monk and scholar Adelard of Bath translates Euclid's *Elements* from Arabic.

1150 Gerard of Cremona uses Arabic numerals in his translation of the Egyptian astronomer Ptolemy's astronomical work, the *Almagest*.

Physics

1135 Muslim philosopher Ibn Bajja (latinized as Avempace) is active. He is the first known commentator on Aristotle in Muslim Spain, and his examination of Aristotle's concepts of motion prove especially fruitful for later scholars.

1200–1299

Astronomy

1220 English-born canon and mathematician John of Holywood *(Sacrobosco)* publishes his *Tractatus de Sphaera/On the Sphere*, an elementary textbook on astronomy.

His textbooks on astronomy and mathematics become standard parts of the university curriculum in Western Europe until the 17th century.

1259 Persian philosopher al-Tūsī, employed by the Mongol Hūlāgū Khan, founds an observatory at Marāghah, Persia.

1276 Chinese astronomer Guo Shou-jing builds a 12 m/40 ft tower gnomon (the stationary arm that projects the shadow on a sundial) at Gao Cheng Zhen, China, with a 36 m/120 ft horizontal stone scale for measuring the length of its shadow at midday. He is thus able to maintain a calendar.

1290 French astronomer William of St-Cloud obtains an accurate measurement of the obliquity of the ecliptic (the angle of the Sun's apparent path around the sky) at Paris, France.

1293 Danish astronomer Peter Nightingale develops an equatorium, an astronomical model for use in calculating and predicting eclipses.

Biology

1235 Albertus Magnus of Bavaria describes the development of different animals including insects. He rejects mythical explanations of natural phenomena and states that evidence should come from observation.

1235 French scholar Thomas de Chantimpré writes his popular encyclopedia of natural history *De natura rerum / On the Nature of Things*.

Chemistry

1249 English scholar Roger Bacon is one of the first sources in Europe to mention gunpowder and how to make it.

Bacon describes in a letter that the proportions are five parts charcoal, five parts sulphur, and seven parts saltpetre. His recipe contains too much saltpetre to be efficient.

Ecology

1273 Complaints in London, England, about pollution produced from the burning of 'seacoal', mined at Newcastle, lead to an eventual prohibition on coal burning throughout England.

Mathematics

1202 Italian mathematician Leonardo Fibonacci writes *Liber abaci / The Book of the Abacus*, which recounts the arithmetic and algebra he had learnt in Arabic countries. It also introduces the famous sequence of numbers now called the Fibonacci sequence.

It is the sequence of integers 1, 1, 2, 3, 5, 8, 13, 21, . . . in which each number is the sum of the previous two. As the numbers get larger, the ratio of successive terms tends to the golden mean.

1225 Italian mathematician Leonardo Fibonacci writes *Liber quadratorum / The Book of the Square*, the first major Western advance in number theory since the work of Diophantus a thousand years earlier.

c. 1230 English-born canon and mathematician John of Holywood *(Sacrobosco)* introduces the decimal number system to England.

Physics

1268 English scholar Roger Bacon, who becomes a Franciscan friar at the age of 33, writes the treatise *On Experimental Science*. He carefully describes the optical phenomenon of light being broken into colours when seen through crystals or droplets of water.

1300–1399

Astronomy

1318 French astronomer Jean de Murs determines the epoch of the spring equinox using the Spanish *Alfonsine Tables*. His explanatory treatise increases the use of the tables in the rest of Europe.

1342 The work of the French-born Hebrew astronomer Levi ben Gerson (Gersonides) is translated into Latin as *De sinibus, chordis, et arcubus / Concerning Sines, Chords, and Arcs*, a treatise on trigonometry introducing the use of the cross-staff for astronomical observations.

1350 French astrologer Jean de Linières compiles a catalogue of stars.

1362 Italian inventor Giovanni de Dondi completes construction of an elaborate astronomical clock.

Called an astrarium, it provides the indication of the position in the sky of all planets known at this time with respect to a Ptolemaic conception of the Solar System, with the Earth at the centre and the Sun and five planets (Mercury, Venus, Mars, Jupiter, and Saturn) rotating around it.

1364 French philosopher and bishop Nicole Oresme writes *Latitudes of Forms*, an early work on coordinate systems.

1377 French philosopher and bishop Nicole Oresme writes a commentary on Aristotle's *De caelo / Concerning the Heavens*, suggesting a heliocentric theory of the universe, and even that there might be 'more than one universe', though he ultimately rejected such ideas.

1391 English writer Geoffrey Chaucer writes *A Treatise on the Astrolabe*.

Biology

1372 Arabic theologian Al-Damīrī writes *The Lives of Animals*, a bestiary of the 931 animals both real and mythical, mentioned in the Quran.

Chemistry

1300 Legend credits Spanish alchemist Arnau de Villanova is credited with the first preparation of pure alcohol.

De Villanova obtains a sample of relatively pure alcohol from the distillation of wine. The resultant liquid contains a much higher percentage of alcohol and is essentially brandy. His technique can be applied to any fermented liquid, such as fermented grain to make whisky. His discovery means that brandy and whisky can now be produced in significant quantities.

c. 1310 An unknown Spanish alchemist, called the false Geber due to his adoption of the name of the famous Arabic alchemist, gives one of the best descriptions of the manufacture of sulphuric acid.

This is the most important chemical discovery of the medieval alchemists and makes possible a wide range of chemical reactions, which are to stimulate the birth of the chemical industry in Europe. Even today, sulphuric acid is one of the most important acids used by the chemical industry.

Gunpowder, or black powder, was the first true explosive. Made from a mixture of 75% potassium nitrate, 15% charcoal, and 10% sulphur it is generally thought to have originated in China, where explosive grenades and bombs were in use by AD 1000. By the early 14th century, gunpowder and cannons were being manufactured in Europe by people such as German alchemist Berthold der Schwarze. *Corbis*

c. **1313** German Grey Friar Berthold der Schwarze is traditionally credited with the independent invention of gunpowder. He is also acknowledged as the first European to cast a bronze cannon.

Although the existence of gunpowder has been known for many centuries, the secret of its composition has not. Several recipes have been tried in Europe and none of these produce an efficient explosive mixture. Schwarze succeeds in perfecting a gunpowder mixture that can be used to fire projectiles from early firearms.

Earth Sciences

1370 Italian writer Giovanni Boccaccio writes that (fossilized) shells found in rocks in Tuscany are the remains of former sea life.

Mathematics

1300 Chinese mathematician Chu Shih-chieh writes *Szu-yuen Yu-chien/The Precious Mirror of the Four Elements*, containing a number of methods for solving equations. He also defines what is now called Pascal's triangle.

1321 Formulae for calculating permutations and combinations are devised by the French-born Hebrew philosopher and mathematician Levi ben Gerson (Gersonides).

Ben Gerson shows how to calculate permutations of *n* objects and how to work out combinations of *n* objects selected *r* at a time.

1335 The English abbot of St Albans, Richard of Wallingford, writes *Quadripartitum de sinibus demonstratis*, the first original Latin treatise on trigonometry.

It concentrates on spherical trigonometry, which is the type used by astronomers.

1343 Jean de Meurs writes a metrical treatise on mathematics, mechanics, and music, *Quadripartitum numerorum/Four-fold Division of Numbers*.

1343 French-born Hebrew philosopher and mathematician Levi ben Gerson (Gersonides) writes *De harmonicis numeris/Concerning the Harmony of Numbers*, a treatise on arithmetic, for the musician Philippe de Vitry.

41

It also discusses trigonometry, but unlike his contemporary Richard of Wallingford, ben Gerson concentrates on plane triangles.

1382 French philosopher and bishop Nicole Oresme's *Le Livre du ciel et du monde/The Book of Heaven and Earth* is published. It is a compilation of treatises on mathematics, mechanics, and related areas.

Oresme will also write two influential books on ratios, in one of which he presents arguments that the 'science' of astrology must be erroneous.

Physics

1335 English scholar William Heytesbury of Oxford, England, produces the first law of uniform acceleration.

Such ideas were to be carried further by Simon Stevinus in 1586 and Galileo Galilei in 1590.

1358 This is generally regarded as the year in which French scholar and philosopher Jean Buridan dies. Buridan spent some 30 years as professor and rector at the University of Paris.

The majority of Buridan's written work are commentaries on the work of Aristotle. In these commentaries, he presents the theory on the nature of the 'impetus' impressed on a moving object. His ideas were taken up by students, and were an influence on Galileo.

1400–1449

Astronomy

1424 Mongolian ruler and astronomer Ulugh Beg, Prince of Samarkand, builds a great observatory, including a 40 m/132 ft sextant, which enables extremely accurate measurements to be made, cataloguing over 1,000 stars.

1437 Astronomers of Samarkand compile the *Tables of Ulugh Beg* for the city's ruler, the Mongol mathematician and astronomer Ulugh Beg; they correct several errors made in the works of Ptolemy.

Biology

1410 Venetian botanist Benedetto Rinio writes his herbal *Liber de simplicibus/Book of Simples*.

Mathematics

1400 Indian astronomer and mathematician Madhava of Sangamagramma proves a number of results about infinite sums. He gives Taylor expansions of trigonometric functions.

He is later credited with discovering the arc-tangent series, not known in the West until rediscovered by Scottish mathematician James Gregory.

Physics

c. 1400 Italian scholar Blasius of Parma writes his *Tractatus de ponderibus/Treatise Concerning Weights*, an essay on statics and hydrostatics.

The Scientific Revolution
1450–1749

The European Scientific Revolution extended philosophy with a new combination of observation, experimentation, and rationality. It resulted in an explosion of scientific discovery that has continued to the present day.

Astronomy

The dawn of a new era in astronomy came in 1543, when a Polish canon, Nicolaus Copernicus (1473–1543), published *De revolutionibus orbium colestium/On the Revolutions of the Heavenly Spheres,* in which he argued that the Sun, not the Earth, is the centre of our planetary system. Danish astronomer Tyco Brahe (1546–1601) increased the accuracy of observations by means of improved instruments allied to his own personal skill, and his observations were used by German mathematician Johannes Kepler (1571–1630) to argue the validity of the Copernican system. Considerable opposition existed, however, for removing the Earth from its central position in the universe; the Catholic Church was openly hostile to the idea, and, ironically, Brahe never accepted the idea that the Earth could move around the Sun. Yet before the end of the 17th century, the theoretical work of Isaac Newton (1642–1727) had established celestial mechanics.

The refracting telescope was invented about 1608, possibly by Hans Lippershey (*c.* 1570–*c.* 1619) in Holland, and was first applied to astronomy by Italian scientist Galileo (1564–1642) in the winter of 1609–10. Immediately, Galileo made a series of spectacular discoveries. He found the four largest satellites of Jupiter, which gave strong support to the Copernican theory; he saw the craters of the Moon, the phases of Venus, and the myriad faint stars of our galaxy, the Milky Way.

Galileo's most powerful telescope magnified only 20 times, but it was not long before larger telescopes were built and official observatories established. Galileo's telescope was a refractor; that is to say, it collected its light by means of a glass lens or object glass. Difficulties with this design led Newton, in 1671, to construct a reflector, in which the light is collected by means of a curved mirror.

Biology

Human dissections were routinely performed from the 14th century and anatomical observation and illustration were promoted by the fervent activity of Belgian physician Andreas Vesalius (1514–1564), whose *De humani corporis fabrica/On the Structure of the Human Body* (1543) is one of the masterpieces of the Scientific Revolution. His book called on anatomists to examine the body itself rather than relying simply on Galen;

43

the illustrations in his work are simultaneously objects of scientific originality and artistic beauty. The rediscovery of the beauty of the human body by Renaissance artists encouraged the study of anatomy by artists such as Italian painter, inventor, and scientist Leonardo da Vinci (1452–1519). Shortly afterwards, the English physician William Harvey (1578–1657) proposed the circulation of the blood and established the utility of experiment in physiology. His little book *De motu cordis / On the Motion of the Heart* (1628) was the first great work on experimental physiology since time of Galen. The eccentric wandering Swiss physician and chemist Paracelsus (1493–1541) added notoriety to the ongoing question of Galen and other ancient authorities by calling publicly for a fresh approach to nature and medicine and for the search for new remedies for disease.

Chemistry

A new era in chemistry began with the researches of Anglo-Irish scientist Robert Boyle (1627–1691), who carried out and systematically recorded experiments on air and other common substances. From Boyle's time onwards, chemistry was allied with the new scientific developments in astronomy and physics.

During the 1700s the phlogiston theory was developed, in part to explain the process of smelting metals, an increasingly important industry. Metals were thought to be composed of a calx combined with phlogiston, which was the same in all metals. When a candle burned in air, phlogiston was given off. It was believed that combustible objects were rich in phlogiston and what was left after combustion possessed no phlogiston and would therefore not burn. Thus wood possessed phlogiston but ash did not; when metals rusted, it considered that the metals contained phlogiston but that its rust or calx did not.

Earth Sciences

European voyages of discovery explored the New World, Africa, Asia, and the Pacific islands. The possibility that the Earth was extremely old arose in the work of natural philosophers such as English physicist Robert Hooke (1635–1703). In the 18th century, Enlightenment naturalists came to view the Earth as a machine, operating according to fundamental laws. Archbishop James Ussher (1581–1656) in his *Sacred Chronology* (1660), arrived at a creation date for the Earth of 4004 BC, but some later naturalists, like George-Louis Buffon (1707–1788), assumed it to be much older.

The integrating of evidence from fossils and strata is evident in the work of Danish naturalist Nicolaus Steno (1638–1686). He was struck by the similarity between shark's teeth and fossil glossopetrae. He concluded that the stones were petrified teeth. On this basis, he posited six successive periods of Earth's history. Steno's work is one of the earliest 'directional' accounts of the Earth's development that integrated the history of the globe and of life.

Ecology and Environment

Scientific work on the impact of human expansion started to emerge at this time; German chemist Georgius Agricola (1495–1555) presented an assessment of the impact of ore-washing in his posthumous 1556 work *De re metallica / On Metalwork*; English diarist John Evelyn (1620–1706) wrote on forest management in a time when timber was the most important resource for fuel and heating, and Swedish biologist Carolus Linnaeus (1707–1778) wrote the ecologically themed 'Oeconomy of Nature', recognizing food webs and niches (referred to as a creature's 'allotted place').

Mathematics

An important advance in mathematics was the discovery of the means of solving cubic equations, credit for this being given to, amongst others, Italian mathematician and physicist Niccolò Fontana (Tartaglia, *c.* 1499–1557). The means of solving quartic equations was discovered soon afterwards. Within another 20 years, the French mathematician François Viète (1540–1603) was improving on the systematization of algebra in symbolic terms and expounding on mathematical (as opposed to astronomical) applications of trigonometry. Ten years after Viète's death, Henry Briggs (1561–1630) in England worked with Scottish mathematician John Napier (1550–1617), the deviser of 'Napier's bones', to produce the first logarithm tables using the number 10 as its base. Simultaneously, the German astronomer Johannes Kepler (1571–1630) was publishing one of the first works to consider infinitesimals, a concept that would lead later to the formulation of differential calculus.

It was in France that the scope of mathematics was then widened by a group of great mathematicians including René Descartes (1596–1650), Pierre de Fermat (1601–1665), and Blaise Pascal (1623–1662). Descartes' greatest contribution to science was in virtually founding the discipline of analytical (coordinate) geometry, in which geometrical figures can be described by algebraic expressions. With Pascal, Fermat investigated probability theory, and in number theory he independently devised many theorems, one of them now famous as Fermat's Last Theorem. French mathematician Abraham de Moivre (1667–1754) formulated game theory, reconstituted probability theory, and set the business of life assurance on a firm statistical basis.

Meanwhile Japanese mathematician Seki Kowa (*c.* 1642–1708) was independently discovering many of the mathematical innovations also being formulated in the West. He worked out new methods for solving algebraic problems, for rectifying arcs of circles, and for approximating the value of π.

In 1665 English physicist and mathematician Isaac Newton (1642–1727) began his work on binomial theorem, which led him to an investigation of infinite series, which in turn led to a study of integration and the notion that it might be achieved as the opposite of differentiation. In Germany, Gottfried Leibniz (1646–1716) reached the same conclusion. Leibniz's work on calculus was greatly admired in Europe, particularly by the great Swiss mathematical family, the Bernoullis, and Swiss mathematicians Leonhard Euler (1707–1783) and Gabriel Cramer (1704–1752).

Physics

Aristotle's ideas on falling bodies were challenged around 1600 by the Flemish scientist Simon Stevinus (1548–1620), who is believed to have dropped unequal weights from a height and found that they reached the ground together. At about the same time Italian physicist and astronomer Galileo measured the speeds of 'falling' bodies by rolling spheres down an inclined plane and formulated the equations that govern the motion of bodies under gravity. This work was brought to a brilliant climax by Newton, who in his three laws of motion achieved an understanding of force and motion, relating them to mass and recognizing the existence of inertia and momentum. Newton's work on motion and gravitation, published in 1687, was essential in promoting the idea of universal laws of nature that could explain all observed effects of force and motion.

In the field of force and motion, important advances were made with the discovery of the law governing the pendulum and the principle of conservation of momentum by Dutch physicist and astronomer Christiaan Huygens (1629–1695) and the determination of the gravitational constant by English natural philosopher Henry Cavendish (1731–1810).

French mathematician and physicist Blaise Pascal (1623–1662) found that pressure is transmitted throughout a liquid in a closed vessel, acting perpendicularly to the surface at any point. Pascal's principle is the basis of hydraulics. Pascal also investigated the mercury barometer invented in 1643 by the Italian physicist and mathematician Evangelista Torricelli (1608–1647) and showed that air pressure supports the mercury column. The law of elasticity was discovered by English physicist Robert Hooke (1635–1703) in 1678 when he found that the stress (force) exerted is proportional to the strain (elongation) produced.

The Scientific Revolution
1450–1749

1450–1479

Astronomy

1450 Austrian astronomer George Peurbach produces his first almanac.

1454 Austrian astronomer George Peurbach completes his *New Theory of the Planets*, supporting the 'sphere' model of the universe, and using new observations to improve the measurement of celestial distances.

1473 Jewish astronomer and historian Abraham ben Samuel Zacuto produces tables of declination of the Sun in his major astronomical work, *Ha-Hibbur ha-Gadol / Rules of the Astrolabe*.

Zacuto's tables are used throughout the Christian and Islamic worlds, and are the basis for the *Regimento do Astrolabio do Quadrante / Regiment for the Astrolabe and Quadrant* prepared for Portuguese mariners under Prince Henry the Navigator. Tables from *Ha-Hibbur ha-Gadol*, along with tables from Zacuto's other book on the influence of the stars, are used by Christopher Columbus on his voyages.

1474 Italian astrologer and astronomer Lorenzo Buonincontro publishes *Commentaria in C Manilii astronomicon / Commentary on the Astronomicon of C Manilius*, a discussion of 1st century-philosopher Manilius' astrological poetry.

1475 German astronomer Regiomontanus publishes *De triangulis planis et sphaericis / Concerning Plane and Spherical Triangles*, a book on spherical trigonometry as applied to astronomy.

In the same year Regiomontanus travels to Rome to advise on possible revisions of the Roman Catholic ecclesiastical calendar. He dies there a year later. A letter found after his death refers to parallax caused by motion of the Earth – indicating he possibly believed in a heliocentric (Sun-centred) universe. His *Epitome* is published posthumously in 1496.

Biology

1458 German philosopher Cardinal Nicholas de Cusa (Nicolaus Cusanus) founds a hospital at his birthplace (Kues, Trier). He bequeaths a library to the hospital, with holdings covering his many interests including diagnostic medicine and experimental science.

From his studies of plant growth, de Cusa concludes that plants absorb nourishment from water 'by the work of the Sun'. De Cusa is the first to suggest the role of sunlight in plant nutrition.

Chemistry

1450 German monk Basil Valentine provided an excellent description of the element bismuth (atomic number 83).

Bismuth is a white, brittle metal, which is commonly mistaken for lead or tin during this period and consequently may have actually been discovered earlier. The element is originally called *Weisse Masse*, a German phrase meaning 'white mass', but is later known as 'wisuth' and 'bisemutum', from which the modern name bismuth is derived.

Earth Sciences

***c.* 1450** German philosopher Cardinal Nicholas de Cusa (Nicolaus Cusanus) devises the first hygrometer for measuring air humidity. It consists of a ball of wool, the weight of which changes as it absorbs moisture from the atmosphere.

1450 Italian mathematician and architect Leone Battista Alberti devises a vane anemometer and makes the first measurements of wind speeds.

1469 Roman writer Pliny the Elder's *Naturalis historia/Natural History* (written in AD 77) is published in Venice; this is the first time the book has been printed.

1480–1509

Chemistry

1509 German chemist Erasmus Eberner of Nuremberg was among the first to investigate zinc (atomic number 30) as a new metal, at Rammelsburg, Germany. The Swiss chemist Paracelsus names it 'zinken' in 1541.

Earth Sciences

1500 German engineer and physician Ulrich Rulein von Kalbe publishes the first mining manual, *Bergbuchlein/Little Book on Mines*.

Mining in Germany expands rapidly in the 16th century and Kalbe's manual describes traditions, ore distribution, alchemy, and astrology as they relate to mining, and guidance on operating a mine.

1503 German Carthusian prior Gregor Reisch publishes the *Margarita philosophia/Pearl of Philosophy*, a university textbook that includes attempts to explain earthquakes.

Mathematics

1482 The first important book on mathematics ever to be printed is Euclid's *Elements*, in a Latin translation by the Italian mathematician and chaplain to Pope Urban IV, Johannes Campanus. It summarizes much of Greek knowledge of mathematics.

1494 Italian mathematician Luca Pacioli publishes *Summa de arithmetica, geometria, proportioni et proportionalita/ The Collected Knowledge of Arithmetic, Geometry, Proportion, and Proportionality* The encyclopedic treatise contains work on theoretical and practical arithmetic, algebra, tables of weights and measures, gaming theory, book-keeping, and a commentary on Euclid's geometry.

1510–1539

Astronomy

1514 Polish astronomer and canon Nicolaus Copernicus writes a short pamphlet on the problems of the Aristotelian Earth-centred view of the universe, the *Commentariolus/ Brief Commentary*, and circulates it privately.

> *Finally we shall place the Sun himself at the centre of the universe.*
>
> Nicolaus Copernicus, Polish canon and astronomer, *De Revolutionibus orbium coelestium*

1533 The ideas of Polish astronomer and canon Nicolaus Copernicus on a Sun-centred (heliocentric) universe, in which the Earth and planets orbit the Sun on circular paths, are presented at a series of lectures in Rome. Pope Clement VII attends some of the lectures.

Nicolaus Copernicus challenged the prevailing views of the workings of the universe.
Oxford Museum of the History of Science

1538 Italian astronomer and physician Girolamo Fracastoro publishes his discovery that the tails of comets always point away from the Sun.

Biology

1517 Naturalists in Amsterdam, the Netherlands, apply the folk skills of taxidermy to science, preserving cassowaries brought from the East Indies.

1538 English naturalist William Turner publishes his three-volume *Libellus de re herbaria novus/ New Letter on the Properties of Herbs*, the first English herbal to take a scientific approach to botany.

Chemistry

1510 The Swiss physician Paracelsus (Theophrastus von Hohenheim) is active. He rejects the theory that an imbalance of humours results in disease, asserting that agents external to the body are the cause.

49

He also suggests that disease agents can be countered by chemical substances, such as mercury and antimony.

The true use of chemistry is not to make gold but to prepare medicines.

Paracelsus, Swiss physician, alchemist, and scientist, attributed remark

---

Earth Sciences

1516 German cartographer Martin Waldseemüller publishes his Carta Marina, a map of the world in chart form, introducing the use of the name 'America' for the southern continent initially.

1530 German chemist and mineralogist Georgius Agricola (Georg Bauer) publishes *Bermannus*, an introduction to mineralogy and the mining business.

Mathematics

1514 Dutch mathematician Vander Hoecke introduces to algebra the + (plus) and − (minus) signs with their modern meanings of addition and subtraction.

1522 English cleric Cuthbert Tunstall publishes *De arte supputandi/On the Art of Computation*, a manual for calculation. This is later developed independently by German mathematician Adam Ries.

1525 German mathematician Christoff Rudolff introduces a symbol resembling $\sqrt{\ }$ for square roots in his *Die Coss*.

It is also one of the first mathematical publications to employ decimal fractions.

Physics

1537 Italian mathematician Niccolò Fontana (Tartaglia) lays the foundations of ballistics in his *Nova scientia/New Science*, attempting for the first time to establish the laws governing falling bodies.

The study of 'falling bodies' makes specific reference to the path of artillery shots. This use of mathematics is later adopted by Galileo.

1540–1569

Astronomy

1540 German astronomer Georg Rheticus persuades the Polish astronomer and canon Nicolaus Copernicus to allow him to publish his *Narrato Prima/First Report*, in which he first describes Copernicus's heliocentric system of the universe.

1542 German Protestant religious reformer Martin Luther attacks Copernicus for expounding his heliocentric concept.

1543 Polish astronomer and canon Nicolaus Copernicus has finally worked out to his satisfaction the details of the heliocentric theory, and they are published in his most important work *De revolutionibus orbium coelestium/On the Revolutions of the Celestial Spheres,* as he lies dying of a cerebral haemorrhage.

Copernicus's view of the Solar System is considered implausible by the vast majority of his contemporaries, but its sophisticated mathematics impresses professional astronomers. Awareness of Copernicus's arguments and methods spreads slowly throughout learned Europe.

1551 English mathematician and philosopher Robert Recorde publishes *The Castle of Knowledge,* an astronomical treatise in which he supports the Copernican heliocentric theory of the universe.

1563 Danish astronomer Tycho Brahe observes a conjunction of the planets Jupiter and Saturn, noting that it occurs one month earlier than predicted by available tables. He sets about producing new, more accurate tables of his own.

1569 While studying at Augsburg, Germany, Danish astronomer Tycho Brahe begins the construction of a 6 m/19 ft quadrant that enables him to plot the positions of celestial objects with unprecedented accuracy.

Biology

1543 Belgian physician Andreas Vesalius writes *De humani corporis fabrica/On the Structure of the Human Body,* a highly illustrated, clearly written study of the human body.

Its seven volumes describe findings from his own dissections of human cadavers; he shows that some of the work of the Greco-Roman physician Galen is based on the dissection of animals, especially the Barbary ape.

1542 German botanist Leonhard Fuchs (after whom the fuchsia is named) publishes *De historia stiripium/On the History of Plants,* a pioneering work of plant classification, in Basel.

1543 Europe's first botanical garden is founded at the University of Pisa, in Italy.

1546 Italian physician Girolamo Fracastoro publishes *De contagione et contagiosis morbis et curatione/On Contagion and the Cure of Contagious Disease,* in which he describes typhus for the first time, and also proposes that diseases are spread through microscopic bodies.

1548 English naturalist William Turner revises and expands his earlier (1538) pioneering work, republishing it as *The Names of Herbes.*

1551 French naturalist Pierre Belon publishes *Histoire naturelle des estranges poissons marins/Natural History of Unusual Marine Animals,* a study of dolphin embryology and one of the founding texts of modern embryology.

1551 Swiss naturalist Conrad Gesner publishes the first volume of his *Historia animalium/History of Animals,* a pioneering, highly illustrated classification of animals.

History of Animals is important for its attempt to distinguish observed facts from myths and folklore. Ultimately five volumes appear covering most vertebrates. Gesner dies before publishing similar works for plants and insects.

1551 English botanist and naturalist William Turner begins publication of his *A New Herball,* a treatise in English on the medicinal properties of plants, using his earlier scientific researches.

1553 French naturalist Pierre Belon writes *De aquatilibus / On Aquatic Animals*, an expansion of his studies on fish and an explanation of comparative anatomy.

1553 Spanish theologian Michael Servetus, in his tract *Christianismi restitutio / The Restoration of Christianity* relates his discovery of the pulmonary circulation of the blood.

The work is printed secretly in 1553 as part of Servetus's efforts to discuss the relationship between the Spirit and regeneration. He is denounced as a heretic against Calvinism and burned at the stake in October 1553.

1555 French naturalist Pierre Belon publishes *L'Histoire de la nature des oyseaux / The Natural History of Birds*, a natural history and classification of birds.

1561 Italian anatomist Gabriello Falloppio writes *Observationes anatomicae / Anatomical Observations*, a pioneering study of anatomy which describes the inner ear and female reproductive organs for the first time.

The book corrects and extends the work of Falloppio's teacher, the Flemish physician Andreas Vesalius. Falloppio is the first to describe the tube extending from each ovary to the uterus (the Fallopian tubes) but does not recognize their function.

Chemistry

1540 German physician Valerius Cordus writes the first description of the preparation of ether.

Theophrastus Bombastus von Hohenheim, who called himself Paracelsus, stood at the crossroads between the mysticism of alchemy and the modern study of chemistry. Although he embraced the mysticism wholeheartedly he saw the purpose of alchemy as being the preparation of medicines with which to treat disease. Claiming to have found 'the elixir of life', he insisted he would live forever, but died aged 47. *Oxford Museum of the History of Science*

He discovers that ether is formed if alcohol and mineral acids, such as nitric acid, are distilled together. Ether becomes an important chemical in the production of medicinal compounds and later becomes one of the first anaesthetics.

1541 Swiss chemist and physician Theophrastus Bombastus von Hohenheim, better known as Paracelsus, is the first European to describe the element zinc (atomic number 30). He calls it 'zinken'. Although the pure form of the element had been produced in India during the 13th century by reducing the zinc-rich mineral calamine with wool, its chemical properties had not been recognized.

Paracelsus is greatly influenced by the tradition of alchemy begun by Arabic writers and extending through the Middle Ages. Although alchemists are popularly believed to hunt for secret formulae to turn lead into gold, scholars like Paracelsus believe their writings hold valuable information on chemical processes.

1556 German chemist and mineralogist Georgius Agricola's (Georg Bauer) pioneering work *De re metallica/On Metalwork*, a highly illustrated treatise on metallurgical and chemical processes, is published posthumously.

Agricola is a physician who studies mining closely and summarizes in this book the practical knowledge he has gained from German miners. It is clearly written and has excellent illustrations of mining machinery.

Earth Sciences

1540 Italian engineer Vannoccio Biringuccio publishes *De la pirotechnia/Concerning Pyrotechnics*, the first clearly expressed, well-illustrated work on ores, metals, and metallurgy.

It has an enormous influence all over Europe. In it Biringuccio discusses minerals, assaying, smelting, and metalworking.

1544 German chemist and mineralogist Georgius Agricola (Georg Bauer) writes *De ortu et causis subterraneorum/On Subterranean Origin and Causes*, discussing the erosive power of water, and the origin of mineral veins as depositions from solution *succus lapidescens* (lapidifying juice).

1546 Flemish cartographer Gerardus Mercator states that the Earth must have a magnetic pole separate from its 'true' pole, in order to explain the deviation of a compass needle from true north.

1546 German chemist and mineralogist Georgius Agricola (Georg Bauer) publishes *De natura fossilum/On the Nature of Fossils*, an expansion of his earlier *Bermannus* (1530), including a summary of the ancient literature on rocks, minerals, and fossils. The book contains the first attempt at a comprehensive mineral classification system, and is instrumental in the development of mineralogy.

1550 German geologist Siegmund von Herberstein publishes *De natura fossilium/On the Nature of Fossils*, discussing mineralogy and fossils.

1550 Italian mathematician Girolamo Cardano publishes *De subtilitate/On Subtlety*, an encyclopedia of science, technology, alchemy, and magic.

He writes that running water is derived from rain, which is in turn derived from the evaporation of seawater – a new idea. He also supports the idea that mountains are worn down through erosion, and that the existence of shells at high elevations is evidence that the land was once submerged.

1565 Swiss naturalist Conrad Gesner, publishes *De Reum Fossilium, Lapidum et Gemmarum/On Fossil Objects, Chiefly Stones and Gems*.

Gesner describes and classifies 'fossils' (anything dug out of the ground) by their resemblance to geometric forms, celestial objects, animals, plants, and artefacts. Using woodcuts, he is the first to include illustrations of the fossils as well. In particular, he illustrates glossopetrae, also known as tonguestones. At the time, they are commonly thought to be the fossilized tongues of snakes and dragons. After comparing them to modern shark teeth, Gesner suggests that glossopetrae are actually fossil shark teeth.

1568 Flemish cartographer Gerardus Mercator devises the cylindrical map projection named after him, for use on sea charts. It enables navigators to plot straight-line courses without having to continually adjust their compass readings.

The problem that Mercator faces is to represent a three-dimensional object, in this case the spherical Earth, on a two-dimensional map. Mercator's projection can be imagined as a cylinder surrounding the Earth with a light at the centre of the Earth casting a shadow on the cylinder. The shadow that results forms the shape of the map. One advantage of the projection is that lines of longitude appear as parallel lines on the final map. Furthermore, a navigator travelling in a constant compass direction moves along a straight line on the map. One drawback of the projection is that surfaces are shown wider than they really are as one moves north or south of the equator. Consequently, Greenland appears larger than Africa when it is in fact 13 times smaller. The cover of Mercator's book of maps shows Atlas holding the world on his shoulders. The name atlas becomes used for similar books of maps.

Ecology

1556 German chemist and mineralogist Georgius Agricola (Georg Bauer) describes how brooks are poisoned and fish are killed or driven away by the practice of ore washing in *De re metallica/On Metalwork*, his treatise on metallurgy.

Mathematics

1545 Italian mathematician and physician Girolamo Cardano publishes a formula that will solve any cubic equation, originally discovered independently by Italian mathematicians Niccolò Fontana (Tartaglia) and Scipione dal Ferro.

1557 In his *The Whetstone of Witte*, English mathematician Robert Recorde introduces the equals sign (=) into mathematics.

The same book is the first in English to use the signs + and − for addition and subtraction.

Physics

c. **1560** Italian scientist and mystic Giambattista della Porta begins work that leads to the publication, in 1589, of *Magia Naturalis/Natural Magic*, outlining a rational approach to the natural world, discussing magnetism, the structure of the eye, and the camera obscura, among other subjects.

Porta is credited with founding several scientific academies. One of these, in Naples, is for the study of the 'secrets of nature' – this leads to all of his works being banned during the years 1592–98.

1570–1599

Astronomy

11 November 1572 Danish astronomer Tycho Brahe observes a bright new star – a supernova – in the constellation Cassiopeia. It becomes known as 'Tycho's Nova'.

1573 Danish astronomer Tycho Brahe publishes his account of his observations of the 1572 supernova in *De nova stella/On the New Star*.

1576 Danish astronomer Tycho Brahe begins to build the island observatory of Uraniborg (named for Urania, the muse of astronomy) financed by King Frederick II of Denmark and Norway.

1577 A great comet appears in the skies over Europe, and Danish astronomer Tycho Brahe observes that the parallax of the comet (its movement against the background stars when seen from different positions) is less than that of the Moon, proving that it must be further away, and therefore not an atmospheric phenomenon as some still believed.

1588 Giovanni Paolo Gallucci's *Theatrum mundi/Theatre of the World* features the first star chart marked with a celestial coordinate system.

    The star positions are taken from the work of Copernicus and mapped by a trapezoidal system of projection commonly used by geographic cartographers of the time. On these are superimposed beautiful new drawings of the mythological figures after whom the constellations are named.

1595 German astronomer Johannes Kepler publishes an astrological calendar of predictions at Graz.

1596 German astronomer Johannes Kepler publishes his *Mysterium Cosmographicum/The Cosmographical Mystery*, an attempt to prove that the orbits of the planets around the Sun are all circles contained within a set of nested Pythagorean solids.

Ubi materia, ibi geometria.
Where there is matter, there is geometry.

Johannes Kepler, German astronomer and mathematician, attributed remark

---

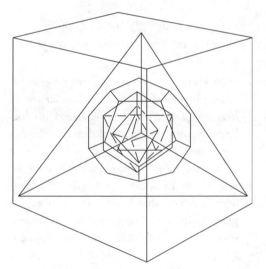

Kepler adopted Plato's five regular solids and suggested that when they were nested in the proper order (from the centre: octahedron, icosahedron, dodecahedron, tetrahedron, and cube) the planetary orbits just fitted between them.

Biology

1574 German naturalist Josias Simler writes his *De Alpibus commentarius/Commentary on the Alps*, a treatise on the natural history of the Alps.

1583 Italian botanist and physician Andrea Cesalpino attempts to classify plants systematically based on variations in their form, in his *De plantis/On Plants*.

1587 The *Physiologus/Naturalist*, the most popular medieval bestiary which repeated many folk stories of real and imaginary animals, is printed for the first time in Rome.

1597 English naturalist John Gerard publishes his *Herball, or Generall Historie of Plantes*, the first plant catalogue, describing over 1,000 species.

55

1599 Italian naturalist Ulisse Aldrovandi publishes the first three volumes of his *Natural History*, methodically listing and describing bird species in the first serious zoological study.

Chemistry

1597 German alchemist Andreas Libau, better known by the Latin name Libavius, publishes his book *Alchemia*, in which he summarizes the greatest achievements of medieval alchemy.

It can be regarded as the first chemical textbook, as it clearly states how to prepare chemicals without using mystical descriptions common to earlier works of alchemy. He is the first to describe the preparation of hydrochloric acid. He also provides clear instructions for the preparation of *aqua regia*, or 'royal water', a powerful mixture of nitric and sulphuric acids that can dissolve gold.

Earth Sciences

1574 Lazarus Ercker, superintendent of mines for Rudolf II of Austria, writes *Beschreibung allerfürnemis Mineralischen Ertzt und Burgswercksarten / Treatise on Ores and Assaying* describing ores, mining techniques, analytical procedures, and materials. The book dominates the field for 150 years.

1576 English scientist Robert Norman discovers the magnetic 'dip' or inclination in a compass needle that is caused by the Earth's magnetic field not running exactly parallel to the surface.

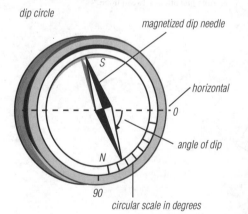

dip circle

magnetized dip needle

S

horizontal

0

angle of dip

N

90

circular scale in degrees

A dip circle is used to measure the angle between the direction of the Earth's magnetic field and the horizontal at any point on the Earth's surface. Magnetic dip was discovered in 1576.

1580 French potter and scientist Bernard Palissy publishes *Discours admirables / Admirable Discourses*, a book based on his Paris lectures on natural history, fossils, and mineralogy.

Among other things, Palissy contends that shells in rocks are fossils of once-living animals that have been petrified in place. He also observes that fossilized species are slightly different from modern species.

1581 English scientist and navigator William Borough writes *A Discourse on the Variation of the Compass or Magneticall Needle*, describing the deviation of the compass from true north depending on longitude.

1581 English scientist Robert Norman writes *The Newe Attractive*, a study of the Earth's magnetism, describing his discovery of the 'dip' in a compass needle (1576).

Mathematics

1572 Italian mathematician Rafael Bombelli becomes the first to use complex numbers to solve equations in algebra in his book *Algebra*.

Complex numbers are expressed as the sum of a real part and an imaginary part (involving the square root of −1).

1591 French mathematician François Viète writes *In artem analyticam isagoge / Introduction to the analytical arts*, in which he uses letters of the alphabet to represent unknown quantities. He uses vowels for the unknowns. The current convention of using consonants near the end of the alphabet comes from Descartes.

1595 German mathematician Bartholomaeus Pitiscus becomes the first to employ the term trigonometry in a printed publication.

Physics

1574 French physician Sebastian Basso publishes *Philosophia naturalis adversus Aristotelem / Natural Philosophy Against Aristotle*, one of the first works concerned with reviving Greek atomic theory.

A quartermaster in the Dutch army, Simon Stevin worked on the mechanics of falling bodies, preceding Galileo's theoretical findings by 18 years. He was also notable for his work in hydrostatics, and developed a system of sluices to flood the country quickly to stop an enemy army.
Oxford Museum of the History of Science

1581 Italian scientist Galileo Galilei discovers that the period of a pendulum is dependent only on its length, not on the arc of its swing.

The time period of a simple pendulum is only truly independent of the arc for small angles. It is also independent of the mass. The time period T for a simple pendulum can be expressed as

$$T = 2\pi\sqrt{(l/g)}$$

where l is the length and g the acceleration due to gravity.

1586 Flemish scientist and engineer Simon Stevinus publishes *Statics and Hydrostatics*, inventing the triangle of forces, and suggesting that objects immersed in liquid experience pressure dependent on the weight of liquid above them. He publishes a report on his experiments with falling bodies, demonstrating that acceleration due to gravity is the same for all objects. This publication is 18 years old before the Italian scientist Galileo Galilei makes the discovery for himself in 1604.

In the previous year Stevin publishes *De Thiende*, containing an early use of decimals.

1586 Italian scientist Galileo Galilei modifies and improves the hydrostatic balance for measuring the density or specific gravity of liquids.

Investigating the Solar System

by Fred W Taylor

EARLY KNOWLEDGE

The ancients knew of just five planets, other than Earth, and to these they attached romantic names and descriptions. Mercury was named after the fleet-footed messenger of the gods, because it moved across the sky so quickly and could only be faintly discerned, close to the Sun. Venus, in contrast, is a brilliant object, the most conspicuous in the sky after the Sun and Moon, casting shadows at night, and sometimes it can be glimpsed during broad daylight. Like Mercury, Venus follows the Sun, known as the evening star when it sets after the Sun, and as the morning star when it rises before the Sun. To the Greeks and Romans these were separate entities, called Hesperos and Vesper, respectively, when trailing the Sun in the evening to appear brilliantly after sunset, and Phosphoros and Lucifer when the planet led the Sun in the morning. The baleful red glow of Mars identified that planet with the god of war, and majestic, slowly moving Jupiter with the king of the gods. Saturn, finally, crawled across the sky so slowly that it suggested to our ancestors the bringer of old age.

THE TELESCOPE AGE: MOONS AND RINGS

The first major discovery after the invention of astronomically useful telescopes, at the beginning of the 17th century, was Galileo Galilei's observation in 1610 of the four large satellites of Jupiter, now known as the Galilean moons in his honour (he called them the 'Medici stars'). They were also studied by the German astronomer Simon Marius, who provided their individual names of Io, Europa, Ganymede, and Callisto.

As telescopes gradually got better, even more exciting discoveries were made. In 1655 Christiaan Huygens observed a satellite orbiting the then most distant known planet, Saturn. Realizing it must be enormous, or he would not have seen it, he called it Titan. Huygens's skill at making telescopes also brought him the answer to a question that had puzzled Galileo 50 years earlier, as to why Saturn had a strange and apparently changing shape. Huygens saw that Saturn was girdled by a ring, and in 1675 Giovanni Cassini observed the dark division in the rings that now bears his name, and surmised correctly that the rings are composed of thousands of tiny satellites, all in independent orbits.

THE DISCOVERY OF URANUS AND NEPTUNE

The discoveries that Earth was no longer alone in having a moon, and that planets could have rings, were momentous enough, but the greatest prize was the discovery of a new planet altogether. Interestingly, the planet beyond Saturn can in fact be observed with the naked eye from Earth, but is so faint and slow moving that it eluded the early, pre-telescope observers. Uranus was discovered in 1781 by William Herschel, after the Titus–Bode law had been used to predict correctly its distance from the Sun. That law fails for Neptune, however, and it was not until 1840 when Friedrich Bessel proposed that the irregularities in the orbit of Uranus were due to the gravitational effects of an unknown planet that astronomers knew where to look for a further, even more distant member of the Solar System. Detailed calculations by Urbain Leverrier in France and John Adams in England preceded the first detection of the new planet by Johann Galle and Johann Encke at the Berlin Observatory on 23 September 1846.

BEYOND NEPTUNE

Reports that Uranus, and now Neptune as well, had discrepancies between their predicted and actual orbits led to a search for another planet still further out. In fact, the reports were in error and all that lies outside the orbit of Neptune is a cloud of planetesimals, relatively small, icy bodies left over from the formation of the planets. These become comets if they are perturbed, by interactions with the giant planets, each other, or even nearby stars, into orbits that approach the Sun. The largest known members of this family are Pluto and Charon, two bodies physically similar to the large icy satellites of the giant planets, which orbit each other once a week and circle the Sun in an eccentric, tilted orbit that passes inside that of Neptune. Pluto was discovered by Clyde Tombaugh in 1930 and shown to be a double object when James Christy detected Charon in 1978.

THE SPACE AGE: THE TERRESTRIAL PLANETS REVEALED

The first space mission to a planet, not crewed of course, was that of *Mariner 2*, which flew past Venus at a distance of about 34,000 km/19,000 mi on 14 December 1962. Its mission was to investigate the microwave emission from Venus which, as measured from Earth, seemed to imply that the surface was scorching hot. The close-up date confirmed the high temperatures.

Mars, too, was initially a disappointment when *Mariner 4* arrived in November 1964. Instead of canals and vegetation, the probe's cameras saw a barren landscape with craters, which looked more like the Moon than Earth. It was not until the arrival of *Mariner 9* in 1971 that it became clear how diverse the surface of Mars is, with giant mountains, deep canyons, plains, polar caps, as well as cratered terrain.

Mariner 10 reached Mercury in March 1973 and remains the only spacecraft to have been to the innermost planet. It mapped a hot, airless, and desolate world dominated by planet-wide cracks and craters and other impact features.

EXPLORING THE GIANT PLANETS

Missions to the outer Solar System naturally came somewhat later than those to Earth's nearest neighbours, but by the end of 1972 *Pioneer 10* and *Pioneer 11* were on their way to Jupiter and Saturn. That early reconnaissance was followed by the two much larger Voyager probes, launched in 1977, which visited the two largest gas giants,

with *Voyager 2* going on to Uranus (in 1986) and Neptune (in 1989) as well. They found incredibly active atmospheres of hydrogen and helium on all four giant planets, with high winds and huge, turbulent eddies. The satellites are also fascinating, especially Titan, which is found to have an atmosphere remarkably like a colder version of Earth, being mainly nitrogen and having a surface pressure actually 50% higher than the mean terrestrial value. ,

LANDING ON MARS AND VENUS

The first probe to land on another planet was *Venera 3* in 1966, and the torrid conditions on Venus were explored by a whole series of Soviet landing spacecraft continuing into the 1980s, complemented by the US *Pioneer* orbiter/probe missions in 1978. The surface and atmosphere of Venus show evidence for massive amounts of volcanic activity, including sulphuric acid clouds that are a major contributor to the greenhouse effect that keeps the surface of Venus so hot.

The US *Viking 1* and *Viking 2* landed on Mars in 1976. The search for life was inconclusive, but evidence was accumulated for a warm, wet phase, more conducive to biological activity, in the early history of the red planet. When *Pathfinder* landed in 1996, it photographed rocks worn smooth by rushing water at its landing site in an ancient channel.

THE FUTURE

Japanese and European missions to Mars are underway, and missions to return samples are to be undertaken by NASA. The focus in the outer Solar System is now on Titan, to be explored by *Cassini* and its Huygens probe in 2004, and on Europa, the Galilean moon of Jupiter, which is believed to harbour an ocean of warm water just below its icy surface. Perhaps most exciting of all, giant space telescopes are being developed to study planets in other solar systems. These will be capable of finding Earth-like planets around Sun-like stars, and analysing their atmospheres for water, ozone, and other signs of life.

1592 Italian scientist Galileo Galilei writes *Della scienza mechanica/On the Science of Mechanics* discussing the problems of raising weights, and falling bodies.

He also constructs the first thermoscope (a thermometer with no measurement scale), which shows temperature changes by variations in the volume occupied by air in a tube. It is not accurate.

1600–1629

Astronomy

1602 Danish astronomer Tycho Brahe's *Astronomia instauratae progymnasmata/Introducing Exercises toward a Restored Astronomy* is published posthumously, giving accurate positions for 777 fixed stars and a description of the 1572 supernova in Cassiopeia.

1603 German astronomer Johann Bayer publishes his *Uranometria* star atlas, the most detailed yet, including the 12 new southern constellations, and introducing the practice of giving the stars Greek letter identifiers.

1604 German Jesuit astronomer Christopher Clavius' *Geometrica practica/Practical Geometry* is published, detailing the principles of accurate engraving for astronomical instruments such as astrolabes.

1606 German astronomer Johannes Kepler publishes his observations of the 1604 nova in the constellation of Ophiuchus in *De stella nova in pede Serpentarii/On the New Star in the Serpent-bearer's Foot*.

1609 Italian scientist Galileo Galilei, having learned of a Dutch telescope, makes his own instruments, including one that magnifies objects 20 times. They are among the first telescopes that can be used for astronomical observation.

1609 German astronomer Johannes Kepler publishes his *Astronomia nova/New Astronomy*, which describes the orbit of Mars accurately and includes his first two laws of planetary motion, which state that all planets move in elliptical orbits around the Sun and that they sweep out equal areas in equal times.

1610 French astronomer Nicholas Pieresc discovers the Orion Nebula.

1610 Italian scientist Galileo Galilei publishes *Sidereus nuncius/The Starry Messenger*, revealing his telescopic discoveries, including the moons of Jupiter, mountains on the Moon, and numerous faint stars in the Milky Way.

Galileo observes four satellites orbiting Jupiter. He names them 'Sidera Medicea', after Cosmo II Medici, Grand Duke of Tuscany. Later he discovers two 'moons' around Saturn, which vanish a year later. In fact, he has seen the planet's rings disappear as they turn to face edge-on to the Earth.

1611 English astronomer Thomas Harriott, Dutch astronomer Johannes Fabricius, German priest Christoph Scheiner, and Italian scientist Galileo Galilei all discover sunspots around the same time. Galileo uses his observations of sunspots to argue for the assertion by the German astronomer Johannes Kepler that the Sun rotates.

1611 German astronomer Simon Marius is the first to observe the Andromeda Nebula. He also observes the four moons of Jupiter and names them Io, Europa, Ganymede, and Callisto.

1615 German astronomer Christoph Scheiner publishes *Solellipticus/The Elliptical Sun*, drawing attention for the first time to the problem of the Sun's elliptical appearance when on the horizon.

1616 Having become convinced of the correctness of the heliocentric system of the Solar System devised by the Polish astronomer Copernicus, Italian scientist Galileo Galilei campaigns for its acceptance; however, the church officials in Rome dislike the controversy this arouses and place Copernicus's work on the index of prohibited books.

1617 In his *Refractiones caelestes/Celestial Refractions*, German astronomer Christoph Scheiner explains the phenomenon of the elliptical form of the Sun near the horizon as a result of refraction of its light in the Earth's atmosphere.

1619 English physician and astronomer John Bainbridge publishes *An Astronomical Description of the Comet of 1618.*

In the same year Bainbridge is appointed Savilian professor of astronomy at the University of Oxford, having earlier translated the 5th- and 2nd-century Greek astro-

nomical works: *Procli Sphaera et Ptolomaei de hypothesibus planetarum/Proclus' 'On Spheres' and Ptolemy's 'On the Hypotheses of Planets'.* He first advances the belief that comets appear as signs of impending disaster, but later in *Antiprognosticon* (1642) he denounces superstitious predictions based on conjunctions of the planets and the appearances of comets.

1619 German astronomer Johannes Kepler publishes *Harmonice mundi/The Harmony of Worlds*, which contains his third planetary law, relating the mean distance of a planet from the Sun to its orbital period.

1624 Under the new, more liberal Pope Urban VIII, Italian scientist Galileo Galilei is allowed to present the Copernican view of the Solar System in his lectures at Florence.

German astronomer Johannes Kepler developed many physical laws, including the ideas that the paths the planets follow around the Sun are ellipses not circles and orbital periods are related to the planets' distance from the Sun. *Oxford Museum of the History of Science*

. . . in my studies of astronomy and philosophy I hold this opinion about the universe, that the Sun remains fixed in the centre of the circle of heavenly bodies, without changing its place: and the Earth, turning upon itself moves round the Sun.

Galileo, Italian scientist, letter to Cristina di Lorena 1615

1626 German astronomer Christoph Scheiner publishes *Rosa ursina sive sol*, detailing his observations of the rotation of sunspots, and calculating the Sun's axis of rotation to be tilted 7.5° from the ecliptic.

1627 German astronomer Johannes Kepler completes the *Rudolphine Tables*, begun by his mentor Danish astronomer Tycho Brahe, for Emperor Rudolph II. The tables give accurate positions of 1,005 fixed stars.

Biology

1602 Italian naturalist Ulisse Aldrovandi publishes the fourth volume of his *Natural History, De animalibus insectus,* focusing on insects.

1603 Italian anatomist Hieronymus Fabricius of Acquapendente discovers that the veins contain valves, and publishes a description in his *De venarum ostiolis / Concerning the Valves of the Veins.*

Fabricius's detailed studies of veins and arteries are crucial for English physician William Harvey's later understanding of blood flow around the body.

1604 Italian anatomist Hieronymus Fabricius publishes *De formata foetu / On the Formation of the Fetus,* the first important study of embryology, in which the placenta is identified for the first time.

1609 English naturalist Charles Butler publishes *The Feminine Monarchie; or a Treatise Concerning Bees,* describing the social structure of bee colonies.

1612 Italian anatomist Hieronymus Fabricius publishes *De formatione ove et pulli / On the Formation of Eggs and Chicks.*

1614 Italian scientist Galileo Galilei records observing 'flies as big as lambs', indicating he was using a microscope by this time.

1618 English horticulturalist John Tradescant visits Russia to collect plant samples, as part of a mission organized by British king James I.

1620 English horticulturalist John Tradescant visits North Africa, returning with many samples of newly discovered species, including gutta percha and many tropical fruits.

1621 The Oxford Physic Garden is opened. The first botanical garden in Britain, it also contains the first rudimentary greenhouse in Britain, a stone greenhouse for preservation of delicate plants.

1628 English physician and anatomist William Harvey publishes *Exercitatio anatomica de motu cordis et sanguinis in animalibus / On the Motions of the Heart and Blood in Animals,* in which he argues that the heart is a pump and that blood circulates from the heart through the aorta and to the heart in the venae cavae.

Harvey explains the functions of the valves in veins but does not demonstrate how blood passes from arteries to veins. The connection is not made until the Italian biologist Marcello Malpighi is able to observe capillary blood vessels under the microscope in the 1660s. In 1651, Harvey's work on embryology is published in *On the Generation of Animals.* It is not until 1959 that Harvey's work on animal locomotion comes to light.

1629 John Parkinson, herbalist to the British king Charles I, publishes a treatise on flowers, *Paradisi in sole Paradisus terrestris / Paradise on Earth.*

Chemistry

1619 German physician Daniel Sennert publishes *De Chymicorum*, which is the first application of Greek atomic theory to chemistry.

He tries to describe chemical change in terms of the interaction between atoms. His main contribution to the field is his distinction between the atoms of an element and the atoms that make up a chemical substance, which he calls 'prima mista'. This is this first time that distinction between different types of atom has been proposed.

1624 Flemish physician Jan Baptista van Helmont publishes his *Supplementum de spadanis fontibus*, outlining his study of gases.

Van Helmont is the first to recognize that more than one air-like substance exists. He calls these substances 'chaos', which if spelled phonetically from the Flemish becomes the word 'gas'. This term is accepted to describe the gaseous state of matter. Helmont is the first to prepare the poisonous gas nitric oxide and to identify and study carbon dioxide, which he produces by burning wood.

Earth Sciences

1600 English physician William Gilbert publishes *De magnete/On Magnetism*, a pioneering study of electricity and magnetism, which distinguishes between electrostatic and magnetic effects.

In it Gilbert suggests that the Earth's magnetic field is similar to the magnetic field of a uniformly magnetized sphere. He believes that magnetism holds the planets in orbit around the Sun, a view which, although not accepted today, is a conceptual advance on previous theories.

1601 The *Naturalis historia/Natural History* by the Roman writer Pliny the Elder is translated into English by the English scholar Philemon Holland.

1622 English mathematician Edmund Gunter discovers that the magnetic needle does not retain the same declination in the same place all the time – the first evidence for variation in Earth's magnetic field.

1622 English philosopher Francis Bacon publishes *Historia naturalis et experimentalis/Natural and Experimental History*, a history of his observations on subjects such as the winds, life, and death.

Mathematics

1606 Dutch mathematician and physicist Willebrord Snel makes the first attempt to measure a degree of the meridian arc on the Earth's surface, and hence determine the shape of the Earth. The same year he translates Flemish scientist Simon Stevin's work on mechanics into Latin. It is published as *Hypomnemata mathematica/Mathematical Memoranda*.

1614 Scottish mathematician John Napier invents logarithms, a method for doing difficult calculations quickly. His results are published in *Mirifici logarithmorum canonis descriptio/Description of the Marvellous Rule of Logarithms*.

Logarithms are also invented independently, six years later, by the Swiss mathematician Joost Bürgi.

1615 German astronomer Johannes Kepler publishes *Nova stereometria doliorum vinarorum/Solid Geometry of a Wine Barrel*, an investigation of the capacity of casks, surface areas, and conic sections.

1617 Dutch mathematician and physicist Willebrord Snel establishes the technique of trigonometrical triangulation to improve the accuracy of cartographic measurements.

1617 In his *Logarithmorum chilias prima/Logarithms of Numbers from 1 to 1,000*, English mathematician Henry Briggs introduces logarithms to the base 10.

They are now known as common logarithms or Briggsian logarithms.

1617 Scottish mathematician John Napier, inventor of logarithms, devises a system of numbered sticks, called Napier's bones, to aid complex calculations. He explains their function in his last work *Rabdologiae/Study of Divining Rods*.

1623 German scholar Wilhelm Schickard makes a 'mechanical clock', a wooden calculating machine that can perform simple additions and subtractions and, with some involvement by the operator, multiplications and divisions.

It works by making use of Napier's 'bones'.

1624 In his *Arithmetica logarithmica/The Arithmetic of Logarithms*, English mathematician Henry Briggs introduces the terms 'mantissa' (for the decimal fractional part after the decimal point) and 'characteristic' (for the integer preceding the decimal point).

He also enlarges the tables of common logarithms (to the base 10) to encompass numbers from 1 to 20,000.

In 1617 Scottish mathematician John Napier published his description of what is arguably the first mechanical calculator – a set of numbered rods, usually made of bone or ivory and therefore known as Napier's bones. Using them, multiplication becomes merely a process of reading off the appropriate figures and making simple additions.

Add the numbers in each of the four horizontal rows.

carry 1 from 12 to 8;

8 × 365 = 2,920

Physics

1604 German astronomer Johannes Kepler publishes *Astronomiae pars optica/Optical Part of Astronomy*, a treatise on optics describing the function of the eye and the way in which light intensity varies with its distance from the source.

In his treatise, Kepler explains atmospheric refraction (the fact that celestial objects appear to be in a different place to their true position due to the 'bending' or refraction of light as it passes through the atmosphere).

1604 Italian scientist Galileo Galilei posits a law of falling bodies, proving that gravity acts with the same strength on all objects, independent of their mass.

This work argues against the Aristotelian view that the rate at which a body falls is proportional to its weight.

1608 Dutch spectacle maker Hans Lippershey of Middelburg, United Netherlands, discovers the magnifying power of two lenses arranged along a tube.

This leads Lippershey to the invention of the refracting telescope. The telescope becomes one of the most important scientific instruments of the time and leads to the further development of the science of astronomy.

1611 Santorio Santorio (Sanctorius), an Italian physician and colleague of Galileo, devises a temperature scale for Galileo's air thermometer, setting melting snow as 0°, and simmering water as 110°.

1620 English philosopher and essayist Francis Bacon publishes the *Novum Organum/New Instrument*, calling for an entirely new scientific method, based on induction, to replace Aristotle's syllogism.

Bacon's ideas on the importance of induction, on experiment, and on collaboration among scientists are the inspiration for the founding of scientific academies in the 17th and 18th centuries.

1621 Dutch mathematician and astronomer Willebrord Snel discovers a law of refraction, relating the angle by which light is refracted at a boundary to the properties of the media it passes between.

The law of refraction is now known as Snel's law. Snel never publishes his findings and it is to be through the work of Christiaan Huygens in 1703 that we first hear of his discovery. This law was discovered and published independently by René Descartes in 1637.

1630–1659

Astronomy

1631 German astronomer Johannes Kepler's *Solemnium/Solemn Service*, is published posthumously. It recounts an imaginary trip to the Moon, and could be considered the world's first science fiction story.

1632 An observatory is founded at Leiden University in United Netherlands.

Its original purpose is to accommodate the quadrant instrument of Willebrord Snel, the Dutch scientist best known for Snel's laws of refraction.

1632 Italian scientist Galileo Galilei publishes *Dialogo sopra i due massimi sistemi del mondo/Dialogue Concerning the Two Chief World Systems*, renewing his attack on the geocentric view of the universe.

1633 Italian scientist Galileo Galilei is tried before the Inquisition at Rome. He formally abjures all teachings contrary to the doctrine of the Catholic Church, a statement widely taken as a forced recantation of his support for the heliocentric theory. He is sentenced to house arrest and turns his attention from astronomy to physics.

The *Dialogue on the Two Chief World Systems: Ptolemaic and Copernican* (1632) is put on the Index of Prohibited Books by the Catholic Church. It remains banned until 1855.

1639 English scientist William Gascoigne invents the micrometer, an improved version of French engineer Pierre Vernier's calliper rule, and uses it on a telescope to measure the distance between stars.

1639 English curate Jeremiah Horrocks is the first person to record a transit of Venus across the Sun.

Kepler had predicted a transit of Venus in 1631, and Horrocks goes on to calculate that these transits occur not singly, but in pairs, eight years apart. In time for the second transit on 24 November 1639, Horrocks prepares apparatus with which he calculates the transit path, angular size, and orbital velocity of Venus, and derives a value for the solar parallax. From the latter Horrocks concludes that the distance of the Sun from the Earth is greater than had been believed previously. His findings were published posthumously in 1662 in *Venus in Sole Visa*.

1642 Polish astronomer Johannes Hevelius observes bright spots on the surface of the Sun and calls them faculae.

1644 Polish astronomer Johannes Hevelius observes the phases of the planet Mercury for the first time. These have particular significance in that they had been predicted by Copernicus as a consequence of his heliocentric theory of the Solar System.

1647 Polish astronomer Johannes Hevelius first charts the lunar surface accurately in his *Selenographia/Moon Map*, a work which also describes his discovery of the Moon's libration in longitude, and includes detailed observations of the Sun.

c. 1650 Polish astronomer Johannes Hevelius builds telescopes in open frameworks with focal lengths in tens of metres.

1651 English scientist William Gilbert's book *A New Philosophy of Our Sublunar World* is published posthumously, proposing, as others had done, theories that the fixed stars are not all at the same distance from Earth.

1651 Italian astronomer Giovanni Battista Riccioli publishes a map of the Moon, giving many features their modern names for the first time.

1652 Italian-born French astronomer Giovanni Cassini observes a comet from his observatory at Bologna, and publishes accurate details of its movements.

1653 Italian-born French astronomer Giovanni Cassini constructs a meridian arc for improved observation of the Sun at the church of San Petronio, Bologna, Italy.

1655 Dutch physicist and astronomer Christiaan Huygens discovers the satellite Titan orbiting the planet Saturn.

1659 Because of his skill at lens grinding and telescope construction, Dutch physicist and astronomer Christiaan Huygens has one of the best instruments available at the time. He discerns the rings surrounding Saturn, and makes the first observations of markings on the surface of Mars.

In *Systema Saturnium* (1659), Huygens explains the phases and changes in the shape of Saturn's rings. His observations and his interpretation are, however, challenged for many years, until wider improvements in telescope technology enable others to confirm his findings. Huygens develops the form of eye-piece which is known by his name. Recognizing the importance for astronomy of accurate timekeeping, Huygens goes on to patent the first pendulum clock, and to invent the cycloidal pendulum for use at sea.

Biology

1637 John Tradescant the younger, son of the celebrated English horticulturalist, visits Virginia to collect samples.

1648 Flemish physiologist Jan van Helmont's collected works, *Ortus medicinae/The Beginning of Medicine*, are published posthumously. In them, he identifies carbon dioxide as distinct from air, and coins the term 'gas' to describe it. He also demonstrates that the increase in weight of a growing tree does not come from the soil. This is one of the first accounts of a quantitative approach to biological problems.

1648 Over the next two years, the English anatomist Thomas Willis meets with others (including Robert Boyle) in Oxford, England. They are the founder members of the Royal Society of London.

1650 French naturalist Jean Brauhin begins publication of his plant catalogue, *Historia plantarum nova et absolutissima/New and Most Complete History of Plants*.

1658 Dutch microscopist Jan Swammerdam records oval particles in the blood of frogs – the first observation of red blood cells.

He pioneers the frog nerve-muscle preparation, which demonstrates muscle contraction when the nerve is stimulated. Much of the work is published in *The Bible of Nature* (1737) more than 50 years after his death.

Chemistry

1642 Seville physician Pedro Barba recounts his use of 'Peruvian bark', a source of quinine, for treating the Countess of Chinchou's malaria – the first documentation of the use of quinine in Europe.

The Incas of Peru had discovered that the bark of the cinchona tree is an effective treatment for malaria. The active ingredient in the bark is later identified as quinine. After the conquest of Peru by the Spanish, the knowledge becomes available and proves to be a useful medicine for the European explorers and settlers who travel to tropical climates.

1646 German chemist Johann Rudolf Glauber publishes *Furni novi philosophici/New Philosophy of Furnaces*, outlining principles for industrial-scale manufacture of chemicals.

The work also includes details of the preparation of nitric acid and *sal mirabile* or sodium sulphate. This latter compound is produced by the reaction between sulphuric acid and common salt, sodium chloride, and becomes known as 'Glauber's salt'. Glauber discusses the medicinal properties of this compound and sells it as a laxative.

1649 Arsenic (atomic number 33) is prepared by two different methods as an element by German pharmacist Johann Schroeder. Arsenic had been described earlier by German scholar Albertus Magnus around 1250.

Arsenic compounds have been known since the 4th century BC, but positive identification of the element is difficult since it exists in several allotropic forms, the primary ones being yellow and grey. Schroeder overcomes this difficulty and publishes two methods for the preparation of the element.

Earth Sciences

1639 English surveyor Richard Norwood attempts to measure a degree of the meridian arc on the Earth's surface, in order to determine its shape.

1644 English scientist Robert Hooke constructs a vane anemometer for measuring wind speeds.

The instrument, known as a pressure plane anenometer, employs a rectangular metal plate attached to a horizontal rod in such a way that it can swing vertically. The force of the wind pushes the plate out at an angle which can be related to wind speed. Similar anenometers are used today.

c. **1650** Duke Ferdinand II of Tuscany constructs a condensation hygrometer. The instrument consists of a chamber filled with ice. Water from the atmosphere condenses on the outside of the chamber and then runs down into a collection glass below. The amount of water collected is related to humidity.

1650 Irish archbishop James Ussher publishes *Annales veteris et novi testamenti / Chronicles of the Old and New Testaments*. In this chronology of world history, he uses the Bible to set the date of creation as 23 October 4004 BC, at nine o'clock in the morning. An English translation appears in 1654.

The idea that the Earth is 6,000 years of age will remain until the Enlightenment, when natural philosophers rely more heavily on observation and reasoning than the Bible to understand the physical nature of the Earth.

1654 Duke Ferdinand II of Tuscany establishes the world's first meteorological office, with Luigi Antinori in charge. Observatories are set up in Parma, Milan, Bologna, Florence, Pisa, Curtigliano, Vallombroua, Innsbrouk, Osnabrouk, Paris, and Warsaw.

The observatories report and record measurements of air pressure, temperature, wind direction, humidity, and local weather. The observatories close in 1667.

Mathematics

1630 English mathematician William Oughtred invents an early form of circular slide rule, adapting the principle behind Scottish mathematician John Napier's 'bones'.

It consists of a pair of graduated scales, so that multiplication can be achieved simply by adding scales together.

1631 In his *Artis analyticae praxis / Practice of the Analytic Art*, published after his death, English mathematician and astronomer Thomas Harriot introduces the symbols > and < for 'greater than' and 'less than'. The symbols, however, are due to the editor of the book, who changed the symbols that Harriot had used.

He also uses a central dot to indicate multiplication, and devises the notation XXX and XXXX to stand for the third and fourth powers of a number (X^3 and X^4). In the same year, English mathematician William Oughtred introduces the symbol × for multiplication in his *Clavis mathematicae / Key to Mathematics*.

1635 What later becomes known as Euler's theorem (not formulated by him until 1752) is discovered by the French mathematician and philosopher René Descartes.

It states that, for a simple polyhedron with no concave faces, $V + F = 2 + E$, where V is the number of vertices (corners), F the number of faces, and E the number of edges.

1637 French mathematician and philosopher René Descartes publishes *La Géométrie / Geometry*, which uses algebra to study geometry.

1639 French architect and engineer Girard Desargues begins the study of projective geometry, which considers what happens to shapes when they are projected onto a non-parallel plane.

Apart from Blaise Pascal, most of his contemporaries ignore Desargues's work.

1642 French mathematician Blaise Pascal, aged only 19, builds a calculating machine to help his father, the Intendant of Rouen, with tax calculations. It performs only additions.

1647 French mathematician Pierre de Fermat claims to have proved a certain theorem, but leaves no details of his proof. Known as Fermat's last theorem, it states that the equation $x^n + y^n = z^n$ has no non-zero integer solutions for x, y, and z when n is greater than 2. This theorem is finally proved to be true by English mathematician Andrew Wiles in 1994.

In the margin of a book Fermat left a note that intrigued and challenged mathematicians for centuries. He stated that a particular theorem could be simply proved, but there was no room to do so in the margin. In 1994, after eight years work, English mathematician Andrew Wiles finally produced a somewhat less-than-marginal 100-page proof of Fermat's last theorem. *Corbis*

1648 English bishop John Wilkins, a member of the Invisible College, publishes *Mathematical Magic*.

Wilkins also writes books on astronomy and will become one of the founders of the Royal Society of London.

1653 French mathematician Blaise Pascal publishes his 'triangle' of numbers. This has many applications in arithmetic, algebra, and combinatorics (the study of counting combinations).

1654 French mathematicians Pierre de Fermat and Blaise Pascal begin to work out the laws that govern chance and probability.

Among their discoveries is the solution of the problem of how to divide the stakes correctly if a game of chance is interrupted and cannot be completed, a solution that had long eluded the Italian mathematicians Cardano, Fra Luca Pacioli, and Tartaglia.

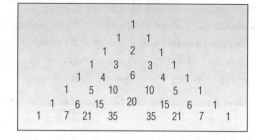

In Pascal's triangle, each number is the sum of the two numbers immediately above it, left and right – for example, 2 is the sum of 1 and 1, and 4 is the sum of 3 and 1. Furthermore, the sum of each row equals a power of 2 – for example, the sum of the 3rd row is 4 = 2^2; the sum of the 4th row is 8 = 2^3.

Together with Pierre de Fermat, Christiaan Huygens, and others, Blaise Pascal founded the modern theory of probability. His triangle of numbers, in which each number is the sum of the two above it, was used by Pascal in his study of the probable outcome of an event such as the toss of a coin. *Corbis*

1657 Dutch scientist Christiaan Huygens's *De ratiociniis in ludi aleae / On Reasoning in Games of Chance* is the first published work on probability theory, outlining for the first time the concept called mathematical expectation.

1659 Publication of the book *Teutsche algebra* by the German mathematician Johann Heinrich Rahn, 71 years after his death, reveals that he had invented the division sign (÷).

Physics

1632 French inventor Jean Rey creates a liquid thermometer by inverting Galileo's air thermometer apparatus.

   The expansion of liquid offers a better standard of measurement, but the thermometer is subject to changes in atmospheric pressure.

1638 Italian scientist Galileo Galilei investigates the motion of falling bodies and publishes *Discorsi e dimostrazioni matematiche intorno a due nove scienze / Discourses and Mathematical Discoveries Concerning Two New Sciences.*

1641 Italian scientist Evangelista Torricelli publishes *De Motu / On Motion*, a study of the physics of motion, which leads him to study with the aged Galileo Galilei in Florence, Italy.

1642 French mathematician Blaise Pascal puts forward the principles of hydraulics, the use of liquids to transmit force.

   Pascal is also to repeat the work of Torricelli and verify that the pressure of the air can be interpreted as a weight.

1643 Italian scientist Evangelista Torricelli, following a suggestion of Galileo Galilei, invents the mercury barometer that records air pressure by the changes in the level of mercury within a glass tube sealed at the top. It is almost exactly the same as those used today.

Torricelli's instrument, to 'show the changes of the air', is based on an earlier, larger, water-filled instrument that was designed to demonstrate the existence of a vacuum.

1645 German physicist Otto von Guericke constructs the first effective vacuum pump and proves that the air presses with equal force in all directions.

Von Guericke uses an apparatus similar to a water pump, but which has been modified to ensure that the working parts fit well enough to be airtight. He successfully uses the pump to create a partial vacuum and demonstrates that candles need air to burn. He also evacuates a metal sphere and measures the weight change before and after the process, thus allowing him to calculate the density of air.

1647 French mathematician Blaise Pascal demonstrates the pressure exerted by the atmosphere, using it to support water and wine 12 m/40 ft up tubes fastened to a ship's mast.

1648 French mathematician Blaise Pascal proves, with his brother-in-law, that the pressure of air decreases with increasing height, by measuring the height of a column of mercury in a barometer carried up a mountain.

1650 Jesuit priest and German scientist Athanasius Kircher conducts an experiment to test whether sound is transmitted in a vacuum. The inadequacy of the vacuum he creates leads him to reach what is now known to be the wrong conclusion.

In the classic school demonstration of this phenomenon, a vacuum is not achieved and the reduction in sound volume is due to a change of 'acoustic impedance'. Sound does travel through the air at reduced pressure but is reflected by the glass.

1658 English physicist Robert Hooke discovers that the period in which a spring vibrates remains constant, and is independent of the amount by which the spring is stretched.

Strictly this is only true if the spring does not reach its 'elastic limit'. Within the elastic limit, then the time period is given by

$$T = 2\pi\sqrt{(m/k)},$$

where m is the mass on the spring and k the 'spring constant'.

1658 English physicist Robert Hooke and the Anglo-Irish physicist Robert Boyle invent an air pump, and begin a series of experiments into the properties of vacuums.

1659 Anglo-Irish physicist Robert Boyle shows that air produces the same elastic pressure in all directions, using his 'pneumatical engine'.

1660–1689

Astronomy

1662 English astronomer John Flamsteed accurately observes a solar eclipse from his home, and corresponds with several leading astronomers on the subject.

Flamsteed goes on to become the first Astronomer Royal in 1675. The Royal Observatory at Greenwich is built and equipped for his observations and he begins observing there in 1676.

1663 Scottish astronomer and mathematician James Gregory, in his *Optica promota / Optics Advanced*, describes the form of compact reflecting telescope that bears his name.

The technology of the time cannot manufacture sufficiently precise mirrors to turn Gregory's design into a working instrument. In 1672, French instrument-maker Cassegrain produces a similar compact telescope design. Both layouts are still in use to day.

1664 Jesuit missionary Adam Schall, once director of the Chinese Imperial Board of Astronomy, is sentenced to execution by the emperor after a bitter dispute with a Chinese astronomer.

Schall and his Jesuit companions were able to predict more accurately than their Chinese rivals the solar eclipse which took place on 21 June 1629. That success gives them the opportunity to reform the calendar then in use, and to produce maps, astrolabes, and other scientific instruments. They are invited to establish an observatory within the royal palace. The Chinese royal scientists are jealous and conspire to have Schall and his companions accused of high treason, false astronomy, and teaching a superstitious religion. But on the day of sentencing, a tremendous thunder storm followed by a great fire in the palace alarms the superstitious judges and results in the sentences being commuted. All of the Jesuits are eventually set free, but Schall is by then old and frail, and dies on 15 August 1666.

1664 Italian-born French astronomer Giovanni Cassini estimates that the planet Jupiter rotates once every nine hours (the modern value is nearer ten).

1666 Italian-born French astronomer Giovanni Cassini determines rotation periods for Jupiter, Mars, and Venus, and is the first to observe the polar ice caps on Mars.

1671 Italian-born French astronomer Giovanni Cassini accepts the invitation of Louis XIV to direct the new Observatoire de Paris, which had been founded at the instigation of Louis's controller of finances, Jean-Baptiste Colbert.

Cassini goes on to discover four satellites of Saturn: Iapetus (in 1671), Rhea (1672), Tethys (1684), and Dione (1684).

1672 Italian-born French astronomer Giovanni Cassini and French astronomer Jean Richer collaborate to measure the distance to Mars at its closest approach to Earth, making simultaneous measurements in Paris, France, and Cayenne, French Guiana, and calculating the distance from parallax. The English astronomer John Flamsteed independently uses a similar method to calculate the distance. Armed with his calculation of the distance to Mars, Cassini calculates other astronomical distances, including the astronomical unit (the distance between Earth and Sun).

1675 Danish astronomer Ole Römer calculates the speed of light from the delay in the expected eclipses of Jupiter's satellites when the planet is farthest from Earth. He estimates that the Sun's rays reach the Earth in 11 minutes. The modern value is 8 minutes, 19 seconds.

Römer also introduces micrometers and reading microscopes into common use in observatories, and invents an automatic planetarium for projecting images of the sky.

1675 Italian-born French astronomer Giovanni Cassini observes the dark division in the rings of Saturn, which now bears his name. This leads him to suggest that the rings are not a solid structure, but rather composed of thousands of tiny satellites, all in independent orbits.

1676 English astronomer Edmond Halley journeys to the Atlantic island of St Helena to map the stars of the southern skies. He is hampered by the poor instruments of the time and the island's weather.

While there, Halley observes a transit of Mercury across the Sun.

1679 English astronomer Edmond Halley publishes his *Catalogus stellarum australium / Catalogue of Southern Stars* in which he gives descriptions of 341 southern stars.

1679 Italian-born French astronomer Giovanni Cassini presents his map of the Moon to the French Academy of Sciences in Paris, France. It is the result of eight years' work, and is the most accurate chart yet published.

1680 English mathematician and physicist Isaac Newton calculates that an inverse-square law of gravitational attraction between the Sun and planets would explain the elliptical orbits discovered by German astronomer Johannes Kepler. He also puts forward a theory that the air resistance encountered by a body increases in proportion to the square of its speed.

1680 English astronomer John Flamsteed publishes his *Doctrine of the Sphere*, in which he gives a new, highly accurate determination of the Sun's eccentricity. Flamsteed invents the conical projection, an important projection of the sphere onto a plane which is used in cartography.

1681 Inspired by the comet of 1680, the Swiss mathematician and physicist Jacques Bernoulli proposes a theory to explain its nature.

The theory holds that comets move in parabolic orbits, with the Sun at the focus. While it represents progress in seeking the natural laws that govern the motions of comets, the detail is not now accepted.

1682 English astronomer Edmond Halley observes the comet that he later (in 1705) concludes returns every 76 years, and which now bears his name.

While being best known for this discovery, Halley also studies magnetic variation over the Earth, and is professor of geometry at Oxford University.

Edmond Halley demonstrated that comets, far from being mysterious harbingers of doom, are astronomical objects following paths around the Sun and subject to the same gravitational laws as the planets. *Oxford Museum of the History of Science*

1682 French philosopher Pierre Bayle publishes anonymously his *Thoughts on the Comet of 1680*, in which he criticizes the long and deeply held view that comets are portents of natural disasters.

1683 Italian-born French astronomer Giovanni Cassini and the French astronomer Nicolas Fatio de Duiller publish a study into the phenomenon of zodiacal light, in which they recognize that it has an astronomical, not meteorological, source.

1687 Polish astronomer Johannes Hevelius publishes *Firmamentum Sobiescianum sive Uranographia*.

73

Hevelius improves the accuracy of measured naked-eye stellar positions down to 1 minute of arc on a routine basis. All the instruments are described by Hevelius in his *Machina Coelestis* (1673). Hevelius introduces 11 new constellations, including Scutum Sobiescanum, Canes Venatici, Leo minor, Lynx, Sextans, Lacerta (the lizard), and the fox with the goose, Vulpecula cum Anser. His *Celestial Atlas*, a catalogue of over 1,500 stars, is finally edited and published posthumously by his second wife Elisabetha in 1690.

Biology

1661 Italian anatomist Marcello Malpighi studies the lung structure of frogs, explaining for the first time how air enters the bloodstream. He also discovers the pulmonary and capillary systems passing blood around the body, confirming the English anatomist William Harvey's theory on the circulation of the blood.

1662 After the death of John Tradescant the younger, the English botanist's collection of plants is incorporated into the collection of Elias Ashmole, and ultimately into the Ashmolean Museum (founded 1683) in Oxford, England.

1663 English naturalists John Ray and Francis Willughby embark on a three-year tour of Europe to study and collect flora and fauna.

1664 English physician Thomas Willis, professor of natural philosophy at Oxford, England, publishes *Anatome cerebri nervorumque descriptio et usus / Use and Description of the Anatomy of the Brain and Nerves*, the most complete description of the brain yet written. It identifies the arteries that supply blood to the base of the brain.

1665 English scientist Robert Hooke publishes *Micrographia*, describing the development of the microscope, and coining the name 'cells' to describe cavities he has found in the structure of cork.

1667 Anglo-Irish scientist Robert Boyle studies bioluminescence of fungi and micro-organisms, and shows that it will not take place unless air is present.

1667 Italian anatomist Marcello Malpighi identifies the lower layer of the skin known today as the Malpighian layer.

1669 Italian anatomist Marcello Malpighi publishes a treatise on the anatomy and development of the silkworm, the first description of the anatomy of an invertebrate.

1670 English naturalists John Ray and Francis Willughby publish their *Catalogue of English Plants*.

1672 English naturalist Nehemiah Grew publishes *The Anatomy of Vegetables Begun*, analysing the structure of bean seeds and inventing much of the terminology for plant embryology.

1674 British physician Thomas Willis publishes *Pharmaceutice rationalis / Rational Pharmacology*, promoting the 'iatrochemical' idea of disease as essentially a chemical problem.

1674 Dutch microscopist Anton van Leeuwenhoek uses single-lens microscopes to study the composition of organisms. He discovers an extensive fauna of minute organisms, which he describes as 'very little animalcules'.

1675 Italian anatomist Marcello Malpighi publishes *Anatome plantarum / Anatomy of Plants*, the first important work on plant anatomy.

1680 Italian scientist and inventor Giovanni Alfonso Borelli's *De motu animalium / Concerning the Motion of Animals*, published posthumously, explains animal movement on physical and mechanical principles.

1682 English botanist Nehemiah Grew's book *Anatomy of Plants* is the first to identify the stamens and carpels as the male and female sex organs, respectively.

1685 The *Historia piscium / Study of Fish*, a classification of fish, mainly the work of English naturalist John Ray, is published under the name of Ray's deceased friend Francis Willughby.

1686 English naturalist John Ray piublishes the first volume of his *Historian Plantarum*, which developed his notion of a natural system of classification for living things.

Chemistry

1661 Anglo-Irish scientist Robert Boyle publishes *The Sceptical Chymist*, in which he proposes a corpuscular or atomic theory of matter, introducing the modern concept of chemical elements, and distinguishing alkali and acid properties.

The *Skeptical Chymist* separates chemistry from medicine, making the subject a science in its own right. From this time onwards, any person who studies chemistry is known as a chemist and not an alchemist. Boyle argues that elements should be identified using experimental techniques and not deduction. He points out that since the elements are the fundamental substances on Earth, they cannot be converted into simpler forms and therefore anything that can be converted is not an element. His work allies chemistry with the emerging ideals of experimental science.

Although he may have believed in the transmutation of gold, Robert Boyle helped transform alchemy into the modern science of chemistry, suggesting for example that an element was a material substance to be identified by experimentation. *Oxford Museum of the History of Science*

1667 German chemist Johann Joachim Becher publishes *Physica Subterranea / Underground Physics*, in which he discusses his theory of general chemistry.

Becher proposes that all solids can be classified into three different types of earth. In particular he mentions *terra pinguis*, or 'fatty earth', which he believes is present in all combustible materials. This concept forms the basis of the phlogiston theory, developed in the 18th century.

1669 German alchemist Hennig Brand discovers the element phosphorus (atomic number 15)

Brand distils a solution of concentrated urine to produce a white, waxy substance that glows in the dark. He calls it 'phosphor' after the Greek word meaning 'light bearer', an ancient name given to the planet Venus when it appears just before sunrise.

1674 English physician John Mayow publishes *Tractatus quinque medico-physici / Five Medico-physical Discussions*, in which he identifies part of the air of the atmosphere is required for combustion and respiration.

Mayow compares breathing with combustion and is the first to report that the volume of inhaled air is reduced after respiration. He also states that oxygen, which he calls *spiritus nitroaereus*, is essential for both combustion and breathing, and suggests that oxygen is carried around the body by blood.

1688 Plate glass is first made in France by a complex process of casting and rolling, followed by lengthy grinding and polishing.

For centuries, the only effective method of producing glass items had been by glassed-blowing techniques, which restricted its use to luxury items. Other techniques such as casting or pressing glass have been improving steadily, until the manufacture of sheet glass is perfected. The French succeed in manufacturing sheets of glass large enough to be used for windows. Their technique is faster and cheaper than glass blowing and soon glass is in common use throughout Europe, especially in the production of windowpanes and mirrors.

Earth Sciences

1660 German scientist Otto von Guericke discovers the sudden drop in air pressure preceding a violent storm – a discovery that will revolutionize weather forecasting. He also suggests comets might return periodically, pre-empting Edmond Halley's work on this subject by 45 years.

1664 German Jesuit schdar Athanasius Kircher publishes *Mundus subterraneus/The Subterranean World*.

Kircher theorizes that there is an enormous fire at the centre of the Earth, which is connected to many smaller fires via a network of channels and tunnels. He states that it is this heat that causes earthquakes as well as volcanic activity. He supports his theory with his own observations of the powerful Calabria earthquake of 1638.

1665 English physicist Robert Hooke writes that there are two kinds of fossils: inorganic rocks and crystals, and petrified animal remains or impressions of animals in rock.

1665 German Jesuit priest and scientist Athanasius Kircher is lowered into the caldera of the volcano Mount Vesuvius in Italy to make notes on its appearance and behaviour.

1667 Danish naturalist Nicolaus Steno publishes his *Elementarum Myologiae Specimen/Sample of the Elements of Myology*, in which he argues that fossils have an organic origin, and are not merely geological curiosities.

1669 Danish naturalist Nicolaus Steno issues *Prodromus*, a dissertation how fossils can be formed from living matter, and outlining basic principles of stratigraphy. He also publishes work on the crystal structure of quartz.

Steno shows that no matter what the shape and size of a quartz crystal is, the angles between the faces are always the same. This observation was instrumental in the development of the science of crystallography.

1671 French astronomer and geographer Jean Picard publishes *Mesure de la terre/Measure of the Earth* that gives the most precise determination of the length of a meridian of latitude since the early Greek measurements. He is among several scientists who take up French cleric Gabriel Menton's proposal of the metre as a standard measure of distance.

1672 Anglo-Irish scientist Robert Boyle publishes *Essay about the Origine and Virtues of Gems*, in which he discusses the physical properties of gems and minerals.

Boyle also experiments with growing crystals. After comparing artificial crystals with natural ones, he concludes that crystals grow from fluids.

1677 English scientist Richard Towneley devises a rain gauge to capture water falling on the roof of his house, and starts to maintain accurate weather records.

1686 English astronomer Edmond Halley publishes a map of the world showing the directions of prevailing winds in different regions – the first meteorological chart.

Ecology

1661 English diarist Samuel Pepys decries the pollution of London, England, calling for a reduction in coal burning, the planting of trees within the city, and the relocation of industry.

1664 English diarist John Evelyn publishes *Sylvia: A Discourse of Forest-Trees and the Propagation of Timber in His Majesty's Dominions*, a report for the Royal Navy on the cultivation of trees.

Evelyn shows a reverence and respect for trees and advocates forest conservation to enable them to serve human needs. As a result of Evelyn's book, tree planting becomes a fashionable activity for the British upper class.

Mathematics

1663 English mathematician Isaac Barrow becomes the first Lucasian Professor of Mathematics at the University of Cambridge, England. He resigns six years later to become the royal chaplain to King Charles II.

1665 English mathematician and physicist Isaac Newton begins work on differential calculus, a fundamental tool in physics for studying rates of change, such as the slopes of curves and the acceleration of moving objects.

1665 English mathematician and physicist Isaac Newton formulates the binomial theorem, a basic tool in solving algebraic equations.

1666 In order to calculate the Moon's orbit accurately, English mathematician and physicist Isaac Newton completes the development of a new type of mathematics, calculus or 'fluxions', to add infinitesimally small elements of the orbit together.

1669 English mathematician John Wallis publishes his *Mechanica/Mechanics*, a detailed mathematical study of mechanics. His *Arithmetica infinitorum/Arithmetic of Infinites* (1655) influences Isaac Newton in his early studies.

1673 German mathematician Gottfried Leibniz demonstrates his incomplete calculating machine to the Royal Society. It is the most advanced yet, capable of multiplication, division, and extracting roots.

1676 German mathematician Gottfried Leibniz invents the differential calculus, a fundamental tool in studying rates of change, independently of Newton.

1679 German mathematician Gottfried Leibniz introduces binary arithmetic (a number system to the base two), in which only two symbols are used to represent all numbers. It is not published, however, until 1701. It will eventually be adopted for use in digital computers.

The imaginary number is a fine and wonderful recourse of the divine spirit, almost an amphibian between being and not being.

Gottfried Leibniz, German mathematician and
philosopher, attributed remark

1683 Japanese mathematician Seki Kōwa publishes a treatise that first introduces determinants, in which functions are represented by a square array of numerical symbols.

Physics

1660 Anglo-Irish scientist Robert Boyle publishes *The Spring of the Air*, outlining his experiments into the nature and properties of air.

1660 English scientist Robert Hooke and Anglo-Irish scientist Robert Boyle study combustion, and suggest that a common substance in all combustible materials may be responsible.

1661 Dutch physicist and astronomer Christiaan Huygens invents a manometer, a device for measuring pressure, in order to ascertain the elastic force of gases. This is a continuation of the work of Robert Boyle in 1659 and 1660.

1662 Anglo-Irish scientist Robert Boyle describes the law that will bear his name, stating that, for a fixed mass of gas in a container, the volume occupied by the gas is inversely proportional to the pressure it exerts.

Boyle explains his findings by suggesting that the gas is composed of particles, which are forced closer together by the applied pressure.

1663 French scientist and mathematician Blaise Pascal's *Traité de l'equilibre des liqueurs / Treatise on the Equilibrium of Fluids* is published posthumously.

It includes Pascal's Law, a formal statement of the fact that force is transmitted through a fluid equally in all directions. This is the basis of all hydraulic systems.

1663 German physicist Otto von Guericke makes a machine for generating static electricity by friction, consisting of a ball of sulphur isolated from earth, and turned by an axle and winch.

1665 Italian physicist Francesco Grimaldi's *Physico-mathesis de lumine / The Physics and Mathematics of Light*, is published posthumously.

The text draws attention to the diffraction of light, and demonstrates its wavelike properties. This view is not widely held but in 1678 Christiaan Huygens publishes *Traité de la lumière / Treatise on Light* which argues strongly in favour of the wave theory.

1665 English physicist Robert Hooke adds a dial to the Italian scientist Evangelista Torricelli's mercury barometer. The needle is operated by a weight floating on top of the mercury, and allows far more accurate recording of pressure changes.

The basic design is still to be seen today.

1673 Dutch physicist and astronomer Christiaan Huygens describes the calculation of equivalent pendulum lengths in *Horologium oscillatorium / On Clock Oscillations*.

While the text deals primarily with pendulums, including the compound pendulum, it also includes a great deal of original mathematics especially concerning cycloids and parabolas. He also derives the law of centrifugal force for uniform circular motion. He visits Newton in England, after which he describes the theory of universal gravitation as 'absurd' for proposing an interaction between bodies unconnected to each other.

1674 English physicist Robert Hooke invents a time-keeping mechanism based on a balance wheel controlled by an oscillating spring instead of a pendulum.

This 'anchor escapement and balance spring' leads to great improvement in the accuracy of clocks of the time.

The Discovery of Boyle's Law

by Peter Lafferty

MEASURING THE SPRING OF THE AIR

In the 1600s, orthodox science held that 'Nature abhorred a vacuum'. In 1643 Italian physicist Evangelista Torricelli (1608–1647) invented the mercury barometer, and suggested that it contained a vacuum. But in general, a vacuum was very rare. Scientists who believed that a vacuum could exist had to explain their scarcity. Irish-born Robert Boyle (1627–1691), who settled in Oxford in 1656, felt that air had an in-built expansive power, or 'spring', which made it expand to fill any vacuum.

THE ACTIVE SPRING

In 1660, Boyle set out to demonstrate and measure this expansive power. Later, describing his experiments, he wrote:

'Diverse ways have been proposed to show both the Pressure of the Air, as the Atmosphere is a heavy Body, and the Air, especially when compressed by outward force, has a Spring that enables it to sustain or resist equal to that as much of the atmosphere, as can come to bear against it, and also to show, that such Air as we live in, and is not condensed by any human or Adventitious force, has not only a resisting Spring, but an active Spring (if I may so speak) in some measure, as when it distends a flaccid or breaks a full-blown bladder.'

Boyle began by demonstrating the 'active' spring of the air, helped by his assistant Robert Hooke (1635–1703), who had made an improved air pump for use in Boyle's experiments. Together they devised an apparatus comprising a container from which air could be extracted, holding a small inner tube containing air trapped and compressed by mercury. When the air was pumped from the outer container, this compressed air expanded, pushing the mercury from the small tube, amply demonstrating the active spring of the trapped air.

THE PASSIVE SPRING

Next, they studied the 'passive' or resisting spring of air when compressed by external pressure. They made a long glass tube 'crooked at the bottom . . . The orifice of the shorter leg . . .

being hermetically sealed'. They pasted strips of paper, carefully marked with a scale in inches, along each arm of the apparatus, poured mercury in the long, open end, and tilted the tube to one side, so that 'the air in the enclosed tube should be of the same laxity (pressure) as the rest of the air about it'.

Then they added more mercury to increase the pressure on the trapped gas, until its volume decreased by half. The additional 'head' of mercury measured 29 in/73.7 cm. Earlier, they had used a Torricellian barometer to measure the atmospheric pressure: it was equivalent to 29 in of mercury. Hence, Boyle concluded, 'this observation does both very well agree with and confirm our hypothesis . . . that the greater the weight is, that leans upon the air, the more forcible is its endeavour of dilation and consequently its power of resistance (as other springs are stronger when bent by greater weights).'

FURTHER EXPERIMENTS

At this point, the glass tube broke, scattering mercury around the laboratory. They constructed a new stronger tube, with a 'pretty bigness', placed it in a wooden box as a precaution against another breakage, and made a series of measurements of the relationship between the volume of the air and the weight of mercury needed to compress it.

They considered the effect of temperature on the results, putting a wet cloth around the tube to cool it; 'it sometimes seemed a little to shrink, but not so manifestly that we dare build anything upon it'. When they heated the closed end with a candle, 'the head had a more sensible operation' but, once again, no conclusion could be drawn. Boyle noted that 'a want of exactness . . . in such experiments is scarce avoidable'.

BOYLE'S LAW

The next step was to investigate the effect of reduced pressure on trapped air, noting the expansion of the trapped gas. Boyle describes the apparatus: 'We provided a slender glass-pipe of about the bigness of a swan's quill.' The tube, with a paper scale marked in inches along its

The Discovery of Boyle's Law *continued*

length and its top end sealed with wax, was inserted into a wide, mercury-filled tube so that about one inch extended above the mercury.

The procedure was to raise the slender tube gradually, reducing the pressure of the air inside. First, the mercury level inside rose until the weight of mercury and the reduced air pressure in the tube balanced the external air pressure. Then the volume of the trapped air gradually increased to double its original volume, when the mercury in the tube was about 14.8 in/37.6 cm above its original level. Atmospheric pressure on that day was 29.5 in/74.9 cm: the trapped air had doubled its volume when the pressure was halved. According to Boyle, this accorded well with 'the hypothesis that supposes the pressures and expansions to be in reciprocal proportions' – a proposition now known as Boyle's Law.

1675 Anglo-Irish scientist Robert Boyle invents a hydrometer for measuring the specific gravity, or density, of liquids.

This simple device becomes important in the wine and brewing industries. Its invention is more remarkable for the fact that in 1670 Boyle had suffered a stroke that left him paralysed down one side.

1675 English mathematician and physicist Isaac Newton proposes a corpuscular theory of light.

This marks the start of a long-running dispute, primarily between Newton and Huygens, regarding the corpuscular and wave theories of light.

1676 French physicist Edmé Mariotte discovers the relationship between volume and pressure in a fixed mass of gas, independently of Anglo-Irish scientist Robert Boyle.

In 1679 Mariotte publishes *De la nature de l'air/On the Nature of Air* in which he expresses the law which is believed to have been known as Mariotte's law in France. This is what we now refer to as Boyle's law.

1678 Dutch physicist and astronomer Christiaan Huygens records his discovery of the polarization of light in his *Traité de la lumière/Treatise on Light*.

The text shows that polarization is responsible for the double refraction in calcite. The text further argues, against Newton, for the wave theory of light, stating that an expanding sphere of light behaves as if each point on the sphere were a source of light – each with the same frequency.

1678 English physicist Robert Hooke discovers the law now named after him – that the extension of an elastic material such as a spring is in proportion to the force exerted on it.

Strictly this is only true provided a point, known as the 'elastic limit', is not exceeded. A simple equation, often called Hooke's law, gives

$$F = kx$$

where F is the applied force, x the extension produced, and k a property of the spring, called the 'spring constant'.

1679 English physicist Robert Hooke proposes an inverse-square law of gravity. Although pre-empting the English mathematician and physicist Isaac Newton's law of gravitation, Hooke does not provide any proof, and does not follow up the idea.

1680 French physicist Denis Papin works with the Anglo-Irish scientist Robert Boyle to develop the condensing pump, publishing their work in *A Continuation of New Experiments*.

1686 The English mathematician and physicist Isaac Newton presents his most important work the *Philosophiae naturalis principiamathematica / Mathematical Principles of Natural Philosophy* to the Royal Society, but they are short of funds and unable to finance its publication. The financial risk of its publication is borne by Newton's friend the English astronomer Edmond Halley. The work is published in 1687.

It presents his theories of motion, gravity, and mechanics, which form the basis of much of modern physics.

Isaac Newton is often cited as the greatest scientist who ever lived. He founded the science of mechanics by setting out three simple laws that described how an object moves when acted on by a force, and set out his universal law of gravitation, which explains why planets move as they do around the Sun and why an apple falls to the ground.
Oxford Museum of the History of Science

1690–1719

Astronomy

1693 English astronomer Edmond Halley compiles tables to assist in the use of the Sun for navigational purposes.

1698 Dutch physicist and astronomer Christiaan Huygens' *Cosmotheoro*, one of the earliest discourses on the possibility of extraterrestrial life on other planets in the Solar System, is published posthumously.

1704 English astronomer Edmond Halley and English mathematician and physicist Isaac Newton denounce the Astronomer Royal, John Flamsteed, for refusing to release data from his sky survey at Greenwich.

1705 English astronomer Edmond Halley conjectures that a comet seen in 1682 is identical to comets observed in 1607, 1531, and earlier; he correctly predicts its return in 1758.

1705 The Berlin Royal Observatory is founded in Germany.

1706 Danish astronomer Ole Römer publishes a catalogue of observations made at his private Tusculaneum observatory with the first telescope attached to a transit circle.

1710 English astronomer Edmond Halley begins a detailed study of the astronomical works of the Greek scientist Claudius Ptolemy (2nd century AD) and from this demonstrates that the 'fixed' stars have proper motions.

1712 English astronomer Edmond Halley and English physicist and mathematician Isaac Newton oversee the unlawful printing of Astronomer Royal John Flamsteed's star catalogue. Flamsteed is furious and burns 300 copies of the published book.

1715 English astronomer Edmond Halley predicts the path of a total solar eclipse across Britain to a high degree of accuracy.

1715 French astronomer Joseph-Nicolas Delisle suggests that the Sun's corona, observed during a total solar eclipse, may be caused by the diffraction of light around the Moon.

His *Mémoires pour servir à l'histoire et au progres de l'astronomie/Memoirs Recounting the History and Progress of Astronomy* (1738) gives the first method for determining the positions of sunspots in heliocentric (Sun-centred) coordinates. In 1753 he organizes a worldwide study of a transit of Venus (1761), the first such systematic study to be made. He used these and other data to find the distance of the Sun from the Earth.

Biology

1691 English naturalist John Ray publishes his *The Wisdom of God Manifested in the Works of Creation*, an exposition on natural theology, showing how nature reflects the divine hand.

Ray, having already made major advances in the study of plants, argues that a study of the natural world gives insights into the mind of the Creator. Ray estimates that there are 10,000 species of insects and probably 18,000 species of plants and concludes that this richness and diversity is evidence of the greatness of God.

1693 English naturalist John Ray, in *Synopsis animalium/Synopsis of Animals*, introduces the first important classification system for animals.

1696 English naturalist Hans Sloane publishes a catalogue of his plant collection, including samples collected on his visit to Jamaica in 1687.

1699 Italian physician Francisco Redi disproves the idea that maggots arise spontaneously in rotten meat, by comparing samples allowed to rot in sealed and open containers.

1710 English naturalist John Ray's *Historia insectorum/Study of Insects*, the first attempt at a systematic classification of insect species, is published posthumously.

1712 French scientist René-Antoine Ferchanlt de Réaumur demonstrates that crayfish can regenerate their claws after they have been removed.

Chemistry

1702 German chemist and physician Wilhelm Homberg is the first to synthesize boric acid, an acid found naturally in the hot lagoons of Tuscany, France.

Homberg produces the compound by reacting borax with iron sulphate. Boric acid is used initially as a preservative, before it is discovered to be toxic. It is later used as a mild antiseptic, in the process for tanning leather, and as a catalyst for the synthesis of certain organic chemicals.

1704 The pigment Prussian blue is first made in the kingdom of Prussia.

It is synthesized by reacting potassium ferrocyanide with iron (II) salts. The first product of the reaction is a white insoluble compound called Berlin white, which is converted to the soluble Prussian blue after oxidation.

1709 English engineer Abraham Darby successfully uses coke instead of charcoal to fuel a furnace for iron smelting, at Coalbrookdale, Shropshire, England. The amounts of charcoal required for smelting have been largely responsible for the high cost of iron goods.

For centuries charcoal has been the main fuel for iron smelting, but there is a shortage in the availability of wood due to the increasing demand for iron. Coke, like charcoal, is mainly composed of carbon, but is stronger and can support a greater weight of iron ore, allowing the use of larger and more cost effective furnaces. The increased availability and cheapness of iron is one of the factors that will start the Industrial Revolution in Britain.

1718 French chemist Etienne Geoffroy presents his *tables des rapports* ('tables of affinities') to the French Academy of Sciences – the first systematic record of the chemical reactivity of elements and compounds.

Geoffroy is the first person to suggest the term 'affinity' to represent the attractive forces between different bodies. He proposes that the completeness of a chemical reaction depends on the relative affinities between different reagents. His publication is a standard reference work until it is superseded in the year 1803 by the theory of chemical equilibrium, put forward by French chemist Claude Louis Berthollet.

Earth Sciences

1690 German scientist Gottfried von Leibniz writes *Protogaea / The Primordial Earth*, a study of the origin of the Earth, including a theory that the Earth was formed from a molten state.

As the Earth cooled and solidified, the oceans drained away exposing dry land. The book is not published until 1749.

1695 English physician John Woodward publishes *An Attempt toward a Natural History of the Earth and Terrestrial Bodies*, reviving the idea of a vast subterranean sea feeding all the world's seas and rivers.

Among other things, Woodward states that subterranean fires control the flow of water, and that earthquakes occur when water flowing deep within the Earth builds up against an obstruction, eventually causing the surface to rupture.

1698 English astronomer Edmond Halley takes command of the naval sloop *Paramour Pink* on a voyage to chart magnetic variations at sea, in order to aid navigation.

1700 English astronomer Edmond Halley publishes charts showing lines of equal magnetic variations in the Atlantic and Pacific oceans, as an aid to navigation.

1702 French scientist Guillaume Amontons, in his experiments on thermometers, notes for the first time that, for a constant mass of air, the change in pressure is proportional to the change in temperature.

Amontons later goes on to suggest that earthquakes are caused by air trapped within the Earth at high pressure and temperature.

1705 The geological papers of English physicist Robert Hooke, most of which are dated 1668, are published posthumously as *A Discourse on Earthquakes*. In them he discusses the long-term geological effects of repeated earthquakes.

1714 An act of Parliament establishes a Board of Longitude in Britain, with the aim of finding a solution to the determination of longitude at sea. A £20,000 prize is offered for the best method.

1715 Royal Naval lieutenant Nathaniel Blackmore uses contour lines to denote areas of equal depth while charting the Bay of Fundy and the waters off Nova Scotia.

Mathematics

1690 Swiss mathematician Jacques Bernoulli uses the word 'integral' for the first time to refer to the area under a curve.

1691 Swiss mathematician Jacques Bernoulli invents polar coordinates, a method of describing the location of points in space using angles and distances.

1694 Swiss mathematician Jean Bernoulli discovers l'Hôpital's rule for determining the correct value of certain limits.

French mathematician Guillaume de l'Hôpital purchased the rights to the rule from Bernoulli.

1706 English mathematician William Jones introduces the Greek letter π to represent the ratio of the circumference of a circle to its diameter in his *Synopsis palmariorum matheseos/ A New Introduction to Mathematics*.

1707 English mathematician and physicist Isaac Newton publishes *Arithmetica universalis/ General Arithmetic*, a collection of his results in algebra.

1707 French mathematician Abraham de Moivre uses trigonometric functions to represent complex numbers for the first time, in the form

$$z = r \left(\cos x + i \sin x \right)$$

where $r \cos z$ is the real part of the complex number, $r \sin z$ is the imaginary part, and i is the square root of -1. He develops his more famous equation on complex numbers in 1722.

1713 Swiss mathematician Jacques Bernoulli's book *Ars conjectandi/The Art of Conjecture* is an important work on probability, and includes his law of large numbers.

1715 English mathematician Brook Taylor publishes *Methodus incrementorum directa et inversa/Direct and Indirect Methods of Incrementation*, an important contribution to Isaac Newton's calculus.

1718 Swiss mathematician Jacques Bernoulli's work on the calculus of variations (the study of functions that yield certain maximum or minimum values) is published posthumously.

1718 French mathematician Abraham de Moivre publishes *The Doctrine of Chances*, an important early work on probability, stimulated by his activities in insurance and acting as a consultant to professional gamblers.

Physics

1695 French physicist and instrument-maker Guillaume Amontons invents a thermometer which uses three different liquids to counteract the change in the reading due to varying atmospheric pressure.

Experiments to Determine the Nature of Light

by Julian Rowe

AND ALL WAS LIGHT

Alexander Pope, the poet, wrote of Isaac Newton's work:

> 'Nature, and Nature's Laws lay hid in Night: God said, **Let Newton be!** and All was **Light**.'

NEWTON DISCOVERS THE SPECTRUM

Isaac Newton (1642–1727), the English physicist and mathematician, would have been remembered as a great scientist for any one of his many discoveries. He made outstanding contributions to mathematics, astronomy, mechanics, and to understanding gravitation and the nature of light.

The first microscopes revealed the world of the very small to scientists for the very first time. They saw in great detail what micro-organisms were really like. In the same way, at the other end of the scale, the first telescopes revealed to astronomers vast numbers of stars and the beauty of the Earth's planets. But there was a problem. Any image formed by the combination of the lenses then available to make microscopes or telescopes was surrounded by a colour fringe. This blurred the outline – an effect that became worse at higher magnifications.

FROM PRISMS AND RAINBOWS . . .

People had long been familiar with the rainbow colours produced when light shone through a chandelier. Now, in order to improve their instruments, scientists needed to know why these colours were formed. Newton ground his own glass lenses and had, since he was an undergraduate student at Cambridge, been interested in the effect of a glass prism on sunlight.

Newton wrote: 'In the year 1666 (at which time I applied myself to the grinding of optick glass or other figures than spherical) I procured me a triangular glass prism, to try the celebrated phaenomena of colours.'

Newton made a small hole in the shutter covering the window in his room to let in a beam of sunlight. He placed the prism in front of this hole and viewed the vivid and intense spectrum of rainbow colours cast on the opposite wall.

Newton now examined these colours individually, using a second prism. He took two boards, each with a small hole in it. The first he placed behind the prism at the window. He positioned the second board so that only a single colour produced by the first prism fell on the hole in it. The second prism was placed so as to cast the light passing through the hole in the second board onto the wall. Newton saw at once that the coloured beam of light passing through the second prism was unchanged. This proved to be true for all the rainbow colours produced by the first prism.

Newton had performed the crucial experiment, because it had been assumed previously that light was basically white, and that colours could be added to it. Now it was clear that white light was a mixture of the colours of the rainbow; the prism simply split them up or refracted the light. A second prism could not 'split' them up further.

. . . TO MODERN TELESCOPES

In his book *Opticks*, published in 1704, Newton described a further experiment in which he used the second prism to recombine the rainbow colours of the spectrum to produce white light. Here for the first time was a simple explanation of the nature of light. The book had a great impact on 18th-century writers. What Newton had to say about what happened when white light passed through a prism could be immediately applied to rainbows, and this captured the imagination of artists and poets.

Because Newton suspected that the colour fringes or chromatic aberration produced by lenses in telescopes could not be avoided, in 1668 he designed a telescope that depended instead on the use of a curved mirror. Light reflected from a surface produces no colour effects, unlike light passing through a lens. Nowadays all large

| **Experiments to Determine the Nature of Light** *continued* |

astronomical telescopes are of the reflecting type.

As a result of his elegant experiments with prisms and light, Newton also speculated about the ultimate nature of light itself. He proposed that light consisted of small 'corpuscles' (small particles) which are shot out from the source of light, rather in the same way that pellets are ejected from a shotgun. He was able to explain many of the known properties of light with this theory, which was widely accepted. Light has proved remarkably difficult to understand; nowadays it is regarded as having both particlelike and wavelike properties.

Amontons invents several other instruments including a water clock to measure longitude while at sea. Around the year 1702 he also puts forward the idea that the increase in pressure of a fixed mass of gas is proportional to the increase in its temperature.

1700 Danish scientist Ole Römer devises a standardized temperature scale, with the freezing point of water at 7.5°, boiling point at 60°, and body temperature at 22.5°.

This particular choice of 'fixed points' does not become generally accepted, and hence we do not see temperatures in °R.

1703 French physicist and instrument-maker Guillaume Amontons publishes papers on thermometers, suggesting the existence of an 'extreme zero' at which air would exert no pressure.

This appears to pre-date the notion of 'absolute zero', used on the absolute temperature scale proposed by Lord Kelvin in 1848.

1704 English physicist Isaac Newton publishes *Opticks*, the result of decades of research, delayed until after the death of English physicist Robert Hooke to avoid a priority argument. In the book, Newton mainly describes experiments on the phenomena of light and colour.

1709 Polish-born Dutch physicist Gabriel Fahrenheit creates a thermometer using the expansion of alcohol with temperature. He devises a new temperature scale to reflect the range and sensitivity of his thermometer.

Fahrenheit selects the temperature of a freezing mixture of ammonium chloride and water to be zero, the freezing point of water to be 32°, and the temperature of the human body to be 96°.

1714 Gabriel Fahrenheit builds a mercury thermometer based on earlier alcohol thermometer designs, but recognizing that mercury has a more even rate of expansion with temperature. The temperature of boiling water is later set at 212°, with 180 equal degrees separating the freezing and boiling points of water. The high accuracy of the Fahrenheit thermometer makes it the first to be useful to scientists and it is still commonly used today; the scale is still the temperature standard used in the USA.

1720–1749

Astronomy

1720 English astronomer Edmond Halley succeeds English astronomer John Flamsteed as Astronomer Royal.

1725 English astronomer John Flamsteed's *Historia coelestis Britannica / Study of the British Heavens*, a mammoth work based on 40 years of observations at Greenwich, is published post-humously.

1725 English instrument-maker George Graham graduates a 2.5 m/8 ft mural circle for improved measurements at the Royal Greenwich Observatory in London, England.

Graham's main concern is in establishing the exact shape of the Earth by means of precision clocks. Using measurements made in the tropics with his instruments and instructions, he corrects Newton's figures for the ratio of the Earth's axes.

1728 English astronomer James Bradley discovers the aberration of starlight – the difference in the angle at which starlight arrives, depending on how the Earth is moving relative to it.

This provides further important evidence for the theory that the Earth revolves around the Sun. Under Bradley's direction the observatory at Greenwich is supplied with new instruments, and with these he catalogues the positions of more than 3,000 stars. In 1748 he completes a 19-year study of the effect of the Moon's orbit on the Earth, discovering the 'nutation' that causes the Earth to wobble in its orbit.

1729 English astronomer John Flamsteed's *Atlas Coelestis / Atlas of the Heavens* is published posthumously.

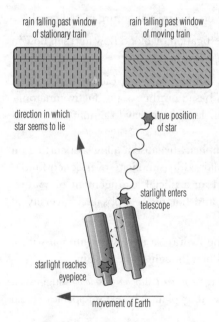

The aberration of starlight is an optical illusion caused by the motion of the Earth. Rain falling appears vertical when seen from the window of a stationary train; when seen from the window of a moving train, the rain appears to follow a sloping path. In the same way, light from a star 'falling' down a telescope seems to follow a sloping path because the Earth is moving. This causes an apparent displacement, or aberration, in the position of the star. The aberration of starlight was first noted in 1728.

1733 Swedish astronomer Anders Celsius publishes a scientific description of the aurora borealis.

1736 Swedish astronomer Anders Celsius participates in an expedition to Lapland to measure the length of a degree along a meridian, close to the pole, so that he can compare the result with a similar expedition to Peru (modern Ecuador) near the Equator. The expeditions confirm Newton's assertion that the shape of the Earth is an ellipsoid flattened at the poles.

1739 North American astronomer and physicist John Winthrop IV publishes his *Notes on Sunspots*, a detailed study of the phenomenon.

1741 Swedish astronomer Anders Celsius persuades the Swedish authorities to donate the resources necessary to build a modern observatory in Uppsala, finished in 1741 and equipped with instruments representing the most modern technology at that time. Together with his assistant Olof Hiorter, he is the first to realize that the aurora has a magnetic origin, after a long period of observations of the inclination of a compass-needle and noting the larger deviations correlating with stronger aurora activity.

Celsius makes observations of eclipses and publishes catalogues of magnitudes for a total of 300 stars using his own photometric system, which consists of using identical transparent glass plates and putting them in the ray of light from a star. He can then compare the magnitudes of the stars by the number of glass plates needed to extinguish the light (the star Sirius, the brightest star in the sky, needs 25 of his plates to be extinguished).

1742 After the death of Edmond Halley, fellow British astronomer James Bradley becomes the third Astronomer Royal.

1748 English benefactor Thomas Lowndes founds a chair of astronomy at Cambridge University, England.

Biology

1727 English botanist Stephen Hales writes *Vegetable Staticks*, in which he describes numerous experiments in plant physiology including work on the uptake and movement of water through sunflowers.

1727 French naturalist Jean-André de Peyssonel discovers the animal nature of red coral.

1733 English botanist Stephen Hales describes his investigation of blood flow and sap flow in *Statical Essays*, an expanded version of his *Vegetable Staticks* (1727)

1735 In his *Systema naturae/System of Nature*, Swedish botanist Carolus Linnaeus (Carl von Linné) introduces a system for classifying plants based on their reproductive anatomy. In later editions he makes use of binomial nomenclature as a way to simplify and stabilize the naming of species.

Linnaeus's system is widely adopted for its simplicity because it allows for sorting on the basis of easily visible features. Critics complained its simplicity overlooked fundamental features of design and thus represented an artificial arrangement of nature. Linnaneus is appointed professor of medicine and botany at Uppsala University in 1741.

1740 Swiss naturalist Abraham Trembley discovers the hydra – a freshwater animal with a simple structure. Because of its appearance, it is first thought to be a plant.

1741 French naturalist Pierre de Maupertuis's *Essai de cosmologie/Essay on Cosmology* suggests a 'survival of the fittest' principle for the first time, to explain why extant species are so well adapted to their environments.

1745 French naturalist Pierre de Maupertuis attacks the currently favoured theory of reproduction, that sperm contains a miniature version of the adult. He argues that characteristics of both parents influence the offspring.

1748 English naturalist John Needham claims to have proved the idea of spontaneous generation, observing micro-organisms coming into existence inside a sealed and sterilized container of water.

Chemistry

1722 French scientist René-Antoine Ferchault de Réaumur gives the first technical description of ironmaking in *L'Art de couvertir le fer forgé en acier et l'art d'adoucir le fer fondu/The Art of Converting Wrought Iron into Steel and the Art of Softening Molten Iron*. He also describes his invention of the cupola furnace.

1723 German chemist Georg Stahl develops his phlogiston theory and discusses its concepts in his *Fundamenta chemicae dogmaticae et experimentalis*. He theorizes that all combustible substances contain a high proportion of a material called phlogiston, which escapes during burning and is replaced by contact with combustible materials.

This work represents the first comprehensive scientific attempt to explain the processes that take place during combustion, and is especially applicable to smelting processes. This fruitful theory stimulates research in combustion and allied phenomenon for decades. The word phlogiston comes from the Greek word for 'burned'.

1731 German chemist Georg Stahl reinforces the phlogiston theory of combustion and chemical action in his *Observationes chemicae/Chemical Observations*.

What remains after burning is without phlogiston and cannot burn any further.

1736 French chemist and botanist Henri Louis Duhamel du Monceau is the first to use the term 'base' to describe those substances that form salts after a reaction with acids. He is also the first to distinguish between caustic soda (sodium hydroxide) and potash (potassium hydroxide), which were thought to be the same substance at this time.

1738 Swiss mathematician Daniel Bernoulli publishes *Hydrodynamica/Hydrodynamics*, a summary of his investigations into the motion of fluids, explaining the forces within the fluid in terms of the movements of individual particles. He also proposes the kinetic theory of gases.

1740 The first large chemical factory is founded in Richmond, England, by Joshua Ward.

It produces sulphuric acid by heating sulphur with potassium nitrate in iron reaction vessels. The resultant acid vapours are condensed into large balls. Chemical plants become a major part of the industrial age and many more chemicals will soon be produced on a large scale.

1742 Swedish scientist Anders Celsius proposes an international fixed temperature scale to the Swedish Academy of Sciences, with 100° set as the freezing point of water, and 0° set as the boiling point; this is later reversed.

The scale is originally called the 'centigrade scale' after the Latin words meaning 'a hundred steps', as there are one hundred equal degrees between the freezing and boiling points of water. However, it is changed to the 'Celsius scale' by international agreement in 1948. The Celsius scale is recognized as the standard temperature scale in most countries of the world.

1747 German chemist Andreas Marggraf is the first to extract sugar from beets.

Marggraf is the first to isolate sugar crystals from sugar beets and prove that this vegetable is a source of sugar. His discovery leads to the development of the sugar industry in Europe. The sugar beet is the world's second most important source of sugar, the first being sugar cane.

1748 Spanish scientist Antonio de Ulloa identifies the element platinum (atomic number 78), although it is not properly investigated until 1750, when English chemist William Brownrigg identifies it as an element.

A silver-white metal was referred to by the French physician and classical scholar Julius Caesar Scaliger in 1557, the metal being found near Mexico. However, Ulloa's is the first definite discovery, in Rio Pinto, Columbia. He called it 'platina' for its resemblance to silver.

1748 French physicist Jean Antoine Nollet observes osmosis and osmotic pressure; however, he cannot explain his observations.

Nollet seals an alcoholic solution with a pig's bladder and then immerses it in a vessel full of water. He observes that the bladder expands over time, indicating that water is passing through it into the alcoholic solution. The passage of water through a membrane is the process known as osmosis. Nollet also notes that liquids crossed the bladder membrane at different rates, the rate of the flow being the osmotic pressure of the fluid.

1749 German chemist Andreas Marggraf isolates formic acid (methanoic acid), which he distils from liquid extracted from red ants.

Formic acid is named after the Latin word *formica*, which means 'ant'. The acid also occurs naturally in bee stings and in certain plants. It becomes a useful chemical reagent in the processing of leather and textiles.

Earth Sciences

1723 James Jurin, the secretary of the Royal Society, London, England, requests the receipt of weather observations from England, Europe, North America, and India.

1735 English lawyer George Hadley describes the circulation of the atmosphere in terms of large-scale convection cells moving hot air away from the Equator and towards the poles.

The convection cells in which warm moist air rises at the Equator and cooler dryer air descends in the subtropics are now known as Hadley cells.

1735 English scientist Benjamin Martin uses the word 'geology' to mean a 'philosophical view' of the 'Globe of the Earth on which we live'.

The word is defined in Samuel Johnson's 1755 English dictionary as 'the knowledge of the state and nature of the Earth'.

1735 Two French scientific expeditions set sail to verify Newton's suggestion that the Earth is not a true sphere but an oblate spheroid. One of them, led by the French naturalist and mathematician Charles-Marie de La Condamine, goes on to explore the River Amazon.

In 1687, Newton had suggested that the rotation of the Earth should cause an equatorial bulge, so that the circumference of the Earth at the Equator is longer than the circumference taking in the two poles. It is reasoned that if the Earth bulges at the Equator then the distance covering a degree of latitude at the Equator should be longer than that covering a degree of latitude near the poles. This is found to be the case.

1739 French cartographer Nicolas de Lacaille leads a survey team measuring a degree of the French meridian between Perpignan and Dunkirk. Their results disprove French astronomer Jacques Cassini's theory that the Earth is flattened at the Equator rather than the poles.

1740 French scientist Pierre Bouguer makes gravity measurements on the La Condamine expedition to South America in order to determine the shape of the Earth. He notices that the Andes mountains do not attract his plumb line as much as he expects.

> The same observation is made on the 1838–43 survey of the Himalayas under George Everest. The observation is key to the understanding of 'isostasy' and will be explained by John Pratt and George Airy in 1855.

1740 Scottish mathematician Colin Maclaurin publishes a widely praised gravitational theory to explain the tides.

1743 French mathematician and physicist Alexis-Claude Clairaut applies hydrodynamic principles to the problem of the shape of the Earth, concluding that it is flattened at the poles, with an equatorial bulge.

1743 Using weather reports from several newspapers, North American politician and scientist Benjamin Franklin observes that a storm has moved northeast from Georgia to Massachusetts. This is the first recorded observation of a storm system.

1745 Russian scientist Mikhail Lomonosov publishes a catalogue of over 3,000 minerals.

1746 French naturalist Jean-Etienne Guettard makes the first mineralogical map (of France and parts of England and Wales). Among his discoveries, he identifies extinct volcanoes in the Auvergne region, and volcanic rocks in large deposits across France.

> Guettard also constructs mineralogical maps of the Switzerland, Poland, the Middle East, and much of North America.

1748 French nobleman Benoit de Maillet publishes *Telliamed*, a theory of the Earth based on widespread observations and diverging greatly from the Bible.

> Maillet states that rocks are laid down by the sea and that there have been numerous cycles of deposition. He estimates that the Earth must be at least 2 billion years old, and puts forward the evolutionary hypothesis that humans originated in the sea, and that mermaids were the halfway stage of the transition.

1749 French naturalist Georges-Louis Leclerc Buffon publishes the first book of his 36-volume *Histoire naturelle, générale et particulière / Natural History, General and Particular*, the first attempt to bring together the various fields of natural history.

Le génie n'est qu'une grande aptitude à la patience.
Genius is only a great aptitude for patience.

Georges-Louis Leclerc Buffon, French naturalist, attributed remark

---

1749 French surveyor Pierre Bouguer publishes his *Figure de la terre determinée / Diagram of the Earth*, an account of his ten-year operations on the La Condamine expedition in measuring a degree of the meridian near the Equator.

Ecology

1722 On Easter Sunday, the Dutch admiral Roggeveen visits a remote island 3,018 km / 2,000 mi off the west coast of South America; he calls it Easter Island. He finds a society of about 3,000 people living in reed huts or caves, almost constantly in a state of civil war and resorting to cannibalism to secure enough food. The island also contains over 600 massive stone statues, of average height about 6 m / 20 ft, staring out to sea.

The first inhabitants of Easter Island are believed to have arrived about AD 400, but no one is sure where they came from. The early islanders created the famous statues, called moai. Today, more than 600 statues are scattered across the island. Most are 3.5–6 m/3.8–6.6 yd tall, although some are as much as 12 m/13 yd in height and weigh over 80 tonnes. It is generally believed that the islanders used trees for rolling the statues across the island, and the decimated habitat could not then support the population. *Corel*

The island is almost completely destitute of trees. Later theories suggest that the trees were used to transport the statues and provide land for farming.

1749 Swedish biologist Carolus Linnaeus publishes his *Oeconomy of Nature.*

Linnaeus had achieved fame in 1735 with his *Systema naturae,* in which he outlined a system for classifying and naming plants. In *Oeconomy*, Linnaeus explains how the world of nature has been neatly organized by God into tidy compartments, assigning each creature to a specific habitation to avoid competition and ensure harmony.

Mathematics

1722 French mathematician Abraham de Moivre proposes an equation that is fundamental to the development of complex numbers.

The equation takes the form

$$(\cos x + i(\sin x))^n = \cos nx + i(\sin x)$$

where $\cos x + i(\sin x)$ is the complex number.

1724 Italian mathematician Jacapo Riccati propounds his equation, an important type of non-linear differential equation.

1727 In correspondence with other mathematicians, Swiss mathematician Leonhard Euler introduces the symbol e for the base of natural logarithms.

It first appears in print in 1736, in Euler's *Mechanica, sive motus scientia analytice exposita.*

1728 Italian mathematician Guido Grandi publishes *Flora geometrica/Geometrical Flowers,* attempting a geometrical definition of the curves of flower petals and leaves.

1730 French mathematician Abraham de Moivre propounds further theorems of trigonometry concerning imaginary numbers and gives Stirling's formula.

1733 French mathematician Abraham de Moivre first describes the normal ('bell-shaped') distribution curve. Later, in 1820, the discovery is credited also to the German mathematician and physicist Carl Gauss.

1736 For the Russian Academy, Swiss mathematician Leonhard Euler solves the long-standing topographical puzzle known as the 'Königsberg bridges problem' (involving seven bridges connected to two islands in a river). He proves that it is impossible to cross all the bridges without crossing one of them twice.

1737 English mathematician Thomas Simpson publishes *Treatise on Fluxions*. Produced originally as a textbook for his private students, it advocates the use of infinite series to solve problems in integration.

1739 French mathematician Jean d'Alembert publishes *Mémoire sur le calcul intégral/Memoir on Intergral Calculus*.

1740 English mathematician Thomas Simpson publishes *Treatise on the Nature and Laws of Chance*.

1742 Scottish mathematician Colin Maclaurin publishes *Treatise on Fluxions*, which treats calculus on the lines of Greek geometry.

1742 German mathematician Christian Goldbach conjectures that every even number greater than or equal to six can be written as the sum of two prime numbers. Goldbach's conjecture has not yet been proved.

Goldbach also conjectures that all odd numbers can be written as the sum of three odd prime numbers, which has been shown to be true for most numbers.

1743 French mathematician Jean d'Alembert expands Newton's work on dynamics in his *Traité de dynamique/Treatise on Dynamics*. He states a principle that the external and inertial forces acting on a solid object in motion are in equilibrium.

1744 French mathematician Jean d'Alembert publishes *Traite de l'equilibre et du mouvement des fluides/Treatise on Equilibrium and on Movement of Fluids*, applying his principle to the motion of fluids.

1746 French mathematician Jean d'Alembert further develops the theory of complex numbers.

1747 French mathematician Jean d'Alembert uses partial differential equations to study the winds in *Réflexion sur la cause générale des vents/Reflection on the General Cause of Winds*.

In the same year he employs differential equations to describe the shape of a vibrating string, producing what today would be called a wave equation.

1748 Maria Gaetana Agnesi of Italy writes *Instituzioni analitiche/Analytical Institutions*. It contains an analysis of a curve that becomes known as 'the witch of Agnesi'.

It resembles the sectional shape of a sheet of flexible material laid over a horizontal cylinder. Its equation is $x^2y = 4a^2(2a - y)$.

1748 Swiss mathematician Leonhard Euler publishes *Analysis infinitorum/Analysis of the Infinite*, an introduction to pure analytical mathematics.

In it he divides functions into two types: algebraic (that use the usual operators of arithmetic and algebra) and transcendental (exponentials and logarithms).

1748 Swiss mathematician Leonhard Euler introduces a formula linking the value of π to the square root of -1.

Physics

1729 English physicist Isaac Newton's *Principia/Principles* is translated into English from Latin by the British scholar Andrew Motte.

Until this time, the work of Newton and the majority of scientific works have only been available in Latin.

1729 English scientist Stephen Gray investigates static electricity and is the first to discover that some materials are conductors of electricity.

Gray classifies some materials as conductors (such as metallic materials) and some as insulators (mostly non-metallic substances). This is still the basic classification used today.

1730 French physicist René-Antoine Ferchault de Réaumur invents a temperature scale in which the freezing point of water in an alcohol thermometer is 0° and the boiling point 80°.

This scale was widely used in Europe but was eventually supplanted by those of Celsius, Fahrenheit, and later, for scientific use, Kelvin.

1733 French physicist Charles Dufay publishes *Six mémoires sur l'électricité/Six Memoirs on Electricity*, the result of his studies of electrostatics. He distinguishes positive and negative electric charges, and notes that electricity can cause a repulsive, as well as an attractive force.

Dufay states specifically that like charges repel while unlike charges attract.

1745 German scientist and dean of the cathedral of Camin, Ewald Georg von Kleist, invents the Leyden jar, a simple capacitor that accumulates and preserves electricity. The following year Dutch scientist Pieter van Musschenbroek makes the same discovery independently.

Following his discovery, van Musschenbroek becomes the first person to record having received a strong 'man-made' electric shock.

1746 North American scientist and politician Benjamin Franklin retires from his business to begin research into electricity, after receiving an 'electrical tube' from a colleague. He devises a new type of capacitor, improving on the Leyden jar.

The 'Franklin pane', as it is known, is a plate of glass coated with conducting foil on either side. This leads to the development of 'parallel plate' capacitors.

1747 Swiss mathematician Leonhard Euler publishes 'Sur la perfection des verres objet des lunettes/On the Perfection of Lenses for Spectacles' – an essay on optics outlining the principle of the achromatic lens.

An achromatic lens is really made from two lenses placed together. This helps to reduce the coloured fringes that otherwise appear around the edges of some objects.

The Industrializing World
1750–1899

This period of great social change was also a time of scientific change. Technology came to play an increasingly important role in scientific methods, as more and more complicated and sensitive instruments were invented.

Astronomy

In the 17th and 18th centuries astronomers were mostly concerned with positional measurements. Uranus was discovered in 1781 by German-born English astronomer William Herschel (1738–1822) and this was soon followed by the discovery of the first four asteroids. In 1846 Neptune was located by Johann Galle (1812–1910), following calculations by British astronomer John Couch Adams (1819–1892) and French astronomer Urbain Leverrièr (1811–1877). Also significant was the first measurement of the distance of a star, when in 1838 German astronomer Friedrich Bessel (1784–1846) measured the parallax of the star 61 Cygni, and calculated that it lies at a distance of about six light years (about half the modern value).

Astronomical spectroscopy was developed, and the spectra of the Sun and stars were successfully interpreted. By the 1860s good photographs of the Moon had been obtained, and by the end of the 19th century photographic methods had started to play a leading role in research.

Herschel investigated the shape of our Galaxy during the latter part of the 18th century and concluded that its stars are arranged roughly in the form of a double-convex lens. Herschel also studied the luminous 'clouds' or nebulae, and made the suggestion that some might be separate galaxies, far outside our own.

Biology

The systematic use of improved microscopes revolutionized the way in which biologists conceived organisms. In the closing years of the 18th century French physician Xavier Bichat (1771–1802), aided only with a hand lens, postulated that the body can be divided into different kinds of units that make up organs. Increasingly, biologists began thinking in terms of smaller functional units, and microscopists noticed regular structures within these units, which we now recognize as cells. By the end of the 1830s, German botanist Matthias Schleiden (1804–1881) and German physiologist Theodor Schwann (1810–1882) systematically developed the idea that all plants and animals are composed of cells.

95

French naturalist Jean Baptiste Lamarck (1744–1829) argued that species change over time but his theory of evolution was based on the idea that acquired characteristics (changes acquired in an individual's lifetime) are inherited by the offspring. 'Lamarckism' continued to be generally accepted until late in the 19th century.

English naturalist Charles Darwin (1809–1882) developed his own theory of evolution and in 1838 he hit upon its mechanism: natural selection. This principle makes use of the fact that organisms produce more offspring than can survive to maturity. In this struggle for existence, those offspring with characteristics best suited to their particular environment will tend to survive. He published his ideas in 1859 in *On the Origin of Species by Means of Natural Selection.*

Meanwhile, unknown to Darwin, Austrian monk Gregor Mendel (1822–1884) was elucidating the laws of modern genetics through his studies of inheritance patterns in pea plants and other common organisms.

Chemistry

The phlogiston theory, which had gained in popularity during the 1700s, was finally put to rest by French chemist Antoine Lavoisier (1743–1794), who put forward the theory of oxygen-based combustion.

English chemist John Dalton (1766–1844) founded the atomic theory in 1803 and set up the first table of atomic weights (relative atomic masses). Swedish chemist Jöns Berzelius (1779–1848) suggested representing each element by a symbol consisting of the first one or two letters of the name of the element. In 1864 English chemist John Newlands (1837–1898) arranged the elements in order of their increasing atomic weights and found that if he wrote them in horizontal rows, and started a new row with every eighth element, similar elements tended to fall in the same vertical columns. Russian chemist Dmitri Mendeleyev (1834–1907) published his version of the periodic table in 1869, which is more or less as we have it today.

Meanwhile the separate branches of chemistry were emerging and organic substances were being distinguished from inorganic ones. Chemists had realized that organic substances were easily converted into inorganic substances by heating or in other ways, but it was thought to be impossible to reverse the process. Then in 1828 German chemist Friedrich Wöhler (1800–1882) succeeded in converting ammonium cyanate (an inorganic compound) into urea. By the middle of the 19th century organic compounds were being synthesized in profusion; a new definition of organic compounds was clearly needed. In 1861 the German chemist Friedrich Kekulé (1829–1886) defined organic chemistry as the chemistry of carbon compounds and this definition has remained, although there are a few carbon compounds (such as carbonates) that are considered to be part of inorganic chemistry.

Earth Sciences

New ideas about the age and origin of the Earth brought momentous social reverberations because they clashed with traditional Judaeo-Christian writings on Creation.

Stratigraphy began to emerge in the 18th century, beginning with the mining schools in Germany, as earth scientists, including Johann Lehmann (1719–1776) and Abraham Werner (1749–1817), sought an understanding of the order of rock formations.

Scottish natural philosopher James Hutton (1726–1797) put forward the theory of a steady-state Earth, in which natural causes had always been of the same kind as the present, acting with precisely the same intensity (uniformitarianism). Scottish geologist Charles Lyell (1797–1875) advocated a revised version of Hutton's theory which argued that both uplift and erosion occurred by uniform natural causes.

The Austrian geologist Eduard Suess (1831–1914) offered an encyclopedic view of crustal movement, the structure and origin of mountain chains, of sunken continents, and the history of the oceans. He disputed whether the division of the Earth's relief into continents and oceans was permanent. This idea was picked up by later scientists and developed into the theory of continental drift.

Ecology and Environment

The study of interactions between various parts of the biosphere became common, as scientists and others began to examine the impact of the industrialization of Europe and North America. The idea of ecology as a distinct discipline developed through the work of German philosopher Ernst Haeckel (1834–1919), who proposed the idea that the living world could be thought of as a unit.

The idea that ecological systems could alter the well-being and development of species, including humans, gained notice through the work of, amongst others, English economist Thomas Malthus (1766–1834) and French zoologist Etienne Geoffroy Saint-Hilaire (1772–1844). Malthus presented a pessimistic view of the future due to the pressure put onto natural systems by population increases. Saint-Hilaire argued that changes in geology and climate altered the development of biological species – an important theory for later evolutionary biologists.

Conservation also gained popularity through state-sponsored nature reserves, and independent bodies, such as the American Forestry Association and the Sierra Club.

Mathematics

Swiss mathematician Leonhard Euler (1707–1783) developed spherical trigonometry and demonstrated the significance of the coefficients of trigonometric expansions; 'Euler's number' (e, as it is now called) has various useful theoretical properties and is used in the summation of particular series. Italian-born French mathematician Joseph Lagrange (1736–1813) advanced mathematical analysis. The publication of his studies of number theory and algebra were models of precise presentation, and his mathematical research into mechanics began a process of creative thought that has not ceased since.

German mathematician Carl Gauss (1777–1855) investigated geometry outside the scope of that described by Euclid, as did Russian mathematician Nikolai Lobachevsky (1792–1856) and Hungarian mathematician János Bolyai (1802–1860). Between them they thus derived non-Euclidean geometry. The ramifications of this were widespread and fast-moving. In Ireland, William Rowan Hamilton (1805–1865) suggested the concept of n-dimensional space; in Germany, mathematician Hermann Grassmann (1809–1877) not only defined it but went on to use a form of calculus based on it. But it was German mathematician Bernhard Riemann (1826–1866) who invented elliptical hyperbolic geometries, introduced 'Riemann surfaces', and redefined conformal mapping (transformations), explaining his innovations with such accuracy that the modern understanding of time and space now owes much to his work.

As the study of geometry expanded rapidly, the importance of algebra also increased accordingly. In algebraic terms, development was furthered by English mathematician Arthur Cayley (1821–1895) who discovered the theory of algebraic invariants as he carried out research into *n*-dimensional geometry. Norwegian mathematician Sophus Lie (1842–1899) made important contributions to contact transformations, and Cayley went on to invent the theory of matrices. German mathematician Felix Hausdorff (1868–1942) is credited with the formulation of topology, a new branch of mathematics.

Physics

In 1801 English physicist Thomas Young (1773–1829) discovered the principle of interference, which could be explained only by assuming that light consisted of waves. This was confirmed in 1821. Newton's discovery of the spectrum remained little more than a curiosity until 1814, when German physicist and optician Joseph von Fraunhofer (1787–1826) discovered that the Sun's spectrum is crossed by the dark lines now known as Fraunhofer lines. An explanation of the lines was provided by the German physicist Gustav Kirchhoff (1824–1887), who in 1859 showed that they are caused by elements in the Sun's atmosphere. With German chemist Robert Bunsen (1811–1899), Kirchhoff discovered elements have unique spectra by which they can be identified.

The velocity of light was first measured accurately in 1862 by French physicist Jean Foucault (1819–1868), who obtained a value within 1% of the modern value. In 1887, German-born US physicist Albert Michelson (1852–1931) and US physicist and chemist Edward Morley (1838–1923) attempted to measure the relative velocity of beams of light travelling in two directions at right angles. They believed this would enable them to determine the speed at which the Earth moves through the 'ether' that was believed to carry light waves. The negative result of the Michelson-Morley experiment eventually helped undermine belief in the existence of the ether.

In 1800 Italian physicist Alessandro Volta (1745–1827) invented the battery. Using this new source of current electricity, Danish physicist Hans Oersted (1777–1851) found, in 1820, that an electric current could deflect a magnetized needle. This discovery of electromagnetism was soon taken up by English physicist Michael Faraday (1791–1867), who pictured a current as surrounded by lines of magnetic force. This concept helped him to discover the principle of the electric motor in 1821 and electromagnetic induction in 1831.

Meanwhile, important theoretical developments were taking place in the study of electricity. In 1827, French physicist André Ampère (1775–1836) discovered the laws relating magnetic force to electric current and also properly distinguished current from tension, or EMF. In the same year, German physicist Georg Ohm (1789–1854) published his famous law relating current, EMF, and resistance. Irish physicist William Thomson (Lord Kelvin) (1824–1907) developed Faraday's work into a full theory of magnetism. Electricity and magnetism were finally brought together in a brilliant synthesis by Scottish physicist James Clerk Maxwell (1831–1879). From 1855 to 1873 Maxwell developed the theory of electromagnetism to show that electric and magnetic fields are propagated in a wave motion and that light consists of such electromagnetic waves. Maxwell's theory predicted that other similar electromagnetic radiations must exist and, as a result, German physicist Heinrich Hertz (1857–1894) produced and detected radio waves in 1888. X-rays and gamma rays were discovered accidentally soon after, by German physicist Wilhelm Röntgen (1845–1923).

The Industrializing World
1750–1899

1750–1769

Astronomy

1750 French astronomer Nicolas de Lacaille leads an expedition to the Cape of Good Hope, the southernmost part of Africa, to observe the southern sky.

Over four years, he is said to have observed over 10,000 stars using just a small refractor. He identifies several new constellations and draws up the first list of 'nebulous stars', dividing them into three classes. Lacaille also studies the Magellanic Clouds.

1750 English philosopher Thomas Wright publishes *An Original Theory or New Hypothesis of the Universe*, describing the shape of the Galaxy and the Sun's location in it.

Wright's hypothesis that our Galaxy is shaped like a disc, along a flat plane, proved influential on German-born English astronomer William Herschel and others, who developed the idea further. Wright also speculates that the stellar nebulae could be galaxies in their own right.

1750 German mapmaker Johann Mayer publishes *Kosmographische Nachrichten und Sammlungen auf das Jahr 1748/Cosmographical Information and Compilations on the Year 1748*, a highly praised collection of astronomical tables.

In 1752, Mayer publishes his *Lunar Tables*, which enable navigators to determine longitude precisely at sea. He also makes improvements in mapmaking and invents the repeating circle, later used in measuring the arc of the meridian. In 1755, he installs a 1.8 m/6 ft diameter mural quadrant at his observatory, allowing him to compile stellar charts with improved accuracy. In 1756, Mayer draws up a catalogue of 1,000 zodiacal stars and, by comparing them with Ole Römer's observations published in 1706, deduces the proper motions of 80 stars.

1755 In his *Allgemeine Naturgeschichte und Theorie des Himmels/Universal Natural History and Theory of the Heavens*, the German philosopher Immanuel Kant proposes a theory for the formation of the Solar System from a primordial nebula and proposes that our Galaxy is just one of many in the universe.

1759 French astronomers Alexis Clairault and Jean Bailly calculate the perihelion (closest point to the Sun) of the orbit of Halley's comet.

French astronomer Charles Messier is the first professional to sight the return of the comet, which has a 76-year orbital period, as predicted by Edmond Halley in 1705. The observations of the return of the comet verify Newton's laws of motion.

99

1759 Inspired by the return of Halley's comet, North American astronomer and physicist John Winthrop IV publishes his *Lectures on Comets*.

1760 To aid his search for more comets, Messier begins the first catalogue of nebulae. It is now known that his catalogue included galaxies and star clusters.

1761 Russian scientist Mikhail Lomonosov deduces that Venus has an atmosphere from watching a transit of the planet across the Sun.

1762 French astronomer Nicole-Reine Lepaute calculates the time of an annular solar eclipse and her tables, which show the eclipse at 15-minute intervals, are widely used by astronomers and navigators around the world.

Biology

1753 Scottish naval surgeon James Lind publishes his *Treatise on the Scurvy*, describing the anti-scorbutic properties of citrus fruit.

The greatest cause of disability and death among sailors on long sea voyages is scurvy, a deficiency disease caused by the lack of vitamin C. Vitamin C is found mainly in perishable foods such as fruit and vegetables, which are always in short supply at sea. Lind correctly reasons that some agent is present in these foodstuffs that is essential for good health and discovers that the regular ingestion of the juice of citrus fruits, such as orange, lemon, and lime, can prevent scurvy.

1754 French naturalist Pierre Louis de Maupertuis outlines a theory of evolution driven by spontaneous generation in his 'Essai sur la formation des corps organisés'/'Essay on the Formation of Organized Bodies'.

1757 Scottish anatomist Alexander Monro distinguishes between the lymphatic system and the blood circulatory system.

Monro also works on the physiology of fish. He is one of a family of Scottish anatomists who influence medical training in Edinburgh for more than 100 years.

1759 German scientist Kaspar Wolff publishes *Theoria generationis*/*Theory of Generations*, containing his observations on the development of plants from seeds and founding modern embryology.

1760 Dutch naturalist and engraver Pieter Lyonnet publishes a monograph on the goat-moth caterpillar, containing details and illustrations of dissections. It is one of the best illustrated books on anatomy ever produced and describes over 4,000 muscles.

1766 Swiss biologist Albrecht von Haller shows that nerves stimulate muscles to contract, and that all nerves lead to the spinal column and brain. His work lays the foundation of modern neurology.

He writes the first specific volumes on physiology, *Physiologie corporis humani*/*Elements of Human Physiology*, with an emphasis on an experimental approach to biological problems.

1767 Italian biologist Lazzaro Spallanzani disproves Georges Buffon's theories of the spontaneous generation of life by preserving organic material in vials sealed by fusing the glass.

1768 Italian biologist Lazzaro Spallanzani studies regeneration in animals and shows that the lower animals have a greater capacity to regenerate lost limbs.

1750 English chemist William Brownrigg identifies the metal platinum (atomic number 78) as a separate and distinct element.

The existence of the element had been known for centuries in South America and had been named 'platina', the Spanish for 'little silver'. In 1748, the Spanish scientist Antonio de Ulloa had compared the properties of 'platina' with those of gold. However, Brownrigg is the first to carry out a thorough scientific investigation of the metal and correctly identify it as an element. He publishes a report of its properties in *Philosophical Transactions*.

1751 Swedish mineralogist Axel Cronstedt discovers the element nickel (atomic number 28).

Cronstedt isolates the pure metal from its ore niccolite, which Swedish miners of the time call 'Kupfernickel' or 'devil's copper' since it resembles copper ore but does not yield any copper. Cronstedt calls the element 'nickel'. He discovers that the metal is weakly attracted to a magnet. Nickel becomes the first metal, other than iron, to be known to be subject to magnetic attraction.

1755 Scottish chemist Joseph Black identifies carbon dioxide, which he calls 'fixed air'.

Black discovers that carbon dioxide can be generated by the decomposition of minerals. It is known that the gas carbon dioxide is generated during combustion and fermentation, but Black is the first to show that the gas can be formed by strongly heating limestone, calcium carbonate. Limestone decomposes to form carbon dioxide and calcium oxide. He calls the gas 'fixed air' because it can be converted, or 'fixed', back to the solid, calcium carbonate, by a reaction with calcium oxide.

1755 Scottish chemist Joseph Black identifies the alkaline-earth element magnesium (atomic number 12).

Magnesium is isolated by English chemist Humphry Davy in 1808, and in a purer form by French chemist Antoine Bussy in 1828.

1756 In *Experiments upon Magnesia, Quicklime, and other Alkaline Substances*, Scottish chemist Joseph Black identifies changes in the alkalinity of carbonates as they take up or lose carbon dioxide.

Black also realizes that carbon dioxide is present in the atmosphere and is the gas exhaled by humans. His experiments involve the careful measurement of changes in weight after reaction and the measurement of the amounts of acids that are required to neutralize alkaline solutions. Black is one of the first scientists to use quantitative methods in the studies of chemical reactions.

1759 German chemist Andreas Marggraf introduces the flame test as a method of distinguishing between chemical compounds.

Marggraf notices that compounds of sodium turn a flame a distinctive yellow colour, whereas potassium compounds, which look very similar to those of sodium compounds, produce a violet-coloured flame. Flame tests develop into a sophisticated method of identifying which elements are present in chemical compounds of unknown compositions.

1765 Swedish chemist Georg Brandt discovers the element cobalt (atomic number 27). This is the first metallic element to be discovered that had not been known of either in antiquity or by the medieval alchemists.

Brandt isolates the element from cobalt-rich ores, which miners had known about for centuries. They had mistaken the blue-coloured ore to be copper ore and, since they could not obtain any copper from it, superstitiously thought that it had been cursed by earth spirits called 'kobolds'. Brandt names the element cobalt after these spirits.

1765 Swedish chemist Karl Scheele identifies prussic acid (hydrocyanic acid). He isolates the volatile, colourless, and extremely poisonous liquid from the pigment Prussian blue.

Prussic acid is used in a variety of industrial chemical processes such as electroplating and case hardening of iron and steel. It is also an effective agent for fumigation and becomes an important reagent in the synthesis of numerous organic chemicals.

1766 English scientist Henry Cavendish discovers the element hydrogen (atomic number 1) and delivers papers to the Royal Society of London, England, on the chemistry of gases.

Cavendish discovers that the action of acids on some metals causes the release of a highly inflammable gas. It is later named 'hydrogène' by French chemist Antoine Lavoisier after the Greek words 'hydro' meaning 'water' and 'gene' meaning 'forming', when it is discovered that water is produced when the element burns. Although other researchers have discovered hydrogen before, Cavendish is the first to study the element fully and report his findings.

1768 English chemist Joseph Priestley discovers carbonated water.

He discovers that passing carbon dioxide through water results in some of the gas dissolving in the liquid to form carbonated water, or soda water as it is more commonly known. This discovery leads to developments that eventually result in the birth of the modern soft drinks industry.

Earth Sciences

1751 French naturalist Nicolas Desmarest suggests the possiblity of an ancient land bridge between France and Britain, and its destruction by ocean currents.

1752 North American scientist and politician Benjamin Franklin performs his most famous experiment, flying a kite during a thunderstorm and charging a Leyden jar to which it is connected. He thereby demonstrates the electrical nature of lightning.

Contrary to popular myth, he is not struck by lightning, although a year later the German scientist Georg Richmann is killed attempting to repeat the experiment. From the results of this experiment, Franklin invents the lightning rod a year later.

1754 German geographer Anton Busching publishes his *Neue Erdbeschreibung/New Geography*. Busching, a founder of modern geography, emphasizes measurements, facts, and statistics rather than descriptive writing.

1755 German philosopher and scientist Immanuel Kant writes three studies on earthquakes.

Kant maintains that earthquakes are a result of hot compressed air in underground chambers beneath high mountain ranges. He states that the immediate cause of earthquakes is an explosive mixing of sulphuric acid, water, and iron.

1760 English geologist John Michell publishes *Conjectures Concerning the Cause and Observations upon the Phaenomena of Earthquakes*.

Having studied ground movements during the 1755 Lisbon earthquake, Michell associates earthquakes with wave motions in the Earth. He states that water vapour in contact with the Earth's subterranean fires is responsible for both earthquakes and volcanic activity.

1760 Russian scientist Mikhail Lomonosov suggests a theory for the formation of icebergs.

1763 Russian scientist Mikhail Lomonosov publishes *On the Layers of the Earth*, a work on mineralogy and geology.

Mathematics

c. **1750** French mathematician Jean d'Alembert works with other mathematicians, including Leonhard Euler, Joseph Lagrange, and Pierre Laplace, on the 'three-body problem', applying calculus to problems of celestial mechanics.

The problem concerns the behaviour of three objects that mutually attract each other by the forces of gravity between them. No general solution has been found, although some particular solutions have been calculated.

1754 French mathematician Jean d'Alembert explains the precession of the equinoxes, and perturbations in the Earth's orbit, using the mathematics of calculus.

1757 Italian-born French mathematician Joseph Lagrange establishes a mathematical society in Italy that will eventually become the Turin Academy of Sciences.

1759 German physicist Franz Aepinus publishes *Tentamen theoriae electriciatis et magnetismi/An Attempt at a Theory of Electricity and Magnetism*, the first work to apply mathematics to problems of electricity and magnetism.

1764 English mathematician and theologian Thomas Bayes publishes 'An Essay Towards Solving a Problem in the Doctrine of Chances'. This includes Bayes's theorem, which is important in statistics.

It expresses the probability of a second event happening knowing that a previous event has already occurred – a type of conditional probability.

1767 German mathematician Johann Lambert proves that the value of π cannot be written exactly as a fraction (or in other words, that it is an irrational number).

Physics

1750 English physicist and teacher John Canton discovers a method of making artificial magnets, a feat that gains him fellowship of the Royal Society.

The ability to create 'artificial' magnets allows for the creation of magnets of various shapes and sizes, of great use to experimental scientists.

1750 Scandinavian physicist Martin Stromer modifies the temperature scale devised by his mentor, Swedish astronomer Anders Celsius. He inverts it, setting freezing point as 0^0C and boiling point as 100^0C, creating the Celsius scale still used today.

1751 German physicist Johann von Segner proposes that capillary effects are caused by a 'surface tension' in liquids. He believes that this tension, which acts like a weak membrane, is caused by attractive forces within the liquid.

Surface tension is the phenomenon responsible for a variety of natural effects, such as the ability of insects to 'walk on water'. Surface tension decreases with temperature and approaches zero at the 'critical temperature'.

1751 North American scientist and politician Benjamin Franklin's experiments on electricity at Philadelphia demonstrate that electricity can magnetize iron needles.

1757 Croatian-born Italian astronomer and mathematician Roger Boscovich, in *Theoris philosophiae naturalis redacta ad unicam legem virium in natura existentium / Theory of Natural Philosophy Reduced to a Single Law of the Strength Existing in Nature*, propounds an atomic theory of matter.

Boscovich proposes that matter can be considered to be the centre of forces of both repulsion and attraction.

1767 English physicist Joseph Priestley publishes his *History and Present State of Electricity*, which suggests that electrical forces follow an inverse-square law, as does gravity.

Priestley is encouraged in this work by North American scientist and politician Benjamin Franklin, but detailed experimental work to confirm his suggestions is not carried out for nearly 20 years.

1770–1789

Astronomy

1772 German astronomer Johann Elert Bode publicizes the Titius–Bode law, first proposed in 1766 by Johann Titius, which states that the distances to the planets are proportional to the terms of the series 0, 3, 6, 12, 24, ...

Uranus is discovered in 1781, after the law is used to correctly predict its distance from the Sun. Neptune, however, is too close to be the next member of the series; Pluto matches more closely. There is some evidence that newly discovered planets orbiting stars other than the Sun may obey the Titius–Bode law, though there are, again, exceptions.

1772 German-born English astronomers William and Caroline Herschel make new discoveries in astronomy, including 8 comets and 14 nebulae.

In 1781 William Herschel discovers the planet Uranus, and in 1784 he observes clouds on Mars. A year later, Herschel argues in his work *On the Construction of the Heavens* that the Milky Way galaxy is composed of individual stars and is not some luminous fluid. In 1786 his *Catalogue of Nebulae*, a catalogue of nearly 2,500 nebulae, is published and in 1789 he completes his 40ft reflecting telescope. With it he discovers a seventh satellite (Mimas) in the Saturnian system and a nebula. He later concludes that the star was condensing out of a surrounding cloud. In 1800 Herschel discovers the existence of infrared solar rays and in 1802 he discovers that some stars revolve around others, forming binary pairs.

1779 German astronomer Heinrich Olbers devises a method of calculating the orbits of comets.

In 1811 he theorizes that pressure from solar radiation always forces the tail of a comet to point away from the Sun. In 1832 Olbers predicts the Earth will pass through the tail of Biela's Comet, causing panic in Europe. In 1801 and 1807, respectively, Olbers discovers the second and fourth asteroids, Pallas and Vesta. He is convinced that asteroids are the remains of a planet formerly lying between Mars and Jupiter that disintegrated.

13 March 1781 German-born English astronomer William Herschel discovers the planet Uranus.

1784 German-born English astronomer William Herschel discovers clouds on Mars.

1784 The first Cepheid variable star is discovered by English astronomer John Goodricke.

Goodricke is the first to calculate the period of Algol, a relatively nearby star (96 light years away) which changes its brightness by more than a magnitude as seen from Earth. Goodricke looks for other variable stars and finds Sheliak or beta Lyrae, which is a double star system with two giant Suns close together. Goodricke also discovers the first truly variable star, Altais, better known as delta Cephei. Many similarly behaved stars are now known and are classified as delta-Cephei stars or Cepheids. Goodricke is admitted to the Royal Society for his discoveries but dies a few days later, at the age of only 21 years.

1788 French astronomer and mathematician Pierre Laplace publishes his laws of the Solar System, demonstrating that planetary orbits are stable.

Working with her brother William, discoverer of the planet Uranus, Caroline Herschel was an astronomer of distinction who discovered no fewer than eight new comets. *Oxford Museum of the History of Science*

Biology

1771 English chemist Joseph Priestley discovers that plants convert carbon dioxide into oxygen.

Priestley burns a candle in an enclosed vessel to convert all the oxygen in the trapped air into carbon dioxide. He then places a live piece of the herb mint into the vessel. He finds that the plant thrives in the jar and, when he tests the air again, it supports the burning of a candle. Priestley is the first to show that plants restore the gas which is used up in respiration and combustion, later recognized as oxygen.

1775 Danish entomologist Johann Fabricius classifies insects according to their mouth structure rather than their wings, in his book *Systema Entomologiae / System of Entomology*.

1776 Swedish chemist Karl Scheele isolates uric acid from kidney stones.

1777 Dutch-born British physician and plant physiologist Jan Ingenhousz discovers two respiratory cycles in plants. He concludes that sunlight is necessary for the production of oxygen by leaves – the first identification of photosynthesis.

Ingenhousz discovers that plants only convert carbon dioxide to oxygen in the presence of light and that the converse is true if the plant is kept in the dark. He publishes his findings in 1779 in *Experiments upon Vegetables, Discovering their Great Power of Purifying the Common Air in Sunshine, and of Injuring it in the Shade and at Night*. This process later becomes known as photosynthesis from the Greek words meaning 'to put together by light'.

Chemistry

1771 English chemist Peter Woulfe discovers picric acid. He obtains the pale yellow solid by reacting indigo dye with nitric acid.

Woulfe names it after the Greek word *pikros* meaning 'bitter'. It is used initially as a yellow dye for silk, but it is soon discovered that the acid is explosive if heated quickly or ignited by a percussion charge. Picric acid-based compounds become the most widely used military explosives until after the end of World War I, when their popularity declines due to their corrosive effect on metal shell cases.

1771 Swedish chemist Karl Scheele identifies the element fluorine (atomic number 9), although it is isolated by French chemist Henri Moissan in 1886.

1772 Scottish medical student Daniel Rutherford publishes his discovery of the element nitrogen (atomic number 7).

Rutherford traps a quantity of air in a sealed container and removes all the oxygen and carbon dioxide. He discovers that a large quantity of another gas remains, which will not support combustion or respiration, but does not know what it is. He has discovered nitrogen. Soon afterwards, French chemist Antoine Lavoisier recognizes nitrogen to be an element.

1772 French chemist Antoine Lavoisier undermines the phlogiston theory by demonstrating that combustion is caused by a reaction with a component of air.

Lavoisier demonstrates that certain elements, such as phosphorus, gain weight when burned in air. According to the phlogiston theory, they should lose weight. He correctly reasons that if the elements are reacting with air then a partial vacuum will result if combustion takes place in a sealed vessel. Lavoisier shows that the weight of air necessary to fill the vacuum is identical to that gained by the element after burning.

1774 Swedish chemist Karl Scheele isolates the element chlorine (atomic number 17).

He thinks that the yellow-green gas is a compound of oxygen. The gas is finally recognized as an element by English chemist Humphry Davy in 1810. The element is named chlorine after the Greek word *chloros* meaning 'green'.

1774 Swedish mineralogist Johan Gahn isolates the element manganese (atomic number 25).

Manganese is first recognized as an element in this year by Swedish chemist Karl Scheele. However, Gahn is the first to isolate the metal, which he accomplishes by the reduction of manganese dioxide using carbon. As the element is very similar to iron in its physical and chemical properties, it is named 'manganese' from the Latin word *magnes* meaning 'magnet'.

1774 English chemist Joseph Priestley discovers the element oxygen (atomic number 8).

Oxygen was discovered earlier by Swedish chemist Karl Scheele, but Priestley publishes first. Priestley notices that a brick-red compound, mercuric oxide, is formed when mercury is heated in air. He uses sunlight concentrated through a lens to heat this compound and discovers that it decomposes to produce mercury and an unknown gas. Substances that burn in the presence of the gas do so brighter and more rapidly than in normal air. Priestley is a believer of the phlogiston theory of combustion and calls the gas 'dephlogisticated air'. It is renamed 'oxygen' by Lavoisier in 1777.

1775 Swedish chemist Torbern Olof Bergman publishes his *Disquisitio de attractionibus electivis/A Dissertation on Elective Attractions.*

It includes tables listing the elements in order of their ability to react and displace other elements in a compound. This ability is also referred to as the affinity of an element. These tables become a standard reference, used well into the next century.

1777 French chemist Antoine Lavoisier presents his 'Memoir' paper to the French Academy of Sciences in which he refutes the phlogiston theory of combustion and puts forward the idea of oxygen-based combustion.

Lavoisier assigns the name 'oxygine' to the gas that had been called 'dephlogisticated air' by Joseph Priestley in 1774. He proposes that air is made up of a mixture of gases, and that one of them, oxygen, is the substance necessary for combustion and rusting to take place. He states that it is the reaction between a burning substance and oxygen that produces combustion and not the loss or gain of phlogiston.

1777 Swedish chemist Karl Scheele discovers that silver nitrate, when exposed to light, results in a blackening effect. This is an important discovery for the development of photography, as soaking in solutions of silver nitrate is the process used for making the first photographic paper.

1778 Swedish chemist Karl Scheele discovers molybdenum (atomic number 42).

Scheele is examining an ore called 'molybdenite', which is thought to be either lead-based or some compound of graphite, when he discovers that it is made up of sulphur and a previously unknown metal. He passes his information to a fellow Swede, mineralogist Peter Jacob Hjelm, who is the first to isolate the element as a pure metal. The new element is named 'molybdenum' after the Greek word *molybdos* meaning 'lead'.

1782 Austrian mineralogist Franz Müller von Reichenstein discovers tellurium (atomic number 52). He extracts a sample of the element from gold ore.

Müller suspects that he has found a new element, but is not confident in his ability to prove his idea. He therefore sends a sample of the material to German chemist Martin Klaproth for confirmation. Klaproth determines that the substance is a new element, which he names tellurium from the Latin word for 'earth'.

1783 Spanish chemists Fausto de Elhuyar and his brother Juan José de Elhuyar discover tungsten (atomic number 74).

They isolate an acidic compound from the mineral wolframite, which yields metallic tungsten when reduced by charcoal. They name the element 'wolfram', but it becomes more commonly known as tungsten from the Swedish words meaning 'heavy stone', after work carried out on tungsten-bearing ores by Swedish chemist Karl Scheele.

1784 English chemist and physicist Henry Cavendish discovers that water is a compound of hydrogen and oxygen.

Cavendish observes that, when hydrogen is burned in a container, a liquid condenses on the cooler parts of the vessel. He establishes that the liquid is water and correctly deduces that it is formed by a combination between hydrogen and the oxygen in the air.

1784 Swedish chemist Karl Scheele discovers citric acid.

Scheele isolates the organic compound from lemon juice as colourless crystals. Citric acid is present in almost all plants and many animals. It becomes mass-produced by the fermentation of cane sugar, or molasses, to which the fungus *Aspergillus niger* has been added, and is widely used as a flavouring agent in soft drinks and confectionery.

1785 French chemist Claude-Louis Berthollet discovers the composition of ammonia. He also introduces chlorine, which he calls 'eau de Javel', as a bleaching agent. Previously, the only bleaching agents were sunlight and urine.

English chemist Joseph Priestley had discovered the gas ammonia in 1774, but had not known its composition. Berthollet determines that ammonia is a compound of nitrogen and hydrogen.

1787 French chemist Antoine Lavoisier, with collaborators, publishes *Méthode de nomenclature chimie / Method of Chemical Nomenclature*, a system for naming chemicals based on scientific principles.

Since the times of the alchemists, there has been no standard way to describe elements, compounds or chemical reactions. Many approaches have been tried throughout the 18th century to devise a common language or system that all chemists could understand no matter what their background or nationality, but none has been successful. Lavoisier's system is both logical and systematic and becomes the accepted method to describe chemistry. It develops into the system in use today.

1789 French chemist Antoine Lavoisier publishes *Traité élémentaire de chimie / Elemental Treatise on Chemistry*.

Lavoisier unifies the subject and proposes his principle of the conservation of matter, stating that in chemical reactions, matter cannot be created or destroyed but can only be changed from one form into another. He states that the total amount of matter at the start of a chemical reaction must equal that at its end. It is from this equality of matter that the term chemical equation is derived.

1789 German chemist Martin Klaproth discovers uranium (atomic number 92).

Klaproth isolates a yellow substance from pitchblende, which he deduces is the compound of a new element, but he is unable to isolate the pure metal. He names the new element 'uranium' after the planet Uranus, discovered eight years earlier.

1789 German chemist Martin Klaproth identifies the element zirconium (atomic number 40). The existence of the orange-coloured gemstone zircon has been known for centuries, before Klaproth proposes that it contains a previously unknown element.

He isolates a new substance from a sample of the mineral, which he names 'zirkon-erde' or 'zirconia'. He has discovered a compound of the element zirconium, though it will be several decades before the pure metal is isolated. The name zirconium is derived from the Persian word *zargun* meaning 'gold-like', a description of the colour of the mineral zircon.

Earth Sciences

1770 American politician and scientist Benjamin Franklin publishes a map of the Gulf Stream.

Franklin notices that American navigators, aware of the Gulf Stream current, cross the Atlantic ocean almost twice as quickly as the English. They use the current on the way east and avoid it going west.

1771 Swiss meteorologist Jean-André Deluc establishes rules for using the barometer to measure altitude and the height of mountains.

Among other accomplishments, Deluc also studies steam in the atmosphere and the effect of temperature and pressure on refraction.

1774 German mineralogist and geologist Abraham Werner publishes *Von den äusserlichen Kennzeichen der Fossilien/On the External Characteristics of Minerals*, a book of mineral identification and classification.

The book is highly regarded and widely influential. Werner goes on to teach earth science at Freiberg.

1774 French naturalist and mathematician Georges de Buffon calculates that the Earth is about 75,000 years old. His calculation is based on the cooling of the Earth from an initially molten state.

1774 The essays on extinct volcanoes by French naturalist Nicolas Desmarest argue for the volcanic origin of basalt, against the theory that all rocks are formed by sedimentation or precipitation from primeval seas.

1776 William Hamilton, British ambassador to the court of Naples, publishes a book of observations of volcanic eruptions at Etna, Stromboli, and Vesuvius.

The book is illustrated by Pietro Fabri.

1778 English clockmaker John Whitehurst writes *Inquiry into the Original State and Formation of the Earth* in which he describes the regular sequence of rock strata in Derbyshire, England, and suggests the igneous origin of 'toadstone' layers (now called dolerites).

1778 In *Epoques de la nature/Epochs of Nature*, French naturalist Georges-Louis Leclerc Buffon, reconstructs geological history as a series of stages – the first to propose such stages. It implies a great age for the Earth.

1780 The Meteorological Society of Mannheim implements a network of weather stations in Russia, Greenland, Europe, and North America.

1781 Swiss geologist Horace de Saussure invents the hair hygrometer to measure humidity. It is based on the principle that hair lengthens when wet.

James Hutton helped lay the groundwork for the development of the modern science of geology. He proposed that rocks were formed by the same processes of erosion and sedimentation still observable, operating continually over vast periods of time. *Oxford Museum of the History of Science*

The hygrometer consists of a human hair boiled in a soda solution. Treated hair is still used in hygrometers today.

1785 At a meeting of the Royal Society of Edinburgh, Scottish natural philosopher James Hutton proposes a theory now known as uniformitarianism: that all geological features are the result of processes that are at work today, acting with uniform intensity over long periods of time.

1786 German mineralogist and geologist Abraham Werner publishes *Kurze Klassifikation und Beschreibung der verschiedene Gebirgsarten/Short Classification and Description of the Various Rocks* in which he classifies rocks according to their sequence of deposition from a global ocean.

Werner is the main proponent of Neptunism: the idea that all rocks, including granites, basalts, and gneisses,

109

were precipitated or deposited from the ocean that covered the entire Earth. Werner dismisses geological processes at work today as recent and insignificant.

1787 German surgeon Johann David Schopf publishes an account of the geology of the eastern USA, from Rhode Island to South Carolina. Schopf is apparently the first to record such observations, but his work is never translated into English and goes unnoticed until the 1840s.

1788 Scottish natural philosopher James Hutton's paper 'Theory of the Earth' expounds his uniformitarian theory of continual change in the Earth's geological features.

Hutton's theory is in contrast to the popular theory of the day, now know as 'catastrophism', that the Earth's surface was formed and shaped by one single catastrophe (the biblical Flood) and that it has not changed significantly since.

Ecology

1789 English naturalist and cleric Gilbert White publishes *The Natural History of the Antiquities of Selborne*.

The book details White's observations of plant and animal life and other curiosities around his parish. White expresses a pious admiration for the order he observes but is in no doubt that nature exists chiefly for humans.

Mathematics

1770 Italian-born French mathematician Joseph Lagrange proves that every whole number can be written as the sum of four square numbers.
Swiss mathematician Leonhard Euler derives the same proof three years later.

1777 In a manuscript, Swiss mathematician Leonhard Euler introduces the symbol i to represent the imaginary number that is the square root of -1; it is not published until 1794.

1784 French mathematician Adrien Legendre introduces his Legendre polynomials, which allow the formulation of polynomial solutions to differential equations. He gives the account in a paper about the orbits of the planets.

1785 French Enlightenment philosopher Jean Caritat, marquis de Condorcet, publishes 'Essai sur l'application de l'analyse à la probabilité des décisions rendues à la pluralité des voix/Essay on the application of analysis to the probability of decisions made by a plurality of voters', a major study of probability in the social sciences.

1788 Italian-born French mathematician Joseph Lagrange publishes *Mécanique analytique/Analytical Mechanics*, an algebraic and analytical exposition of Newtonian mechanics.

Physics

1772 Swiss mathematician and physicist Leonhard Euler expounds the principles of mechanics, optics, acoustics, and astronomy in *Lettres à une princesse d'Allemagne/Letters to a German Princess*.

This work gives a 'layperson's' explanation of the physical science of the time, to which Euler has made many original contributions.

1775 Italian physicist Alessandro Volta invents the electrophorus, a device for generating and storing static electricity; it later leads to modern electrical condensers.

The principle of the electrophorus is later used in the electrostatic-generating Wimshurst machine.

1777 French scientist Charles Coulomb invents the torsion balance in which forces are measured by the amount of twist induced in a metal wire.

1785 French scientist Charles Coulomb publishes *Recherches théoriques et expérimentales sur la force de torsion et sur l'élasticité des fils de métal / Theoretical and Experimental Research on the Force of Torsion and on the Elasticity of Iron Threads*, in which he makes the first precise measurements of the electric forces of attraction and repulsion between charged bodies.

Coulomb's experimental work confirms the suggestions made by Priestley in 1767, and the inverse square law becomes known as Coulomb's law.

1786 The gold-leaf electroscope is invented by English physicist Abraham Bennet. It indicates the presence of an electric charge by the mutual repulsion of two gold leaves.

1787 French physicist Jacques Charles demonstrates that different gases expand by the same amount for the same increase in temperature.

This leads to the statement that, for a fixed mass of gas at constant pressure, the volume is directly proportional to the absolute temperature. This is now known as Charles's law.

1787 German physicist Ernst Chladni demonstrates sound patterns on vibrating plates.

These patterns, which come to be known as 'Chladni's figures', are traditionally produced by sprinkling fine powder onto a metal plate which is then 'bowed' with a cello bow.

1790–1809

Astronomy

1796 French mathematician and physicist Pierre Simon Laplace publishes *Exposition du système du monde / Account of the System of the World*, in which he enunciates the 'nebular hypothesis', which forms the basis of modern theory, proposing that the Solar System formed from a cloud of gas.

The *Exposition* consists of five books: the first is on the apparent motions of the celestial bodies, the motion of the sea, and also atmospheric refraction; the second is on the actual motion of the celestial bodies; the third is on force and momentum; the fourth is on the theory of universal gravitation, including an account of the motion of the sea and the shape of the Earth; and the fifth gives a historical account of astronomy and includes the nebular hypothesis.

1799 French mathematician and physicist Pierre Simon Laplace discovers the invariability of planetary mean motions, and proves that the eccentricities and inclinations of planetary orbits to each other always remain small, constant, and self-correcting.

These and many other of his earlier results form the basis for his *Traité de Mécanique Céleste / Celestial Mechanics*, published in 1799.

1 January 1801 Italian astronomer Giuseppe Piazzi discovers the first asteroid, Ceres.

German mathematician Carl Gauss calculates its orbit, allowing other astronomers to locate it.

May 1801 German astronomer Johann Bode becomes the first cartographer to draw boundaries between adjacent constellations, publishing *Uranographia*, a catalogue of 17,240 stars, which also contains 20 star maps.

1802 English scientist William Hyde Wollaston discovers dark lines in the solar spectrum. They are caused by the absorption of specific wavelengths of the Sun's radiation by gaseous elements in the solar atmosphere.

Wollaston notes the presence of only seven dark lines. Ten years later, Joseph von Fraunhofer finds many more while examining the spectrum of solar light passing through a thin slit, and accurately measures the positions of 324 of the 500 or so lines he can see in his solar spectrum. Dark lines in solar and stellar spectra are now known as Fraunhofer lines.

Biology

1790 German writer and polymath Johann Goethe publishes his *Versuch die Metamorphose der Pflanzen zu erklären/Attempt to Explain the Metamorphosis of Plants*, a scientific study of plant forms.

1791 Italian physiologist Luigi Galvani announces his observations on the muscular contraction of dead frogs, which he argues are caused by electricity.

The emphasis on electricity as a source of animation is brought to full effect in Mary Shelley's *Frankenstein, or the Modern Prometheus* (1818).

1794 Erasmus Darwin, English naturalist, physician, and grandfather of Charles Darwin, publishes *Zoonomia or the Laws of Organic Life*, expressing his ideas on evolution, which he assumes to have an environmental cause.

1798 English physician Edward Jenner publishes at his own expense the results of his experiments and his discovery of a vaccination against smallpox in *Inquiry into the Causes and Effects of Variolae Vaccinae*.

Jenner describes how the pus from the blisters on the hand of a woman suffering from the mild disease cowpox is scratched into the skin of the boy James Phipps. Several weeks later Jenner scratches pus taken from a victim of the much more serious smallpox, into James's skin. The boy does not develop smallpox. The work quickly establishes vaccination (immunization) as a powerful weapon in the fight against disease.

1801 Dutch botanist Christiaan Hendrik Persoon publishes *Synopsis fungorum/An Overview of Fungi*, a description of various fungi. It is the founding text of modern mycology.

1802 German naturalist Gottfried Treveranus entitles his book *Biologie*, and the word, taken from the Greek *bios* meaning 'life', and *logos* meaning 'study', is widely accepted.

1803 English botanist and entomologist Adrian Hardy Haworth publishes the first volume of *Lepidoptera Britannica/Butterflies of Britain*, the first complete description of several hundred British moths and butterflies.

1804 Swiss plant physiologist Nicolas-Theodre de Saussure publishes *Recherches chimiques sur la végétation/Chemical Research on Vegetation*, in which he demonstrates that plants require nitrogen from the soil, and increase in weight through the absorption of water and carbon dioxide.

1809 French biologist Jean-Baptiste de Lamarck publishes *Philosophie zoologique/Zoological Philosophy*, in which he theorizes that organs improve with use and degenerate with disuse and that these environmentally adapted traits are inheritable. He also proposes a progressive theory of evolution.

According to Lamarck, these acquired characteristics are the basis of evolutionary change.

1809 French physiologist François Magendie describes experiments which test the effects of plant alkaloids on animals (and himself). The work shows that many of the compounds have useful medicinal properties.

Later work is the basis for establishing experimental physiology in France. The work is continued by his assistant and successor Claude Bernard.

Chemistry

1762 Scottish chemist Joseph Black introduces the term 'latent heat' to describe the characteristic energy of a substance that is required to change its physical state. He observes that additional heat that must be supplied to turn ice into water without changing its temperature – demonstrating that ice absorbs heat without changing temperature when it melts.

Black reasons that the energy is being taken in by the ice to change it from a solid to a liquid. Black calls this the 'latent heat' of the ice, from the Latin word for 'hidden', as there is no temperature rise. This concept is to prove useful in the development of the steam engine.

1783 French chemist Nicolas Leblanc invents an inexpensive process for the manufacture of soda (sodium hydroxide and sodium carbonate) from salt (sodium chloride). This leads to the development of the soap industry in France and provides raw materials for the glass, porcelain, and paper-pulp industries.

1790 Nitrogen is first named such by French chemist Jean Chaptal.

French chemist Antoine Lavoisier had previously named the element 'azote', a name derived from the Greek word *zoe* meaning 'life', because of the element's inability to sustain life. However, Chaptal renames the element when it is discovered that it is present in nitre, potassium nitrate. The name is derived by combining the words 'nitre', from potassium nitrate and the Greek word *genes* meaning forming, literally 'nitre-forming'.

1791 English clergyman William Gregor discovers titanium (atomic number 22).

Gregor finds samples of titanium ore in sand from the beaches in Cornwall, England. Although he cannot identify it, he reports the observation of a new substance, which he calls a 'reddish-brown calx'. German chemist Martin Klaproth determines that the ore contains a previously unknown metallic element, which he names 'titanium' after the giants of Greek mythology, the Titans.

1794 Finnish chemist Johan Gadolin discovers the rare-earth element yttrium (atomic number 39).

The term 'earth' is used to describe any oxide that is resistant to heat and insoluble in water. Common earths such as iron oxide and silicon dioxide make up a large

113

proportion of the Earth's crust. Gadolin discovers a new 'earth' in a quarry near the town of Ytterby, Sweden. It is so unlike those previously known that he calls it a 'rare earth' and names it 'ytrria' after the town. Yttria is a mixture of oxides of several unknown metals, notably the element yttrium.

1796 English chemist Smithson Tennant is first to prove that diamonds consist entirely of carbon.

Tennant carries out a series of quantitative experiments in which he accurately measures the amount of carbon dioxide formed when a diamond is completely burned. His observations show that the only way the measured amount of carbon dioxide could be produced during combustion is if the diamond is composed only of carbon.

1797 French chemist Louis-Nicolas Vauquelin discovers chromium (atomic number 24).

Vauquelin isolates an unknown substance from a mineral from Siberia. He recognizes it as the compound of a new element and isolates the pure metal in 1798. He names it 'chromium' after the Greek word *chromo* meaning 'colour', because the element forms brightly coloured compounds.

1798 French chemist Louis-Nicolas Vauquelin discovers beryllium (atomic number 4).

Vauquelin discovers an unknown oxide while studying the gemstones beryl and emerald but is unable to isolate the new element. This is accomplished by German chemist Friedrich Wöhler in 1828. The new metal is called beryllium after the gemstone beryl.

1799 French chemist Joseph-Louis Proust discovers that the elements in a compound combine in definite proportions.

Proust carries out a series of experiments on the chemical composition of a range of substances including iron oxides and copper carbonate. He shows that the elements making up a particular compound are always in the same proportion by weight, no matter how the compound is synthesized. This is the first stating of the law of definite proportions and, in this form, is also known as Proust's law.

***c.* 1800** English chemist and physicist William Hyde Wollaston develops a method to make platinum malleable.

The properties of platinum make it ideally suited for making laboratory apparatus. However, these properties also make the metal difficult to work into the required shapes. Wollaston is the first to develop a method that allows platinum to be easily worked. Wollaston arranges to have his method published after his death and his techniques form the basis of modern powder metallurgy.

1800 Italian physicist Alessandro Volta invents the voltaic pile made of discs of silver and zinc – the first electric battery.

Volta uses bowls of salt solution connected by metal arcs to generate electricity. The arcs are constructed so that different metals contact separate bowls, usually copper on one end and nickel or zinc on the other. This is the first source of a continuous current of electricity, which proves to be of enormous use to scientists. His invention is also known as the 'voltaic cell'. The unit of electrical potential is now known as the volt.

1800 English chemist Humphry Davy publishes *Researches, Chemical and Philosophical, Chiefly Concerning Nitrous Oxide*, detailing the effects of nitrous oxide.

The gas was initially discovered in 1772 by English chemist Joseph Priestley, who called it 'diminished nitrous air'. Davy is the first to determine the properties of the gas and renames it 'nitrous oxide'. He discovers that the gas induces an intoxicated state when inhaled, causing weeping and uncontrollable laughing, earning nitrous oxide the nickname 'laughing gas'. More importantly, Davy reports feeling no pain under the influence of the gas. Nitrous oxide eventually becomes the first true chemical anaesthetic.

1800 English chemist William Nicholson and English anatomist Anthony Carlisle are the first to electrolyse water. They build their version of a voltaic cell and use it to pass an electric current through a slightly acidic solution of water.

Nicholson and Carlisle discover that bubbles of gas are formed at the two wires that are inserted into the solution. They have successfully broken up water into its individual elements, hydrogen and oxygen, the first example of electrolysis. This discovery is the first demonstration that electricity can be used to produce a chemical reaction and founds a new branch of science, which becomes known as electrochemistry.

1800 German physicist Johann Ritter shows that water is composed of two parts of hydrogen to one of oxygen.

He repeats William Nicholson's electrolysis of water and collects the two gases given off in separate containers for the first time. He observes that the volume of hydrogen is twice that of oxygen, which confirms the chemical composition of the water molecule as H_2O.

1800 German physicist Johann Ritter develops electroplating.

Ritter discovers that passing an electric current through an aqueous solution of copper sulphate results in the deposition of copper metal onto one of the wires, or electrodes, which have been inserted into the solution. The process of depositing metal using an electric current becomes known as electroplating and leads to many commercial applications, such as silver plating of tableware and gold plating of jewellery.

1801 English chemist and physicist John Dalton formulates the law of partial pressure of gases.

Dalton proposes that the pressure that a mixture of gases applies, for a given temperature and volume, will be the sum of the pressures that each component gas would exert if it alone were present. The pressure that each component gas can apply is called the partial pressure of the gas. This empirical relationship becomes known as 'Dalton's law' and is found to describe the behaviour of real gases at low pressures and high temperatures.

1801 English chemist Charles Hatchett discovers niobium (atomic number 41).

Hatchett discovers the new element while analysing an unusual mineral from Connecticut. He names the new element 'columbium' after the USA, which is sometimes called Columbia at this time. However, his claim is disputed and the issue is not settled until 1846, when an international commission recognizes his discovery of a new element and renames it 'niobium' after Niobe, the daughter of Tantalus in Greek mythology.

1801 Spanish mineralogist Andrés del Rio discovers vanadium (atomic number 23).

He names the new element 'erythronium', from the Greek word *erythros* meaning 'red', after the red colour of the compound he is working on. His discovery is disputed and he is persuaded to withdraw his claim. The element is rediscovered in 1830 by Swedish chemist Nils Sefström. Vanadium compounds form brightly coloured solutions, which prompt Sefström to name the element 'vanadium', after Vanadis, the Swedish goddess of beauty.

1802 Swedish chemist Anders Ekeberg discovers tantalum (atomic number 73) while studying minerals found in Finland and Sweden.

Ekeberg finds the task of isolating compounds of the new element so difficult that he names it 'tantalum' after the character from Greek mythology, Tantalus.

1803 English chemist and physicist John Dalton proposes his atomic theory of matter. He revives the theory put forward by the Greek philosopher Democritus (460–370 BC) that elements are made up of minute indestructible particles, called atoms.

However, while the ancient theory suggests that elements differ from each other by shape, Dalton proposes that the difference is weight. He states that the individual atoms of an element are identical, but that atoms of different elements have different weights. Dalton devises a system of chemical symbols to describe the elements and arranges them into a table according to their relative atomic weights in 1808.

1803 English scientist William Hyde Wollaston discovers the element palladium (atomic number 46) while refining platinum.

Wollaston names the metal after Pallas, the recently discovered asteroid. It is later used as a catalyst for chemical reactions, and is used to form alloys such as white gold, a mixture of palladium and platinum.

1803 French chemist Claude-Louis Berthollet publishes *Essai de statique chimique/Essay on Chemical Statistics*, in which he notes that the completeness of chemical reactions depends partly on the mass and partly on the temperature of the reacting substances. He is the first to propose that chemical affinity is not the only factor determining how substances react.

Berthollet also states that some chemical reactions are reversible. This is the first serious attempt at understanding the physics of how chemicals react and can be considered to be the start of the branch of science known as physical chemistry.

1803 Swedish chemist Jöns Jakob Berzelius and Swedish mineralogist Wilhelm Hisinger discover cerium (atomic number 58).

They name the new element after the recently discovered asteroid 'Ceres'. Cerium is also discovered independently this year by German chemist Martin Klaproth. It is the most abundant of the rare-earth metals.

1803 Swedish chemist Jöns Jakob Berzelius extends the concept of electrical polarity to include chemical elements.

Berzelius carries out electrolysis experiments on a wide range of solutions. His observations lead him to the conclusion that the theory of electrical polarity can be extended to include the elements. He classifies each element by its electrical tendency and arranges them into a series, oxygen being the most electronegative element, potassium the most electropositive. Berzelius is the first to suggest that acids and bases have opposite electrical charges.

1804 English chemist and physicist William Hyde Wollaston discovers the element rhodium (atomic number 45) while refining crude platinum ore from South America.

Wollaston names the new metal after the Greek word *rhodon* meaning 'rose'. The metal is used in alloys to harden platinum and palladium.

1804 English chemist and physicist John Dalton proposes the law of multiple proportions. He states that when two elements combine to form more than one compound, the weight of one element combines with a fixed weight of the other in a ratio of small whole numbers. The law provides strong support for his atomic theory.

1804 English chemist Smithson Tennant discovers and isolates the elements osmium (atomic number 76) and iridium (atomic number 77).

Tennant discovers the new metals in residues of platinum ore that have been treated with *aqua regia*, a very powerful acid. The unpleasant odour coming from one of the new compounds prompts Tennant to call the parent element 'osmium', from the Greek word *osme* meaning 'smell'. The other new element he names 'iridium', from the Latin word *iris* meaning 'rainbow', as an acknowledgement of the many coloured compounds of this element.

1805 German chemist Friedrich Wilhelm Saturner isolates the painkiller morphine.

The only painkiller known at this time is laudanum, an extract from immature opium plants. Saturner succeeds in isolating the active ingredient from laudanum, which he finds to be much more effective for lessening pain and inducing sleep. He names the chemical after *morphos*, the Greek word meaning 'sleep'.

1805 French chemist and physicist Joseph-Louis Gay-Lussac determines the relative proportions of hydrogen and oxygen in water by measuring the proportions of the gases that combine.

Around three years later, Gay-Lussac proposes the law that now bears his name. This states that, for chemical reactions between gases, the volumes of reactants and products measured under the same conditions occur in simple numerical ratios.

1806 French chemist Louis Nicolas Vauquelin discovers the amino acid asparagine. He isolates a substance from asparagus, which he names 'asparagine' after the vegetable.

Vauquelin is the first scientist to isolate an amino acid. This is not considered an important discovery at the time. However his discovery is reassessed when it is later realized that it belongs to a set of compounds, known as 'amino acids', that are essential for life.

1807 English chemist Humphry Davy isolates the elements potassium (atomic number 19) and sodium (atomic number 11).

Davy constructs an electric battery consisting of over 200 metal plates, the most powerful battery of the time, and uses it to electrolyse molten caustic potash or potassium hydroxide. The result is the deposition of metallic potassium at the negative electrode of the cell. Similarly, he isolates sodium metal from molten caustic soda or sodium hydroxide. He names one element 'potassium' after potash and the other 'sodium' after caustic soda. Davy is the first to use electrolysis to isolate new elements.

1808 English chemist Humphry Davy isolates the alkaline-earth metals magnesium (atomic number 12), calcium (atomic number 20), strontium (atomic number 38), and barium (atomic number 56).

After his success in isolating potassium and sodium in the previous year, Davy repeats his experiment on a series of molten chemicals and isolates the four new metals. He names them 'magnesium' after Magnesia, a district of Thessaly, 'calcium' from the Latin word *calx* meaning 'lime', 'strontium' in honour of the Scottish village of Strontia, and 'barium' from the Greek word *barys*, which means 'heavy'.

1808 English chemist Humphry Davy, and French chemists Joseph-Louis Gay-Lussac and Louis-Jacques Thérard, independently isolate the element boron (atomic number 5) by reacting boric acid with potassium.

31 December 1808 French chemist and physicist Joseph-Louis Gay-Lussac publishes *The Combination of Gases*, in which he announces that gases combine chemically in simple proportions of volumes.

Gay-Lussac states that the relationship between the combining gases and the contraction in their original volume can be expressed in small integer numbers. This relationship becomes known as 'Gay-Lussac's law'.

Earth Sciences

1793 Irish chemist, mineralogist, and geologist Richard Kirwan, (as a Neptunist geologer he believes that the Earth was initially covered by water, from which all rocks were deposited) attacks the Scottish natural philosopher James Hutton's Plutonist theory (that the origin and nature of all rocks are explicable by heat) as 'fanciful and groundless'. This is the incentive Hutton needs to finally publish his ideas as a full book, which he does in 1795.

1795 Scottish natural philosopher James Hutton publishes *Theory of the Earth with Proofs and Illustrations.*

Hutton's writing is not very clear and the book is read by very few people. His ideas will be made popular by John Playfair's 1802 book, and later by Charles Lyell.

1798 English scientist Henry Cavendish calculates the average density of the Earth at 5.48 times the density of water using a torsion balance developed by English physicist John Michell.

Because most rocks on the surface of the Earth are much less dense than 5.5, Cavendish's result suggests that there must be a very dense, probably metallic, rock deep within the Earth.

1799 English surveyor William Smith publishes his first geological map, a map of the strata around Bath, England.

1799 The British Mineralogical Society is founded.

1800 Scottish geologist James Hall publishes results of his experiments. By melting basaltic glass and allowing it to cool slowly, he shows that minerals crystallize upon slow cooling.

Hall's experiments support the Scottish natural philosopher James Hutton's theory that some rocks originally crystallized from a molten liquid (Plutonism) and were not precipitated from water (Neptunism).

1801 French mineralogist René-Just Häuy publishes *Traité de minéralogie / Treatise on Mineralogy*, a classification system, including a theory of the crystal structure of minerals that establishes him as one of the founders of crystallography.

1802 English chemist John Dalton discovers that the amount of water vapour required to saturate the air varies with temperature. This leads to the concept of relative humidity. Dalton also shows that the aurora borealis is a magnetic phenomenon.

1802 French biologist Jean-Baptiste de Lamarck publishes *Hydrogeologie / Hydrogeology*.

In this book Lamarck demonstrates a uniformitarian (believing the Earth forms a steady-state system, in which terrestrial causes had always been of the same kind and degree as at present) approach to such phenomena as the transitory nature of shorelines and the uplift involved in mountain building. Lamarck notices that fossils in lower and therefore older rocks are less evolved than those in more recent rocks, and advocates the idea that living things change over time.

1802 Scottish geologist and mathematician John Playfair publishes *Illustrations of the Huttonian Theory of the Earth*. The book is read fairly widely.

1803 English meteorologist Luke Howard creates a system of classifying clouds based on their shapes. He describes heap-like clouds as 'cumulus', broad layers of clouds as 'stratus', feathery wisp-like clouds as 'cirrus', and rain-bearing clouds as 'nimbus'. The classification is still in use today.

1803 French scientist Constantin-François Chasseboeuf (comte de Volney) publishes *Tableau du Climat du Sol des Etats Unis d'Amérique/ Tables of Climate and Soil of the USA*.

1806 Scottish geologist James Hall publishes the results of his experiments using a pressure chamber, which he makes out of an old cannon. The first high pressure–high temperature apparatus, his chamber reaches almost 300 bars and 1,000°C/ 1,832°F.

1807 Persuaded by Swiss geodesist Ferdinand Hassler, US president Thomas Jefferson establishes a Survey of the Coast to create precise maps of the US coastline.

Actual surveys do not begin until 1816, but will be suspended in 1818. The programme is re-established in 1836 as the Coast Survey (headed by Hassler), and renamed the Coast and Geodetic Survey in 1878.

1807 The Geological Society of London is founded as a scientific organization to foster research and discussion on geological topics, to maintain a library and map collection, and to maintain a specimen collection.

Its first transactions are published in 1811, and it is still very active today.

1809 English scientist William Hyde Wollaston develops the first precise reflection goniometer, used to measure the angles between crystal faces on very small specimens.

The instrument is used not only to characterize known minerals, but also to identify or distinguish new minerals.

1809 French mineralogist René-Just Häuy publishes *Tableau comparatif/ Comparative Tables*, a classification system of minerals.

1809 US geologist William Maclure publishes *Observations on the Geology of the United States*, the first geological map of the eastern USA.

Ecology

1798 English economist and cleric Thomas Malthus publishes *Essay on the Principle of Population*, in which he argues that population increases geometrically while food production increases only arithmetically, resulting in competition; this insight is later a key idea for the English naturalist Charles Darwin.

The motivation for Malthus's work can be gleaned from its full title *Essay on the Principle of Population as it affects the Future Improvement of Society with Remarks on the Speculations of Mr Godwin, M Condorcet and other writers*. Godwin and Condorcet are radical thinkers who argue that the human condition could be improved by restructuring society. As a conservative, Malthus is reluctant to entertain the idea that social reform, especially along the lines currently being practised in France, is desirable. His conclusion is a pessimistic one.

Thomas Malthus believed that, because population tends to increase more rapidly than food supplies, wars and disease would inevitably kill off the extra population. Malthus's ideas were influential on Charles Darwin in his formulation of the idea of natural selection. Rapid population growth in the 20th century, especially in underdeveloped countries, led to renewed interest in Malthus's theories. *Corbis*

1799–1804 German explorer and geographer Alexander von Humboldt publishes his *Personal Narrative*, accounts of his travels in Latin America.

Humboldt explores the idea that plants can be grouped in geographical zones. He embraces a holistic view of nature and attempts a blend of romanticism and science.

1802 English theologian and philosopher William Paley publishes *Natural Theology*. Paley argues that the natural world shows clear evidence of design and purpose and that, therefore, God exists.

Paley finds no contradiction between the work of scientists seeking to explore the workings of nature and the idea of a Christian God. In his view, the structure of living organisms can only be understood using the notion that they have been designed for a purpose. In this work, Paley revives the clock analogy: if we chance upon a watch, we must assume a watchmaker; therefore, examining the even more delicate structure of organisms implies a Creator for the natural world.

April 1803 US ornithologist John James Audubon becomes the first person in the USA to band birds for scientific purposes, when he tags some phoebes.

Mathematics

1794 French mathematician Adrien Legendre publishes *Eléments de géométrie*, an influential account of geometry.

1797 Italian-born French mathematician Joseph Lagrange publishes *Théorie des fonctions analytiques / Theory of Analytical Functions*, which introduces the modern notation (such as dy/dx) for derivatives.

1797 Norwegian mathematician Caspar Wessel presents the vector representation of complex numbers.

It first appears in a report by Wessel in 1787. His paper was published in Danish in 1799. The representation, of a line at an angle to a pair of coordinate axes, is similar to the Argand diagram of 1806.

1799 German mathematician Carl Friedrich Gauss proves the fundamental theorem of algebra: that every algebraic equation has as many solutions as the exponent of the highest term.

Thus a cubic equation has three solutions, a quartic has four solutions, and so on. Not all solutions have to be real: solutions can be complex numbers that involve 'imaginary numbers' (numbers that are a multiple of the square root of -1).

1799–1825 French mathematician and physicist Pierre-Simon Laplace publishes the five-volume *Traité de mécanique céleste / Celestial Mechanics*, which applies calculus to the motions of celestial bodies and Isaac Newton's theories of the Solar System to show how its stability is implicit in the law of gravitation.

1801 German mathematician Carl Gauss publishes *Disquisitiones arithmeticae / Discourses on Arithmetic*, which deals with relationships and properties of integers and leads to the modern theory of algebraic equations.

The book also publicizes the use of the symbol i to represent the square root of -1, as first proposed by Leonhard Euler in 1777.

1806 French mathematician Jean Argand introduces the Argand diagram as a way of graphically representing a complex number.

The number is plotted as a point on a plane whose coordinates are the real part (horizontal axis) and imaginary part (vertical axis) of the complex number.

1806 French mathematician Adrien Legendre develops a method of finding the best approximate values from a set of observed data, called the method of least squares. It can be used, for example, to draw the 'best' straight line through a series of non-colinear points.

1807 French mathematician Jean-Baptiste-Joseph Fourier discovers his method of representing continuous functions by the sum of a series of trigonometric functions.

He thereby founds what becomes known as Fourier analysis.

1808 French mathematician Sophie Germain works on Pierre de Fermat's last theorem and proves that his equation does not hold for the case in which n is equal to 5. This comes to be known as Germain's theorem.

1809 German mathematician Carl Gauss describes the least-squares method for minimizing errors in calculations in *Theoria motus corporum coelestium / Theory of the Movement of Heavenly Bodies*.

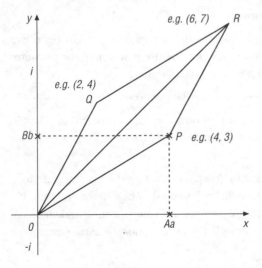

A complex number can be represented graphically as a line whose end-point coordinates equal the real and imaginary parts of the complex number. This type of diagram is called an Argand diagram after the French mathematician Jean Argand (1768–1822) who devised it.

A similar method had been proposed three years earlier by French mathematician Adrien Legendre.

Physics

1792 Italian physicist Alessandro Volta demonstrates the electrochemical series.

1798 English physicist and chemist Henry Cavendish calculates Newton's universal gravitational constant, G, using a torsion balance of the type invented by Coulomb. He is able to estimate that the Earth is 5.5 times as dense as water.

Cavendish was a student at Cambridge University, England, and the Cavendish Laboratory at Cambridge now bears his name. However, Cavendish himself never actually graduated.

1798 US-born British physicist and inventor Benjamin Thompson, Count Rumford, demonstrates experimentally the theory that heat is the increased motion of particles.

1800 English physicist Thomas Young publishes *Outlines and Experiments Respecting Sound and Light*, in which he proposes a wave theory of light.

The following year Young carries out a series of experiments in which he observes that light passing through two closely spaced slits produces alternating bands of light and dark in the area of overlap. This 'interference pattern' can only be explained by treating light as a wave.

1801 German physicist Johann Ritter discovers ultraviolet rays when he notes that silver chloride decomposes faster in invisible than visible light. He also discovers electroplating.

1802 French chemist and physicist Joseph-Louis Gay-Lussac demonstrates that all gases expand by the same fraction of their volume when their temperature is increased by the same amount.

French physicist Jacques Charles had discovered this empirical relationship in 1787, but did not publish his findings. However, the relationship is named 'Charles' law' in recognition of this earlier work. The existence of a common thermal expansion coefficient for gases allows the establishment of a new temperature scale.

1802 German physicist Johann Ritter invents the dry voltaic cell.

1802 Italian jurist Gian Domenico Romagnosi observes that an electric current deflects a magnetic needle. The observation is announced in an obscure newspaper; the phenomenon is later rediscovered by the Danish physicist Hans Oersted.

1802 Scottish physicist John Leslie accurately explains capillary action.

24 August 1804 French physicists Jean Biot and Joseph-Louis Gay-Lussac ascend to a height of 4,000 m/13,000 ft in a hydrogen balloon to study the effects of altitude on fluctuations in the Earth's magnetic field.

16 September 1804 French chemist and physicist Joseph-Louis Gay-Lussac ascends to a height of 7,016 m/23,018 ft in a hydrogen balloon, a record unbeaten for 50 years. He takes measurements of the Earth's magnetism, temperature, air pressure, and chemical composition, and determines that at that altitude the air contains the same percentage of oxygen (21.49%) as on the ground.

1807 English physician and physicist Thomas Young enunciates 'Young's modulus', a measurement of the elasticity of a material defined as tensile stress divided by tensile strain.

This is a constant for a given material obeying Hooke's law.

1808 French physicist Etienne Malus discovers the polarization of light by reflection.

1810–1829

Astronomy

1817 German astronomer Friedrich Bessel develops the 'Bessel function', a mathematical function that explains the movement of three stellar objects whose gravitational effects influence each other.

1818 German astronomer Friedrich Bessel's book *Fundamenta astronomiae/Fundamental Principles of Astronomy* records the positions of 50,000 stars on the basis of James Bradley's observations; it is the most accurate star catalogue to date.

1819 German astronomer Johann Encke discovers the short-period comet (Encke's comet), which returns every 3.29 years.

1825 English astronomer John Herschel invents the actinometer for measuring the Sun's light energy.

Herschel's actinometer is basically a bowl of water and a thermometer, the latter used to make a measurement of the rate at which the bowl warms up. Herschel finds a value for the solar constant, which is about half the modern figure.

1826 English astronomer John Herschel begins to measure the parallax of stars – the apparent shift in a star's position relative to a neighbour caused by the Earth's revolution around the Sun. He thus proves conclusively that the Earth revolves around the Sun.

Biology

1813 Swiss botanist Augustin Pyrame de Candolle publishes *Théorie élémentaire de la botanique/ Elementary Theory of Botany*, in which he coins the word 'taxonomy', introduces the idea of homology in plants, and argues that the basis of plant classification should be anatomy and not physiology.

1815–1822 French biologist Jean-Baptiste de Lamarck publishes *Histoire naturelle des animaux sans vertèbres/Natural History of Invertebrate Animals*, in which he maintains that museum collections should form the basis for revising biological classification schemes.

1821 The first international congress on biology is held.

1822 US army surgeon William Beaumont begins the study of digestion when he treats French-Canadian trapper Alexis St Martin who has been wounded in the stomach by a shotgun blast. The wound has failed to heal and the remaining opening (fistula) allows examination of digestive processes.

A ten-year study of 238 observations provides the basis of an understanding of gastric physiology. Beaumont goes on to use artificial fistulas as a tool in experimental physiology.

Pioneering Experiments on the Digestive System

by Julian Rowe

AN ARMY MARCHES ON ITS STOMACH

On 6 June 1822 at Fort Mackinac, Michigan, USA, an 18-year-old French Canadian was accidentally wounded in the abdomen by the discharge of a musket. He was brought to the army surgeon, US physician William Beaumont (1785–1853), who noted several serious wounds and, in particular, a hole in the abdominal wall and stomach. The surgeon observed that through this hole in the patient 'was pouring out the food he had taken for breakfast'.

The patient, Alexis St Martin, a trapper by profession, was serving with the army as a porter and general servant. Not surprisingly, St Martin was at first unable to keep food in his stomach. As the wound gradually healed, firm dressings were needed to retain the stomach contents. Beaumont tended his patient assiduously and tried during the ensuing months to close the hole in his stomach, without success. After 18 months, a small, protruding fleshy fold had grown to fill the aperture (fistula). This 'valve' could be opened simply by pressing it with a finger.

DIGESTION . . . INSIDE AND OUTSIDE

At this point, it occurred to Beaumont that here was an ideal opportunity to study the process of digestion. His patient must have been an extremely tough character to have survived the accident at all. For the next nine years he was the subject of a remarkable series of pioneering experiments, in which Beaumont was able to vary systematically the conditions under which digestion took place and discover the chemical principles involved.

Beaumont attacked the problem of digestion in two ways. He studied how various substances were actually digested in the stomach, and also how they were 'digested' outside the stomach in the digestive juices he extracted from St Martin. He found it was easy enough to drain out the digestive juices from his fortuitously wounded patient 'by placing the subject on his left side, depressing the valve within the aperture, introducing a gum elastic tube and then turning

him . . . on introducing the tube the fluid soon began to run.'

A TYPICAL EXPERIMENT

Beaumont was basically interested in the rate and temperature of digestion, and also the chemical conditions that favoured different stages of the process of digestion. He describes a typical experiment (he performed hundreds), where (a) digestion in the stomach is contrasted with (b) artificial digestion in glass containers kept at suitable temperatures, like this:

(a) 'At 9 o'clock he breakfasted on bread, sausage, and coffee, and kept exercising. 11 o'clock, 30 minutes, stomach two-thirds empty, aspects of weather similar, thermometer 298°F, temperature of stomach $101\frac{1}{28}$ and $100\frac{3}{48}$. The appearance of contraction and dilation and alternative piston motions were distinctly observed at this examination. 12 o'clock, 20 minutes, stomach empty.'

(b) 'February 7. At 8 o'clock, 30 minutes a.m. I put twenty grains of boiled codfish into three drachms of gastric juice and placed them on the bath.'

'At 1 o'clock, 30 minutes, p.m., the fish in the gastric juice on the bath was almost dissolved, four grains only remaining: fluid opaque, white, nearly the colour of milk. 2 o'clock, the fish in the vial all completely dissolved.'

ALL A MATTER OF CHEMISTRY

Beaumont's research showed clearly for the first time just what happens during digestion and that digestion, as a process, can take place independently outside the body. He wrote that gastric juice: 'so far from being inert as water as some authors assert, is the most general solvent in nature of alimentary matter – even the hardest bone cannot withstand its action. It is capable, even out of the stomach, of effecting perfect digestion, with the aid of due and uniform degree of heat (100°Fahrenheit) and gentle agitation . . . I am impelled by the weight of evidence . . . to conclude that the change effected by it on the aliment, is purely chemical.'

Our modern understanding of the physiology of digestion as a process whereby foods are gradually broken down into their basic components follows logically from his work. An explanation of how the digestive juices flowed in the first place came in 1889, when Russian physiologist Ivan Pavlov (1849–1936) showed that their secretion in the stomach was controlled by the nervous system. By preventing the food eaten by a dog from actually entering the stomach, he found that the secretions of gastric juices began the moment the dog started eating, and continued as long as it did so. Since no food had entered the stomach, the secretions must be mediated by the nervous system.

Later, it was found that the digestion that takes place beyond the stomach is hormonally controlled. But it was Beaumont's careful scientific work, which was published in 1833 with the title *Experiments and Observations on the Gastric Juice and Physiology of Digestion*, that triggered subsequent research in this field.

1823 Italian microscopist Giovanni Battista Amici proves the existence of sexual processes in flowering plants, by observing pollen approaching the plant ovary.

1824 French chemists Jean Dumas and C Prevost show that sperm is essential to fertilization.

1824 Swiss botanist Augustin Pyrame de Candolle begins his 17-volume classification of plants *Prodromus systematis naturalis regni vegetabilis / Treatise on the Classification of the Plant Kingdom*. Completed in 1873 by his son Alphonse, it replaces Lamarck's classification system and serves as the model for future systems.

1826 English administrator Stamford Raffles founds the Royal Zoological Society in London, England.

1826 Estonian embryologist Karl von Baer begins the study of the mammalian ovary. He shows that the Graafian follicle contains the microscopic ovum (egg).

Later he points out that the embryos of different species are similar early on in development and that differences can only be distinguished during later stages. He states that during the development of the embryo of a higher animal, the early stages are similar to the developmental stages of a lower animal. The observation is the foundation of conflicting theories of evolution later proposed by Ernst Haeckel and Walter Garstang.

1827 Estonian embryologist Karl von Baer provides detailed and systematic investigation of developmental processes in mammals and humans in *Epistola de ovo mammalium et hominis generis / On the Mammalian Egg and the Origin of Man*.

1827–1838 US ornithologist John James Audubon publishes the first volume of his multi-volume work *Birds of America*.

Chemistry

1811 French chemist Bernard Courtois discovers the element iodine (atomic number 53) when he accidentally adds too much sulphuric acid to seaweed. The reaction produces a violet vapour that, when condensed, forms dark purple crystals.

He calls his discovery 'substance X' and, suspecting that it is a new element, gives it to English chemist Humphry Davy for confirmation. Davy names the new element 'iodine' after the Greek word for violet, *iodes*.

1813 French chemist and physicist Pierre-Louis Dulong discovers the explosive nitrogen trichloride.

Dulong synthesizes the explosive while studying compounds of nitrogen and chlorine and loses an eye and two fingers during his research on this compound.

1813 Swedish chemist Jöns Jakob Berzelius introduces the modern system of chemical symbols.

The concept of atoms is now becoming established and it is necessary to introduce an acceptable system to describe them. Berzelius proposes that letters be used, based on the first letter of the name of the element, such as O for oxygen. The first two letters are to be used in the case of duplication, such as Ca for calcium. Latin is adopted to describe elements that are known by different names in different languages, such as gold, Au. These symbols are also used to describe chemical compounds, such as sulphuric acid, H_2SO_4.

1814 Swedish chemist Jöns Jakob Berzelius publishes his *An Attempt to Establish a Pure Scientific System of Mineralogy through the Use of the Electrochemical Theory of Chemical Proportions*, a very short treatise, devoted not to classifying minerals, but rather to the principles by which minerals can be treated as chemical compounds.

In this same year, Berzelius publishes his *Theory of Chemical Proportions and the Chemical Action of Electricity*, establishing himself as one of the founders of modern chemistry.

1815 French chemist Joseph-Louis Gay-Lussac is the first to describe the composition of an organic radical.

Gay-Lussac discovers 'cyanogen', C_2N_2, a highly poisonous gas, which he prepares from the equally poisonous compound prussic acid and publishes his findings in *Recherche sur L'Acide Prussique/ Research on Prussic Acid*. Lussac demonstrates that the 'cyano group', the name used to describe a carbon/nitrogen grouping, is very stable and remains unchanged through various chemical reactions. Such groupings come to be known as organic radicals and their discovery is an important step in the understanding of organic chemistry.

1817 French chemists Joseph Pelletier and Joseph-Bienaimé Caventou discover chlorophyll, a green pigment responsible for the colouration of all green plants.

They name it after the Greek words meaning 'green leaf'. It is discovered later that chlorophyll is essential for the process of photosynthesis in plants.

1817 German chemist Friedrich Stromeyer discovers cadmium (atomic number 48).

He discovers that an impurity in zinc carbonate turns the compound an orange colour when heated. He successfully isolates this impurity, which is the yellow coloured oxide of an unknown metal, similar in its chemical properties to zinc. He names the element 'cadmium' after the ancient Greek name for zinc carbonate, *kadmeia*.

1817 Swedish chemist Jöns Jakob Berzelius discovers selenium (atomic number 34) while investigating a red deposit found on the floor of a sulphuric acid factory.

He works on the problem with Swedish mineralogist Johan Gahn. The residue they eventually isolate from the red deposit shows unique properties and Berzelius recognizes it as the compound of an unknown element. He names the new element 'selenium', after the Greek word *selene* meaning 'moon', because he associates it with the element tellurium, which is named after the Earth.

1817 Swedish chemist Johan Arfwedson discovers lithium (atomic number 3).

Arfwedson isolates the new element from the mineral 'perlite', expecting the substance to be a compound of sodium. He is surprised to find that he has discovered a previously unknown element. He names the new metal 'lithium', from the Greek word *lithos* meaning 'stone', to indicate that he has obtained the element from a mineral.

1819 French chemist Pierre Dulong and French physicist Alexis Petit show that the specific heat of an element is inversely proportional to its atomic weight.

The determination of the atomic weight of an element is a difficult and time-consuming process at this time. However, it is considerably easier to determine the specific heat of an element. The relationship found by Dulong and Petit provides a simpler method for the determination of atomic weight using specific heat measurements. The relationship becomes known as the Dulong–Petit law and considerably helps to improve and expand the table of atomic weights.

1819 German chemist Eilhardt Mitscherlich proposes the theory of isomorphism after observing the crystallization of phosphates and arsenates.

Mitscherlich observes that compounds with similar compositions crystallize out of solution at the same time and correctly deduces that this is because they have similar molecular structures. He further states that if two compounds crystallize out together and the structure of only one is known, then it can be assumed that the molecular structure of the other is similar to the first. His theory helps to explain the observed similarity in crystalline structure between different chemical compounds.

1820 French chemist Joseph Pelletier and French physicist Joseph Caventou isolate the anti-malarial drug quinine.

They successfully isolate the active ingredient of the drug from the bark of the cinchona tree from Peru. In this same year, they also isolate the alkaloids brucine, cinchonine, colchicine, strychnine, and veratrine.

1821 Caffeine is discovered by German chemist Friedlieb Runge. He extracts the compound from coffee beans and names it after the French name for coffee, *café*.

The chemical is a member of the alkaloid group and is also found in tea and cocoa. It acts as a mild stimulant and a diuretic.

1823 English physicist and chemist Michael Faraday liquefies chlorine for use in bleaches and water purification.

Faraday is the first to use a combination of cooling and pressure to liquefy chlorine gas. The temperature required to liquefy chlorine at atmospheric pressure is -37°C/-34.6°F, but by causing the pressure of the gas to increase, Faraday achieves the task using only a bucket of ice. He goes on to use his technique to liquefy a number of other gases, notably carbon dioxide. His work can be regarded as the start of the branch of physics known as cryogenics, the study of science at low temperature.

1823 French chemist Michel-Eugène Chevreul is the first to study the chemical nature of fats.
Chevreul discovers that all fats are combinations of glycerol with organic or fatty acids. He isolates the most common of these, in particular palmitic, stearic, and oleic acids. These compounds are the main ingredients of soap and candles and his discovery allows the industrial manufacture of these items. This is one of the first chemical discoveries to have an immediate major impact on the daily lives of people.

1823 Swedish chemist Jöns Jakob Berzelius isolates silicon (atomic number 14).
Although silicon makes up more than 25% of the Earth's crust, the element has never been isolated in a pure form due to the chemical stability of its compounds. Berzelius heats silicon tetrafluoride with potassium and purifies the residue by repeated

washing. What remains after this process is a pure form of silicon. Berzelius names the element after the Latin word for flint, *silex*.

1825 Danish physicist Hans Oersted isolates aluminium (atomic number 13) in powdered form.

Oersted uses the action of potassium on compounds of aluminium to obtain tiny beads of the pure metal. However, his method is so difficult that the credit for discovery of the element is usually attributed to German chemist Friedrich Wöhler. In 1828, he heats a mixture of aluminium and potassium chlorides together in a platinum crucible and obtains sufficient quantities of aluminium to form a bar of the metal. Aluminium is named after the Latin word *alumen* meaning 'alum'.

1825 English physicist and chemist Michael Faraday discovers benzene, a product of the distillation of whale oil.

Benzene, an aromatic hydrocarbon, is a colourless, flammable liquid that causes severe irritation of the mucous membranes. It is not in itself an important chemical at the time. However, it will be of significant importance to later scientists, notably German chemist Friedrich Kekulé, in the understanding of the chemical structure of organic compounds.

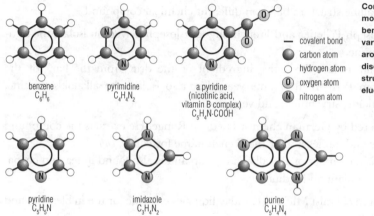

benzene
C_6H_6

pyrimidine
$C_4H_4N_2$

a pyridine
(nicotinic acid,
vitamin B complex)
$C_5H_4N \cdot COOH$

— covalent bond
● carbon atom
○ hydrogen atom
Ⓞ oxygen atom
Ⓝ nitrogen atom

pyridine
C_5H_5N

imidazole
$C_3H_4N_2$

purine
$C_5H_4N_4$

Compounds whose molecules contain the benzene ring, or variations of it, are called aromatic. Benzene was discovered in 1825 and its structure was first elucidated in 1865.

1826 French chemist Antoine-Jérôme Balard discovers bromine (atomic number 35).

Balard is studying brine collected from a salt marsh near Montpellier, when he obtains a dark brown liquid that has a very foul smell. His analysis of the liquid shows that it contains a new element with properties similar to those of iodine and chlorine. It is named after the Greek word *bromos*, meaning 'stench'.

1826 French physiologist Henri Dutrochet carries out the first quantitative experiments on osmosis – the passage of a solvent through a semipermeable membrane.

Dutrochet rediscovers and names osmosis. French physicist Jean Antoine Nollet had discovered this phenomenon first in 1748, but had been unable to explain his findings. Dutrochet expands upon this earlier work and shows that the pressure involved during the diffusion process, the 'osmotic pressure', is proportional to solution concentration.

1827 Russian chemist Gottfried Wilhelm Osann discovers the platinum-metal element ruthenium (atomic number 44).

Ruthenium is isolated more purely by Russian chemist Karl Klaus in 1844. Osann names the element after a Latin name for Russia, Ruthenia.

1828 German chemist Friedrich Wöhler synthesizes urea from ammonium cyanate. It is the first synthesis of an organic substance from an inorganic compound.

Wöhler's work conclusively disproves the current consensus that only living organisms can produce organic compounds. Scientists now begin searching for new ways to synthesize this highly important class of chemicals. This discovery marks the beginning of modern organic chemistry.

1828 Swedish chemist Jöns Jakob Berzelius discovers the element thorium (atomic number 90).

Berzelius discovers the new metal while studying a mineral that originated from Norway. Consequently, he names the element 'thorium' after Thor, the Scandinavian god of thunder. The metal is silver in appearance, but turns black on exposure to air. It is later found to be radioactive and eventually will be used as a nuclear fuel.

Organic chemistry just now is enough to drive one mad.
It gives one the impression of a primeval, tropical forest full of the
most remarkable things, a monstrous and boundless thicket, with no
way of escape, into which one may well dread to enter.

Friedrich Wöhler, German chemist, letter to Berzelius, 28 January 1835

1829 Scottish chemist Thomas Graham formulates 'Graham's law', which states that the rate of diffusion of a gas is inversely proportional to the square of its molecular weight.

This relationship allows mixtures of gases to be separated by measuring their rates of diffusion.

Earth Sciences

1813 English surveyor Robert Bakewell publishes *An Introduction to Geology*, one of the first textbooks of geology. The book is very successful and goes through many editions, including three in the USA.

1813 US chemistry and mineralogy professor Thomas Cooper concludes that the Earth must have a metallic core.

Cooper bases his hypothesis on experiments by English physicist Henry Cavendish and English astronomer Nevil Maskelyne that show the mean density of the Earth to be around 5.5 times the density of water, and on his own analyses of the densities of crustal rocks. Because crustal rocks have a density of about 2.5–3.0, the interior of the Earth must be denser.

1815 English surveyor William Smith publishes *Geological Map of England and Wales with Part of Scotland*. Smith makes use of characteristic fossil assemblages to identify and classify different rock formations.

Though versions of some of his geological tables have been in circulation for many years, this provides a full synthesis of his views regarding the stratigraphic layering of rocks across England and Wales.

1816 German meteorologist Heinrich Brandes at the University of Breslau, Poland, begins using reports from the Meteorological Society of Mannheim to create a series of synoptic (large scale) weather maps. Among other things, the maps show the movement of high and low pressure systems over Europe.

Discovering the Interior of the Earth

by Margaret W Carruthers

UNDERSTANDING

Many geology textbooks include a cross-section through the Earth revealing its four layers: a solid metallic inner core, a liquid metallic outer core, a solid rocky mantle, and a thin rocky crust. Some diagrams go on further to show that the continental crust is much thicker than the oceanic crust, that there are different layers within the mantle, that the crust and uppermost mantle are grouped together as the 'lithosphere', and that both the liquid outer core and the solid mantle rocks are convecting.

How have we come to this understanding of the interior of the Earth if it is physically inaccessible beyond a depth of a few kilometres? For the most part, the Earth's interior has been mapped remotely using a combination of gravimetry, seismology, and experimental petrology.

GRAVITY

The first major stride towards understanding the overall structure of the Earth was made in 1798 when British physicist Henry Cavendish determined the value of the gravitational constant, allowing him to calculate the average density of the Earth. He found this was much greater than that of the bulk of the rocks that make up the continents. Cavendish's experiment therefore showed that the interior must be composed of materials much denser than those on the surface, and in 1813 US mineralogist Thomas Cooper concluded that the Earth must have a metallic core.

Because the force of gravity varies slightly according to the distribution of masses, gravity measurements have also been used to determine the Earth's structure on regional scales. For instance, gravity measurements in the Andes, made by French scientist Pierre Bouguer on the La Condamine expedition in 1740, indicated a mass deficiency beneath the mountains. This led the English geophysicist George Airy to conclude that tall mountain ranges are balanced by a keel of deep crustal roots that displaces the dense

mantle. Gravity measurements near the Atlantic and Pacific oceans, made by John Hayford of the US Coast and Geodetic Survey in the first decade of the 20th century, showed a mass surplus below the oceans. This was an indication that the rocks beneath the oceans were denser than those of the continents, thus oceanic crust and continental crust were distinct.

Along with geomorphological studies (for example those of US geologist Grove Gilbert in 1890), the discovery that mountains are balanced by deep roots gave rise to isostasy, the concept that the crust can adjust to the addition or release of a load, championed by US geologist Clarence Dutton. Isostasy showed that the Earth was not entirely rigid as many believed, but that at some depth it was plastic.

SEISMOLOGY

The detailed mapping of distinct layers in the Earth has been carried out through analysing the behaviour of seismic (earthquake) waves. In 1828 French mathematician Simeon Poisson discovered that two types of motion waves can propagate through elastic materials: compressional waves (P waves), which compress and extend the material as they pass through, and shear waves (S waves), which move the material up and down, back and forth. He also showed that the waves move at different speeds from each other and at different speeds through different materials. With the improvement of seismometers in the mid- to late 19th century, geologists were able to record the arrival time of seismic waves with increasing accuracy and precision.

In 1900 Irish geologist Richard Oldham distinguished P waves, S waves, and surface waves (seismic waves that move over the surface of the Earth rather than through it) on seismograms. He recognized that because the velocity of seismic waves varied with the physical properties of the rocks (most notably density, which is related to temperature and composition), the ability to distinguish the arrival times of different waves travelling through the Earth would allow geophysicists to 'image' the Earth's interior. In 1906 Oldham put his new technique to work when he recognized a seismic discontinuity (a place where the behaviour of seismic waves changes) at great depth in the Earth and interpreted it as evidence for a dense core. In 1914 US geophysicist Beno Gutenberg calculated the depth of the core/mantle boundary (also known as the Gutenberg discontinuity) at 2,880 km/1,790 mi; and in 1926 British geophysicist Harold Jeffreys used a combination of geophysical data to show that the core is a liquid. Ten years later, Danish geophysicist Inge Lehmann recognized seismic evidence for a solid inner core within the liquid outer core. Seismic imaging of the core continues; in 1996 geophysicists announced that they had found seismic evidence that the inner core is spinning within the outer core.

The quantitative study of the thickness of the crust began in 1909 when Croatian physicist Andrija Mohorovičić discovered the seismic discontinuity of the crust/mantle boundary (the Mohorovičić discontinuity, or Moho). Seismic probing of the Moho has since revealed such characteristics as the deep roots of high mountain chains and the marked difference in thickness between continental and oceanic crust.

Since the 1980s geophysicists have been able to process enormous amounts of digital seismic data from numerous seismic stations around the world. The results include images of smaller-scale heterogeneities within the crust, mantle, and core, as well as global, three-dimensional computerized axial tomography (CAT) scans of the Earth.

GEOCHEMISTRY AND EXPERIMENTAL PETROLOGY

Geophysics has provided an enormous amount of information relating to the physical properties of the Earth's interior, but with geophysical data alone it would be impossible to work out its chemical composition, the rocks and minerals that characterize the different layers of the Earth. Deducing the chemical layering of the Earth has required sampling and experimental work.

Although sampling of the Earth's interior is extremely limited, it has provided some important information. As early as the Middle Ages, as mining increased in Germany, miners noticed that rocks got warmer with depth. In his book *Mundus subterraneus* (1665) German Jesuit and scientist Athanasius Kircher noted that the Earth's temperature increases with depth.

Scottish geologist James Hutton understood that contorted rocks had been buried to great depth where, with high temperatures and pressures, they were lithified (turned to stone) and folded. Geologists now routinely examine surficial rocks that have been exhumed from depth (by

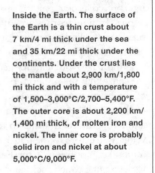

crust between 7 km / 4 mi thick (oceanic) and 35 km/22 mi thick (continental)

mantle 2,900 km / 1,800 mi thick

outer core 2,200 km / 1,370 mi thick

inner core 2,500 km / 1,550 mi diameter

12,756 km / 7,926 mi equatorial diameter (12,714 km / 7,900 mi polar diameter)

Inside the Earth. The surface of the Earth is a thin crust about 7 km/4 mi thick under the sea and 35 km/22 mi thick under the continents. Under the crust lies the mantle about 2,900 km/1,800 mi thick and with a temperature of 1,500–3,000°C/2,700–5,400°F. The outer core is about 2,200 km/ 1,400 mi thick, of molten iron and nickel. The inner core is probably solid iron and nickel at about 5,000°C/9,000°F.

Discovering the Interior of the Earth *continued*

erosion and/or tectonics) in order to understand the physical conditions under which they formed and which therefore characterize the deep crust.

Attempts at drilling into the deep crust have been moderately successful. The deepest drill hole, which was began in 1967, is 13 km/8 mi into the continental crust of the Kola Peninsula in Russia and has provided some interesting insights into the relationships between the local seismic and compositional structure of the continental crust.

We do have samples of the mantle in spite of our inability to drill into it. When some volcanoes erupt, they eject pieces of mantle xenoliths, and pieces of olivine- and pyroxene-rich rocks thought to originate in the mantle. Since the 1960s, when it became apparent that mid-ocean ridge basalts were derived from the mantle, geochemists have been analysing the detailed chemistry of basalts in order to calculate the composition of the mantle rocks from which they originate.

The other solid source of knowledge of the Earth's interior comes, strangely enough, from outside the Earth: from meteorites. Since the 1920s geochemists, such as H S Washington of the Geophysical Laboratory in Washington, DC, have based their models of the Earth on the composition of various meteorites, which are interpreted as being pieces of other planetary bodies. Iron meteorites are analogous to core rocks, and pallasites (stony irons) are used as estimates of the lower mantle.

In the late 19th century geochemists began to realize that the minerals present on the surface of the Earth were not necessarily representative of those stable deep in the Earth at high pressure and temperature. In order to relate the seismic velocity structure to the petrological (rock) and mineral composition of the Earth, geologists had to determine which rocks and minerals were

likely to be present at certain pressures and temperatures.

Although the roots of experimental petrology coincided with those of modern geology itself, the field blossomed in the early 20th century. In 1908 British physicist Percy Bridgman began developing high-pressure experimental apparatus and subsequently discovered many high-pressure 'polymorphs', minerals with the same composition but different molecular structure. In documenting the temperatures and pressures of the 'phase transitions' of one polymorph to another, Bridgman paved the way for understanding the mineralogy of the deep Earth. Experimental results were soon incorporated into models of the Earth. By the 1950s geochemists, such as the American Francis Birch and the Australian Alfred Ringwood, were relating their findings directly to those of geophysicists who continued to discover seismic discontinuities in the crust and mantle.

The invention of the diamond anvil cell in 1958 marked a major breakthrough in high-pressure research. Because diamond is the hardest substance known and because it is transparent to many forms of electromagnetic radiation, including visible light and X-rays, diamond cells allowed geochemists to perform experiments at increasing pressures and temperatures, while being able to observe the results. In 1975 mineral physicists David Mao and Peter Bell of the Geophysical Laboratory in Washington, DC, announced that they had reached pressures approaching those of the inner core.

As geologists and meteoriticists search for more samples, geophysicists continue to image in greater and greater detail, and geochemists push the boundaries of technology to simulate deep planetary interiors, it is clear that exploration of the Earth's interior is far from over.

1816 Parker Cleaveland, professor of mathematics and natural philosophy at Bowdoin College, Maine, USA, publishes his *Elementary Treatise on Mineralogy and Geology*.

The book gives the first well-researched assemblage of US mineral occurrences and localities. As a result of the book, he is made the first US member of the Geological Society of London.

1818 English cleric and geologist William Buckland of Oxford University, England, finds the bones of an animal he later names *Megalosaurus*, in the village of Stonesfield, England. *Megalosaurus* is the second dinosaur to be found in Europe, and the first to be named.

1818 US chemist Benjamin Silliman founds the *American Journal of Science*. Still in print, the journal discusses geological topics and provides a means of international communication among geologists.

1819 English cleric and geologist William Buckland becomes reader in geology at Oxford University. Reader in mineralogy since 1813, Buckland is a very popular lecturer, and inspires many future geologists, including Charles Lyell.

1819 English geographer and geologist George Bellas Greenough, first president of the Geological Society of London, publishes *A Geological Map of England and Wales* at a scale of 6 miles to the inch.

1819 The American Geological Society is founded in New Haven, Connecticut, USA. Accepting geologists of all social, economic, educational, and professional backgrounds, the society serves to bring together ideas and observations from around the country. In 1840 it will become the Association of American Geologists.

1821 US meteorologist William Redfield discovers that during a hurricane, trees are toppled toward the northwest in Connecticut, and toward the southeast 80 km/ 50 mi further west, evidence that winds in tropical storms move in a cyclonic or spiralling path.

1822 German mineralogist Friedrich Mohs devises a hardness scale, now known as the Mohs hardness scale, used to describe the resistance of a mineral to being scratched. The scale is still used today as a reliable tool in mineral identification.

The scale is based on ten minerals that increase in hardness from one to ten: talc, gypsum, calcite, fluorite, apatite, orthoclase, quartz, topaz, corundum, and diamond.

1822 English geologists William Conybeare and William Phillips publish a highly influential textbook of stratigraphy, *Outlines of the Geology of England and Wales*. The book is used as a reference for English stratigraphy and serves as a starting point for stratigraphy in the rest of Europe and in the USA.

1823 English cleric and geologist William Buckland publishes *Reliquiae Deluvianae/Relics of the Flood*, discussing the large fossil vertebrates found in Kirkdale Cave in Yorkshire, England. The book helps to promote vertebrate palaeontology.

1824 English cleric and geologist William Buckland publishes the first description of a dinosaur, *Megalosaurus*, which he found in Stonesfield, England, in 1818.

1825 English surgeon and palaeontologist Gideon Mantell describes the dinosaur *Iguanodon*.

1826 Swiss engineer Ignatz Venetz and mining superintendent Johann de Carpentier take Swiss-born US naturalist, geologist, and teacher Jean Louis Agassiz out into the field to show him evidence that glaciers once covered the landscape. Agassiz is convinced. He goes on to do more research and finally writes a book on the subject in 1841.

1827 English geologist and politician George Poulett Scrope publishes *Memoir on the Geology of Central France*, in which he discusses the volcanoes of the Auvergne, France.

Scrope discusses observations that support the idea that volcanoes are built of layer upon layer of lava and ash, and are not bulges produced by catastrophic uplifts. The book is also influential in that it is Charles Lyell's glowing review of it that brings Lyell onto the geological scene.

1828 French mathematician Siméon-Denis Poisson shows that there are two types of motion waves that can propogate through an elastic material: P waves and S waves. He also shows that these waves move at different speeds through the materials. The observation is the basis for the seismological study of the Earth.

1829 Scottish geologist William Nicol develops a polarizing microscope and a method of grinding sections of rock to transparent thinness. The microscope allows the observer to identify minerals by their interaction with polarized light, and is still a major tool of petrography.

Ecology

1810 Scottish-born US ornithologist Alexander Wilson observes a flock of passenger pigeons 400 km/250 mi long, which he estimates contains 2 billion birds. The passenger pigeon is extinct by 1914.

1825 French zoologist Etienne Geoffroy Saint-Hilaire attributes change in species to change in the geological and climatological environment. He suggests that major changes in the environment cause saltation (large change) in species, whereas minor changes in the environment cause only small and gradual change in species.

Mathematics

1811 French mathematician Siméon-Denis Poisson publishes *Traité de mécanique/Treatise on Mechanics*, which discusses the application of mathematics to magnetism, electricity, physics, and mechanics.

He is best remembered for Poisson's ratio, which is an elastic constant given (for a material being stretched) by the contraction per unit thickness divided by the extension per unit length.

c. **1820** German mathematician Carl Gauss reintroduces the normal distribution curve ('Gaussian distribution') – a basic statistical tool.

1820 French mathematician Charles-Thomas de Colmar develops the first mass-produced calculating machine – the 'arithmometer'.

1821 French mathematician Augustin-Louis Cauchy publishes *Cours d'analyse/A Course in Analysis*, which sets mathematical analysis on a formal footing for the first time.

1822 French mathematician Jean-Victor Poncelet systematically develops the principles of projective geometry in *Traité des propriétés projectives des figures/Treatise on the Projective Properties of Figures*.

Poncelet also reintroduces the abacus into Europe (from Russia).

1822 French mathematician Augustin-Louis Cauchy formulates the basic mathematical theory of elasticity; he defines stress as the load per unit area of the cross-section of a material, agreeing with Poisson's conclusions of 1811.

1822 French mathematician Joseph Fourier publishes *Théorie analytique de la chaleur/Analytical Theory of Heat*, introducing a technique now known as Fourier analysis, which has widespread applications in mathematics, physics, and engineering.

Using this technique, any continuous function can be represented as an infinite sum of sine curves and cosine curves.

1823 English mathematician Charles Babbage begins construction of a large 'difference engine', a machine for calculating logarithms and trigonometric functions. He had already developed a small 'difference engine' between 1819 and 1822.

The new machine is never completed through lack of funds, and Babbage goes on to work on his 'analytical engine', which also is not completed. However, these machines represent the first instance of automated number processing – the first computers.

1824 Norwegian mathematician Niels Abel proves that equations involving powers of x greater than four cannot be solved using a standard formula like that used for quadratic equations.

1824 German mathematician and astronomer Friedrich Bessel further develops results on a class of integral functions, now called Bessel functions, that arise in many areas of physics. He first discovered the functions in 1817.
The functions have so many applications that tables of their values are published.

Charles Babbage designed the first mechanical devices capable of automated number processing. His designs had many similarities to those later used in electronic computers – his analytical engine, for example, had a 'mill' where the actual calculations took place and a 'store' where the data and results were kept. *Corbis*

1824 Swiss mathematician Jakob Steiner begins his innovative work on synthetic geometry; he makes particular contributions to the theory of stereographic projection and the treatment of polar curves.

1827 German mathematician Carl Gauss introduces the subject of differential geometry that describes features of surfaces by analysing curves that lie on them – the intrinsic-surface theory.
In essence it is the application of calculus to geometry.

1828 Norwegian mathematician Niels Abel begins the study of elliptic functions, which are doubly periodic.
At about the same time Abel produces a complete proof that a quintic (an equation of the fifth degree) cannot be solved.

1828 English mathematician George Green publishes *Essay on the Application of Mathematical Analysis to the Theory of Electricity and Magnetism*, in which he applies mathematics to the properties of electric and magnetic fields.
It introduces a theorem that enables volume integrals to be calculated in terms of surface integrals. Green also makes extensive use of potential functions, a term he coins.

1829 Russian mathematician Nikolay Lobachevsky develops hyperbolic geometry, in which a plane is regarded as part of a hyperbolic surface shaped like a saddle. Austrian mathematician János Bolyai publishes a treatise on non-Euclidean geometry in 1832. It is the beginning of non-Euclidean geometry.

135

1829 French mathematician Evariste Galois invents group theory, which helps use ideas of symmetry to solve equations. His work was not published until 1846, 14 years after his death, and even then is not well understood. But today group theory is fundamental to modern algebra and has applications in quantum theory.

Physics

1810 Scottish physicist John Leslie is the first to create artificial ice – he freezes water using an air pump.

1811 Italian physicist Amedeo Avogadro proposes that equal volumes of all gases contain the same number of particles if they are at the same temperature and pressure.

This statement becomes known as 'Avogadro's hypothesis'. Avogadro is the first to make the distinction between single atoms and the groups of atoms that are present in chemical compounds. He called these groups 'molecules' from the Latin words meaning 'small masses'. The hypothesis is not accepted by scientists for another 50 years.

1811 Scottish physicist David Brewster discovers the law now named after him that describes the behaviour of polarized light.

Brewster shows that complete polarization occurs when the reflected and refracted rays are perpendicular. When this occurs, the angle of incidence is known as the Brewster angle. For glass, the Brewster angle is 57°.

1814 German physicist Joseph von Fraunhofer plots more than 500 absorption lines (Fraunhofer lines) and discovers that the relative positions of the lines are constant for each element. His work forms the basis of modern spectroscopy.

Fraunhofer labels the lines from 'A' to 'H', corresponding to the bright lines of emission spectra. The C and F lines, for example, are the red and blue lines in the hydrogen spectrum.

1815 French physicist Augustin-Jean Fresnel shows that light consists of transverse waves – he thus explains the diffraction of light.

Fresnel diffraction can also explain much of the nature of shadows, including the coloured fringes around shadows produced by white light.

1817 English physician and physicist Thomas Young explains the polarization of light – the alignment of light waves so that they vibrate in the same plane (light waves normally vibrate in all planes at right angles to the direction of travel).

1819 Danish physicist Hans Oersted discovers electromagnetism when he observes that a magnetized compass needle is deflected by an electric current.

Oersted's observation is taken up by André Ampère in 1820.

1820 French physicist André Ampère formulates Ampère's law, which states the relationship between a magnetic field and the electric current that produces it.

This allows him to develop an instrument that uses the deflection of a needle to measure the flow of electricity.

1821 English physicist Michael Faraday builds an apparatus that transforms electrical energy into mechanical energy – the principle of the electric motor. Faraday's device uses a magnet and an electric current to produce continuous mechanical rotation.

Faraday first comes to science when, working as an apprentice bookbinder, he happens upon scientific books. In 1813 he applies for the position of assistant to

Humphry Davy. Within just 14 years of this initiation he will have succeeded Davy as professor at the Royal Institution.

1821 German physicist Thomas Seebeck discovers thermoelectricity – the conversion of heat into electricity – when he generates a current by heating one end of a metal strip comprising two metals joined together.

This is the discovery of the thermocouple, and the process becomes known as the Seebeck effect. James Cumming of Cambridge University, England made the same discovery but he does not publish it until 1823.

1823 English engineer William Sturgeon demonstrates the world's first electromagnet by wrapping a current-carrying wire around an iron bar. The 200 g/7 oz magnet can support 4 kg/9 lb and is an essential invention in the development of electric motors and the electric telegraph.

In 1836 Sturgeon develops the first moving-coil galvanometer.

1824 French scientist Sadi Carnot publishes *Réflexions sur la puissance motrice du feu et sur les machines propres à développer cette puissance/Thoughts on the Motive Power of Fire, and on Machines Suitable for Developing that Power*, a pioneering study of thermodynamics in which he explains that a steam engine's power results from the difference in temperature between the boiler to the condenser. He also describes the 'Carnot cycle' whereby heat is converted into mechanical motion and mechanical motion converted into heat – the basis of the second law of thermodynamics.

1824 German scientists (and brothers) Wilhelm Eduard Weber and Ernst Heinrich Weber publish 'Wellenlehre, auf experimente gegründet/Theory of waves, founded on experiments', in which they investigate wave motion and acoustics.

1826 Italian physicist Leopoldo Nobili invents the astatic galvanometer, which measures small electric currents without interference from the Earth's magnetism.

This is the most sensitive galvanometer of the time.

1827 French physicist André Ampère publishes *Electrodynamics*, in which he formulates the mathematical laws governing electric currents and magnetic fields. It lays the foundation for electromagnetic theory.

1827 German physicist Georg Ohm formulates Ohm's law, which states that the current flowing through an electric circuit is directly proportional to the voltage, and inversely proportional to the resistance.

Strictly speaking, this is not true for all conductors. Some are said to be non-ohmic, and the validity of the law relies on temperature remaining constant.

1827 Scottish botanist Robert Brown is the first to report the observation of the continuous motion of tiny particles in a liquid solution – now known as Brownian motion.

Brown is studying a suspension of pollen grains in water using a microscope, when he notices that the grains are moving in a random manner. At first he thinks that the observed movement is because the pollen grains are alive and represents some aspect of their living state. However, he repeats the experiment with non-living dye particles and observes the same random movements.

1829 French mathematician Gustave Gaspard de Coriolis writes *Du calcul de l'effet des machines/On the Calculation of Mechanical Action*, in which he is the first to use the term 'kinetic energy'.

The term is taken from the Greek word *kinos*, meaning 'to move'. Coriolis is also the first to make use of the term 'work' in the sense that it is accepted in physics today.

1830–1849

Astronomy

1834–1838 English astronomer John Herschel, at the Cape of Good Hope, catalogues the locations of 68,948 stars in the southern hemisphere.

1837 German astronomer Friedrich Struve publishes *Micrometric Measurement of Double Stars*, a catalogue of over 3,000 binary stars.

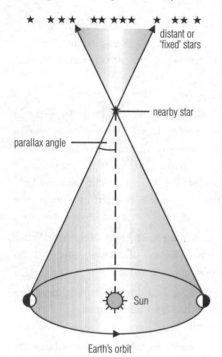

The parallax of a star, the apparent change of its position during the year, can be used to find the star's distance from the Earth. The star appears to change its position because it is viewed at a different angle in July and January. By measuring the angle of parallax, and knowing the diameter of the Earth's orbit, simple geometry can be used to calculate the distance to the star. The technique was developed by the German astronomer Friedrich Bessel.

11 October 1838 Using the method of parallax, German astronomer Friedrich Bessel calculates the star 61 Cygni to be 10.3 light years away from Earth. It is the first determination of the distance of a star other than the Sun.

1840 German astronomer Friedrich Bessel explains the irregularities in the orbit of Uranus as due to the gravitational effects of an unknown planet (later identified as Neptune).

1840 US astronomer John William Draper takes the first photograph of the Moon.

1840 US astronomer William Cranch Bond erects the first astronomical observatory in the USA.

In 1847 a 37.5 cm/15 in telescope, equal in size to the largest in the world, is installed. With it, Bond makes elaborate studies of sunspots, of the Orion nebula, and of the planet Saturn. He and his son develop the chronograph for automatically recording the positions of stars, and pioneer the use of the chronometer and the telegraph for determining longitude. They also make the first practical use in the USA of photography in astronomy.

1842 German astronomer Friedrich Bessel suggests that perturbations to the motion of Sirius are due to the existence of a companion star.

1842–1845 Irish astronomer William Parsons (Lord Rosse) builds the 180 cm/72 in reflecting telescope 'Leviathan'.

Used to observe nebulae, its size is not exceeded until the 250 cm/100 in Hooker telescope is built in 1917 at the Mount Wilson Observatory in California. This in turn remains the largest telescope in the world until 1948, and it is still the largest plate glass mirror ever cast, with a mass of 4,080 kg/9,000 lb. In early 1845 Parsons observes the nebula M51, and finds that it is a spiral galaxy of stars. Other nebulae seem to be unresolvable clouds of gas, but in others he is able to pick out individual stars. By the end of 1850 he has identified 14 spiral galaxies.

1843 German astronomer Samuel Schwabe discovers that sunspots and the effects of solar disturbances have a cycle of about 11 years.

Schwabe makes his discovery after a decade of searching for the hypothetical (and non-existent) planet Vulcan, thought by some at the time to orbit nearer to the Sun than Mercury.

23 September 1846 German astronomer Johann Galle discovers the planet Neptune on the basis of French astronomer Urbain Le Verrier's calculations of its position.

October 1846 Just one month after the planet Neptune is discovered, the British astronomer William Lassell discovers Triton, Neptune's largest moon.

He uses a 61 cm/24 in reflecting telescope of extraordinary precision, which he designs and builds himself. In 1995 the mirror, which is preserved in Liverpool Museum, is optically tested, and found to be accurate to within a fraction of a wavelength of light, a remarkable achievement for its time. Lassell goes on to discover two satellites of Uranus, Ariel and Umbriel, in 1851.

Biology

1831 Scottish botanist Robert Brown discovers and names the nucleus in plant cells.

Earlier he spends five years classifying the more than 4,000 plant species he collects during an expedition to Australia in 1801. In so doing he establishes the distinction between gymnosperms and angiosperms.

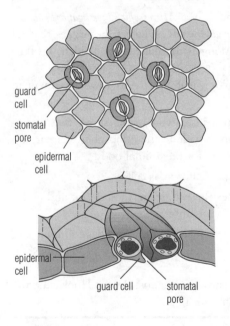

guard cell

stomatal pore

epidermal cell

epidermal cell

guard cell

stomatal pore

The stomata, tiny openings in the epidermis of a plant, are surrounded by pairs of crescent-shaped cells, called guard cells. The guard cells open and close the stoma by changing shape. Stomata were first identified in 1832.

27 December 1831–2 October 1836 English naturalist Charles Darwin undertakes a five-year voyage, to South America and the Pacific, as naturalist on the *Beagle*. The voyage convinces him that species have evolved gradually but he waits over 20 years to publish his findings.

1832 French plant physiologist Henri Dutrochet finds the small openings on the surface of leaves called stomata. Later it is found that gas exchange in plants occurs through them.

1834 French chemist Jean-Baptiste Boussingault discovers that plants absorb nitrogen from the soil, and carbon from carbon dioxide in the air.

Later he suggests the use of iodine salts for the treatment of goitre. Iodine is an essential element of the hormone thyroxine, the lack of which leads to the disease. His suggestion is not taken up for many years.

1835 HMS *Beagle* visits the Galapagos Islands off the coast of Peru. English naturalist Charles Darwin observes varieties of finch and giant tortoise. He later uses these observations as supporting evidence for his proposition that species change through time, as a result of natural selection, adapting to different environmental circumstances.

1835 Czech histologist Johannes Purkinje observes ciliary motion using a compound microscope.

New microscopic observations follow: in 1837 he describes cells (Purkinje cells) in the cerebral cortex, and describes nerve cells, their nuclei, and dendrites; in 1838 he observes cell division; in 1839 he uses the term 'protoplasm' for the first time, applying it to embryonic tissue. In 1837 Purkinje outlines cell theory, paving the way for Schleiden and Schwann's more comprehensive proposals in 1839.

1836 German physiologist Theodor Schwann discovers pepsin, the first known animal enzyme to be isolated.

1837 French plant physiologist Henri Dutrochet proves that chlorophyll is essential for photosynthesis in plants.

Since its discovery in 1817, scientists have been trying to find out what the green coloration in plants is used for. Dutrochet is the first to show conclusively that photosynthesis only occurs in plant cells that contain chlorophyll.

1838 German botanist Matthias Jakob Schleiden publishes the article 'Contributions to Phytogenesis', in which he recognizes that cells are the fundamental units of all plant life. He is thus the first to formulate cell theory.

1838 Dutch chemist Gerard Johann Mulder coins the word 'protein'.

1839 German physiologist Theodor Schwann publishes *Microscopic Investigations on the Accordance in the Structure and Growth of Plants and Animals*, in which he argues that all animals and plants are composed of cells. Along with Matthias Schleiden, he thus founds modern cell theory.

The theory rapidly comes to dominate biology, leading to Virchow's statement in 1855 that 'all cells arise from pre-existing cells'.

1840 Over the next decade, German organic chemist Justus von Liebig shows that carbohydrates in plants are synthesized from carbon dioxide in the atmosphere. He argues that carbohydrates and fats are a source of energy for the animal body.

1840 Swiss embryologist Rudolf Kölliker identifies spermatozoa as cells.

1842 Polish-German physician Robert Remak names the ectoderm, mesoderm, and endoderm (inner, central, and outer) layers of cells in an embryo.

1842 Fearing an early death, English naturalist Charles Darwin writes a 35-page summary of his theory of evolution, with instructions that it should be published posthumously if necessary.

Darwin discusses his views with two close friends, the scientist Robert Hooke and the geologist Charles Lyell, but he does not publish until 1859.

What can be more curious than that the hand of a man, formed for grasping, that of a mole for digging, [. . .] and the wing of a bat, should all be constructed on the same pattern, and should include the same bones, in the same relative positions?

Charles Darwin, British naturalist, *On the Origin of Species* (1859)

1842 German botanist Carl von Naegeli describes cell division during pollen formation.
Later he distinguishes between the structural tissues of plants and their meristematic tissues.

1842 German physician Robert Remak discovers that the early embryo consists of three layers: ectoderm, mesoderm, and endoderm.
Earlier, he finds the myelin sheath surrounding the axon of neurones.

1844 German physician Robert Remak discovers the nerve cells in the heart now known as Remak's ganglia.

1846 Italian botanist Giovanni Battista Amici establishes the circulation of sap in plants.

1846 English palaeontologist Richard Owen publishes *Lectures on Comparative Anatomy and Physiology of the Vertebrate Animals*, one of the first textbooks on comparative vertebrate anatomy.

Chemistry

1830 German naturalist Karl von Reichenbach discovers paraffin wax. He obtains the white waxy substance while experimenting with wood tar.
Reichenbach finds the substance to be inert to most chemical reactions. He names it paraffin from the Latin words *parum* and *affinis*, which can be loosely translated as meaning 'not very reactive'. The substance becomes known as paraffin wax.

1831 German chemist Justus von Liebig and US chemist Samuel Guthrie independently discover chloroform.
It is soon discovered that the chemical has anaesthetic properties and is first used for this purpose by Scottish physician James Simpson in 1847. It does not become widely accepted until 1853, when it is given to Queen Victoria by her physician during the birth of her eighth child, Prince Leopold.

1831 Swedish chemist Jöns Jakob Berzelius discovers isomerism – the relation of two or more compounds of the same element with the same number of atoms but which differ in structural arrangement and properties.
French chemist Joseph-Louis Gay-Lussac asks Berzelius to investigate a report that two substances, silver fulminate and silver isocyanate, which have different chemical properties, have the same chemical formula. Berzelius finds that they do indeed have the same chemical formula despite showing completely different chemical behaviours. He classifies these substances as being 'isomers', a word derived from the Greek meaning 'equal parts'. This is the first indication that the arrangement of atoms in a molecule is just as important as what number and type are present.

1832 French chemist Pierre-Jean Robiquet isolates the analgesic codeine from opium.
Morphine has already been isolated from opium, but Robiquet suspects that there is something else present in the drug. He discovers that the methyl derivative of opium can also be isolated from opium. This is named 'codeine' and proves to be a gentler and less addictive form of morphine and is initially used as a sedative. Codeine is still used as a component of modern painkillers.

1833 French chemist Anselme Payen discovers the first enzyme, diastase. He separates it from a malt extract, and finds that it is able to accelerate the conversion of starch to glucose.
Payen names the substance after the Greek word meaning 'separate', since it is able

butane $CH_3(CH_2)_2CH_3$

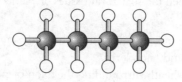

methyl propane $CH_3CH(CH_3)CH_3$

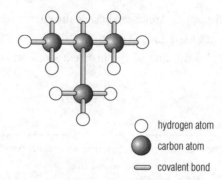

○ hydrogen atom

● carbon atom

⬭ covalent bond

The chemicals butane and methyl propane are isomers. Each has the molecular formula C_4H_{10}, but with different spatial arrangements of atoms in their molecules.

to separate large starch molecules into individual glucose units. It is because of this discovery that each enzyme has the suffix '-ase' in its name.

c. 1834 French chemist Anselme Payen discovers cellulose, the basic component of plant cells

Payen isolates a substance from wood that can be broken down chemically to form glucose, but which is not starch, a plant compound already known to do this. Since he has obtained the substance from the cell walls of the plant, he names it 'cellulose'. At this time, sugars are given individual names, such as starch and cane sugar. After Payen's discovery, sugars become named by convention with the suffix '-ose', so that starch becomes known as amylose and cane sugar becomes known as sucrose.

1834 French chemist Jean Dumas formulates the law of substitution by showing that halogens can replace hydrogen in organic compounds to form the organohalogens.

The reactions of these compounds are some of the most important in organic chemistry and they produce chemicals that have numerous important industrial applications. Organohalogens are used in the production of dyes, drugs, pesticides, and plastics.

1834 German chemist Friedlieb Runge is the first to obtain the simple phenol, carbolic acid.

Runge dissolves coal tar using aqueous acidic solutions and then neutralizes the resultant liquid, which causes the formation of a layer of oil. Runge finds that, through distillation, he can separate the oil into a number of organic compounds, one of which is carbolic acid. This chemical is used extensively as an antiseptic agent, usually in the form of carbolic soap.

1835 Swedish chemist Jöns Jakob Berzelius first uses the word 'catalysis' to describe the phenomenon of the increased rate or occurrence of a chemical reaction, which is induced by the addition of another material that remains unchanged after the reaction.

The material which is unaffected by the chemical reaction is known as the 'catalyst'.

He names the process 'catalysis' after the Greek phrase meaning 'to break down'.

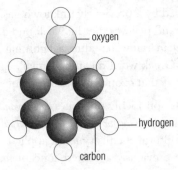

oxygen

hydrogen

carbon

The phenol molecule with its ring of six carbon atoms and a hydroxyl (OH) group attached. Phenol was first extracted from coal tar in 1834. It is used to make phenolic and epoxy resins, explosives, pharmaceuticals, perfumes, and nylon.

1836 Irish chemist Edmund William Davy discovers the gas acetylene (ethyne).

He prepares the gas by adding water to potassium carbide. It is found to be poisonous and highly flammable, but can be safely stored in pressurized containers. It becomes used as a fuel for high-temperature welding torches and is still used for this purpose today.

1837 German chemist Robert Bunsen synthesizes the first organometallic compound, cacodyl.

The compound is a foul smelling, toxic liquid, which is spontaneously flammable on contact with air. It is the first in a class of substances where a metal atom is bonded to an organic molecule or group. Bunsen works on cacodyl for six years, during which time he nearly dies from arsenic poisoning and loses an eye during an explosion.

1838 Italian chemist Rafaelle Piria synthesizes salicylic acid, the basis of aspirin.

Piria prepares the acid from an extract obtained from willow bark. The acid is later used in the preparation of the painkiller and anti-inflammatory drug acetylsalicylic acid, commonly known as aspirin.

1839 Swedish chemist Carl Mosander discovers the metallic element lanthanum (atomic number 57).

Mosander obtains the oxide of the new element from the residue of a sample of cerium nitrate that he heats with nitric acid. He names the new rare-earth compound 'lanthano' from the Greek word meaning 'hidden'. When the metal is isolated from the oxide, it is called 'lanthanum'.

1839 US inventor Charles Goodyear invents vulcanized rubber by adding sulphur and then heating it.

Natural rubber is limited to mainly waterproofing applications as it becomes sticky in warm conditions and brittle in cold. Goodyear is trying to improve the properties of natural rubber by the addition of sulphur, when he accidentally drops the mixture onto a hot stove. He finds that heating has made the mixture flexible when cold and dry when warm. He perfects the technique, which becomes known as 'vulcanization', from Vulcan, the Roman god of fire and the forge. The most famous use of vulcanized rubber is the manufacture of rubber tyres for automobiles.

1840 German chemist Christian Schönbein discovers ozone.

Schönbein notices a peculiar smell in the vicinity of electrical equipment, which he discovers is being caused by a previously unknown gas. He succeeds in making the gas by electrolysing water and believes that he has found a new element. Schönbein shows that the gas is present in rainwater after a lightning strike and names his discovery 'ozone', after *ozo*, the Greek word meaning 'smell'. Later, Irish chemist and physicist Thomas Andrews proves that ozone is an allotrope of oxygen.

1840 German chemist Justus von Liebig publishes *Die organische Chemie in ihrer Anwendung auf Agrikulturchemie und Physiologie/Organic Chemistry in its Application to Agriculture and Physiology*, which establishes agricultural chemistry as an applied science.

Liebig dismisses the previously accepted concept that 'humus' provides plants with nourishment and argues instead that carbon dioxide from the air and water and ammonia from the soil are used by plants to grow. Liebig argues that chemical analysis of plants should determine the substances present in fertilizers and that mineral as well as organic fertilizers can be used to regenerate depleted soils.

1840 Swiss-born Russian chemist Germain Hess proposes the law of constant heat formation

Hess carries out a series of experiments that accurately measure the amount of heat involved in a number of chemical reactions. He discovers that the total amount of heat that is generated or absorbed in producing a chemical is the same no matter how many steps are involved or what route is taken to achieve it. This observation becomes known as 'Hess's law' and marks the beginning of the science of thermochemistry.

1841 Swedish chemist Jöns Jakob Berzelius is the first to describe the chemical allotropy of carbon.

Berzelius converts a sample of charcoal into graphite and correctly recognizes that the two materials are different forms of the same element. The existence of an element in more than one stable form is known as 'allotropy'.

1841 French chemist Eugène-Melchior Péligot isolates uranium (atomic number 92).

The element was discovered by German chemist Martin Klaproth in 1789, but had not been isolated in its pure form. Péligot obtains metallic uranium by the reduction of uranium tetrachloride with potassium metal. Uranium is a silvery-white, heavy metal, which is radioactive and tarnishes readily in air.

1842 English agronomist John Bennet Lawes patents superphosphate – the first artificial fertilizer – by treating phosphate rock with sulphuric acid. He opens the first factory for the production of artificial fertilizer.

The only method to replenish minerals lost from soils due to intensive cultivation is the use of organic fertilizers such as animal manure. Lawes realizes that it is possible to use chemicals to replenish the soil, if it can be determined what elements are removed. Lawes experiments on his own farm and patents his method for producing 'superphosphate'. This is a mineral-based fertilizer, made by reacting phosphate-rich rock with sulphuric acid.

1842 US physician Crawford Williamson Long is the first to use ether as an anaesthetic. He does not publish his findings until 1849.

Long administers the chemical to a patient before removing a tumour from his neck. The patient experiences no pain during the operation. This is the first successful use of a chemical anaesthetic. US physician Oliver Wendell Holmes later suggests that the use of chemicals for this purpose be called anaesthesia from the Greek words meaning 'no sensation'.

c. **1843** English physicist Michael Faraday develops electroplating by coating metals with nickel.

1843 German chemist Justus von Liebig develops melamine, the starting material for the first synthetic resins.

Melamine is formed when dicyandiamide is heated at high pressure. When this compound is reacted with formaldehyde, the resulting substance is a chemically stable and insoluble resin, which can be moulded into any shape. It becomes used in the manufacture of household items, dishes, kitchen utensils, and as a coating agent for wood and textiles. Melamine resins are commonly known by the trade names 'Formica' and 'Melmac'.

1843 Swedish chemist Carl Mosander discovers the rare earth metals erbium (atomic number 68) and terbium (atomic number 65).

He separates the rare earth yttria, discovered in 1794, into three components, which he designates as 'yttria', 'erbia', and 'terbia'. These last two fractions are the oxides of new elements, but they are so alike in their chemical properties that they are frequently mistaken for each other. The positive identification of terbia is not confirmed until 1873. The elements are named 'erbium' and 'terbium' in honour of the Swedish town Ytterby, where the original mineral was mined.

1844 Estonian chemist Karl Klaus isolates the platinum group element ruthenium (atomic number 44).

Klaus isolates the metal while studying waste residues from a platinum refinery in St. Petersburg. Ruthenium is similar in appearance to platinum but is harder and more brittle. It becomes used as an alloying agent to harden palladium and platinum.

1845 German chemist Adolph Kolbe is the first to synthesize an organic compound from inorganic materials.

Kolbe synthesizes acetic acid from carbon disulphide using a number of chemical steps to achieve his aim. The organic compound urea was earlier produced from the inorganic compound ammonium cyanate, in 1828. However this reaction only involves a change in the physical state of the two compounds, which are isomers, and does not involve any synthetic steps. Kolbe is the first to show that organic chemicals can be synthesized from purely inorganic sources. This discovery is a major breakthrough for synthetic chemistry.

1845 German-Swiss chemist Christian Schönbein accidentally discovers the explosive compound nitrocellulose, or guncotton, when he uses a cotton apron to mop up a spillage of sulphuric and nitric acid in his laboratory.

It has been known since 1838 that cotton treated with nitric acid produces the explosive compound nitrocellulose. However, no process for obtaining the material had been perfected. The acid combination Schönbein uses, along with washing excess acid out of the material, becomes the standard method for producing nitrocellulose. It is a smokeless explosive and becomes known as 'guncotton' since it replaces black powder as the explosive used in firearms.

1846 Italian chemist Ascanio Sobrero is the first to prepare the powerful explosive nitroglycerine.

Sobrero synthesizes the compound by adding glycerine to a combination of nitric and sulphuric acids. He tests a drop of the compound by heating it in a test tube and is so shocked by the force of the explosion that he immediately stops working on the substance. He becomes so concerned that the explosive might be used in warfare that he only publishes his results a year later to a very small and select group of scientists.

22 May 1848 French chemist Louis Pasteur discovers molecular asymmetry when he discovers two mirror-image forms of tartaric acid that polarize light in opposite directions.

Pasteur is studying two forms of tartaric acid, one from grape juice and the other from a commercial process. The compounds have identical chemical compositions and structure, but show differences in their properties. He notices that the crystals of the compounds are asymmetric in shape and that half of those produced by the commercial source are the mirror image of those from the grape juice. Pasteur suggests that the difference in chemical properties is associated with the arrangement of atoms at a molecular level. This is an important breakthrough in the field of stereochemistry.

Earth Sciences

1830 US scientist Alexander Dallas Bache establishes the first US magnetic observatory at his house in Philadelphia.

Bache finds that variations in the direction of the magnetic field are related to longitude, and consequently that declination can be used as a guide to longitude.

July 1830–April 1833 Scottish geologist Charles Lyell publishes the first volume of his three-volume work *Principles of Geology*, in which he argues that geological features are the result of presently observable processes acting over millions of years. It creates a new time frame for other sciences such as biology and palaeontology.

Charles Lyell argued that geological processes occurred slowly over long periods of time and not in a series of cataclysmic upheavals as was then believed. Darwin was a friend of Lyell's and a keen reader of his work. The vastness of time revealed by Lyell provided the backdrop for Darwin's theory of evolution to unfold. *Oxford Museum of the History of Science*

A scientific hypothesis is elegant and exciting insofar as it contradicts common sense.

Charles Lyell, Scottish geologist, attributed remark, quoted by S J Gould *Ever Since Darwin* 1978

1831 Having been slighted by the Geological Society for 24 years, essentially because he is a professional rather than a gentleman geologist, English engineer and surveyor William Smith is awarded the Society's first Wollaston medal.

1833–1837 Swiss palaeontologist Jean-Louis Agassiz publishes *Recherches sur les poissons fossiles/Researches on Fossil Fish*, in which he classifies nearly 1,700 fossil fish.

1834 English geologist Henry de la Beche publishes *Researches in Theoretical Geology*.

Among other things, de la Beche discusses the Earth's internal heat and the idea that it began as an igneous fluid. He also discusses the millions of years required to accumulate the layers of rock visible on the Earth's surface.

1835 English geologist Adam Sedgwick works out the sequence of fossil-bearing rocks in north Wales. He calls the oldest period the Cambrian, which is later dated at around 600 million years BC.

1835 English geologist Henry de la Beche becomes the first director of the Geological Survey of Great Britain. The Survey is still involved in mapping the country and regions abroad, and assessing natural resources.

1835 French mathematician Gustave Gaspard de Coriolis describes the inertial forces acting on a rotating body that act at right angles to the direction of rotation. The Coriolis force causes wind and current systems to rotate to the right in the northern hemisphere and to the left in the southern.

Coriolis reasons that air and water at the Equator move with the Earth as it spins. Points on the Equator move westward at about 1,600 kph/1,000 mph. Points north or south of this line move at a slower surface velocity since the distance around the Earth at these points is less. The result is that if wind moves from the south or north towards the Equator, then it arrives moving more slowly than the ground beneath. The result of this is that the winds are deflected in the opposite direction to which the Earth rotates: they move eastward. The Coriolis force explains the movement of ocean currents and air north and south of the Equator.

1836 English cleric and geologist William Buckland publishes *Geology and Mineralogy Considered with Reference to Natural Theology*, the sixth of the eight Bridgewater Treatises commissioned by Francis Henry, the eighth Earl of Bridgewater. The book is the last British geology textbook to be written from the Diluvialist (world flood) point of view.

1837 English geologist John Phillips publishes *A Treatise on Geology*, in which he discusses geological time.

Phillips states that to understand the history of the Earth, one must examine the stratigraphy from bottom (oldest) to top (youngest). He also suggests that one could begin to estimate the age of the Earth by calculating the time taken to form the crust.

1837 English naturalist Charles Darwin states that atolls are formed by the subsidence of extinct volcanoes. He publishes his ideas in 1842.

1837 US geologist James Dwight Dana publishes *System of Mineralogy*, which goes through five editions in his lifetime and becomes the standard textbook of mineralogy.

The book is now in its 32nd edition.

1838 An expedition led by US explorer Charles Wilkes sets sail from Norfolk, Virginia, with seven scientists including a geologist, botanist, marine biologist, and naturalist. The expedition sails around the world via Cape Horn, Manila, Singapore, Cape Town, and New York, returning in 1842.

1838 German mathematician Carl Gauss publishes *Allegmeine Theorie des Erdmagnetismus/ Unified Theory of the Earth's Magnetism*, in which he demonstrates that 95% of the Earth's magnetic field originates in the interior of the Earth, rather than outside the Earth.

1838 Scottish geologist Charles Lyell publishes his textbook *Elements of Geology*.

The first volume is a text on physical geology, while the second is on historical geology. Like his earlier *Principles*, it goes through many editions and is highly influential.

1838 German physicist Karl Steinheil discovers that the Earth acts as a conductor and can replace the return wire in telegraph systems.

1839 Scottish geologist Roderick Murchison publishes *The Silurian System*, a geological treatise that establishes the Silurian period of the early Palaeozoic rocks, which is today dated between roughly 439 to 408 million years ago.

1840 English geologist Adam Sedgwick and Scottish geologist Roderick Murchison publish 'On the Physical Structure of Devonshire' in which they identify and name the 'Devonian' period.

The Devonian period of geological time is today accepted as lasting from roughly 408 to 360 million years ago.

1840 English geologists Adam Sedgwick and John Phillips establish the three main stratigraphic groups, the Palaeozoic, Mesozoic, and Cenozoic. The groups reflect the progression of life through the stratigraphic record from the Palaeozoic ('early life') to Mesozoic ('middle life') to Cenozoic ('recent life').

1841 French geologists Jean-Baptiste Elie de Beaumont and Ours Dufrénoy produce a geological map of France.

1841 Scottish geologist Roderick Murchison visits the Ural Mountains where he finds a sequence of fossiliferous rocks above the Carboniferous rocks. He calls the sequence 'Permian' after a town in the region. The Permian period is today dated between roughly 290 to 245 million years ago.

1841 Swiss naturalist Jean-Louis Agassiz describes the motion and behaviour of glaciers in *Etudes sur les glaciers/Studies of Glaciers*. He argues that Europe was covered by great sheets of ice in the geologically recent past.

Agassiz argues that such features as U-shaped valleys and loose deposits of boulders and sediments were produced by glaciers now long gone.

1841 German astronomer Friedrich Bessel deduces the elliptical distortion of the Earth – the amount it departs from a perfect sphere – to be 1/299.

1842 British palaeontologist Richard Owen coins the term 'dinosaur' (meaning 'terrible lizard') to describe the great reptile-like animals that lived on earth during the Mesozoic Era (about 230 to 65 million years ago).

1842 The Geological Survey of Canada is established by Canadian geologist William Logan.

1845 Scottish geologist Charles Lyell publishes *Travels in North America*.

An account of his 13-month-long field trip from Niagara Falls to Savanna, Georgia, the book serves to educate and interest British geologists in the geology of the USA.

1845–1858 German naturalist and explorer Alexander von Humboldt lays the basis of modern geography with the publication of *Kosmos/Cosmos*, in which he arranges geographic knowledge in a systematic fashion.

1847 The Association of American Geologists becomes the American Association for the Advancement of Science (AAAS) based in Boston, Massachusetts. The AAAS, publisher of the journal *Science*, is still very active today and promotes science of all disciplines.

1848 Irish engineer and seismologist Robert Mallet describes an earthquake as 'the transit of a wave of elastic compression in any direction from vertically upward to horizontally in any azimuth through the surface and crust of the Earth'. His idea is that earthquakes are caused by sudden ruptures or flexures of the crust.

1848 The first wind and current charts for the North Atlantic are compiled by US naval officer and hydrographer Matthew Fontaine Maury.

As the head of the US Navy's Depot of Charts and Instruments, Maury has access to numerous ships' logs with their wind and current reports. He understands their significance in navigation and decides to compile them for future expeditions.

1849 Following the invention of the telegraph in the 1830s, US scientist Joseph Henry, secretary of the Smithsonian Institution, establishes a network of meteorological stations linked by telegraph. Synoptic weather maps are displayed daily in the Smithsonian.

1849 Irish engineer and seismologist Robert Mallet publishes the first modern catalogue of earthquakes.

1849 English geologist Henry Clifton Sorby shows that minerals can be identified by their optical properties.

Ecology

1836 US philosopher, and poet Ralph Waldo Emerson publishes his essay 'Nature', one of the central works of the US literary and philosophical movement known as transcendentalism.

The essay quickly becomes a manifesto for romantic thought in the USA. Emerson stresses the importance of self-reliance and argues for transcendentalist ideas that the world of nature can be read as a source of spiritual truth.

1848 English philosopher John Stuart Mill publishes *The Principles of Political Economy*. Mill argues that population growth and increasing wealth cannot continue indefinitely.

Mill suggests that a stationary state should be reached where economic growth is zero but humanity still advances morally and intellectually. Mill argues that 'solitude in the presence of natural beauty' provides thoughts that are good for the individual and the social whole.

Mathematics

***c.* 1830** English mathematician Charles Babbage creates the first accurate actuarial tables for use in insurance calculations.

1833 Charles Babbage designs his 'analytical engine' in England; never fully built it is a prototype of the modern computer, using levers, rods, and gears to perform calculations.

> *The whole of the developments and operations of analysis are now capable of being executed by machinery . . . As soon as an Analytical Engine exists, it will necessarily guide the future course of science.*
>
> Charles Babbage, English mathematician, *Passages from the Life of a Philosopher* 1864

1834 Irish mathematician William Rowan Hamilton introduces 'canonical equations', now known as 'Hamiltonians', based on the translation of Lagrangian functions.

They have various applications involving energy, momentum, and time in classical mechanics.

149

1835 Belgian mathematician Adolphe Quetelet publishes *Sur l'homme et le développement de ses facultés/A treatise on Man and the Development of his Faculties,* in which he presents the idea of the 'average man' in whom measurements of various traits are normally distributed around a central value.

1836 French mathematician Jean-Victor Poncelet publishes *Cours de mécanique appliquée aux machines/A Course in Mechanics Applied to Machines;* it introduces the use of mathematics to machine design.

1837 French mathematician Siméon-Denis Poisson publishes *Recherches sur la probabilité des juge-ments/Researches on the Probabilities of Opinions,* in which he establishes the rules of probability and describes the Poisson distribution for a discrete random variable.

1843 English mathematician Arthur Cayley is the first person to investigate spaces with more than three dimensions, as used in the phrase 'geometry of *n* dimensions' in the title of his paper of that year.

1843 Irish mathematician William Rowan Hamilton invents quaternions, which generalize the complex numbers to a non-commutative four-dimensional system.

Quaternions are generalized complex numbers that are composed of one real number and three real multiples of symbols representing the square root of -1.

1844 French mathematician Joseph Liouville finds the first transcendental numbers – numbers that cannot be expressed as the roots of an algebraic equation with rational coefficients.

Liouville was unable, however, to show that π (the ratio of any circle's diameter to its circumference) and e (the base for natural logarithms) are transcendental.

1845 English mathematician Arthur Cayley publishes *Theory of Linear Transformations.* He studies compositions of linear transformations.

The formulation of a method for representing logic using mathematical formulae (symbolic logic) by George Boole led to the development of Boolean algebra, which is fundamental to the design of modern digital computer circuits. *Corbis*

1845 French mathematician Augustin Cauchy proves a fundamental theorem of group theory, in the context of permutation groups, subsequently known as Cauchy's theorem.

1846 French mathematician Evariste Galois's research on the resolvability of algebraic equations is published posthumously.

1847 English mathematician George Boole publishes *The Mathematical Analysis of Logic,* in which he shows that the rules of logic can be treated mathematically. Boole's work lays the foundation of computer logic.

1847 Indian-born English mathematician Augustus de Morgan proposes two laws of set theory that are now known as de Morgan's laws. Now applied mainly in Boolean algebra, they concern the duality of binary operations.

Physics

29 August 1831 English physicist Michael Faraday discovers electromagnetic induction – the production of an electric current by change in magnetic intensity (and also the principle of the electric generator). US scientist Joseph Henry makes the same discovery independently of Faraday, and shortly before him, but does not publish his work.

October 1831 English physicist Michael Faraday makes the first transformer.

The transformer is later to prove itself in the electric power industry's choice of alternating rather than direct current for power distribution.

1832 French inventor Hippolyte Pixii builds the first magneto or magneto-electric generator. A magnet rotates in front of two coils to produce an alternating current, making it the first induction electric generator, and the first machine to convert mechanical energy into electrical energy.

July 1832 US scientist Joseph Henry discovers the phenomenon of self-induction – the production of electric current when a conductor is disconnected from a battery.

The reason for this is that the change of current through a coil produces a change in magnetic flux, which in turn induces an electromagnetic force in the coil itself.

1833 English physicist Michael Faraday announces the basic laws of electrolysis: that the amount of a substance deposited on an electrode is proportional to the amount of electric current passed through the cell, and that the amounts of different elements are proportional to their atomic weights.

1834 Estonian physicist Heinrich Lenz formulates the electromagnetic law that an induced current is in a direction that opposes the change that produces it.

1834 French physicist Benoît-Pierre Clapeyron uses Sadi Carnot's thermodynamic principles to formulate a relationship between vapour pressure and heat of vapourization.

1834 French physicist Jean-Charles Peltier discovers the Peltier effect: the emission or absorption of heat by an electric current depending on the direction of current. In the 1960s the effect is used for refrigeration.

1839 French physicist Edmund Becquerel discovers the photovoltaic effect when he observes the creation of a voltage between two electrodes, one of which is exposed to light.

1839–1855 English physicist Michael Faraday publishes *Researches in Electricity*, summarizing his discoveries.

1840 Italian optician Giovanni Battista Amici introduces the oil-immersion technique for observing specimens under the microscope. By immersing the objective lens in a drop of oil placed on top of the specimen light aberrations are minimized and magnifications up to 6,000 times are achieved.

1840 English physicist James Joule publishes *Production of Heat by Voltaic Electricity*, in which he states his law that the amount of heat produced per second in any conductor by an electric current is proportional to the product of the square of the current and the resistance of the conductor.

This is now expressed by the equation:

$$P = I^2R$$

where I is current and R is resistance.

1842 Austrian physicist Christian Doppler publishes *Über das farbige Licht der Doppelsterne / On the Coloured Light of Double Stars*, in which he describes how the frequency of sound and light waves changes with the motion of their source relative to the observer – the 'Doppler effect'. He also theorizes that the wavelength of light from a star will vary according to the star's velocity relative to Earth.

This becomes an invaluable tool for astronomers when measuring both the rotational speed of stars and binary stars.

1842 German physicist Julius Robert Meyer publishes 'Bermerkungen über die kräfte der unbelebten natur / Remarks on the forces of inanimate nature', in which he recognizes that heat and mechanical energy are aspects of the same phenomenon, and that the total amount of this energy in any closed system is conserved – the first law of thermodynamics.

1843 English physicist James Joule determines the value for the mechanical equivalent of heat (now known as the joule), that is the amount of work required to produce a unit of heat.

Joule carries out a series of measurements on falling masses attached to a paddle wheel in water. He finds that a mass of 328 kg/722 lb falling through a vertical distance of 0.3 m/1 ft would raise the temperature of 0.45 kg/1lb by 0.56°C/1°F.

1845 English physicist Michael Faraday discovers diamagnetism. He also discovers that a magnetic field rotates the plane of polarization of a light beam – the Faraday effect.

1847 English physicist James Joule states the law of conservation of energy – the first law of thermodynamics.

1847 German physicist Hermann von Helmholtz presents his paper 'On the Conservation of Force' to the Physical Society of Berlin. In it he sets forth the mathematical principles of the law of conservation of energy.

1848 Irish physicist William Thomson (Lord Kelvin) devises the absolute temperature scale. He defines absolute zero as $-273°C/-459.67°F$, where the molecular energy of molecules is zero. He also defines the quantities currently used to describe magnetic forces: magnitude of magnetic flux, β, and H the magnetizing force.

1848 French physicist Armand Fizeau shows how the shift in wavelength of the light from stars can be used to measure their relative velocities independently repeating the 1842 work of Christian Doppler.

1849 French physicist Armand Fizeau measures the velocity of light to within 5% of the value accepted today.

Fizeau uses a rotating, toothed wheel for the process; this is the first measurement based on wholly terrestrial observations.

1850–1869

Astronomy

1852 English astronomer Edward Sabine demonstrates a link between sunspot activity and disturbances in the Earth's magnetic field.

Sabine sets up magnetic monitoring stations from Canada to Van Diemen's Land (Tasmania). The vast network not only makes possible the first global models of the field, but also demonstrates the worldwide character of so-called 'magnetic storms', when disturbances of typically up to 1% are superimposed on the magnetic field. These are found to occur most frequently during the years with the most sunspots.

1855 English astronomer John Russell Hind discovers the first dwarf nova, U Geminorum. The star brightens by a factor of nearly 40 in a matter of days before returning to its normal brightness.

1859–1862 German astronomer Friedrich Argelander publishes *Bonner Durchmusterung/Bonn Survey*, a star catalogue listing over 324,000 stars and their magnitudes in the northern hemisphere.

c. **1860** By observing sunspots, the English astronomer Richard Carrington discovers that the Sun rotates faster at the equator than at the poles.

c. **1862** English astronomer Warren de la Rue takes stereoscopic photographs of the Sun and Moon.

1862 US astronomer Alvan Clark observes the companion star of Sirius – the first white dwarf to be discovered.

1863 English astronomer William Huggins uses the spectra of stars to show that they are composed of the same elements that exist on the Earth and in the Sun.

First visually and then photographically he explores the spectra of stars, nebulae, and comets. He is the first to show that some nebulae, including the great nebula in Orion, have pure emission spectra and thus must be truly gaseous, while others, such as that in Andromeda, yield spectra characteristic of stars, and are therefore galaxies. He is also the first to attempt to measure the radial velocity of a star from the Doppler shift of its spectral lines.

1868 English astronomer Norman Lockyer and the French astronomer Pierre-Jules Janssen announce, independently, a method of spectroscopically observing solar prominences without waiting for an eclipse to block out the Sun's glare.

They each use the invention to discover a new element, not known on Earth at the time, named by Lockyer helium. Through spectroscopic observations Lockyer goes on to discover that solar prominences are due to storms in the outer layer of the Sun (which Lockyer names the chromosphere).

Biology

1851 German botanist Wilhelm Hofmeister publishes *On the Germination, Development and Fructification of the Higher Cryptogamia and on the Fructification of the Coniferae,* in which he establishes the alternation of generations in ferns and mosses, and the relationships between the conifers (gymnosperms) and flowering plants (angiosperms). The work helps to frame a unified view of the plant kingdom.

1852 French physiologist Claude Bernard proposes that living processes depend on a constant internal environment (homeostasis).

1852 Polish-German physician Robert Remak discovers that the growth of tissues, including diseased tissues, involves both the multiplication and division of cells.

1854 German biologist Christian Ehrenberg publishes *Microgeology* and establishes micro-palaeontology, the study of fossil micro-organisms.

1856 French chemist and microbiologist Louis Pasteur establishes that micro-organisms are responsible for fermentation, thus laying the groundwork for the discipline of micro-biology.

1856 German botanist Nathaniel Prongshemi is the first to notice fertilization when he observes sperm entering the ovum in plants.

Later he describes plastids, structures in the cytoplasm of plant cells containing either chlorophyll or starch.

1856 German scientist Theodore Bilharz discovers the parasitic worm that causes schistoso-miasis (bilharzia).

1858 German pathologist Rudolf Virchow publishes *Die Cellularpathologie in ihrer Begründung auf physiologische und pathologische Gewebenlehre / Cellular Pathology as Based upon Physiological and Pathological Histology*. In it he expands his ideas on the cell as the basis of life and disease, establishing cellular pathology as essential in understanding disease.

A portrait of the English naturalist Charles Darwin, copy by John Collier (1850-1934) in the National Portrait Gallery, London. The theory of evolution which Darwin first expressed in *The Origin of the Species* resulted from his discoveries as naturalist on board the *Beagle* when he was in his early 20s. Darwin spent the rest of his life studying the results of that expedition, and formulating his theory of the evolutionary process. © *Billie Love Historical Collection*

1858 The ideas of English naturalists Charles Darwin and Alfred Russel Wallace are jointly presented to the Linnaean Society of London, England.

Darwin is awarded priority based on his previous twenty years of research. The two men remain on good terms.

1859 Charles Darwin publishes *On the Origin of Species by Means of Natural Selection*, which expounds his theory of evolution by natural selection, and by implication denies the truth of biblical creation and God's hand in nature. It sells out immediately and revolutionizes biology.

Darwin's argument is two-fold: (1) descent with modification is the key pattern in life's history, and (2) natural selection drives adaptation and evolutionary change. He goes out of his way to accommodate moderate religious views. In the public sphere, the core argument over evolution focuses on the 'ape theory' – how did humans appear? One compromise view is that human bodies may have evolved but God intervened at some point to inject soul and mind. This is the view held even by some of Darwin's closest colleagues. Darwin intentionally avoids discussion of human origins in his book.

1860 At the British Association for the Advancement of Science meeting in Oxford, Darwin's theory of evolution is widely discussed, especially its implications for the origin of the human species. Conservative English cleric Archbishop Samuel Wilberforce joins forces with scientists opposed to Darwin's theories. English biologist Thomas Henry Huxley aggressively defends Darwin, earning the nickname 'Darwin's bulldog', and is an important advocate of Darwin's theory of evolution through natural selection.

Both sides claim victory in this 'debate', which serves as one of the first of many debates about Darwinian evolution. This debate focuses on three general claims: that life's history is a story of evolution, that evolution occurred via natural selection, and that human history is part of that story. Though Darwin himself tries to avoid the issue, most public discussion of evolution focuses on the 'ape theory' of human origins. Huxley openly describes 'man's place in nature' among the apes using arguments based on anatomical similarities.

1861 German zoologist Max Schultze defines the cell as consisting of protoplasm and a nucleus, a structure he recognizes as fundamental in both plants and animals.

1861 English entomologist Henry Walter Bates publishes his paper 'Contributions to the Insect Fauna of the Amazon', in which he describes over 1,000 different types of insect that he has collected.

1862 French biologist Louis Pasteur publishes his *Germ Theory of Disease*, in which he gathers together the current theories that contagious diseases are caused by micro-organisms.

Pasteur publishes this work to give credibility to the study of the causes of infectious diseases. Acceptance of this theory becomes the basis for the transformation of medical practice, leading to the development of vaccines for many diseases and to aseptic procedures in surgery.

1862 German botanist Julius von Sachs proves that starch is produced by photosynthesis.

Sachs is the first to discover that chlorophyll, the substance that gives plants their green colour, is not distributed evenly throughout plant cells. He locates a discrete area in each plant cell where chlorophyll is located. These areas become known as 'chloroplasts'. He also finds that starch is present inside chloroplasts. This proves for the first time that starch is a product of photosynthesis.

1862 English entomologist H(enry) W(alter) Bates is the first to describe a type of biological mimicry whereby a harmless and vulnerable species resembles a dangerous or unpalatable species that is ignored by potential predators. This is known today as Batesian mimicry.

Batesian mimicry provides evidence for Darwin's theory of adaptation via natural selection. Here, birds act as selective agents by preying on individual butterflies that poorly resemble members of the poisonous species.

1863 English biologist Thomas Henry Huxley publishes *Evidence as to Man's Place in Nature*, in which he argues that humans are anatomically similar to gorillas and chimpanzees and that this similarity is greater than the similarity of the apes to other great apes.

On the question of human origins, Huxley is the most aggressive proponent of the Darwinian view. English anthropologist Edward Tylor's subsequent *Primitive Culture* (1871) allows Darwin to set human cultures on a progressive scale – from savage to civilized – to further develop the study of human cultures.

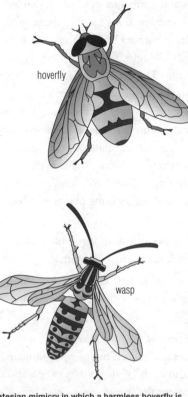

hoverfly

wasp

Batesian mimicry in which a harmless hoverfly is coloured like an unpleasant wasp in order to confuse a predator. A predator that has tried to eat a wasp will avoid the hoverfly.

1864 English philosopher Herbert Spencer coins the term 'survival of the fittest' when discussing the process of natural selection in *Principles of Biology*.

1865 French biologist Louis Pasteur reports his discovery that heating wine kills the micro-organisms that cause it to turn to vinegar. The process becomes known as pasteurization.

Pasteur is asked to investigate why a large proportion of French wine becomes sour with age. He discovers that two types of yeast cells are present, one of which produces lactic acid and causes the wine to spoil. His solution is to devise a method of gentle heating that is hot enough to kill yeast cells, but not hot enough to damage the wine. The method is eventually used to kill harmful bacteria in milk.

1865 English surgeon Joseph Lister first uses phenol (carbolic acid) as an antiseptic during surgery, to kill germs.

Lister is aware of Louis Pasteur's germ theory and insists that all instruments and doctors' hands are washed in phenol solutions immediately before surgery takes place. The death rate of patients after surgery drops dramatically. Solutions used to kill bacteria become known as antiseptics from the Greek words *anti* meaning 'against' and *septic* meaning 'rotting'. By the 1880s the use of antiseptics in surgery becomes a widely accepted practice.

1866 Austrian monk and botanist Gregor Mendel publishes 'Versuche uber Pflanzen-hybriden/Experiments on Plant Hybridization', in the *Proceedings of the Brunn Natural History Society*, in which he describes the inheritance of different characteristics in pea plants and proposes general methods for predicting patterns of inheritance, from one generation to the next. The work establishes the fundamental laws of heredity and is the basis of modern genetics. Mendel sends the paper to eminent biologists but is ignored.

Mendel's work did not attract much immediate attention from other biologists, possibly because of its place of publication (a relatively obscure journal), and because Mendel was using mathematical concepts in his study of genetics. Most biologists of the period did not approach biological problems mathematically. Mendel's work also seems to have been viewed as outdated, relevant only to a then-stale debate about hybridization.

1866 German embryologist Ernst Haeckel proposes a third category of living beings intermediate between plants and animals. Called Protista, it consists mostly of microscopic organisms such as protozoans, algae, and fungi.

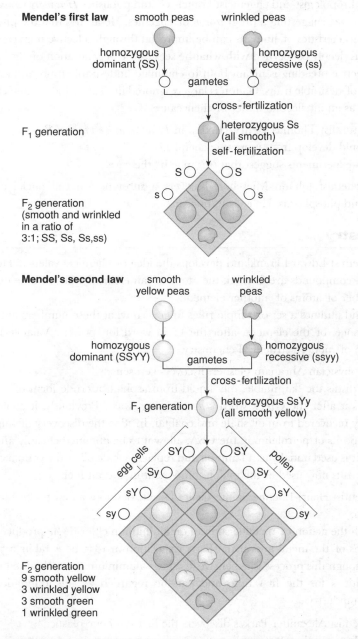

Mendel's first law

smooth peas — wrinkled peas

homozygous dominant (SS) — homozygous recessive (ss)

gametes

cross-fertilization

F₁ generation

heterozygous Ss (all smooth)

self-fertilization

S   S
s   s

F₂ generation (smooth and wrinkled in a ratio of 3:1; SS, Ss, Ss,ss)

Mendel's second law

smooth yellow peas — wrinkled peas

homozygous dominant (SSYY) — homozygous recessive (ssyy)

gametes

cross-fertilization

F₁ generation

heterozygous SsYy (all smooth yellow)

egg cells   SY   SY   pollen
             Sy   Sy
             sY   sY
             sy   sy

F₂ generation
9 smooth yellow
3 wrinkled yellow
3 smooth green
1 wrinkled green

Mendel's laws explain the proportion of offspring having various characteristics. When pea plants with smooth yellow peas are crossed with plants with wrinkled green peas, the first-generation offspring all have smooth yellow peas. The second-generation offspring, however, contain smooth yellow, wrinkled green, smooth green, and wrinkled yellow peas. This can be understood by tracing the passage of alleles Y, S, s, y throughout the generations. S and Y are dominant genes.

The same year he postulates the biogenetic law that ontogeny recapitulates phylogeny – that the evolutionary development of the species is reproduced in the development of the animal embryo.

Walter Garstang and Gavin de Beer provide evidence that discredits the theory in the first half of the 20th century.

1869 English anthropologist and eugenicist Francis Galton publishes *Hereditary Genius, its Laws and Consequences*, suggesting a programme of eugenics. He argues that the physical and mental characteristics of humans can be improved through selective marriages.

Galton is deeply concerned with what he sees as the degeneration of British human stock. Selective breeding is meant both to eliminate undesirable traits and expand the frequency of desirable traits. Galton is largely ignored until after 1900, when eugenicists claim him as an intellectual father of their cause.

1869 English naturalist Thomas Henry Huxley, in *Protoplasm, the Physical Basis of Life*, argues that life could develop from inorganic chemicals.

Modern experiments suggest that this may be the case.

1869 Swiss biochemist Johann Miescher discovers a substance in cell nuclei containing nitrogen and phosphorus. He calls it 'nuclein'.

Chemistry

1852 English chemist Edward Frankland develops the idea of chemical valence. He suggests that, when compounds are formed, the atoms of an element will combine only with a fixed number of atoms of another element.

Frankland outlines a set of simple rules to govern what these numbers are and calls it the 'valence' of the element, after the Latin word for 'power'. Valance forms the foundation of modern structural chemistry.

1853 Canadian physician Abraham Gesner discovers kerosene.

Gesner distils the flammable, oily liquid from asphalt, a crude form of petroleum, and names it after the Greek word *keros* meaning 'wax'. Previously, it could only be expensively recovered from oil shale and coal tar. In 1859 the discovery of huge underground deposits of petroleum in the USA, allow it to be obtained cheaply and in large amounts. It is used initially as a lamp oil and replaces whale oil as the primary fuel used in lighting. It is now used as the major component in aircraft fuel.

1854 French scientist Henri-Etienne Sainte-Claire Deville develops a new process for making aluminium.

Through the action of metallic sodium on aluminium chloride he produces marble-sized lumps of the metal and at the Paris Exposition in 1855 he exhibits a 7 kg/15 lb ingot. Although this process is expensive, it allows aluminium to be obtained in reasonable quantities for the first time and can be regarded as the foundation of the aluminium industry.

1855 English chemist Alexander Parkes discovers the first synthetic plastic.

Parkes discovers that if pyroxylin, or partially nitrated cellulose nitrate, is dissolved in a mixture of ether, alcohol, and camphor, it will produce a hard solid material after it is recovered by evaporation. Parkes finds that the solid becomes malleable upon heating and can be worked into complex shapes. This material is the first example of a synthetic plastic. He names his product 'Parkesine', but fails to make it a commercial success.

1856 English chemist William Perkin synthesizes the first artificial dye.

While still a chemistry student studying a method to synthesize the anti-malarial drug quinine, Perkin treats aniline with potassium dichromate and notices a purple trace in the resulting product. He obtains a bright purple solution when he adds alcohol and suspects that he has synthesized a dye. He sets up a factory to produce 'aniline purple' dye, which becomes known as 'mauve'. The factory is an instant commercial success and represents the beginning of the synthetic organic chemical industry.

1858 English chemists William Perkin and B F Duppa synthesize the amino acid, glycine.

Previous preparations of amino acids had involved isolating compounds from a natural source. This is the first instance of the manufacture of an amino acid by a purely synthetic route. Glycine is the simplest amino acid and is found in most proteins.

1858 German chemist Friedrich Kekulé publishes *Uber die Konstitution und die Metamorphosen der chemischen Verbindungen und über die chemische Natur des Kohlenstoffs / On the Constitution and Changes of Chemical Compounds and on the Chemical Nature of Carbon*, in which he shows that carbon atoms can link together to form long chains, the basis of organic molecules.

This work is of fundamental importance to the understanding of the molecular structure of organic compounds.

1858 German chemist Friedrich Kekulé and Scottish chemist Archibald Scott Couper independently devise systems of chemical symbols to represent the valence state of an atom.

Kekulé suggests that the valence number of an element represents the number of 'hooks' that can attach to other atoms. Couper proposes that valences should be represented by a series of dashes; for example, an oxygen molecule O_2 would be written as O=O. The system is used to represent the molecular structure of a molecule and helps to avoid the confusion caused by the existence of isomers.

1858 German-born British chemist Johann Griess is first to discover how to synthesize azo dyes.

Griess discovers that the addition of nitrous acid to primary aromatic compounds obtained from coal tar results in the production of yellow-coloured synthetic dyes. These organic dyes have nitrogen forming a key group, called the azo group, in their chemical structure. The synthesis of over half the commercial dyes in use today is based on research carried out on the chemical reactions used by Griess to produce azo dyes.

1859 German chemist Adolph Kolbe discovers a method to synthesize salicylic acid, the active ingredient in aspirin, from inorganic chemicals.

Kolbe develops a chemical reaction between the common industrial compounds carbon dioxide and phenol, which produces salicylic acid. The chemical reaction becomes known as the 'Kolbe reaction' and leads to cheap mass production of the drug for the first time.

1859 German chemist Albert Niemann isolates cocaine.

The native peoples of Peru and Boliva had been chewing the leaves of the coca plant, *Erythroxylum coca*, for centuries because of its mildly intoxicating properties and to counter the effects of strenuous labour. Niemann is the first to isolate the active ingredient, as a white crystalline powder, from leaves of the plant. The drug is many times more potent in this form and is highly addictive.

1860 German chemist Robert Bunsen and German physicist Gustav Kirchhoff discover the element caesium (atomic number 55).

This alkali metal is the first new element to be found using the spectroscopic techniques developed by Bunsen and Kirchhoff in the previous year. They identify the element by the presence of unknown characteristic blue lines in the spectrum of a sample of mineral water from Durkheim, Germany. They name the element 'caesium' from *caesius*, the Latin descriptive word for 'blue sky'.

1860 Italian chemist Stanislao Cannizzaro revives Avogadro's hypothesis to distinguish between atomic and molecular weight.

The First International Congress of Chemistry is held in Karlsruhe, Germany, to discuss the continuing confusion over the classification of the fundamental properties of elements and compounds. Cannizzaro presents a speech in which he explains the principles behind Avogadro's hypothesis and how to apply it. He shows that the molecular weight of any volatile compound can be accurately determined and that specific heat can be used to calculate this property with non-volatile substances. Most scientists accept the hypothesis.

1861 Belgian chemist Ernest Solvay patents a method for the economic production of sodium carbonate (washing soda) from sodium chloride, ammonia, and carbon dioxide.

Solvay's method becomes known as the 'Solvay process'. Used to make paper, glass, and bleach and to treat water and refine petroleum, it is a key development in the Industrial Revolution. The first production plant is established in 1863. The process proves to be so economical that it soon replaces all other methods of sodium carbonate production.

1861 English chemist and physicist William Crookes discovers the element thallium (atomic number 81).

Crookes discovers the element spectroscopically, by noticing unknown prominent green lines in the spectrum obtained from selenium-rich pyrites used in sulphuric acid production. He names the element 'thallium' after *thallos*, the Latin word for 'green shoot'. Crookes isolates the element as a pure metal in 1862. The metal and all of its soluble compounds are poisonous and are used as a rodenticide.

1861 German chemist Robert Bunsen and German physicist Gustav Kirchhoff discover the element rubidium (atomic number 37).

They identify the element by the presence of unknown characteristic red lines in the spectrum of a sample of the mineral lepidolite. They name the alkali metal 'rubidium' from *rubidus*, the Latin descriptive word for 'deepest red'.

1861 Russian chemist Aleksander Butlerov coins the term 'chemical structure' to describe the chemical nature of a molecule.

Butlerov proposes that the arrangement of atoms in a molecule is an important factor in determining how a compound will react. He attributes the nature of isomers – compounds with the same empirical formula but different chemical properties – to the different arrangements of their atoms.

1861 Scottish chemist Thomas Graham develops dialysis, based on the principles of osmosis.

Graham develops a method of separating particles suspended in a colloidal solution using a semi-permeable membrane. Particles in solution diffuse through this barrier at different rates. Graham classifies those particles with high diffusion rates as 'crystalloids' and those with low rates as 'colloids'. Although a slow method, dialysis proves to be efficient in the separation of compounds from colloidal solutions, and establishes the science of colloidal chemistry.

1862 German physician and biochemist Ernst Hoppe-Seyler prepares haemoglobin in crystalline form.

Haemoglobin is a protein present in the red blood cells. Hoppe-Seyler successfully isolates a sample of the protein in crystalline form. He names it haemoglobin from *haemo*, the Greek word for blood, and 'globin', a shortened version of 'globumin', the class name for this type of protein.

1863 English chemist John Newlands devises the periodic table of the elements.

Newlands notices that when elements are arranged according to atomic weight, the chemical properties of each set of seven elements tend to repeat. He calls this the 'law of octaves' and arranges the elements into vertical columns of seven. This shows that the chemical behaviours of the horizontal rows are approximately the same. This is the first time that the periodic nature of the elements has been proposed. The table is published in 1864, but is not accepted because the positions of the elements are inaccurate.

1863 German chemist Adolf von Baeyer synthesizes barbituric acid, the first barbiturate drug.

Baeyer makes the acid while studying derivatives of uric acid. He is reputed to have named the acid in honour of a female friend of his, whose name was Barbara. The acid is the primary compound for the synthesis of the barbiturate class of drugs. These compounds are commonly used today in the manufacture of sleeping pills and sedatives.

1863 German chemist J Willbrand discovers the chemical explosive trinitrotoluene (TNT).

Willbrand prepares the explosive by progressively nitrating toluene with nitric acid. The chemical is ideally suited for use as a military explosive since it melts at 82°C/179.6°F and does not explode at temperatures less than 240°C/464°F, which enables it to be melted and poured into shell casings and moulds using steam. It is resistant to shock and will not normally explode without being ignited with a detonator. Despite these advantages, TNT is not used until 1904. It becomes the most commonly used military explosive of the 20th century.

1863 German mineralogist Ferdinand Reich and his assistant Hieronymus Richter discover the element indium (atomic number 49).

Reich isolates a yellow compound from a zinc ore he had been studying and suspects that he has found a new element. He decides to analyse his sample spectroscopically but is colour-blind and has to ask his assistant Richter to carry out the experiment. Richter identifies the element by observing the presence of two unknown dark blue lines in the spectrum. They name the new element indium after the brilliant indigo colour of its spectral lines.

1864 Norwegian chemist and mathematician Cato Guldberg and Norwegian chemist Peter Waage propose the chemical law of mass action.

Guldberg and Waage state that the direction in which a chemical reaction will proceed not only depends on the mass of reagents present, but also on their concentration in a given volume. The higher the number of reacting particles present, the more likely the possibility of collisions between them. This relationship becomes known as the law of mass action, but is not noticed for many years since it was initially published in Norwegian.

1865 German chemist Friedrich Kekulé proposes the molecular structure of benzene.

Kekulé suggests that the six carbon atoms present in the benzene molecule form a hexagonal ring structure. This proposal is contrary to the current opinion that carbon only forms long chain compounds, but it explains the chemical stability of the compound and is universally accepted. Kekulé's theory introduces the concept of ring structures to organic chemistry.

161

Becoming a wealthy man on the proceeds of his invention, dynamite, Alfred Nobel left a fund of $9,200,000 after his death for the establishment of annual prizes (the Nobel prizes) in five fields, Peace, Literature, Physics, Chemistry, and Physiology or Medicine. Nobel actually believed that his invention would put an end to war by making it too horrific to contemplate. *Corbis*

1867 Swedish chemist Alfred Nobel patents the blasting explosive dynamite.

Nobel discovers that the highly explosive chemical nitroglycerine can be absorbed in the porous material 'kieselguhr', a silica-rich earth. The resulting material can be handled easily and will not explode unless ignited by heat or percussion. He calls his invention 'dynamite' and it becomes widely used in engineering and mining projects, as well as having military applications. After his death, Nobel bequeaths over $9 million to the foundation of annual prizes for recognition of scientific achievement. These become known as the Nobel prizes.

1868 English chemist William Henry Perkin synthesizes the red dye alizarin inexpensively.

Alizarin is a red dye, which can be obtained from the root of the common madder plant. Perkin is the first to develop a method of producing the dye artificially from anthraquinone. The dye is used to colour cotton, silk, and wool.

1868 English chemist William Henry Perkin synthesizes the first artificial perfume, coumarin. This perfume was previously obtained from an extract from the tonka tree, native to Guyana.

Perkin develops a laboratory procedure that enables him to synthesize unsaturated acids and he uses this method to synthesize coumarin artificially for the first time. His process becomes known as the 'Perkin reaction' and is an important breakthrough in the field of organic synthesis.

1868 French astronomer Pierre Janssen and English astronomer Norman Lockyer discover helium (atomic number 2).

Janssen is examining the spectrum of the solar prominences during a total eclipse of the Sun, when he notices an unknown bright yellow spectral line. He consults with Lockyer, who determines that the line belongs to a new element, unknown on Earth. The element is named 'helium' from the Greek name for the Sun, *helios*.

1869 Irish physical chemist Thomas Andrews proposes the concept of 'critical temperature'.

Andrews suggests that every gas has a precise temperature above which it cannot

Mendeleyev drew on observations of the properties of the chemical elements to draw up a table based on their atomic weights and the recurring patterns, or periods, exhibited by those properties. He announced that the gaps he left in the table would be filled by elements not yet discovered and accurately described the properties of the missing elements.
Oxford Museum of the History of Science

be liquefied, even at higher pressures. This concept helps scientists to develop methods that allow the liquefaction of all known gases.

1869 Russian chemist Dmitry Mendeleyev develops the periodic table of the elements. He leaves gaps for elements yet to be discovered.

As Newlands did before him in 1863, Mendeleyev classifies the elements according to their atomic weights and notices that they exhibit recurring patterns, or periods, of properties. However, he places more reliance on the valence of the element in determining its place in his table and allows the length of each period to increase beyond seven. He places gaps in his table where a known element does not fit and predicts the chemical properties of these unknown elements. He also rearranges those elements with atomic weights that he considers to be wrong.

There will come a time, when the world will be filled with one science, one truth, one industry, one brotherhood, one friendship with nature . . . this is my belief, it progresses, it grows stronger, this is worth living for, this is worth waiting for.

Dmitry Mendeleyev, Russian chemist, in Y A Urmantsev
The Symmetry of Nature and the Nature of Symmetry 1974

1869 US inventor John Wesley Hyatt, in an effort to find a substitute for the ivory in billiard balls, invents celluloid. The first artificial plastic, it can be produced cheaply in a variety of colours, is resistant to water, oil, and weak acids, and quickly finds use in making such things as combs, toys, and false teeth.

In the USA, a prize has been offered for a material that could replace ivory for the manufacture of billiard balls. To win this prize, Hyatt investigates the earlier work of English chemist Alexander Parkes, who produced a synthetic plastic from partially nitrated cellulose nitrate in 1855. Hyatt improves upon the process and names his material 'celluloid'.

Earth Sciences

c. **1850** French naturalist Antonio Snider-Pellegrini suggests that the similarities between European and North American plant fossils could be explained if the two continents were once in contact.

1850 English engineer Jacob Brett lays the first long-distance submarine telegraph cable, between Dover in England and Calais in France.

It not only fosters communication between continents, but forces scientists to begin taking a detailed look at the topography of the ocean floor.

1850 Irish engineer and seismologist Robert Mallet publishes the first seismic map of the Earth, showing earthquake foci around the world.

1853 US naval officer and hydrographer Matthew Maury organizes an international meteorological conference in Brussels. It calls for international cooperation in collecting and sharing oceanographic data.

1854 English Admiral Robert FitzRoy heads the new meteorological department of the British Board of Trade. He provides barometers for ships, creates weather maps, and in 1861 establishes a storm warning service for ships.

1854 US naval officer and hydrographer Matthew Maury maps the depth of the North Atlantic to 4,000 fathoms (7,300 m/23,950 ft).

Maury had directed the first systematic efforts at sounding the North Atlantic, and subsequently produces the first bathymetric chart of the ocean. The map includes the high region in the middle of the Atlantic, which he refers to as 'Middle Ground', the first observation of what is today called the mid-ocean ridge.

1855 English amateur scientist John Pratt and English astronomer George Airy propose two different hypotheses to explain observations that large mountains have a lower gravitational pull than expected and therefore exhibit a mass deficiency of some sort.

Airy suggests that mountains are of the same composition as the rest of the crust, but have deep roots that descend into the mantle, thereby displacing dense mantle with less dense crust. Pratt suggests that the base of a mountain is the same as everywhere else, but that high mountains are composed of less dense rock than lower hills and plains.

1855 Italian Luigi Palmieri invents an electromagnetic seismograph. It is probably the first able to detect seismic waves that are imperceptible to humans, and allows geophysicists to register earthquakes in very distant locations.

1855 US naval officer and hydrographer Matthew Maury publishes *The Physical Geography of the Sea*, the first major work on oceanography.

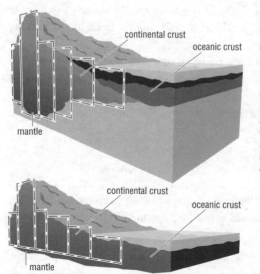

continental crust

oceanic crust

mantle

continental crust

oceanic crust

mantle

Isostasy explains the vertical distribution of the Earth's crust. George Airy proposed that the density of the crust is everywhere the same and the thickness of crustal material varies. Higher mountains are compensated by deeper roots. This explains the high elevations of most major mountain chains, such as the Himalayas. John Pratt hypothesized that the density of the crust varies, allowing the base of the crust to be the same everywhere. Sections of crust with high mountains, therefore, would be less dense than sections of crust where there are lowlands. This applies to instances where density varies, such as the difference between continental and oceanic crust.

The book is extremely popular and influential, going through many editions and printings. Maury is today known as a pioneer of physical oceanography.

1856 French hydraulic engineer Henri Darcy develops an equation to describe the flow of groundwater; it is now known as Darcy's law.

1856 US engineer and physicist Alexander Bache estimates the depth of the Pacific Ocean as 4 km/2.5 mi.

Tide-recording stations set along the Californian Pacific coast in San Francisco and San Diego detect very long wavelength 'tidal' waves (now known as tsunamis) from a Japanese earthquake. Bache uses the information to calculate the velocity of the waves travelling from Japan to California and, using the knowledge that wave speed depends upon depth, calculates the mean depth of the ocean.

1856 US meteorological researcher William Ferrel publishes 'Essay on the Winds and Currents of the Ocean'. Ferrel explains the circulation paths in the atmosphere and ocean with respect to Coriolis's theory (1835) of relative motion on a rotating sphere.

1857 Austrian geologist Eduard Suess argues that mountains and continents are formed by the movement of portions of the Earth's crust and not simply by vertical uplift.

1858 French naturalist Antonio Snider-Pelligrini publishes *La Création et ses mystères dévoilés/ Creation and its Mysteries Unveilled*, in which he suggests that Africa and South America were once one continent. His theory states that on the biblical fifth day of creation the continents were one, with a fissure running down the middle. On the sixth day, the deluge occurs and gases vent from the fissure, causing the continents to split.

1860 English geologist John Phillips, in an attempt to come up with a reasonable estimate of geological time, calculates the time it took to form the entire sedimentary section of the Weald, England. He calculates a period of 96 million years by using measurements of the sedimentary load of the Ganges River. This method of estimating geological time will be common for the rest of the century.

1861 The first *Archaeopteryx* fossil is found in upper Jurassic limestones of Solnhofen, Bavaria (now in Germany). *Archaeopteryx* has many reptilian features and is 55–190 million years old. It is still the oldest known bird.

1862 English geologist Joseph Jukes publishes 'On the Mode of Formation of Some of the River-Valleys of the South of Ireland' and 'Upon the Carboniferous Slates and Devonian Rocks and the Old Red Sandstone of the South of Ireland', in which he suggests that the structure of underlying rock formations influences the pattern of river valleys in southern Ireland.

1862 English meteorologists James Glaisher and Robert Coxwell begin making balloon flights over England. They carry a variety of weather instruments and are able to begin to characterize the lower atmosphere.

1862 Irish physicist William Thomson (Lord Kelvin) estimates the age of the Earth as about 100 million years (as little as 10 million and not older than 400 million).

Kelvin's estimate is based on the latest principles of thermodynamics and French mathematician and physicist Pierre-Simon Laplace's nebular hypothesis (1796). He essentially calculates how long it should have taken for the Earth to cool from its initial state to its present temperature. In 1897 Kelvin revises his estimate down to 20–40 million years. Kelvin's age of the Earth contradicts geologists' estimates, which

165

are much greater and are based on measurements of sedimentation rates and rock thicknesses.

1862 Scottish-born German astronomer Johann von Lamont discovers the electrical current within the Earth's crust.

1863 Canadian geologist William Logan, director of the Geological Survey of Canada, publishes *Geology of Canada*.

1863 English geologist Charles Lyell publishes *The Geological Evidence for the Antiquity of Man*, in which he provides evidence for ancient human history.

1863 The National Academy of Sciences is established in the USA, with US scientist Alexander Bache as its first director.

1864 Scottish amateur scientist James Croll publishes his hypothesis that the beginning of an ice age is linked to the eccentricity of the Earth's orbit. He later suggests that one might be able to calculate when previous ice ages occurred by examining times of extreme eccentricity.

1865 Irish physicist William Thomson (Lord Kelvin) attacks the geological principle of uniformitarianism (the theory that the Earth forms a steady-state system, in which terrestrial causes have always been of the same kind and degree as at present), stating it is contrary to physical laws.

Kelvin's argument is that the processes on Earth cannot have acted in a steady and completely uniform manner over time because the Sun, which drives processes such as evaporation and precipitation, and therefore erosion, has not had an unchanging brightness and heat output over geological time.

1867 US geologist Clarence King begins his geological, geographical, and natural resources survey of 210,000 sq km/80,000 sq mi along the 40th parallel from 105°W to 120°W.

The survey is one of four US government surveys, headed by King, F V Hayden, George Wheeler, and John Wesley Powell, employed to create maps and assess the natural resources west of the Mississippi River. The surveys eventually publish many volumes of reports of geological and palaeontological interest containing much that is previously unknown to the world.

1868 English meteorologist Alexander Buchan begins the use of weather maps for forecasting, with the publication of a map showing the movement of a cyclonic depression across North America and Europe.

1868 English naturalist Thomas Henry Huxley makes the first classification of the dinosaurs, creating the order *Ornithoscelida* and two suborders.

Ecology

1852 Eight pairs of English sparrows (*Passer domesticus*) are imported into Brooklyn, New York. Within 100 years their range extends across North America.

1853 The Forest of Fontainebleau in France becomes the first designated nature reserve.

1854 English physician John Snow traces a local epidemic of cholera and typhoid to a communal pump in Broad Street, London, England. He discovers that the well's water supply is being contaminated by a leakage from a neighbouring sewage tank.

Snow had noted that approximately 500 deaths have been reported from a single area in the city. He observes that nearly all the deaths have taken place near to the pump in Broad Street; people in the area who have escaped the disease use water from another well. Snow concludes that the disease is water-borne. The day after he informs the parish authorities the pump handle is removed. Following Snow's intervention, the death rate declines.

1866 German embryologist Ernst Haeckel coins the word 'oecology' in his work *Generelle Morphologie,* to describe the relationships between plants, animals, and their environment.

Haeckel takes the word from the Greek root *oikos,* meaning 'household', and uses it to convey the idea that the living world can be thought of as a unit resembling a family of individuals living together. The new word helps to define the field and focus the study of the modern science of ecology.

Mathematics

1850 Russian mathematician Pafnuty Chebyshev publishes *On Primary Numbers,* in which he further develops the theory of prime numbers.

Chebyshev also writes about integrals, probabilities, and quadratics, as well as technical subjects such as gearing and map-making.

1851 A posthumous book by the Czech mathematician Bernard Bolzano, *Paradoxien des Undendlichen/Paradoxes of the Infinite,* introduces ideas about infinite sets into mathematics.

The idea does not catch on until the later work of Georg Cantor and Richard Dedekind in 1872.

1851 French mathematician Joseph Liouville publishes a second work on the existence of specific transcendental numbers (that is, irrational numbers such as π that cannot be expressed by an algebraic function). These are now known as Liouville numbers.

1852 English mathematician James Sylvester establishes the theory of algebraic invariants – algebraic-equation coefficients that remain unchanged when the coordinate axes are altered or rotated.

Two years earlier, he invented the term 'matrix', though it did not appear in print until the work of Arthur Cayley, published in 1858.

1854 German mathematician Bernhard Riemann formulates his concept of non-Euclidean geometry in *On the Hypotheses forming the Foundation of Geometry.* He also gives a lecture entitled 'Über die Hypothesen welche der Geometrie zu Grunde liegen/On the Hypotheses that Lie at the Foundation of Geometry', in which he shows the connection between geometry and our assumptions about the universe.

1854 English mathematician George Boole publishes *The Laws of Thought,* in which he outlines his system of symbolic logic now known as Boolean algebra, which is now much used by computer programmers.

1854 English mathematician Arthur Cayley makes important advances in group theory.

Cayley attempts to define abstract mathematical groups and defines various operations with them.

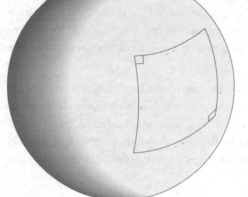

In the 1850s, Bernhard Riemann showed that, in spherical geometry (a form of non-Euclidean geometry), although a parallelogram may have two opposite angles of 90°, it does not necessarily follow that the other two angles are also 90°.

1857 French mathematician Augustin Cauchy completes his lifelong mathematical research, having made great advances in analysis, probability, and group theory.

1858 English mathematician Arthur Cayley introduces single-letter notation for matrices (rectangular arrays of numbers) and studies their properties in *A Memoir on the Theory of Matrices*.

1858 German mathematician August Möbius describes a strip of paper that has only one side and one edge. Now known as the Möbius strip, it also remains in one piece when cut down the middle.

German mathematician Johann Listing discovers it in the same year.

The Möbius strip has only one side and one edge. It consists of a strip of paper connected at its ends with a half-twist in the middle.

1859 German mathematician Bernhard Riemann makes a conjecture about a function called the zeta function, an important special function in number theory. Riemann's hypothesis is still unproved, although is known to be true in millions of cases, and is an important key to understanding prime numbers.

1864 French mathematician Joseph Bertrand publishes *Treatise on Differential and Integral Calculus*.

1865 German mathematician and physicist Julius Plückner makes advances in projective geometry, invents line geometry, and notes the interchangeability in theorems of projective geometry of the terms 'line' and 'point' (the principle of duality).

Physics

1851 Irish physicist William Thomson (Lord Kelvin) publishes his essay 'On the Dynamical Theory of Heat', which states that energy in a closed system tends to become unusable waste heat – the second law of thermodynamics.

However, Kelvin is not the first to make the connection between 'energy' and 'waste heat'. In the previous year German physicist Rudolf Clausius made a similar statement.

1852 English physicist James Joule and Irish physicist William Thomson (Lord Kelvin), discover that an expanding gas cools. This becomes known as the Joule–Kelvin effect, and is used in the refrigeration industry.

Strictly speaking, not all gases cool when they expand; indeed Joule and Kelvin find that this is not the case for hydrogen. For all gases there exists a temperature, called the inversion temperature, above which expansion will have a net heating effect and below which expansion will have a net cooling effect.

1853 French physicist Jean Foucault uses a rotating mirror to measure the speed of light in air. When he repeats the measurement in 1862, he obtains a result that is within 1% of today's accepted value.

1857 German mathematical physicist Rudolf Clausius develops the mathematics of the kinetic theory of heat and demonstrates that evaporation occurs when more molecules leave the surface of a liquid than return to it, and that the higher the temperature, the greater the number of molecules that will leave.

In the same year he publishes *Über die Arder Bewegung welche wir Wärme nennen/On the Kind of Motion which we call Heat*, which presents the first probabilitistic arguments and introduces the notion of 'mean free path'.

1859 German chemist Robert Bunsen and German physicist Gustav Kirchhoff discover that each element emits a characteristic wavelength of light.

They find that elements emit a unique spectrum of light when heated to incandescence. Parts of the full spectral range are missing, leaving a series of lines called 'spectral lines' that act as a 'fingerprint' for the element. They also discover that the vapour of an element will absorb its characteristic spectral pattern if it is cooler than the source of the light shining through it. This is the beginning of elemental spectroscopic analysis.

1860 French physicist Jean Foucault uses his rotating mirror method to measure the speed of light in air and in water. He finds the speed of light to be lower in water than in air. This result was consistent with the wave theory of light but not with the particle theory of light. By then, however, almost all scientists had already come to accept the wave theory on other grounds.

1864 Scottish physicist James Clerk Maxwell introduces mathematical equations that describe the electromagnetic field, and predict the existence of radio waves.

In 1873 he publishes *A Treatise on Electricity and Magnetism*, in which he provides a mathematical model of electromagnetic waves and identifies light as consisting of such waves. Maxwell's theory is the starting point for Einstein's work on relativity.

1867 German scientist Hermann von Helmholtz publishes *Handbook of Physiological Optics*, which develops the idea, contrary to Immanuel Kant, that concepts of time and space are due to experience.

1868 Swedish physicist Anders Ångström expresses the wavelengths of Fraunhofer lines in units of 10^{-10} m, a unit that came to be known as the angstrom.

This is now an obsolete unit that has been replaced in the SI system by the nanometre.

1868 French physicist Louis du Hauron patents a technique for taking colour photographs. Photographs are taken through green, orange, and violet filters and then printed on sheets of gelatin containing red, blue, and yellow pigments – the complementary colours of the negatives. A colour photograph is produced by superimposing the three positive transparencies.

This is similar to the process still used today.

1870–1889
Astronomy

1871 Italian astronomer Giovanni Schiaparelli observes a network of lines on Mars, which he calls *canali* ('channels'). The word is mistakenly translated as 'canals' and leads to widespread speculation that they were constructed by intelligent beings.

1872 US astronomer Henry Draper develops astronomical spectral photography and takes the first photograph of the spectrum of a star – that of Vega.

The *Henry Draper Catalogue* is a listing of the positions, magnitudes, and spectral types of stars in all parts of the sky. It is the first to use the present alphabetical system of classifying stars by their spectral type. Modern versions have raised the number of stars included to about 400,000.

1874 US astronomer Henry Draper photographs the transit of Venus.

1877 US astronomer Asaph Hall discovers Deimos and Phobos, the two moons of Mars.

In response to a suggestion by Henry Madan of Eton, England, Hall names the satellites after 'Fear' (Phobos) and 'Flight' (Deimos), the attendants of Mars mentioned in Homer's *Iliad*. Hall's observations give him the information he needs to work out the mass of Mars from an analysis of the motion of the tiny moons. It is 0.1076 times that of the Earth, very close to the currently accepted value of 0.1074.

1880 US astronomer Henry Draper photographs the Orion Nebula, the first photograph of a nebula.

1881 German-born US physicist Albert Michelson develops an interferometer to measure distances between stars.

The 'Michelson interferometer' later becomes important in high resolution spectroscopy and for determining atomic length standards, since displacements of a fraction of the wavelength of light can be measured. In 1887 Michelson and Morley famously use the device to provide evidence against the existence 'ether' surrounding Earth and acting as the medium in which light waves are propagated. These ether-drift experiments precede Einstein's creation of the special theory of relativity, and Michelson receives a Nobel prize in 1907 for his work in optics.

1881 US astronomer Edward Barnard is the first to discover a comet using photography.
Barnard goes on to take the first photograph of the Milky Way in 1889.

1881 English astronomer William Huggins and US astronomer Henry Draper take the first photograph of the spectrum of a comet.

1882 English astronomer Ralph Copeland observes the transit of Venus, in Jamaica.

1885–1890 British astronomer David Gill photographs over 450,000 stars of 11th magnitude or brighter in the southern hemisphere, in South Africa.

The plates are catalogued by Dutch astronomer Jacobus Kapteyn to produce the *Cape Photographic Durchmusterung* star catalogue in 1900.

1887 The Paris Observatory in France enlists 18 observatories around the world to make a photographic map of 10 million stars, and to catalogue all stars of 12th magnitude or brighter.

The map, known as the *Carte du ciel*, is completed in 1961.

1888 Danish astronomer Ludvig Dreyer publishes the *New General Catalogue of Nebulae and Clusters of Stars*, which lists over 8,000 nebulae.

This number is increased to 13,000 by 1895; stellar objects are known by their catalogue designations, such as NGC1898.

1888 US physicist Henry Rowland publishes the *Photographic Map of the Normal Solar Spectrum*, an 11 m/35 ft long spectrogram of the Sun, which serves as a standard reference for astronomers.

Biology

1872 German botanist Ferdinand Cohn publishes *Untersuchungen über Bakterien / Researches on Bacteria*. The first major work on bacteriology, he classifies bacteria on the basis of their morphology and physiology.

Earlier, he concludes that plant cells and animal cells are essentially the same in structure.

1873 In the first modern example of biological pest control, British-born US entomologist Charles Riley exports the acarid *Rhizoglyphus phylloxerae* to France to destroy aphids.

1875 German botanist Ferdinand Cohn describes the production of endospores by the bacteria *Bacillus subtilis*. The discovery leads to the final rejection of the doctrine of spontaneous generation and becomes important in developing sterilization techniques.

1875 German embryologist Oskar Hertwig discovers that fertilization occurs with the fusion of the nuclei of the sperm and ovum.

1875 British physician Richard Caton is the first to demonstrate the electrical activity of the brain.

The work leads to the development of the electroencephalogram (EEG), which is a trace of the brain's electrical signals recorded through electrodes placed on the scalp. EEGs are used to diagnose different brain disorders.

1875 German cytologist Eduard Strasburger publishes *Über Zellbildung und Zellteilung / On Cell Formation and Cell Division*, in which he describes the process of mitosis.

Strasburger deduces that the nucleus is responsible for heredity.

1877 German physician Robert Koch demonstrates new techniques of staining and fixing bacteria with heat for microscopic examination.

1877 Russian biologist Ilya Metchnikov discovers the process of phagocytosis whereby certain amoeba-like cells engulf other cells or foreign bodies such as dust or bacteria.

1877 English amateur entomologist Eleanor Ormerod begins to publish the *Annual Report of Observations of Injurious Insects*. Used by agriculturalists worldwide, 170,000 of the pamphlets are printed yearly.

1877 French chemist and microbiologist Louis Pasteur discovers that certain bacteria die when cultured with another type of bacteria, suggesting that the latter gives off a toxic substance. Later, these 'microbial antagonisms' were identified as antibiotics and in the 1940s applied as medical therapy.

1878 German-born Brazilian zoologist Fritz Müller proposes a second type of mimicry (besides Batesian mimicry described 1862) whereby several distasteful species evolve similar appearances. This condition is now called Müllerian mimicry. He suggests this similarity increases their protection from predators, who need only learn one pattern rather than many.

It is advantageous to the mimics, which each need to lose fewer individuals in order to teach potential predators to avoid them. Predators also benefit because less effort is required to avoid two or more similar unpalatable species than to avoid two or more unpalatable species that do not look alike.

1880 English naturalist Charles Darwin and one of his sons, Francis, write *The Power of Movement in Plants*, in which they describe the phototropic response of plants.

They conclude that the shoot tip is particularly sensitive to light and suggest that 'when seedlings are freely exposed to a lateral light some influence is transmitted from the upper part to the lower part, causing the latter to bend'. In the 1920s, experiments by Fritz Went suggest that the 'influence' is a growth-regulating substance called auxin (indolyl-3-acetic acid).

1880 German cytologist Eduard Strasburger announces that new cell nuclei arise from the division of old nuclei.

1881 German botanist Wilhelm Pfeffer publishes *Pflanzenphysiologie: Ein Hanbuch des Stoffwechsels und Kraftwechsels in der Pflanz / The Physiology of Plants: A Treatise Upon the Metabolism and Sources of Energy in Plants*, which becomes the basic handbook on plant physiology.

1881 French chemist and microbiologist Louis Pasteur vaccinates sheep against anthrax. It is the first infectious disease to be treated effectively with an antibacterial vaccine, and his success lays the foundations of immunology.

1882 German anatomist Walther Flemming publishes *Zellsubstanz, Kern und Zelltheilung / Cell Substance, Nucleus and Cell Division*, in which he describes how animal cell nuclei are derived from the division of pre-existing nuclei.

Flemming does not know of Gregor Mendel's work and does not relate his studies to genetics.

1882 German physician Robert Koch announces the discovery of *Mycobacterium tuberculosis*, the bacillus responsible for tuberculosis. This is the first time a micro-organism has been definitively associated with a human disease.

1883 German physician Robert Koch isolates the cholera bacillus.

1883 Russian physician Openchowski freezes parts of the cerebral cortex of dogs using a probe frozen by the evaporation of ether. He discovers that adjacent tissues are little damaged. His experiments are the forerunner of cryosurgery.

1884 German bacteriologists Edwin Klebs and Friedrich Löeffler discover *Corynebacterium diptheriae*, the diphtheria bacillus.

The discovery is the result of using a new medium of thickened serum on which the bacterium is cultured.

1885 English anthropologist and eugenicist Francis Galton proves the permanence and individuality of fingerprints, and devises an identification system based on them.

1886 German biologist August Weismann states that reproductive cells, or 'germplasm' cells, remain unchanged from generation to generation, and that they contain some hereditary substance.

Weismann defines reproduction as the halving of the chromosome number in the gametes and the later combination of chromosomes from two individuals. He suggests that variation arises from the combination of different chromosomes.

1887 English naturalist H G Seeley classifies dinosaurs into two groups, those with birdlike pelvises, the *Ornithischia*, and those with reptile-like pelvises, the *Saurischia*.

His proposals are meant as refinements and corrections to English biologist T H Huxley's system, devised in 1868.

1888 The word 'chromosome' is first used, by German anatomist Wilhelm von Waldeyer.

Chemistry

1871 Austrian physicist Ludwig Boltzmann describes the general statistical distribution of energies among the molecules in a gas.

Boltzmann bases his description on an earlier study carried out by Scottish physicist James Clerk Maxwell, who in 1859 proposed a model to explain the distribution of velocities of gas molecules. The two theories are combined to form the 'Maxwell–Boltzmann distribution law', which describes the statistical distribution of energy in a classical gas. The plot of energy distribution becomes known as the 'Boltzmann distribution'.

1871 German chemist Adolf von Baeyer synthesizes the dye fluorescein.

It crystallizes as a deep red powder, but when added to alkaline solutions, it imparts to them an intense green fluorescence. This colour is so strong that it is still observable even in dilutions of one part per 40 million, and is named 'fluorescein' because of this effect. It becomes used as a chemical marker in water and as an indicator dye in analytical instruments.

1874 German chemist Othmar Zeidler reports the synthesis of dichlorodiphenyl-trichloroethane, DDT.

Zeidler produces the chemical using a reaction between chloral hydrate and chlorobenzene, to which he adds sulphuric acid. He routinely reports his synthesis of this colourless, odourless compound and has no idea of its properties as an insecticide.

1874 Dutch chemist Jacobus van't Hoff and the French chemist Joseph-Achile Le Bel, independently propose a three-dimensional shape for organic molecules based on a tetrahedral carbon atom.

Chemical formulae had always been thought of in terms of two dimensions. Van't Hoff and Le Bel suggest a three-dimensional approach to the subject. A tetrahedrally shaped carbon atom allows the formation of three-dimensional molecular structures that are mirror images of each other. This is used to explain the occurrence of optical isomers. The three-dimensional representation of a chemical formula becomes known as 'stereochemistry' from the Greek words meaning 'solid chemistry'.

1875 French chemist Paul-Emile Lecoq de Boisbaudran discovers the metal gallium (atomic number 31). Russian chemist Dmitry Mendeleyev had predicted the existence of gallium in his periodic table of the elements and its discovery results in the universal acceptance of the table.

Lecoq de Boisbaudran uses spectroscopic analysis to identify two unknown violet spectral lines, present in the spectrum of a zinc ore from the Pyrenees region, as belonging to a new element. He names the element after the Latin name for France, 'Gallia'.

1876 US physicist Josiah Willard Gibbs publishes 'On the Equilibrium of Heterogeneous Substances', which lays the theoretical foundation of physical chemistry.

Gibbs applies the laws of thermodynamics to chemical reactions and proposes his concept of chemical equilibrium, where the components of a system come to a point when there is no further change in different phases, such as a liquid and a gas. He derives an equation to explain changes in pressure, temperature, and concentration at the equilibrium point. This becomes known as the 'phase rule' and marks the beginning of chemical thermodynamics.

1877 German botanist Wilhelm Pfeffer publishes *Osmotische Untersuchungen: Studien zur Zellmechanik/Osmotic Research: Studies on Cell Mechanics*, in which he describes how osmotic pressure depends on the size of molecules.

Pfeffer separates solutions containing proteins of varying sizes with a semi-permeable membrane and measures the osmotic pressure across the barrier. He shows that large protein molecules are unable to cross the membrane and that the osmotic pressure is caused by the diffusion of smaller molecules. From his observations he becomes the first scientist to be able to calculate the molecular weight of a protein molecule.

1877 US chemist James Crafts and French chemist Charles Friedel develop the 'Friedel–Crafts' reaction.

They discover that aluminium chloride acts as a versatile catalyst in chemical reactions involving the attachment of alkyl and acyl groups to aromatic carbon ring compounds. This discovery finds numerous applications in chemical synthesis.

1878 French chemist Hilaire de Chardonnet invents rayon, the first artificial fibre.

Chardonnet develops a process to make fibres out of nitrocellulose. He treats cotton with a mixture of nitric and sulphuric acids to produce nitrocellulose, which he dissolves in alcohol and ether. He forces this solution through a series of glass tubes and allows the solvent to evaporate, resulting in the production of fibres. He calls the material 'rayon' from the French word meaning 'ray of light'. It is the first of the 'rayon' class of artificial textiles.

1878 Swiss chemist Jean de Marignac discovers the rare-earth element ytterbium (atomic number 70).

De Marignac separates a new component from the rare earth erbia (see 1843), which he calls 'ytterbia' after the Swedish town of Ytterby, where the original mineral had been mined. The pure metal is isolated in 1937 and is named 'ytterbium'.

1878 French chemists Jacques Soret and Marc Delafontaine discover the rare-earth metal holmium (atomic number 67). It is discovered independently the following year by Swedish chemist Per Cleve, who names it after his home town of Stockholm.

Cleve separated holmium chemically from erbium oxide, in which holmium oxide was present as an impurity.

1879 Swedish chemist Lars Nilson discovers the rare-earth element scandium (atomic number 21).

Nilson separates an unknown oxide from the rare-earth minerals gadolinite and euxenite. He names the new element 'scandium' after Scandinavia. Later this year, Swedish chemist Per Teodor Cleve identifies scandium with 'eka-boron', an element whose existence had been predicted by Dmitry Mendeleyev in his periodic table.

1879 Swedish chemist Per Teodor Cleve discovers the rare-earth element thulium (atomic number 69).

Cleve isolates an unknown rare earth from 'erbia' (see 1843). He names the oxide 'thulia' after 'Thule', the ancient name for Scandinavia. When the element is obtained in its pure metal form, it is called 'thulium'.

1879 US chemist Ira Remsen and his German student Constantin Fahlberg discover the artificial sweetener saccharin; it is 500 times sweeter than sugar.

They synthesize the organic compound orthobenzoyl sulfimide as part of a routine series of preparations while studying derivatives of toluene. Fahlberg accidentally tastes the compound and notices the intense sweetness of its flavour. It is given the name 'saccharin' from the Latin word meaning 'sweet' and is later used as an artificial sweetener.

1879 French chemist Paul-Emile Lecoq de Boisbaudran discovers the rare-earth element samarium (atomic number 62).

Boisbaudran notices a series of unknown spectral lines, while carrying out spectroscopic analysis of the mineral samarskite. He attributes the lines to a new element, which he names 'samarium' after the mineral. Samarskite had been named after a Russian mining official called 'Samarsky', who had been the first to discover the mineral.

1880 French physicists Pierre and Paul-Jacques Curie discover that electricity is produced when pressure is placed on certain crystals including quartz – the 'piezoelectric' effect.

The Curies notice that a quartz crystal generates electricity when compressed. They also observe the converse reaction, where a crystal contracts and expands when an electric potential is placed across it. This behaviour becomes known as piezoelectricity from the Greek word *piezo* meaning 'pressure'. This property of quartz is eventually used in the manufacture of quartz oscillators for timing devices such as watches and clocks.

1880 Swiss chemist Jean de Marignac discovers the rare-earth element gadolinium (atomic number 64).

Marignac isolates an unknown rare earth from the mineral samarskite. French chemist Paul-Emile Lecoq de Boisbaudran had independently separated the same metallic oxide from a sample of the mineral yttria (see 1843). The mineral is renamed 'gadolinite' after the discoverer of the first rare-earth mineral in 1794, Finnish chemist Johan Gadolin.

1883 German chemist Adolf von Baeyer formulates the structure of the dye indigo.

Indigo is a dark blue dye, which can be obtained from any plants of the genus *Indigofera* or *Isatis*. Baeyer is the first to determine the molecular structure of the dye, allowing it to be synthesized artificially. Commercial manufacture begins in the 1890s. The dye is commonly used to colour cotton work clothes, such as denim jeans.

1884 French chemist Henri-Louis Le Châtelier identifies that any change to a system at chemical equilibrium will cause the system to adjust itself in order to minimize this disturbance and return to an equilibrium position.

This rule becomes known as 'Le Châtelier's principle'. It proves useful to scientists for determining what conditions are necessary for efficient chemical reactions.

1884 French chemist Hilaire de Chardonnet, patents his process for making artificial silk.

Chardonnet perfects his method of making artificial fibres from nitrocellulose. In 1891, he opens a factory in Besançon, France, to commercially produce the fibre. The material becomes known as 'Chardonnet silk' as it closely resembles natural silk in appearance.

1884 French scientist Paul Vieille adds stabilizers to nitrocellulose (cellulose nitrate) to make smokeless explosive powder.

Vieille adds solvents to nitrocellulose to produce a colloidal gelatinous substance. This material is much more stable than nitrocellulose and can be safely moulded into any required shape. It becomes designated by the French army as 'Powder B' and is the first of a series of smokeless explosives. Stable and reliable, it quickly replaces black powder as the propellant charge used in artillery shells.

May 1884 Swedish chemist Svante Arrhenius proposes ionic dissociation. He suggests that those compounds that form electrolytic solutions break up in solution to produce charged particles or 'ions'.

These charged fragments have opposite charges and are the reason that electrolytic solutions are able to conduct electricity. The theory is not well received as it depends on the concept that atoms could carry an electric charge. This suggestion is at odds with the strongly held consensus view that atoms are indestructible and unchanging.

1885 Austrian chemist Carl von Welsbach discovers the rare-earth elements neodymium (atomic number 60) and praseodymium (atomic number 59).

Welsbach isolates two unknown rare earths while studying compounds of didymia, the oxide of the supposed rare-earth element, didymium, which had been proposed by Swedish chemist Carl Mosander in 1839. Welsbach proves that didymia is actually a mixture of the metallic oxides of two new elements. He names them neodymium and praseodymium from the Greek words *neo* meaning new, *prasios* meaning 'green', and *dymium* meaning 'twin'.

1885 French horticulturist Pierre Millardet develops the Bordeaux mixture, a blend of copper sulphate and hydrated lime. The first successful fungicide, it rapidly achieves worldwide usage.

From the mid-1860s, wine production in France had been devastated by the twin problems of insect infestation and the spread of the fungus *Plasmopara viticola*. Millardet notices that a blend of copper sulphate, lime, and water, used by farmers in the Médoc region of France to discourage the theft of their grapes, controls the spread of the fungus. He studies the mixture for three years before publishing his work.

1886 French chemist Henri Moissan isolates the halogen element fluorine (atomic number 9).

Chemists had known of the existence of a highly reactive gaseous element for decades and had named it 'fluorine' from the Latin word *fluo*, which means 'flow'. However, they had been unable to isolate the element despite almost a century of study. Moissan achieves this goal by electrolysing a solution of potassium hydrogen fluoride in hydrogen fluoride. He cools the solution to $-50°C/-58°F$ to reduce the reactivity of the pale yellow fluorine gas that is liberated by the process. He receives a Nobel prize in 1906 for this work.

1886 French chemist Paul-Emile Lecoq de Boisbaudran discovers the rare-earth element dysprosium (atomic number 66).

Boisbaudran realizes that he has found a new element while studying a rare-earth ore containing holmium. However, he fails to isolate the element either as a compound or in its pure state. This does not happen until the 1950s, when the development of better separation techniques allows the isolation of the pure metal. He names the element 'dysprosium' from the Greek word *dysprositos*, whose literal translation is 'difficult to get at'.

1886 German chemist Clemens Winkler discovers germanium (atomic number 32).

Winkler finds an unknown component in a sample of the silver ore that he is studying. Winkler succeeds in isolating a compound of the new element from this fraction, which he further refines to obtain the pure metal. He names the element 'germanium' after 'Germania', the Latin name for Germany. It is soon recognized that germanium is 'eka-silicon', the third element predicted in 1869 by Russian chemist Dmitry Mendeleyev on the basis of his periodic table.

1886 German chemist Hermann Hellriegel discovers nitrogen fixing – this becomes important for agriculture and leads to the planting of nitrogen-fixing crops to replenish soils that have been depleted of nitrates.

Hellriegel shows that certain plants of the pea family, leguminosae, have root nodules that contain bacteria that can fix nitrogen from the air. These nitrogen-fixing bacteria are able to combine nitrogen with other constituents to form nitrate compounds.

1886 Swedish chemist Svante Arrhenius introduces the idea that acids are substances that dissociate in water to yield hydrogen ions, H^+, and that bases are substances that dissociate to yield hydroxide ions, OH^-.

Arrhenius's theory explains the properties of many common acids and bases through their ability to yield ions in solution. It leads to a further classification of these compounds into 'strong' and 'weak' categories depending on the concentration of ions in solution. This theory becomes known as 'Arrhenius theory'.

1886 US chemist Charles Hall and French chemist Paul Héroult independently develop the same method for the production of aluminium by the electrolysis of aluminium oxide.

The method becomes known as the 'Hall–Héroult process' and produces aluminium at a much lower cost than was possible before, paving the way for the widespread use of the metal. Aluminium is second only to steel as the most common structural material in use today.

1887 Swedish chemist Alfred Nobel invents ballistite, the first nitroglycerine-based smokeless blasting powder.

The explosive is composed of 60% nitroglycerine and 40% nitrocellulose. For this process, Nobel uses partially nitrated nitrocellulose, which is more soluble than the conventional material. He dissolves the mixture in alcohol and ether and then evaporates the solvent to obtain the smokeless explosive. It is safe to handle and can be cut into flakes. It is used as a propellant in firearms and as a blasting powder in mining operations.

1888 French chemist Henri-Louis Le Châtelier publishes *Loi de stabilité de l'equilibre chimique/ Law of Stability of Chemical Equilibrium* in which he outlines his principles of chemical equilibrium.

Le Châtelier expands upon his earlier work on 'Le Châtelier's principle' and discusses what effect changing conditions, such as temperature and concentration, will have on a chemical system at equilibrium. His work is independent of that carried out by US physicist Josiah Willard Gibbs, although many of the principles he discusses are the same.

1889 English chemist Frederick Abel and Scottish chemist James Dewar invent the smokeless explosive cordite.

They dissolve a mixture of nitroglycerine and nitrocellulose in petroleum jelly to form a gelatinous mass. This is then squeezed into cords that are carefully dried and cut to the required amount. They call their invention 'cordite' after the shape of the final product. The material is in breach of a patent filed by Swedish chemist Alfred Nobel, but the inventors ignore this restriction and proceed with manufacturing their product. It becomes the standard explosive used by the British Army.

1889 Swedish chemist Svante Arrhenius performs the first systematic study of activation energy.

Arrhenius proposes that molecules require a given amount of activation energy before they can react with other substances. Once 'activated', a reaction will generate enough energy to be self-sustaining and can lead to explosive chain reactions if the spread of energy through the reacting compounds is fast enough.

Earth Sciences

1870 US president Ulysses S Grant establishes the meteorological division of the Signal Service, better known as the Weather Bureau.

1871 US palaeontologist Othniel Charles Marsh discovers the first pterodactyl skeleton, an extinct flying lizard, now classified as a pterosaur.

7 December 1872–26 May 1876 The British ship *Challenger* undertakes the world's first major oceanographic survey. Under the command of the Scottish naturalist Wyville Thomson, the crew collect marine animals and water samples, dredge and core samples of the ocean bottom, and make hundreds of temperature and depth measurements.

The survey initiates the new science of oceanography.

1873 Irish physicist William Thomson (Lord Kelvin) reforms the mariner's compass by mounting thin cylindrical bars on silk thread.

1873 The International Meteorological Organization (now the World Meteorological Organization) is founded to increase communication and data sharing among nations around the world.

1873 US geologist James Dwight Dana revises and enlarges the theory that the Earth is contracting while cooling from an initially molten state.

The theory includes the 'geosynclinal' theory of mountain building, in which mountains begin as continental and oceanic depressions filled with sediments that become squashed and elevated as the Earth contracts.

1874 Austrian meteorologist Julius von Hann discovers that 90% of atmospheric water vapour is found below 5,500 m/18,000 ft, thus showing that mountain ranges can act as barriers to the transport of water vapour.

1876 US palaeontologist Charles Walcott discovers, in New York State, the first American trilobite that has its appendages still preserved.

In 1890 he will find the oldest known fossil of the time at the base of the Cambrian, and then in 1899 he will find even older fossils in pre-Cambrian rocks in Montana.

1877 The Como Bluff palaeontological site is discovered in Wyoming. It contains a large number and variety of dinosaur remains, including the first specimens of *Stegosaurus*, *Brontosaurus* (now known as *Apatosaurus*), and *Allosaurus*.

1878 The complete skeletons of several dozen *Iguanodon* are discovered in a coal mine in Belgium. They provide the first evidence that some dinosaurs travelled in herds.

1879 English astronomer George Darwin suggests that the Moon is made of mass that was somehow expelled from the Earth, leaving the scar of the Pacific Ocean basin. The idea persists for some time and is included in a number of contemporary theories of how the Earth's crust came to be.

1879 English geologist Charles Lapworth resolves a geological controversy by proposing that the lower Palaeozoic era be divided into three systems, rather than two. The new 'Ordovician' system covers many of the rocks that English geologist Adam Sedgwick had claimed were Cambrian and Robert Murchison had claimed were Silurian. The Ordovician period is today dated between roughly 510 and 439 million years ago.

1879 The US Congress consolidates the western surveys and establishes the United States Geological Survey (USGS), with US geologist Clarence King as its first director.

1881 English cleric Osmond Fisher publishes *Physics of the Earth's Crust*, the first book of its kind.

In it he suggests that the interior of the Earth is fluid and that there are convection currents rising beneath the oceans, especially at the Mid-Atlantic Ridge, and falling beneath the continents.

1882 US geologist Clarence Dutton publishes Monograph 2 of the US Geological Survey's *Tertiary History of the Grand Canyon District*. He argues that the Colorado River cut through the strata of the Colorado Plateau over a very short period of time (geologically speaking), and that the rocks then buoyed up isostatically as a result of the erosion of rocks above. This description of the Grand Canyon essentially ends the controversy over whether or not rivers carve the valleys they run through.

1883 Austrian geologist Eduard Suess publishes the first volume of his *Das Antlitz der Erde / Face of the Earth*, in which he postulates the existence of an ancient supercontinent in the southern hemisphere called Gondwanaland, and discusses how various processes are responsible for the present features of the Earth's surface.

Like many at the time, Suess believes that the Earth contracted as it cooled from an initially molten state. This contraction, he believes, resulted in the formation of mountain ranges and the collapse and subsidence of the ocean basins.

1883 British metallurgist Robert Abbott Hadfield patents manganese steel, the first special alloy of steel, and one that is exceptionally hard.

1883 Italian geologist Michele Stefano de Rossi and Swiss geologist Francois Forel produce the Rossi–Forel intensity scale, used to describe qualitatively the surface effects of earthquakes.

It is the first well-known seismicity scale and is used to determine isoseismal lines. It is eventually replaced with the Mercalli scale.

1883 US geologist G K Gilbert publishes an article in the *Salt Lake City Tribune*, warning the city of a devastating earthquake sometime in the future.

The article is reprinted in the *American Journal of Science* in the next year. It is the first article on earthquakes by a US Geological Survey scientist. In the article, Gilbert discusses his theory of mountain building by way of faulting rather than folding, and his contention that earthquakes are a result of friction between two masses of rock trying to slide past one another. He uses this theory to predict that, owing to a buildup of stress along the fault, the next segment to experience an earthquake would

179

Grove Karl Gilbert was one of the outstanding geologists of the late 19th and early 20th centuries. His work introduced ideas regarding erosion, river development, and glaciation that now form part of modern theories of physical geology. His seismic gap theory, however, was not accepted at the time of publication, and had to wait 25 years until the work of Henry Fielding Reid for widespread acceptance. *Corbis*

be the one that had moved the least recently. This is now known as the 'seismic gap' theory.

26–28 August 1883 Krakatoa volcano, Indonesia, erupts in one of the most catastrophic volcanic events in human history. The climactic explosion on the second day is heard nearly 4,800 km/ 3,000 mi away. Over 36,000 people in Sumatra and Java are drowned by an ensuing tsunami (tidal wave) 35 m/115 ft high, and dust, which is thrown 80 km/ 50 mi into the air, drifts around the world, causing spectacular sunsets for over a year.

Some citizens in the USA mistake the brilliant red sunsets for fires, and fire engines are called out. The ash that enters the atmosphere starts to restrict the levels of sunlight reaching the Earth, and global temperatures fall slightly.

1888 Canadian geologist John Dawson publishes *Geological History of Plants*, based on his observations of plant fossils in Canada, which is to be the standard textbook on palaeobotany for several decades.

1888 Swedish geologist Alfred Elis Törnebohm presents the theory that mountain ranges are the result of overthrusting, in which the upper surface of a fault plane moves over the rocks of the lower surface.

1889 German physicist Ernst Von Rebeur Paschwitz suggests that the seismic waves registered on pendulums in Germany are the result of a powerful earthquake in Tokyo. Geophysicists subsequently detect earthquakes around the world, regardless of whether or not the areas noticeably affected are inhabited.

1889 US geologist Clarence Edward Dutton develops a method of determining the epicentre of earthquakes and precisely measuring the speed of seismic waves.

Ecology

1872 Angus Smith, one of Britain's earliest alkali inspectors, publishes *Air and Rain*, in which he shows that levels of sulphate are much higher in industrial towns than in the countryside. He coins the phrase 'acid rain'.

1875 John Warder helps found the American Forestry Association, an early conservation organization, in Chicago, Illinois.

1881 English amateur entomologist Eleanor Ormerod publishes *Manual of Injurious Insects, with Methods of Prevention and Remedy*, in which she explains the importance of economic entomology within the context of agricultural studies.

In 1900 Edinburgh University recognizes her professional scientific standing by awarding her an honorary LLD – the first such award by the university to a woman.

Mathematics

1871 German mathematician Karl Weierstrass discovers a curve that, while continuous, has no definable derivative at any point.

1872 German mathematician Richard Dedekind publishes a paper that demonstrates how irrational numbers (those that cannot be written as a whole number fraction) may be defined formally, and gives a strict definition of an integer (whole number). He made his discovery in 1858.

1873 In his *Sur la fonction exponentielle / On the Exponential Function*, French mathematician Charles Hermite demonstrates that e (the base of natural or Naperian logarithms) is a transcendental number (that is, an irrational number that is not the root of a polynomial equation with rational coefficients).

1874 German mathematician Georg Cantor is the first person rigorously to describe the notion of infinity. He shows that infinities come in different sizes, and proves that 'almost all' numbers are transcendental.

1880 French mathematician Jules-Henri Poincaré publishes important results on analytic functions called automorphic functions, a subject of great importance in modern mathematics.

Science is built up with facts, as a house is with stones. But a collection of facts is no more a science than a heap of stones is a house.

Jules-Henri Poincaré, French mathematician, *Science and Hypothesis* 1905

1881 English mathematician John Venn introduces the idea of using pictures of circles to represent sets, subsequently known as Venn diagrams.

Overlapping circles, for example, represent intersecting sets, while a circle inside another one is a subset.

1881 US scientist Josiah Willard Gibbs develops the theory of vector analysis and uses it to extend James Clark Maxwell's mathematical analysis of electromagnetic waves.

1882 German mathematician Ferdinand von Lindemann proves that π is a transcendental number. A consequence of this is that it is impossible to construct a square with the same area as a given circle using a ruler and compass. This had been a classic mathematical problem ('squaring the circle') dating back to ancient Greece.

1888 German mathematician Richard Dedekind publishes *Was sind und was sollen die Zahlen? / The Nature and Meaning of Numbers*. He gives a rigorous foundation for arithmetic, later known as the Peano axioms after the Italian mathematician Giuseppe Peano, one of the pioneers of mathematical logic.

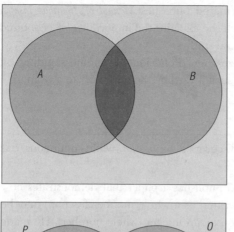

Sets and their relationships are often represented by Venn diagrams. The sets are drawn as circles – the area of overlap between the circles shows elements that are common to each set, and thus represent a third set. Here the top diagram is a Venn diagram of two intersecting sets and the bottom diagram is a Venn diagram showing the set of whole numbers from 1 to 20 and the subsets P and O of prime and odd numbers, respectively. The intersection of P and O contains all the prime numbers that are also odd.

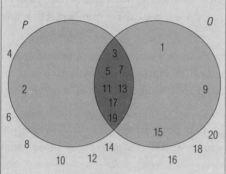

O = set of odd numbers
P = set of prime numbers

Physics

1871 English physicist John Strutt (Lord Rayleigh) explains that the sky is blue because of the preferential scattering of light in the blue region of the spectrum by particles in the atmosphere.

A red sky in the evening or morning is due to light in the red part of the visible spectrum being preferentially scattered as sunlight passes through the atmosphere at an angle.

1873 Dutch physicist Johannes van der Waals introduces the idea of weak attractive forces between molecules.

The van der Waals forces between planes of atoms in graphite allow the planes to slide, and it is this that makes it a good lubricator.

1873 English electrician Willoughby Smith confirms that the electrical conductivity of selenium increases with the amount of illumination; it proves to be an important discovery in the development of television.

1873 German physicist Ernst Abbe discovers that to distinguish two separate objects under the microscope the distance between them must be more than half the wavelength of the light that illuminates them. The discovery becomes important in the later development of electron and X-ray microscopes.

It also becomes important in the design of radio telescope dishes.

1874 Austrian physicist Ludwig Boltzmann develops the basic principles of statistical mechanics when he demonstrates how the laws of mechanics and the theory of probability, when applied to the motions of atoms, can explain the second law of thermodynamics.

The equation

$$S = k \log W$$

(where S is entropy, k the Boltzmann constant, and W is the 'number of states' of the system) is inscribed on his tombstone.

1874 Irish physicist George Johnstone Stoney coins the word 'electron' for the hypothetical unit of atomic charge and estimates its value. Stoney's estimate of 10^{-20} coulomb can be compared favourably with the currently accepted value of 1.6×10^{-19} coulomb.

The term electron is from the Greek word for amber – the Greeks noticed that chaff from their fields was attracted to amber jewellery. This is now explained by the fact that the amber is charged by friction.

1876 German electrical engineer Werner von Siemens develops a selenium cell in which electrical resistance decreases when it is illuminated.

This can be considered to be the first light dependent resistor (LDR). LDRs are now used in many applications for which light level acts as a trigger.

1877 French physicist Louis-Paul Cailletet liquefies oxygen.

Cailletet achieves this by compressing the oxygen to 300 atmospheres (approximately 3.04×10^8 Pa), using liquid sulphur dioxide to reduce the temperature to $-29°C$, then suddenly reducing the pressure. The rapid expansion causes the oxygen to condense at atmospheric pressure. Cailletet later uses this method to liquefy nitrogen and air.

1879 Austrian physicist Josef Stefan discovers the fourth-power law of blackbody radiation, which states that the energy radiating from a blackbody is proportional to the fourth power of its temperature.

This law, called the Stefan–Boltzmann law, is often expressed by the equation

$$E = \sigma T^4,$$

where σ is the Stefan–Boltzman constant, E is energy, and T is temperature.

1879 English-born US electrical engineer Elihu Thomson shows how induction coils can be used to increase current and step down voltage – the basic principle of the transformer developed a few years later.

The development of the transformer is to prove vital for the mass distribution of electrical energy.

1881 Dutch physicist Johannes van der Waals develops a version of the gas law, now known as the van der Waals equation, which takes into account the size and attraction of atoms and molecules.

It is often expressed by the equation,

$$(P + a/V^2)(V - b) = RT$$

where P, V, and T are the pressure, volume, and temperature (in Kelvin), respectively; R is the gas constant; and a and b are constants for that particular gas.

1881 French physicist Marcel Deprez shows that increasing the voltage allows electric power to be carried over longer distances with less loss due to resistance.

Present day power transmission uses very high voltages to reduce power loss. The

183

electricity supply companies will often transform voltages to 500,000 volts for long-distance transmission.

1882 French physicist Jacques-Arsène d'Arsonval invents the d'Arsonval galvanometer to measure weak electric currents. Using the torque created when an electric current passes through a coil located between the poles of a magnet, it long remains the most commonly used galvanometer.

1882 Scottish physicist Balfour Stewart postulates the existence of an electrically conducting layer of the outer atmosphere (now known as the ionosphere) to account for the daily variation in the Earth's magnetic field.

1882 English engineer James Wimshurst invents an electrostatic generator. It uses two counter-rotating plates and is capable of producing very high voltages. Electrostatic generators are later used to accelerate charged particles in nuclear bombardment experiments.

1882 US physicist Henry Augustus Rowland invents the concave diffraction grating, in which 20,000 lines to the inch are engraved on spherical concave mirrored surfaces. The grating revolutionizes spectrometry by dispersing light and permitting spectral lines to be focused.

1883 Irish physicist George Francis FitzGerald suggests that electromagnetic (radio) waves can be created by oscillating an electric current. A later demonstration of such waves by the German physicist Heinrich Hertz leads to the development of wireless telegraphy.

This, in turn, leads to a series of developments resulting in radio, television, and mobile telephones.

1883 US inventor Thomas Alva Edison observes the flow of current between a hot electrode and a cold electrode in one of his vacuum bulbs. Known as the 'Edison effect', it is later found to result from the thermionic emission of electrons from the hot electrode, and is the principle behind the working of the electron tube, which is to form the basis of the electronics industry.

Even today the vast majority of computer monitors and television screens rely on thermionic emission.

1884 English physicist Oliver Lodge discovers electrical precipitation of dust.

Electrical precipitation is now used to clean the emissions from power stations and other industries.

1885 French inventor Etienne Lenoir develops the electrical spark plug.

Spark plugs are to become a vital component in the development of the internal combustion engine.

1885 Italian physicist Galileo Ferraris develops a rotating magnetic field by placing electromagnets at right angles to one another and supplying them with alternating current 90° out of phase. The principle is used in self-starting electric motors and polyphase motors.

May 1885 Croatian-born US physicist Nikola Tesla sells his polyphase system of alternating current (AC) dynamos, transformers, and motors to US industrialist George Westinghouse, who begins a power struggle to establish AC technology over US inventor Thomas Alva Edison's direct current (DC) systems.

Since AC voltages can be 'stepped up' by transformers and experience reduced power loss at high voltages, AC transmission becomes the norm.

1886 US astronomer and physicist Samuel Pierpont Langley begins the first systematic aerodynamic research. He measures lift and drag on models of wings and other objects, which he attaches to a counterweighted beam, mounted on a pivot, that may be rotated at a speed of up to 112 kph/70 mph.

1887 German physicist Heinrich Hertz discovers the photoelectric effect, in which a material gives off charged particles when it absorbs radiant energy, when he observes that ultraviolet light affects the voltage at which sparking between two metal plates takes place. Later work on this phenomenon leads to the conclusion that light is composed of particles called photons.

The photoelectric effect is to become an area of study for Einstein, and his explanation of the effect is to lead to his Nobel Prize for Physics.

1887 US physicist Albert Michelson and US chemist Edward Morley fail in an attempt to measure the velocity of the Earth through the 'ether' by comparing the speed of light in two directions. Their failure eventually helps discredit the idea of the ether and leads to the conclusion that the speed of light is a universal constant, a fundamental premise of Einstein's theory of relativity.

However, Einstein appears to have been only indirectly aware of this result at the time he publishes his work on special relativity.

1888 German physicist Heinrich Hertz produces electromagnetic waves and demonstrates that they are reflected in a manner similar to light waves. Hertz's experiments now support Maxwell's theory that light is also electromagnetic in nature.

1890–1899

Astronomy

1890 The first version of the *Henry Draper Star Catalogue* is published. Produced by astronomers at Harvard College Observatory, it lists the position, magnitude, and type of over 10,000 stars, and begins the alphabetical system of naming stars according to temperature. Subsequent editions increase the listing to 400,000 stars.

1891 The 'blink' comparator is invented. It permits the discovery of objects in the Solar System by comparison of two photographs, taken a few hours apart, of the same region of the sky. Stars remain fixed, while planets and asteroids move or 'blink'.

1892 US astronomer Edward Emerson Barnard discovers Jupiter's fifth moon, Amalthea.

Barnard uses the 91 cm/36 in refractor at Lick Observatory. Amalthea is the first moon of Jupiter to be discovered since the time of Galileo, and the last moon of any planet to be discovered by direct visual observation, as opposed to photography. Amalthea is Jupiter's fifth largest moon, but only $^1/_{15}$ the size of Europa, the fourth largest.

1895 Russian scientist Konstantin Tsiolkovsky publishes *Gryozy o zemle i nebe/Dreams of Earth and Sky*. The first book about space travel, it discusses the possibility of space flight using liquid-fuelled rockets, and the idea of designing spacecraft with a closed biological cycle to provide oxygen from plants for long flights.

The beginning of scientific space travel is considered by many to date from this point. His other innovations include multi-stage rockets, rocket fuels, biological and

185

technical problems of weightlessness, the use of solar energy on Earth and in space, the role of orbital stations, and the design of spacesuits.

1897 US astronomer Alvan Clark completes construction of the 102 cm/40 inch Yerkes optical refracting telescope in Wisconsin.

1898 The asteroid Eros is discovered by German astronomer Gustav Witt.

1899 US astronomer William Pickering discovers Phoebe, the ninth satellite of Saturn.

Pickering notes that it revolves around Saturn in a retrograde direction.

Biology

1892 Dutch geneticist Hugo Marie de Vries, through a programme of plant breeding, establishes the same laws of heredity discovered by Gregor Mendel in 1865.

Earlier, he uses the term 'plasmolysis' to describe the shrinkage of cell content from the plant cell wall with the loss of turgor pressure. De Vries's work on water relations advances plant physiology.

1892 German embryologist and anatomist Oscar Hertwig establishes the science of cytology by suggesting that the processes that go on inside the cell are reflections of organismic processes.

1892 Russian microbiologist Dimitry Ivanovsky publishes 'On Two Diseases of Tobacco', in which he announces that mosaic disease in tobacco is caused by micro-organisms too small to be seen through a microscope.

Ivanovsky observes the disease is still transmissible after passing through ceramic filters that are impermeable to bacteria. He puts the result down to a bacterial toxin. However, this is the first evidence for the existence of viruses.

c. 1895 Dutch microbiologist Martinus Beijerinck develops a method of creating pure cultures of micro-organisms, which are invaluable for research purposes.

1895 Belgian bacteriologist Emilie Van Ermengem isolates the bacterium *Clostridium botulinum*, which causes botulism.

1897 British physician and bacteriologist Ronald Ross, a medical officer in the British army in India, discovers the malaria parasite in the gastrointestinal tract of the *Anopheles* mosquito, and realizes that the insect is responsible for the transmission of the disease.

For a number of years, Ross had failed to vindicate the idea that malaria could be transmitted by drinking water into which the bodies of mosquitoes had fallen. Ross observes the malarial parasite (*Plasmodium falciparium*) invade the salivary glands of the mosquito and experimentally shows that malaria can be transmitted between birds through mosquito bites.

1897 Danish veterinarian Bernhard Bang discovers the bacillus *Brucella abortus* (Bang's bacillus), which is responsible for contagious abortion in cattle and brucellosis in human beings.

1897 Russian physiologist Ivan Pavlov publishes *Lectures on the Work of the Digestive Glands*, which discusses the gastrointestinal secretions involved in digestion and results in the idea of the conditioned reflex.

1898 Dutch microbiologist Martinus Beijerinck discovers a new form of infectious agent, which he calls a 'contagious living liquid'. This is now taken as the beginning of the discipline of virology.

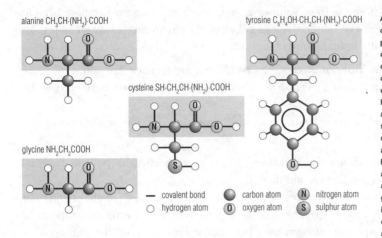

alanine CH$_3$CH·(NH$_2$)·COOH

tyrosine C$_6$H$_4$OH·CH$_2$CH·(NH$_2$)·COOH

cysteine SH·CH$_2$CH·(NH$_2$)·COOH

glycine NH$_2$CH$_2$COOH

— covalent bond
○ hydrogen atom
● carbon atom
Ⓞ oxygen atom
Ⓝ nitrogen atom
Ⓢ sulphur atom

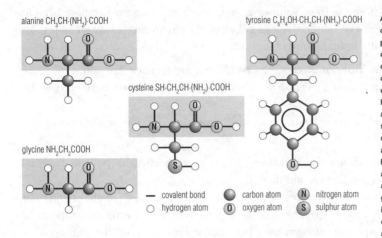

Amino acids are natural organic compounds that make up proteins and can thus be considered the basic molecules of life. There are 20 different common amino acids. They consist mainly of carbon, oxygen, hydrogen, and nitrogen. Each amino acid has a common core structure (consisting of two carbon atoms, two oxygen atoms, a nitrogen atom, and four hydrogen atoms) to which is attached a variable group, known as the R group. In glycine, the R group is a single hydrogen atom; in alanine, the R group consists of a carbon and three hydrogen atoms.

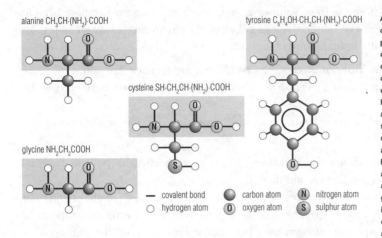

1898 German bacteriologists Friedrich Löffler and Paul Frosch show that it is possible to infect a cow with foot-and-mouth disease by inoculation with a cell-free extract. The work helps to establish the discipline of virology.

1899 English zoologist Edwin Lankester publishes 'The Significance of the Increased Size of the Cerebrum in Recent as Compared with Extinct Animals', in which he argues that the ability to learn is inherited and plays a significant role in human evolution.

1899 The amino acid cysteine is discovered to be a component of protein.

Chemistry

1891 US chemist Herman Frasch patents the Frasch process for the recovery of sulphur from underground deposits. Superheated water is pumped into the deposits, melting the sulphur and bringing it to the surface. The process allows the exploitation of deposits that would otherwise be prohibitively expensive. The liquid sulphur is more than 99% pure. The process is first used successfully in Texas and Louisiana in 1894.

1891 US inventor Edward Goodrich Acheson discovers carborundum (silicon carbide). It is the hardest synthetic material known until 1929 and is used as an industrial abrasive.

Trying to create artificial diamonds, Acheson strongly heats a mixture of coke and clay in an iron vessel. He notices that tiny hexagonal shaped crystals have been produced and that these are almost as hard a material as diamond. He has produced silicon carbide, but mistakenly thinks that it is a compound of carbon and alumina. He names the compound 'carborundum' after 'carbon' and 'corundum', a mineral made up from fused alumina.

1892 British chemists Charles Cross, Edward Bevan, and Clayton Beadle dissolve cellulose to create viscose rayon. In 1905, the British textile company Courtaulds buys the rights and begins commercial production.

Cross, Bevan, and Beadle treat cellulose with sodium hydroxide and carbon disulphide to form the compound cellulose xanthate. This is dissolved in caustic soda to produce a viscous yellow liquid, which they call 'viscose'. They develop a method of injecting this liquid into a bath of sulphuric acid, which recovers the cellulose present in the form of a fibre. The material – a member of the rayon class of artificial textiles – is initially used in the manufacture of 'artificial silk'.

1892 French chemist Henri Moissan invents the first electric-arc furnace. He uses it to create many new compounds and to vaporize substances that had been considered to be impossible to melt.

Moissan also develops a commercially viable procedure for the production of the flammable gas acetylene.

1894 English physicist John Strutt (Lord Rayleigh) and Scottish chemist William Ramsay isolate the element argon (atomic number 18).

Rayleigh collaborates with Ramsey after noticing that the atomic weight of nitrogen obtained from air is always greater than that prepared from nitrogen compounds. Ramsey passes atmospheric nitrogen over hot magnesium, until all the nitrogen has reacted, and collects whatever gas remains for spectroscopic analysis. He is able to detect a series of unknown spectral lines from a heated sample of the residue gas, showing the presence of a new element. They name the gaseous element 'argon' after *argos*, the Greek word for 'inactive'.

1895 Scottish chemist William Ramsay isolates the rare gas helium (atomic number 2). Its existence had been established in 1868 from the solar spectrum.

Ramsey collects a gas produced by the uranium mineral cleveite, and examines it spectroscopically. He discovers that the gas produces spectral lines that are identical to those observed in the solar spectrum by Janssen and Lockyer in 1868. Ramsey has isolated helium and shown that the element can exist on Earth.

1896 The first significant organic chemical herbicide, Sinox, is developed in France.

It is discovered that sodium orthodinitrocresylate is a selective killer of vegetation, and can be used to control the spread of weeds in fields of cereal crops. The chemical becomes known by the name 'Sinox' and is used as a herbicide well into the 20th century.

1897 French chemist Paul Sabatier develops nickel catalysis, making large-scale hydrogenation reactions commercially viable and eventually leading to the production of edible fats, such as margarine, from inedible plant oils.

Sabatier accidentally discovers that nickel acts as a catalyst for the reaction that adds hydrogen to unsaturated carbon double bonds in compounds such as ethylene. This process is known as 'hydrogenation' and had previously only been possible using an expensive platinum catalyst. He is awarded a share in the Nobel Prize for Chemistry in 1912 for his discovery.

1898 Scottish chemist William Ramsay and English chemist Morris Travers discover the rare gases neon (atomic number 10), krypton (atomic number 36), and xenon (atomic number 54).

They distil a quantity of liquid air to isolate the fraction that they know is associated with the rare gas argon (see 1894). After careful examination, they discover the presence of three unknown gaseous elements. They name the new elements 'krypton', 'neon', and 'xenon' after the Greek words for 'hidden', 'new', and 'stranger'. These gases have a valency of zero and are chemically stable. Along with helium and argon, they become known as the 'noble gases' or rare gases.

Marie Curie and her husband Pierre in their Paris laboratory. They received the Nobel Prize for Physics in 1903 for the discovery of radioactivity. In 1911 Marie Curie became the first person to be awarded the Nobel prize twice, when she was awarded the chemistry prize for her discovery of radium. Some of her notebooks are so radioactive that they still cannot be handled. *AEA Technology*

1898 French chemists Pierre Curie and Marie Curie discover the radioactive elements polonium (atomic number 84) and radium (atomic number 88). Radium is discovered by Marie Curie in pitchblende and is the first element to be discovered radiochemically.

Marie Curie notices that some uranium ores produce more radiation than can be accounted for by the presence of uranium alone. Working on this assumption, the Curies purify the uranium mineral pitchblende in a deliberate attempt to find these more active radioactive elements. Their first success is the discovery of 'polonium', which they name after Marie Curie's native country, Poland. They use the same methods to discover a second radioactive element, which they name 'radium' after the Latin word for 'ray', because it is even more radioactive than polonium.

1898 Scottish chemist and physicist James Dewar liquefies hydrogen for the first time.

He invents a special flask to contain the liquid gas. It is a double walled, silvered glass vessel that contains a vacuum between the walls. The idea is developed to produce the modern thermos flask. In 1899, he succeeds in solidifying hydrogen, leaving helium as the only known substance that has not been made solid.

1899 French chemist André Debierne discovers the radioactive element actinium (atomic number 89).

The discovery of new radioactive elements from the uranium mineral pitchblende in 1898 prompts scientists to investigate this mineral more carefully. One such scientist is Debierne, who treats pitchblende with ammonium hydroxide and discovers an undiscovered radioactive element in a precipitate of the residues. He acknowledges the earlier work of the Curies by naming the element 'actinium' after *aktis* and *aktinos*, the Greek words for 'ray' and 'beam', which make up the name 'radium'.

1899 German chemist Emil Fischer shows that proteins are polymers, or large molecules, comprised of amino acids. His wide-ranging work brings organization to carbohydrate chemistry.

Earth Sciences

1890 US geologist G K Gilbert publishes Monograph 1 of the US Geological Survey's *Lake Bonneville*.

After first identifying the former presence of a deep lake in the area that is now Salt Lake, Gilbert goes on to show that after the lake had drained, the crust experienced isostatic rebound in response to the release of the weight of water. He then uses his measurements of the rebound to estimate the geophysical properties of the crust.

1891 British meteorologist W H Dines invents a pressure tube wind gauge that is still in use.

1891 Russian mathematician Evgraf Fyodorov and German mathematician Arthur Schoenflies mathematically derive 230 possible space groups of crystals, working independently.

The work becomes the basis for determining crystal structures when, about 20 years later, the X-ray diffraction of crystals can be analysed. Fyodorov also invents the universal stage, which allows mineralogists to study mineral grains in any orientation under the microscope.

1892 US geologist Clarence Dutton publishes the paper 'On Some of the Greater Problems of Physical Geology', in which he advances the idea of isostasy, whereby lighter material in the Earth's crust rises to form continents and mountains, while heavier material sinks, to form basins and oceans.

1892 US geologist G K Gilbert publishes *The Moon's Face*, in which he argues that the craters on the Moon are not volcanic as most scientists think, but are formed by meteorite impact.

Gilbert's article, which includes the results of very simple experiments as well as telescope observations, is considered to be the first paper in planetary geology. Seen as catastrophic by many, his theory is not accepted by geologists until the 1960s.

1893 US geologist Charles Walcott studies the palaeozoic sediments east of the Sierra Nevada mountain range and calculates that they took 17.5 million years to accumulate. He estimates the age of the Earth at 55 million years.

1899 Irish geologist and physicist John Joly announces his calculation of the age of the oceans at somewhere between 90 and 99.4 million years.

Joly's calculation is based on the ocean's salt content, and the assumption that the ocean began as fresh water and that the salt, supplied by rivers, has increased steadily over time.

1899 The third volume of Scottish geologist James Hutton's book *Theory of the Earth with Proofs and Illustrations*, lost for more than 100 years, is published after being found in the library of the Royal Society.

Ecology

1891 US Congress passes the Forest Reserves Act, which establishes the first national forests in order to preserve their water, timber, wildlife, and other resources. Over 13 million acres are set aside over the next 2–3 years.

John Muir (right) was one of the founders of the environmental movement. An explorer, naturalist, and writer, he campaigned for the conservation of land, water, and forests in the USA. He was instrumental in bringing about the Yosemite National Park Bill in 1890, establishing both Yosemite and Sequoia national parks. He also helped persuade President Theodore Roosevelt (on the left of the picture) to set aside 59,900,000 hectares of forest reserves. *Corbis*

1892 Scottish-born US naturalist, explorer, and writer John Muir founds the Sierra Club to help conserve Yosemite and other natural wonders in California.

Muir's life-long dedication to the beauties of the wilderness began as a result of civil war draft dodging when, aged 26, he fled Wisconsin for the Canadian wilderness. Muir suggests that nature was not created solely for human use and that humans should not elevate themselves above the rest of creation. Consequently, Muir rejects the anthropocentrism of traditional Christianity. Muir is convinced, however, that areas of natural beauty should be preserved for their own sake and for the benefit of future generations. He becomes a champion of the establishment of national parks and the importance of the wilderness experience.

1893 At a meeting of the British Association for the Advancement of Science (BAAS), the physiologist John Burdon-Sanderson tells the BAAS that 'oecology' is one of the three great divisions of biology along with physiology and morphology. At a meeting of the International Botanical Congress in this same year, the modern spelling 'ecology' is established.

1898 The 19,480 sq km/7,500 sq mi Kruger National Park is established (in part) in South Africa.

1899 Mount Rainier National Park is created in the state of Washington. The 95,264 ha/ 235,404 acre park encompasses a volcano, which contains 41 glaciers on its summit.

Mathematics

1890 US inventor and statistician Herman Hollerith uses punched cards to automate counting the US census. The holes, which represent numerical data, are sorted and tabulated by an electric machine, the forerunner of modern computers. In 1896

191

Hollerith forms the Tabulating Machine Company, which later changes its name to International Business Machines (IBM).

1892 US inventor William Seward Burroughs patents and manufactures the 'arithmometer', the first adding machine that prints out each entry as well as the total.

1892 French mathematician Jules-Henri Poincaré publishes the first of three volumes of *Les Méthodes nouvelles de la mécanique céleste*/*New Methods in Celestial Mechanics*, which describes the interactions of objects in space due to gravity.

Poincaré makes prolific contributions to most branches of mathematics, particularly probability and topology. He is also a philosopher and popularizer of science.

1895 French mathematician Jules-Henri Poincaré publishes his first work on topology, sometimes referred to as 'rubber sheet geometry'. He is the originator of algebraic topology.

1896 The prime number theorem is proved independently by mathematicians Jacques-Salomon Hadamard of France and Charles-Jean de la Vallée-Poussin of Belgium. This theorem gives an estimate of the number of primes there are up to a given number.

1899 German mathematician David Hilbert publishes *Grundlagen der Geometrie*/*Foundations of Geometry*, which provides a rigorous basis for geometry.

Physics

1891 German electrical engineer Oskar von Miller develops a cable that can transmit alternating current (AC) at 25,000 volts over long distances.

This can be considered to be the first step towards the current power transmission at over 400 kV.

1892 German-born US electrical engineer Charles Steinmetz discovers the law of hysteresis, which, by explaining why all electrical devices lose power when magnetic action is converted to heat, allows engineers to improve the efficiency of electric motors, generators, and transformers through design, rather than trial and error.

1892 Presaging an aspect of Einstein's theory of relativity, Irish physicist George Francis FitzGerald suggests that a body contracts when in motion and that this contraction affects scientific measurements.

The effect still carries the name 'FitzGerald contraction'.

1892 Russian scientist Konstantin Tsiolkovsky pioneers the study of aerodynamics by building a series of wind tunnels to measure air resistance on moving vehicles.

1893 British physicist Oliver Heaviside shows theoretically that as the velocity of an electric charge increases, so does its effective mass. It presages Einstein's special theory of relativity

It will be another 12 years before Einstein presents his work in this area.

1893 German physicist Wilhelm Wien states that the wavelength at which the most energy is emitted by a hot body is inversely proportional to the absolute temperature of the body.

Wien's law, as it becomes known, can be expressed by the equation

$$\lambda_{max} T = \text{a constant}$$

where λ is wavelength and T is temperature. Taking the wavelength at which the most energy is emitted by the Sun to be 500 nm and the constant to be 2.9 mK then the surface temperature of the Sun is calculated to be 6,000 K (5,727°C/10,340°F) at its surface.

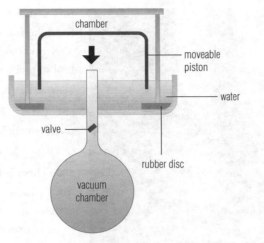

The cloud chamber devised by Charles Wilson was the first instrument to detect the tracks of atomic particles. It consisted originally of a cylindrical glass chamber fitted with a hollow piston, which was connected, via a valve, to a large evacuated flask. The piston falls rapidly when the valve is opened, reducing the pressure in the chamber, and water vapour condenses along the tracks of any particles in the chamber.

1893 German-born US electrical engineer Charles Steinmetz develops a mathematical method for making calculations about alternating current (AC) circuits. By allowing the performance and efficiency of electrical equipment to be predicted, it leads to the rapid development of devices using AC.

1895 Scottish physicist Charles Wilson develops the first cloud chamber. He builds it to duplicate the effects of clouds on mountain tops, but later realizes its potential in nuclear physics.

The first 'antiparticle', the positron, is to be discovered by the physicist Carl Anderson using the Wilson cloud chamber.

1895 US physicist Wallace Sabine discovers that the acoustics of a building depend on its volume and the absorptivity of the materials used in its construction; he thus founds the science of architectural acoustics.

This is now part of the design of any theatre or concert hall. Many concert halls now have moveable panels or ceilings to allow the acoustics to be matched to the performance.

1895 Russian physicist Aleksandr Popov constructs a lightning detector to register atmospheric electrical disturbances, and suggests that it can be used to detect radio waves.

However it is not the first detector of electromagnetic waves. In 1890, Edouard Branly of the Catholic University in Paris developed a 'coherer', a glass tube filled with copper filings, which would conduct when subjected to a source of radio waves.

1895 Russian physicist Aleksandr Popov demonstrates the transmission of radio waves, at the University of St Petersburg, Russia.

By then Oliver Lodge in England, Frederick Trouton in Ireland, Guglielmo Marconi in Italy, and others had also demonstrated radio wave transmission.

1895 German scientist Wilhelm Röntgen experiments with a Crookes tube and notices that a sheet of paper coated with barium platinocyanide becomes fluorescent. Realizing that

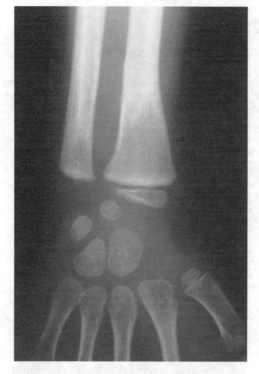

Röntgen's accidental discovery of the penetrating power of X-rays revolutionized medicine, allowing doctors to see inside the human body without cutting it open. Only later would the potential of X-rays to cause cancer become evident. *Ecoscene*

this is due to some unknown emission from the tube, he names it the X-ray.

X-rays soon become an invaluable tool in medicine.

1895 New Zealand-born British physicist Ernest Rutherford discovers that magnetic fields can be used to detect electromagnetic or radio waves.

1896 British-born US electrical engineer Elihu Thomson deliberately exposes a finger to X-rays and subsequently provides the first description of the development of X-ray burns.

1896 French physicist Antoine-Henri Becquerel reports to the Académie des Sciences his discovery that uranium is radioactive.

This is a 'lucky' discovery for Becquerel since, at the time, he is investigating fluorescence. The full impact of his work will, however, have to wait a few more years until Marie and Pierre Curie publish their work. The three of them eventually share the Nobel Prize for Physics in 1903.

1897–99 English physicist Joseph John Thomson demonstrates the existence of the electron, the first known subatomic particle. It revolutionizes knowledge of atomic structure by indicating that the atom can be subdivided.

The assumption of a state of matter more finely subdivided than the atom of an element is a somewhat startling one.

J(oseph) J(ohn) Thomson, English physicist, Royal Institution Lecture 1897

1897 German-born US electrical engineer Charles Steinmetz publishes *Theory and Calculation of Alternating Current*, which outlines his method of making calculations concerning alternating current (AC) circuits.

Steinmetz is the first to make use of the complex number i, the square root of minus one, in engineering calculations. He is granted over 200 patents for his discoveries in the field of electrical engineering.

The Discovery of X-rays

by Julian Rowe

THE BONES IN HIS WIFE'S HANDS

On 8 November 1895, German physicist Wilhelm Röntgen (1845–1923) discovered X-rays – by accident. This key discovery in atomic physics transformed medical diagnosis, and later provided a powerful tool in cancer therapy.

Röntgen was investigating the effects of electricity discharged through gases at low pressures to produce a beam of cathode rays (or electrons). Röntgen used a Crookes tube: an improved vacuum tube invented by English chemist and physicist William Crookes (1832–1919). Röntgen directed a narrow beam of rays from the tube, which was covered with black cardboard, onto a screen in a darkened room. He noticed a faint light on a nearby bench, caused by fluorescence from another screen.

THE UNKNOWN RAYS

Röntgen knew that cathode rays could only travel a few centimetres through air. This other screen was about a metre away from his Crookes tube; he realized that he had discovered a new phenomenon. For six weeks, he worked feverishly to find out all about the new rays. Because their nature was unknown, he called them X-rays. On 2 December 1895, he made the first X-ray photograph, of his wife's hand. It shows that the bones in her hands were deformed by polyarthritis.

Röntgen announced his discovery on 28 December 1895. He accurately described many of the properties of X-rays. They are produced at the walls of the discharge tube by cathode rays; like light, they travel in straight lines and cast shadows; most bodies are transparent to them in some degree; they can blacken photographic plates, and cause some chemical substances to fluoresce; unlike cathode rays, they are not deflected by magnetic fields.

THE USES OF X-RAYS

All this created tremendous scientific excitement. In 1901, Röntgen received the first Nobel Prize for Physics for his discovery. Initially, X-rays were used to examine the bones of the skeleton: effectively the start of radiology and the use of X-rays for medical purposes. Fractures could be seen clearly, as could embedded foreign bodies such as bullets or needles. Accidentally swallowed objects were accurately located. For the first time, surgeons knew in advance of operating exactly what to do.

THE BARIUM MEAL

Solid objects and bones could be clearly visualized on photographic film, recently developed by George Eastman (1854–1932) in the USA. However, it was hard to distinguish the soft tissues of the body. US physician Walter Bradford Cannon (1871–1945), who researched digestive problems and pioneered the use of X-rays in studying the alimentary canal, invented the 'barium meal' in 1897. Barium salts are relatively impervious to X-rays; by swallowing a solution of them before radiology, the patient's digestive system is selectively thrown into relief on the X-ray film. By the 1930s, X-ray examination was routine medical practice, particularly in examining the lungs for tuberculosis.

THE DAMAGING EFFECTS

As X-rays traverse living tissue, they also damage it. This danger to health was only understood much later, and both Röntgen and his assistant suffered X-ray poisoning. The destructive nature of radiation can be harnessed therapeutically. By focusing X-radiation on a tumour, its tissue is destroyed. When a cancer is not widely disseminated within the body, radiation therapy is a powerful weapon in the fight against cancer. Other forms of radiation, from radioactive sources for example, can be similarly employed.

TOMOGRAPHY

Traditional radiology now accounts for no more than half of medical photography. In tomography, an interesting extension of the medical uses for X-rays, they are used to photograph a selected plane of the human body. Tomography was first demonstrated successfully in 1928. In 1973, British engineer Godfrey Hounsfield (1919–) invented

The Discovery of X-Rays *continued*

computerized axial tomography (CAT), in which a computer assembles a high-resolution X-ray picture of a 'slice' through the body (or head) of a patient from information provided by detectors rotating around the patient. CAT can detect very small changes in the density of living tissue.

At first, such scanners were used to diagnose diseases of the brain; now, improved machines can scan an entire body. With Allan Macleod Cormack (1924–1998), the South African physicist who independently developed the mathematical basis for computer-assisted X-ray tomography, Hounsfield received the 1979 Nobel Prize for Physiology or Medicine for the development of CAT.

FROM X-RAYS TO RADIOACTIVITY

Röntgen's discovery of X-rays led directly to the discovery of radioactivity by French physicist

Antoine-Henri Becquerel (1852–1908). In 1896 Becquerel was searching for X-rays in the fluorescence seen when certain salts absorb ultraviolet radiation: from the Sun, for example. He placed his test salt on a photographic plate wrapped in black paper, reasoning that any X-radiation given off by a fluorescing salt would fog the film. By chance, he left a test salt in a drawer near a wrapped plate. When the plate was developed, it had fogged.

The salt could not have fluoresced in the dark; some radiation other than X-rays must be responsible. His test salt was potassium uranyl sulphate, a salt of uranium; Becquerel had discovered radioactivity.

1898 German physicist Wilhelm Wien demonstrates that a magnetic field can deflect a beam of ions. His work lays the foundations of mass spectroscopy.

1899 New Zealand-born British physicist Ernest Rutherford discovers and names alpha and beta radiation from radioactive substances.

Rutherford is able to distinguish between the two types of radiation by their penetrating power. The low penetration particles he names alpha particles and the higher penetrating particles he names beta particles.

1899–1900 Serbian-born US inventor Nikola Tesla discovers stationary terrestrial waves, demonstrating that the Earth can act as a conductor of certain electromagnetic frequencies.

The 20th Century
1900–2000

Science in the 20th century progressed at a phenomenal rate, with the development of the computer aiding all branches of science.

Astronomy

By the end of the 19th century, photography had begun to take over from naked-eye observations of the universe and spectroscopy was being used to determine the composition, motion, and distance of stars. In 1916 German-born US physicist Albert Einstein (1879–1955) published his general theory of relativity. Essentially a theory of gravitation, it marked the greatest theoretical advance in our understanding of the universe since Newton's *Principia* and, like Newton's theory, it had far-reaching implications for astronomy.

In 1923 US astronomer Edwin Hubble (1889–1953) verified 18th-century English astronomer William Herschel's suggestion that there might be separate galaxies, far outside our own – the concept of an expanding and evolving universe.

Radio astronomy – beginning in the 1930s with experiments by US radio engineer Karl Jansky (1905–1950) and US radio astronomer Grote Reber (1911–) – probed the structure of the universe, demonstrating in 1954 that an optically visible distant galaxy was identical with a powerful radio source known as Cygnus A. Later analysis of the comparative number, strength, and distance of radio sources suggested that in the distant past these, including the quasars discovered in 1963, had been much more powerful and numerous than today. This fact suggested that the universe has been evolving from an origin, and is not of infinite age as expected under a steady-state theory. The discovery in 1965 of microwave background radiation was evidence for the enormous temperature of the giant explosion, or Big Bang, that brought the universe into existence.

Although the practical limit in size and efficiency of optical telescopes has apparently been reached, the siting of these and other types of telescope at new observatories in the previously neglected southern hemisphere has opened fresh areas of the sky to search. Australia has been in the forefront of these developments. The most remarkable recent extension of the powers of astronomy to explore the universe is in the use of rockets, satellites, space stations, and space probes. Even the range and accuracy of the conventional telescope may be greatly improved free from the Earth's atmosphere. When the USA launched the Hubble Space Telescope into permanent orbit in 1990, it was the most powerful optical telescope yet constructed, with a 2.4 m/94.5 in mirror.

Biology

Biochemistry became more and more dominant in 20th-century biology. Molecular biologists and chemists have been concerned with determining the structures of many large biological molecules, such as the muscle protein myoglobin, using X-ray crystallography, a technique pioneered by English chemist Dorothy Hodgkin (1910–1994).

The development and understanding of genetics has also had a huge impact, particularly towards the end of the 20th century. By 1900, Dutch botanist and geneticist Hugo de Vries (1848–1935), English geneticist William Bateson (1861–1926), and others were recognizing the importance of the pioneering work of Austrian monk Gregor Mendel (1822–1884). The study of mutations was pioneered by geneticists such as US geneticists Thomas Hunt Morgan (1866–1945) and Herman Muller (1890–1967). In 1953 English molecular biologist Francis Crick (1916–   ) and US biologist James Watson (1928–   ) determined the structure of DNA. In the mid-20th century the chemical link between the gene and the expression of that gene in the organism was beginning to be unravelled. This link proved to be the production by the gene of a specific protein, called an enzyme, which has a particular effect in the cell in which it is produced. In 1941, US biologists George Beadle (1903–1989) and Edward Tatum (1909–1975), established the gene-enzyme hypothesis, showing that one gene was responsible for the production of one specific enzyme ('one gene, one enzyme').

In genetic engineering, the splicing and reconciliation of genes is used to increase knowledge of cell function and reproduction, but it can also achieve practical ends. Gene splicing was invented in 1973 by the US scientists Stanley Cohen (1922–   ) and Herbert Boyer (1936–   ), and patented in the USA in 1984. In 1995 a team at the Institute for Genomic Research in Gaitherburg, Maryland, USA, unveiled the first complete genetic blueprint (for a free-living bacterium). The Human Genome Organization was established in 1988 in Washington, DC, with the aim of mapping the complete sequence of human DNA. A working draft of the genome was completed in 2000. At the beginning of the 21st century, biology is one of the most rapidly expanding scientific disciplines, as its implications for improving human life are realized.

Chemistry

The 20th century was a time of tremendous advancement for the synthesis and manipulation of chemicals. In organic chemistry, German organic chemists Kurt Alder (1902–1958) and Otto Diels (1976–1954) found a method to synthesize cyclic carbon compounds, an essential step in drug development. German chemist George Wittig (1897–1987) developed the Wittig reaction, a route to produce unsaturated hydrocarbons, and in the 1960s US chemist Charles Pedersen (1904–1990), French chemist Jean Marie Lehn (1939–   ), and US chemist Donald Cram (1919–   ) discovered and developed crown ethers, cryptands, and cryptates, versatile organic reagents with broad applications in biochemistry and organic synthesis.

US chemist Elias J Corey (1928–   ) developed retrosynthesis, a powerful tool for building complex molecules from smaller, cheaper, and more readily available ones. Modern chemists use retrosynthesis to design everything from insect repellents to better drugs. Significant advances also occurred in the field of structural analysis. German physicist Max van Laue (1879–1960) developed X-ray crystallography to work out the structures of large molecules.

This field was advanced by English chemists and X-ray crystallographers Rosalind Franklin (1920–1958) and Dorothy Hodgkin (1910–1994), among others.

In 1912, Polish-born US biochemist Casimir Funk (1884–1967) isolated vitamin B from yeast. A year later US biochemist Elmer McCollum (1884–1967) discovered vitamin A and found vitamin D in 1920. Hungarian-born US biochemist Albert Szent-Györgyi (1893–1986) isolated vitamin C from cabbages. Vitamin E was isolated by US chemists soon afterwards. Once the vitamins had been isolated in pure form, their structures could be determined. The structure of vitamin B_1 was determined in 1934 and, by the 1940s, most of the vitamins we know today had been found, isolated, and synthesized in laboratories.

German organic chemist Hermann Staudinger (1881–1965) pioneered the concept of macromolecules and his theories formed the foundation of polymer science. Synthetic polymers were made which closely resemble natural rubber; the leader in this field was US organic chemist Wallace Carothers (1896–1937), who also invented nylon. German organic chemist Karl Ziegler (1898–1973) and Italian chemist Giulio Natta (1903–1979) worked out how to prevent branching during polymerization, so that plastics, films, and fibres can now be made more or less to order. Work on the make-up of proteins had to wait for the development of chemical techniques such as chromatography and electrophoresis.

German chemists Fritz Haber (1868–1934) and Carl Bosch (1874–1940) developed industrial techniques using high pressures, catalysts, and high temperatures to manufacture chemicals that could not be produced economically in the 19th century, notably nitrogen fixation, which led to the development of modern artificial fertilizers and explosives. Pioneers such as US chemist William Burton (1865–1964), French-born US inventor Eugene Houdry (1892–1962), and US chemical engineer Warren K Lewis (1882–1975) developed techniques for the conversion of petroleum oil and natural gas into chemicals (petrochemicals). This route allowed chemicals to be produced more cheaply and in greater quantities than ever before. Petroleum also provided the chemical industry with a variety of previously unavailable feedstocks, which led to the development of plastics, synthetic rubber, and synthetic fibres.

Earth Sciences

The most significant advance in earth science of the 20th century was the theory of plate tectonics. A theory of continental drift was put forward in 1912 by German meteorologist Alfred Wegener (1880–1930), who suggested that continents rupture, drift apart, and eventually collide with one another. Wegener's theory explained why the shape of the east coast of the Americas and that of the west coast of Africa seem to fit together like pieces of a jigsaw puzzle; evidence for the drift came from the similarity of distant rock deposits, which indicated that continents have changed position over time. In the early 1960s scientists discovered that most earthquakes occur along lines parallel to ocean trenches and ridges, and in 1965 the theory of plate tectonics was formulated by Canadian geophysicist John Tuzo Wilson (1908–1993).

In 1900 the Earth was estimated to be about 90 million years old. In 1907 US chemist Bertram Boltwood (1870–1927) used the recent discovery that some forms of lead were the products of radioactive decay of uranium to demonstrate that some rocks were as old as 2,200 million years. English geologist Arthur Holmes (1890–1965) pioneered the use of this new tool of radioactive decay methods for rock-dating.

Towards the close of the 20th century three-dimensional imaging of Earth's structure was made possible through the use of global seismic tomography. Developed by geophysicists at Harvard University, this method is analogous to CAT scans of the human brain. With seismic tomography it is possible to follow the fate of great slabs of the Earth's crust as they descend into the mantle, and it is also possible to find the sources of sustained volcanism such as that which created the Hawaiian Islands in the Pacific.

In recent years Earth observation satellites have measured continental movements with unprecedented accuracy. The surface of the Earth can be measured using global positioning geodesy (detecting signals from satellites by Earth-based receivers), satellite laser ranging (in which satellites reflect signals from ground transmitters back to ground receivers), and very long-baseline interferometry, which compares signals received at ground-based receivers from distant extraterrestrial bodies. These techniques can measure distances of thousands of kilometres to accuracies of less than a centimetre. Movements of faults can be measured, as can the growth of tectonic plates.

Ecology and Environment

The study of ecology took a new urgency and a new theoretical structure in the 20th century. Great leaps in the understanding of processes on a habitat and global level occurred. Plant ecology was advanced by Danish biologist Eugenius Warming (1841–1924), who developed the theory of climax communities and US ecologist Frederic Clements (1874–1945), who stressed the role of climate in determining the climax community. Animal ecology was furthered by English zoologist Charles Elton (1900–1991) who proposed the ideas of food chains, population pyramids, and ecological niches, and US biologist Raymond Lindeman (1915–1942) who consolidated these ideas into his theory of trophic levels.

Scientists from many disciplines have come together to produce theories and models of the interactions between the biosphere, atmosphere, and lithosphere, particularly relating to climate change and pollution. Large-scale international efforts have discovered the thinning of the ozone layer, the greenhouse effect, global warming, and acid rain. However, individual scientists have also contributed to the public understanding of ecological and environmental problems, for example, US biologist Rachel Carson's (1907–1964) influential work *Silent Spring* (1962), and Dennis Meadow's *Limits to Growth* (1972). Ecology promises to be one of the most exciting, and important, sciences at the beginning of the 21st century.

Mathematics

New branches of mathematics have been developed in the 20th century, which are of great practical importance. Probably the most important of these is statistics, in which much pioneering work was done by English statistician Karl Pearson (1857–1936).

The 20th century was the century of algebra, as the very language of much of pure mathematical proof. It is also a story of the development of very general and abstract theories, rather than the solution of individual problems, and the provision of very abstract settings for such theoretical accounts. In consequence of this broadening of scope and scale, mathematical work is increasingly done by teams of collaborators rather than by individual mathematicians. One notable exception to this came with the announcement in 1994 by English mathematician Andrew Wiles (1953–   ) that he had solved Fermat's last theorem. The theorem's proof had eluded mathematicians for over 300 years.

Chaos theory, which attempts to describe irregular, unpredictable systems, arose in

the 1960s. The central discovery in 1961 by US meteorologist Edward Lorenz (1917–   )
observed that random behaviour can arise in systems whose mathematical description
contains no hint whatever of randomness. Polish-born French mathematician Benoit
Mandelbrot (1924–   ) produced the first fractal images in 1962, using a computer that
repeated the same mathematical pattern over and over again.

Physics

The principal focus of 20th-century physics was to determine the inner structure of the
atom. It began with the discovery of the electron in 1897–99 by English physicist
J J Thomson (1856–1940). The charge and mass of the electron were then found by Irish
mathematical physicist John Townsend (1868–1937) and US physicist Robert Millikan
(1868–1953). Meanwhile, radioactivity was detected by French physicist Henri Becquerel
(1852–1908) in 1896. Three kinds of radioactivity were identified (alpha, beta, and
gamma) by New Zealand-born British physicist Ernest Rutherford (1871–1937), and in
1911 he produced the nuclear model of the atom, proposing that it consists of electrons
orbiting a nucleus. In 1913, Danish physicist Niels Bohr (1885–1962) showed that the
electrons must move in orbits at particular energy levels around the nucleus.

In 1923, French physicist Louis de Broglie (1892–1987) described how electrons
could behave as if they made up waves around the nucleus. This discovery was developed into a theoretical system of wave mechanics by Austrian physicist Erwin
Schrödinger (1887–1961) in 1926. It showed that electrons exist both as particles and
waves. In 1927 German physicist Werner Heisenberg (1901–1976) showed that the
position and momentum of the electron in the atom cannot be known precisely, but
only found with a degree of probability or uncertainty (his uncertainty principle).

A series of discoveries of nuclear particles was made, starting in 1932 with the discovery of the positron by US physicist Carl Anderson (1905–1991) and the neutron by
English physicist James Chadwick (1891–1974). This work was aided by the development
of particle accelerators, the first built by English physicist John Cockcroft (1897–1967)
and Irish physicist Ernest Walton (1903–1995) in 1932. It led to the discovery of nuclear
fission by German radiochemist Otto Hahn (1879–1968) in 1938 and the production of
nuclear power by Italian-born US physicist Enrico Fermi (1901–1954) in 1942.

Much of modern physics has been concerned with the behaviour of elementary
particles. US physicists Richard Feynman (1918–1988) and Julian Schwinger (1918–
1994), and Japanese physicist Sin-Itiro Tomonaga (1906–1979) developed quantum electrodynamics (QED), which describes the interaction of charged subatomic particles in
electric and magnetic fields. By 1960 the existence of around 200 elementary particles
had been established, some of which did not behave as theory predicted.

Pakistani physicist Abdus Salam (1926–1996) and US physicists Steven Weinberg
(1933–   ) and Sheldon Glashow (1932–   ) demonstrated that at high energies the electromagnetic and weak nuclear forces could be regarded as aspects of a single combined
force, the electroweak force. This was confirmed in 1983 by the discovery of new particles predicted by the theory. In the 1980s a mathematical theory called string theory
was developed, in which the fundamental objects of the universe were extremely small
stringlike objects. There are many unresolved difficulties, but some physicists think that
string theory, or some variant of it, could develop into a 'theory of everything' that
explains space-time, together with the elementary particles and their interactions,
within one comprehensive framework.

The 20th Century
1900–2000

1900–1909

Astronomy

1903 Russian scientist Konstantin Tsiolkovsky publishes his classic 'Investigations of Space by Means of Rockets', in which he outlines the use of liquid-propelled rockets to escape the Earth's gravity.

1905 US astronomer Percival Lowell, after a study of the orbit of Uranus, predicts the existence of the planet Pluto.

Lowell first analyses the discrepancies between the observed and calculated positions of Uranus after making allowance for the perturbations of Neptune. Pluto is discovered by Clyde Tombaugh in 1930.

1905–1907 Danish astronomer Ejnar Hertzsprung discovers that there is a relationship between the colour and absolute brightness of stars and classifies them according to this relationship. It is used to determine the distances of stars and forms the basis of theories of stellar evolution.

1906 The asteroid Achilles is discovered by German astronomer Max Wolf. It is the first of the asteroid family known as Trojans, which orbit in positions that form an equilateral triangle with Jupiter and the Sun.

1906 US astronomer Percival Lowell publishes *Mars and its Canals*, in which he argues that the canal-like markings on Mars are irrigation canals built by intelligent creatures.

During 15 years of intensive study of Mars, whose surface markings he draws in intricate detail, Lowell publicizes the existence of the network of several hundred dark, straight lines and their intersection in a number of 'oases'. Lowell concludes that the bright areas are deserts and the dark ones patches of vegetation, and that water from the melting polar cap flows down the canals toward the equatorial region to supply the vegetation there. The canals, as it turns out, are mostly chance alignments of dark patches which the eye, at the limit of resolution, tends to join together by lines.

1908 US astronomer George Ellery Hale discovers that sunspots have magnetic fields.

1909 English astronomer John Evershed discovers what is now known as the Evershed effect, a wavelength shift and asymmetry of spectral line profiles due to a horizontal, radial outflow of gas across a sunspot penumbra. Explaining this effect has remained a classic problem in solar magnetohydrodynamics ever since.

Biology

1900 Dutch geneticist Hugo de Vries, German botanist Carl Correns, and Austrian botanist Erich Tschermak von Seysenegg, simultaneously and independently, rediscover the Austrian monk Gregor Mendel's 1865 work on heredity.

Breeding experiments using the evening primrose prompt de Vries to coin the term 'mutation' to describe the spontaneous formation of varieties of the plant. He does not realize that the varieties are the result of changes in the number of chromosomes and not of changes in genes. Nevertheless his work shows that evolution need not necessarily be a process of slow change. In England, geneticist William Bateson becomes an advocate of Mendelism. He sees that Mendel's work supports his theory that species evolve discontinuously, in a sequence of 'jumps'.

1900 The Kral Collection of micro-organisms is established in Prague, Czechoslovakia; it is the first collection of pure cultures of micro-organisms used for research purposes.

June 1900 The Galton–Henry fingerprint classification system is published (devised 1885). Scotland Yard adopts it in 1901 in place of the Bertillon anthropometric system; other law enforcement agencies change over soon after.

1901 Austrian immunologist Karl Landsteiner discovers the ABO blood group system.

Later, Landsteiner finds other blood antigens: the MNP system in 1927 and the rhesus factor system in 1940. Like the ABO system, both are inheritable.

1901 English biochemist Frederick Gowland Hopkins isolates the amino acid tryptophan.

1902–1904 US geneticist Walter Sutton and the German zoologist Theodor Boveri found the chromosomal theory of inheritance when they show that cell division is connected with heredity.

1903 Russian physiologist Ivan Pavlov describes learning by conditioning. Whilst researching digestion Pavlov observes that a dog salivates when food is in its mouth – an example of an unconditional reflex. However, he notices that when accustomed to a feeding routine, dogs salivate at meal times even before food is given to them. If a bell always rings before food appears, the dog soon salivates at the sound of the bell even in the absence of food. The work establishes that reflexes can be conditioned (trained).

1904 Russian physiologist Ivan Pavlov receives the Nobel Prize for Physiology or Medicine for his work on the physiology of digestion.

1905 Danish botanist Wilhelm Johannsen introduces the terms 'genotype' and 'phenotype' to explain how genetically identical plants differ in external characteristics.

1905 English neurophysiologist K Lewis shows that nerve impulses do not change with the strength or nature of the stimulus promoting their propagation. His friend English neurologist Edgar Adrian confirms the result, thereby establishing the 'all or none' nature of the nerve impulse.

Later Lewis studies the electrical activity of the brain. The work leads to development of the electroencephalogram (EEG) which displays the pattern of electrical discharge by the brain. The method is soon used to diagnose brain disorders.

1906 English biologist William Bateson introduces the term 'genetics'.

Bateson discovers that some characteristics are not inherited independently, thereby contradicting Mendel's laws. The contradiction is the result of linkage where genes are close together on a chromosome and inherited *en bloc*.

1908 English geneticist Reginald Punnett, troubled by the relationship between the frequency of alleles and the frequency of their expression in phenotypes, puts the problem to his friend, the mathematician Geoffrey Hardy. Hardy's solution is the equation

$$p^2 + 2pq + q^2 = 1,$$

thereby making his single but vital contribution to population genetics.

The same year, the German physician Wilhelm Weinberg proposed the same equation independently. The formula, now known as the Hardy–Weinberg law, allows for complex predictions based on the mathematical consequences of Mendelian inheritance. In 1910, Punnett is the first to set out a cross between parents with contrasting characteristics as a table. The table is called a Punnett square.

1909 Danish botanist Wilhelm Johannsen introduces the term 'gene' – a derivative of genesis – to explain the section on a chromosome that controls the phenotype of an organism.

1909 French bacteriologist Charles-Jules-Henri Nicolle discovers that typhus is transmitted by the body louse.

Nicolle notices that individuals that do not wash or change clothes regularly transmit typhus at much greater rates. He deduces that the carrier of typhus must be transported within clothes or on skin. He proves the body louse to be the carrier through a series of experiments.

Chemistry

1900 French chemist François Grignard discovers organometallic compounds that become known as Grignard reagents and find important uses in organic synthesis.

Grignard discovers a series of magnesium-based organometallic compounds that help to catalyse reactions involving the combination of carbon containing groups to organic molecules. These compounds become widely used in the production of organic chemicals, notably in new approaches to alcohol synthesis. He is awarded a share in the Nobel Prize for Chemistry in 1912 for his discovery.

1900 German physicist Friedrich Dorn discovers the radioactive noble gas radon (atomic number 86).

Dorn is studying the radioactive element radium, when he discovers that the material gives off a gas that is also radioactive, which he calls 'radium emanation'. On closer inspection, it is found that the gas is not only a new element, but also the final member of the noble gas class of elements. It is renamed 'radon' after the element radium.

1901 French chemist Eugène Demarçay discovers the rare-earth element europium (atomic number 63).

Demarçay isolates the oxide of the element from a compound of samarium. He names the new element 'europium' after the continent of Europe. It is eventually discovered that europium is a highly efficient neutron absorber and it is used in the control systems of nuclear reactors.

1901 Dutch chemist Jacobus van't Hoff receives the first Nobel Prize for Chemistry for his discovery of the laws of chemical dynamics and osmotic pressure.

1902 Austrian chemist Richard Zsigmondy invents the ultramicroscope.

Colloidal solutions are composed of particles that are too small to be seen using a conventional microscope, but large enough to scatter light shining through the solution. The ultramicroscope projects an intense beam of light through a colloidal solution and allows the observation of the scattered light using a microscope positioned perpendicular to the projected beam. He discovers that he can distinguish the motion of individual particles as bright flashes against a dark background. Zsigmondy is awarded the Nobel Prize for Chemistry in 1925 for his invention of the ultramicroscope and his subsequent study of colloids.

1902 German chemist Emil Fischer is awarded the Nobel Prize for Chemistry for his synthesis of sugars and purines.

1902 German chemist Friedrich Ostwald patents the industrially important process of producing nitric acid by the catalytic oxidation of ammonia.

Ostwald uses a platinum gauze catalyst to convert ammonia gas into nitrogen dioxide, which is dissolved in water to form nitric acid. His procedure becomes known as the 'Ostwald process' and becomes the primary industrial method for the manufacture of nitric acid.

1902 Polish-born French physicist Marie Curie and French chemist André-Louis Debierne isolate radium as a chloride compound.

They began the process of refining the uranium mineral pitchblende in 1898. The task takes them four years to complete, after which time they have isolated only 4 oz of the chloride compound from over 7 tonnes of the mineral.

1903 Swedish chemist Svante Arrhenius receives the Nobel Prize for Chemistry for his theory of electrolytic dissociation.

1904 English chemist Frederick Kipping synthesizes the first long-chain silicon compound.

Kipping's compounds are structurally equivalent to saturated hydrocarbons, except that their molecules are composed of Si–Si bonds instead of C–C bonds. This class of compounds become known as 'silicones'. They are used extensively during World War II to produce non-freezing hydraulic fluids for aircraft landing gears and as general-purpose water repellents; they still find use in these applications today.

1904 Scottish chemist William Ramsay receives the Nobel Prize for Chemistry for his involvement in the discovery of rare gases in air and their locations in the periodic table.

1905 German chemist Adolf von Baeyer receives the Nobel Prize for Chemistry for his discovery and work with indigo and other hydroaromatic dyes.

1906 English biochemist Frederick Gowland Hopkins is first to suggest the existence of vitamins.

Hopkins proposes that foods contain trace quantities of substances essential to life in addition to carbohydrates, fats, minerals, and water. He calls these compounds 'accessory factors' and suggests that the diseases scurvy and rickets are caused by a lack of these factors in the diet. These essential compounds become known as vitamins. Hopkins shares the Nobel Prize for Physiology or Medicine in 1929 for proposing the vitamin concept.

1906 German physicist Walther Nernst formulates the third law of thermodynamics, which states that matter tends towards random motion and that energy tends to dissipate at a temperature above absolute zero ($-273.15°C/-459.67°F$).

Nernst's theorem – which becomes known as 'Nernst's law' – states that a system can never reach the temperature of absolute zero by any means because the extraction of energy from the system will always become progressively more difficult. Nernst's law is found to be a powerful tool for predicting chemical equilibrium and determining the theoretical feasibility of many chemical reactions. The theory earns Nernst the Nobel Prize for Chemistry in 1920. Even using today's modern techniques, absolute zero has not been reached.

The History of Plastics

by Keith B Hutton

INTRODUCTION

The development of humankind from prehistoric times up to the modern age has been closely associated with the materials used to fabricate tools and artefacts. The earliest materials were wood and stone, which were replaced by copper and iron when it was discovered how metals could be obtained from mineral ores. Advances in metallurgy led to the development of metallic alloys, such as bronze and steel, which soon replaced pure metals for making tools. However, it was to take until the middle of the 19th century before the development of the first truly synthetic material, plastic. The word 'plastic' is derived from the Greek word *plastikos*, which means 'able to be moulded'. The strict definition of a plastic is a synthetic non-metal material that can be shaped into almost any form.

CELLULOSE-BASED PLASTICS

The first synthetic plastics were produced from cellulose, a natural material present in the cell walls of plants. In 1855 English chemist Alexander Parkes (1831–1890) discovered that dissolving partially nitrated cellulose in a mixture of ether, alcohol, and camphor resulted in a hard solid that could be shaped when hot, but retained its final form when cold. Parkes presented his material at the Great International Exhibition of 1862 in London and it became known as 'Parkesine'. He suggested that it could be used as a substitute for natural materials such as ivory and tortoiseshell, which were becoming increasingly scarce. In 1866 he founded the Parkesine Company to exploit his discovery commercially, but the high cost of raw materials led to the company failing a few years later.

A rise in the popularity of the game of billiards in the USA, using ivory balls, led to a crisis in the ivory trade. The demand was so high that hundreds of elephants were killed each day in an attempt to supply enough material to make the required amount of billiard balls. A prize of $10,000 was offered to anyone who could produce a cheap alternative to ivory. In 1866 US inventor John Hyatt (1837–1920) accidentally spilt a bottle of collodian, a solution of nitrocellulose similar to that used by Parkes, and noticed that it congealed to form a tough film. His initial efforts to make billiard balls were hampered by the fact that the material was unstable and the balls were prone to explode on impact. However, he discovered that the addition of camphor stabilized the material, which he called 'celluloid'. He won the prize and patented the material in 1869. Celluloid could be produced cheaply in a variety of colours and was used in the manufacture of combs, toys, and false teeth. It was later used in the production of photographic films.

THE FIRST SYNTHETIC FIBRES

It was not long before the knowledge of plastics was applied to find a replacement for natural fibres such as silk and cotton. The first synthetic fibres were made from cellulose derivatives and were collectively known as rayon, from the French meaning 'ray of light'.

French chemist Hilaire de Chardonnet (1839–1924) had noticed that silkworms produced a liquid that rapidly hardened on contact with air to form silk fibres, and he reasoned that forcing a plastic material through a series of small holes would similarly produce fibres. Chardonnet perfected a technique of forcing a nitrocellulose solution through a series of thin glass tubes to produce the first synthetic fibres. The material closely resembled natural silk in appearance and consequently became known as 'Chardonnet silk'. He presented the material at the Paris Exposition of 1889 and produced it commercially from a factory in Besançon, France, in 1891. Unfortunately the material was highly flammable, which led to it being withdrawn.

VISCOSE AND CELLOPHANE

In 1892 English chemists Charles Cross and Edward Bevan discovered an alternative method for producing artificial silk that was significantly less flammable than Chardonnet silk. Cellulose again formed the basis of the material and a solution based on caustic soda was used to produce a viscous yellow liquid, which they called 'viscose'. This was injected into a neutralizing acid bath through a series of small holes to produce synthetic fibres.

However, French chemist Jacques Brandenberger (1872–1954) discovered the most important use for viscose in 1908. He was searching for a plastic that could be applied to cloth to protect it from liquid spillages and developed a process for forcing viscose through narrow slots to produce plastic sheets. He perfected the process to allow the mass production of a transparent, flexible plastic film, which he called 'cellophane'. This became widely used as a waterproof wrapping material and marked the beginning of the popularization of plastic materials.

PLASTIC RESINS

Cellulose-based plastics lose their shape when heated, but in 1909 US chemist Leo Baekeland (1863–1944) patented a material that retained its shape once hardened, even if heat was reapplied. He called this material 'Bakelite' and it was the first plastic to be made without using cellulose. Bakelite was resistant to almost all acids and solvents and could be added to materials, such as softwoods, to make them more durable. It was widely used in World War II in the manufacture of lightweight components for weapons and equipment, and is still used as an electrical insulator in many household applications today.

The discovery of the non-stick plastic resin Teflon in 1938 was the result of the cooling under pressure of the then recently developed refrigerant gas Freon. US chemist Roy Plunkett (1910–1994) had accidentally left a sample overnight and discovered the next day that the gas had formed a chemically inert plastic to which fats could not stick. Teflon became the world's most widely used non-stick coating.

THE SCIENTIFIC APPROACH TO PLASTICS

The development of plastics owed a great deal to the persistence of German chemist Hermann Staudinger (1881–1965). Plastics belong to a class of compounds known as polymers, which were thought to be composed of disorderly conglomerates of small molecules. Staudinger was virtually the only person who thought that polymer molecules were composed of tens of thousands of atoms held together with ordinary chemical bonds. In 1922 he coined the word 'macromolecule' to describe the huge molecular chains that make up natural rubber and proposed a method for producing a polymer molecule by

the reaction of hundreds of smaller molecules called 'monomers'. His ideas were universally rejected until Swedish chemist Theodore Svedberg (1884–1971) developed the ultracentrifuge in 1924 and conclusively proved the existence of macromolecules. The acceptance of Staudinger's theories led to methods of synthesizing polymers without requiring the use of natural materials such as cellulose and can be considered as the foundation of the modern plastics industry.

MODERN PLASTICS

The discovery of how polymer molecules are constructed led to the rapid development of many new plastics. The Naugatuck Chemical Company in Canada was the first to produce polystyrene commercially in 1925. US chemist Waldo Semon (1898–1999) developed the manufacture of polyvinyl chloride, PVC, in 1926 to produce the rubber-like plastic Koroseal.

Canadian-born British chemist Michael Perrin (1906–1988) discovered one of the most important new plastic materials in 1935 when he successfully synthesized polyethylene. His method required pressures of over 30,000 atmospheres and it took until 1939 before full-scale production began. However, it became one of the key materials of World War II, and was used as a lightweight insulating material for communication cables and portable radars. The use of polyethylene allowed radar equipment to be made light enough to be fitted into aircraft.

In 1953 German chemist Karl Ziegler (1898–1973) developed a catalyst that allowed polyethylene to be produced at atmospheric pressure. This form of the plastic was considerably cheaper to produce and led to the material becoming the most used plastic by volume in the world today.

MODERN SYNTHETIC FIBRES

The first totally synthetic fibre, nylon, was produced by US company DuPont in 1938. This was the culmination of years of research by US chemist Wallace Carothers (1896–1937). The material was lightweight and strong and was initially used to replace animal hair in toothbrushes and silk in the manufacture of stockings, and was later used to make parachutes.

This discovery led the way to further developments. In 1941 British chemists Rex Whinfield and James Dickson produced the first polyester-based synthetic fibre, Terylene, which was independently developed in the USA as Dacron. Blends of this material with natural fibres proved to be more popular than the pure material and are still widely used in clothing manufacture.

The adaptability of the plastic industry to changing clothing trends was reflected by the development of the elastic polymer Lycra in 1958. This fibre retains its shape after stretching and is used, for example, in the production of sportswear.

THE NEXT GENERATION OF PLASTICS

Research is continuing on the development of plastics with useful properties. In 1977 Japanese researcher Hideki Shirakawa (1936–) and US researchers Alan McDiarmid (1929–) and Alan Heeger (1936–) made the first electrically conducting plastic. This led to the development of the first plastic battery in 1981. Their discovery also allowed the development of semiconducting plastics and the production of the first light-emitting polymers in 1993. These are more energy efficient than conventional light-emitting diodes and give off less heat.

In an increasingly environmentally aware society, the plastics industry is developing biodegradable materials. The first practical example of this was the development of Biopal in 1990. This plastic is degraded by exposure to sunlight.

1906 Russian botanist Mikhail Tsvet publishes his work on developing chromatography to separate plant pigments.

Pigments are composed of similar organic chemicals and are difficult to separate by conventional methods. In a technique he first used in 1901, Tsvet passes solutions of pigment mixtures through a column of powdered aluminium oxide and notices that each component is attracted to the powder with a different degree of strength. As the solution is washed through the tube, individual compounds separate out into different coloured fractions. He calls the technique 'chromatography' from the Greek phrase meaning 'writing in colour'. Chromatography develops into an important separation technique for complex chemical mixtures.

1906 French chemist Henri Moissan receives the Nobel Prize for Chemistry for his discovery of fluorine and for the introduction of the Moissan electric furnace.

1907 French chemist Georges Urbain and Austrian chemist Carl von Welsbach independently discover the rare-earth element lutetium (atomic number 71).

Urbain shows that the rare-earth ytterbia (see 1878) can be separated into two components, 'neoytterbia' and 'lutecia', the oxides of ytterbium and lutetium respectively. Welsbach carries out the same separation independently and names the two elements 'aldebaranium' and 'cassiopeium'. However, Urbain is given credit for the discovery and names the new element 'lutetium' after Lutetia, the ancient Roman name for the city of Paris. This is the 14th element to be discovered in the mineral from Ytterby, Sweden.

1907 German chemist Emil Fischer publishes *Researches on the Chemistry of Proteins*, in which he describes the synthesis of amino acid chains in proteins.

Fischer develops a chemical procedure for combining amino acids together to form an amino acid chain 18 units in length, which he compares with protein molecule fragments that have been broken down by digestive enzymes. These fragments are called 'peptides' after the Greek word meaning 'to digest' and his synthetic protein chain is an example of a 'synthetic polypeptide'. He discovers that his synthetic molecules have the same chemical properties as naturally produced peptides.

Hollywood actress sunbathing in cellophane (1932). Cellophane, developed in 1908, was believed to allow tanning without burning.
Bettmann/CORBIS

1907 German chemist Eduard Buchner receives the Nobel Prize for Chemistry for his work on enzymes from 1897, when he demonstrated that fermentation could occur outside yeast cells.

1908 French chemist Jacques Brandenberger invents cellophane.

Brandenburger develops a process for producing sheets of the cellulose-based synthetic plastic viscose, by forcing the liquid though a series of thin slots. He perfects his method and patents his process in 1911, calling the material 'cellophane'. His method produces thin transparent sheets of plastic film, which are initially used during World War I to make eyepieces for gas masks. The material is better known for its later use as a transparent wrapping material.

1908 German chemist Fritz Haber develops a process for combining hydrogen with nitrogen from air to form ammonia, the 'Haber process'.

At this time, most nitrate compounds, used in the manufacture of artificial fertilizers and explosives, are not produced synthetically because methods for the industrial production of ammonia are too expensive. Haber advocates the use of high pressures and an osmium catalyst to facilitate a reaction between hydrogen and atmospheric nitrogen that produces ammonia. The procedure is so economical that it quickly replaces all other industrial methods for ammonia production. Haber is awarded the Nobel Prize for Chemistry in 1918 for his discovery.

1908 New Zealand-born British physicist Ernest Rutherford is awarded the Nobel Prize for Chemistry for his investigations concerning the disintegration of elements and the chemistry of radioactive substances.

1909 Belgian-born US chemist Leo Baekeland patents the plastic Bakelite; its insulating and malleable properties, combined with the fact that it does not bend when heated, ensures it has many uses.

Baekeland reacts phenol with formaldehyde to obtain a resin that is insoluble in water and other solvents and is a non-conductor of electricity. The material adopts the shape of the container it is in while hot, but cools to form a hard, easily machinable solid that remains hard when reheated – the first thermosetting synthetic plastic material. He markets the material under the tradename 'Bakelite'. This discovery

The pH levels of some common substances. The lower the pH, the more acidic the substance; the higher the pH, the more alkaline the substance. The pH scale was devised in 1909.

marks the beginning of the modern plastics industry.

1909 Danish biochemist Søren Sørensen devises the pH scale for measuring acidity and alkalinity.

Sørensen originally uses pH to denote the negative logarithm of the hydrogen ion concentration in a solution. Since hydrogen ions are always present in water-based solutions, the scale can be applied equally to alkaline solutions. However, uncertainties in the measurement of hydrogen ion concentration lead to a redefinition of the pH number to base it on standard instrument measurements. The scale is used to monitor the extent and rate of chemical reactions and has developed into an important measurement for soil assessment and pollution monitoring.

1909 German chemist Friedrich Ostwald wins the Nobel Prize for Chemistry for his work on catalysis, chemical, equilibrium, and reaction velocities.

Earth Sciences

1900 French meteorologist Léon de Bort discovers that the Earth's atmosphere consists of two main layers: the troposphere, where the temperature continually changes and is responsible for the weather, and the stratosphere, where the temperature is invariant.

1900 Irish geologist Richard Oldham distinguishes P waves, S waves, and surface waves on seismographs.

Because the velocity of each wave depends on the properties of the materials it travels through, the ability to distinguish the arrival times of these different types of waves will allow geophysicists to 'image' the interior of the Earth.

1900 Swedish chemist Per Teodor Cleve publishes *The Seasonal Distribution of Atlantic Plankton Organisms*, which serves as a basic text on oceanography for many years.

1902 French meteorologist Léon de Bort convinces other meteorologists of his observation (from uncrewed balloons) that air temperature stops decreasing above an altitude of 8,840 m/29,000 ft and instead increases. This is the discovery of the layer of the atmosphere called the stratosphere.

1902 Italian seismologist Giuseppe Mercalli creates a qualitative earthquake intensity scale to describe the surface effects of earthquakes. It is replaced by the modified Mercalli scale in 1931 and then by the more quantitative Richter scale.

1903 French physicists Pierre Curie and Albert Laborde announce that radium salts are continuously releasing heat. The discovery is important to geologists because it suggests a heat source that could be keeping the interior of the Earth warm.

Irish geologist and physicist John Joly suggests that as radioactivity may affect the Earth's thermal gradient, its discovery may undermine the 1862 calculation of Irish physicist William Thomson (Lord Kelvin) – based on the Earth's temperature – of the planet's age as being between 20 and 400 million years old.

1904 New Zealand-born British physicist Ernest Rutherford suggests that ores can be dated by measuring the amount of helium trapped in radioactive minerals. Two years later he performs the experiment and comes up with ages of at least 400 million years.

1904 In a lecture at St Louis, Missouri, French mathematician Jules-Henri Poincaré proposes a theory of relativity to explain Michelson and Morley's failed experiment to determine the velocity of the Earth.

1905 English physicist John Strutt (Lord Rayleigh) measures the helium content of a radium bromide salt and then computes the age of the crystal as 2 billion years. Strutt goes on to investigate radiogenic helium and its use in dating minerals.

1905 The International Association of Seismology (now the International Association of Seismology and Physics of the Earth's Interior) is founded.

1906 By studying the arrival times of seismic waves that pass all the way through the Earth, Irish geologist Richard Oldham suggests that the Earth has a core.

Oldham notices that the waves are arriving at some stations later than they should. He postulates that they are being slowed down as they pass through the centre of the Earth. That is an indication that the centre might have a different composition from the rest of the Earth.

1906 French physicist Bernard Brunhes observes that the magnetic minerals in some lava flows are oriented reversely to the Earth's magnetic field. This suggests that the field periodically reverses polarity.

Brunhes's observation will become key to the development of the theory of plate tectonics in the 1960s.

18 April 1906 San Francisco, California, experiences the worst earthquake in US history, estimated to measure 8.3 on the Richter scale. The quake ruptures water and gas lines and ignites a fire that burns for three days, destroying 28,000 buildings, around two-thirds of the city, and causing $350–400 million in damage. An estimated 2,500 people are killed and 250,000 are left homeless.

The earthquake is instrumental in increasing the study of earthquakes and earthquake hazards. Far from any active volcanoes, it poses something of a problem to geologists who generally associate earthquakes with volcanism. The quake spurs a series of studies regarding the impact of earthquakes on different rock types and on potential earthquake hazards.

1907 German scientists Richard Anschütz and Max Schuler perfect the gyrocompass, which always points to true north (as opposed to magnetic north).

The compass is non-magnetic; using a gyroscope, it aligns itself parallel to the

211

Earth's axis of rotation. The gyrocompass is especially useful for surveying underground.

1907 US chemist Bertram Boltwood uses the ratio of lead and uranium in rocks to determine their age. He estimates his samples to be 410 million to 2.2 billion years old.

Boltwood assumed that lead is a product of the radioactive decay of uranium, and suggests that the age of the Earth might be determined by measuring the amount of radiogenic lead trapped in uranium ores. He calculates the geological ages of almost 50 minerals.

1909 Croatian physicist Andrija Mohorovičić discovers a seismic discontinuity. Located about 30 km/18 mi below the surface, it forms the boundary between the crust and the mantle.

Mohorovičić notices that seismic waves increase in velocity at this boundary, which, he suggests, indicates an increase in density of the rocks, probably as a result of changes in composition. The boundary between the crust and mantle is now known as the 'Moho'.

1909 US explorer Robert Peary and his associate Mathew Henson become the first humans to reach the North Pole.

Peary began his quest for the North Pole in 1886 when he started his exploration and mapping of Greenland. He showed by 1891 that Greenland was an island that did not extend to the North Pole. After Peary's trek to the pole, controversy begins when another of his associates, Fredrick Cook, claims to have reached the pole in 1908.

1909 US geologist and secretary of the Smithsonian Institution in Washington, DC, Charles D Walcott discovers fossils of soft-bodied organisms in the Cambrian Burgess Shale of the Canadian Rockies. The discovery provides unprecedented evidence pertaining to the rapid evolution of life that started in the Cambrian period.

Ecology

1900 Less than 30 bison remain wild in the USA. Action by cattle farmers and conservationists to protect them on government reserves, however, saves them from extinction.

1903 US president Theodore Roosevelt establishes the first US national wildlife refuge on Pelican Island, off the east coast of Florida. By 1929, 87 federal refuges will be established.

1904 The Society for the Protection of Birds, formed in Britain in 1891, becomes the Royal Society for the Protection of Birds (RSPB).

This society grew out of the Fur and Feather Group set up in 1889 in Manchester by Emily Williamson. The Society opposes the feather trade, which results in the death of birds merely to supply feathers for fashion garments. By 1901 the RSPB is employing watchers to guard threatened species and promoting measures to conserve birds and their habitats.

1905 US politician, forester, and conservationist Gifford Pinchot becomes the first chief of the US forest service. Under Pinchot's leadership, the first US national forests are established.

Pinchot views the forests as a resource that must be properly managed for human use, embracing a utilitarian philosophy of improving forests for human welfare. His original contribution is to bring the ideas of progressive agriculture to the management of public lands.

1905 US conservationists combine with nature lovers to found the National Audubon Society.

13 May 1908 In the USA 44 state and territorial governors attend an environmental conservation conference, convened by President Theodore Roosevelt, at the White House.

1909 Danish biologist Eugenius Warming publishes *Plantesamfund/The Oecology of Plants: An Introduction to the Study of Plant Communities.*

Warming reviews the influence of such factors as light, heat, humidity, and soil type on the growth, distribution, and physiology of plants. Warming explores the notion of successional development of plant communities and shows how existing formations can be disturbed and give rise to new ones. He advances the notion that habitats develop towards a final or 'climax' formation. Warming resists, however, the notion that a plant community has a life of its own not explicable in materialistic terms.

August 1909 Seattle, Washington, hosts the first National Conservation Congress, representing 37 states.

Mathematics

1900 German mathematician David Hilbert poses 23 problems at the International Congress of Mathematics, as a challenge for the 20th century. Most of these problems are now solved.

1900 English mathematician Karl Pearson develops the χ^2 (chi-squared) test to determine whether a set of observed data deviates significantly from what would have been predicted by a null hypothesis (that is, totally at random). It is a fundamental concept in statistics.

1902 US physicist J Willard Gibbs publishes *Elementary Principles in Statistical Mechanics*, in which he develops the mathematics of statistical mechanics.

It employs statistical methods to evaluate data relating to a large number of entities, and is used in physics and chemistry (Gibbs applied it to thermodynamics).

1906 French mathematician Maurice Fréchet introduces functional calculus, which involves functions on metric spaces. He formulates the abstract definition of compactness.

1906 Russian mathematician Andrei Markov studies random processes that are subsequently known as Markov chains.

Each chain is a sequence of separate random events (or variables) whose probabilities depend on previous events in the chain.

1908 German mathematician Ernst Zermelo publishes *Untersuchungen über die Grundlagen der Mengenlehre/Investigations on the Foundations of Set Theory*, which forms the basis of modern set theory and overcomes the difficulties inherent in the earlier theory of his compatriot Georg Cantor.

1908 French mathematician Jules-Henri Poincaré publishes *Science et méthode/Science and Method.*

Physics

1900 French physicist Henri Becquerel demonstrates that beta particles are fast-moving electrons.

However, the name 'beta particle' is restricted to fast moving electrons that originate from within the nucleus.

213

Max Planck solved a problem that had been plaguing physicists for decades by suggesting that energy came in discrete packets, which he called quanta, rather than being infinitely variable. **This was such a radical idea that Planck did not pursue it, although Albert Einstein and others would develop the idea a few years later.** *Oxford Museum for the History of Science*

1900 French physiologist Paul Villard discovers gamma rays.

1900 German physicist Max Planck suggests that black bodies (perfect absorbers) radiate energy in packets or quanta, rather than continuously. He thus begins the science of quantum physics, which revolutionizes the understanding of atomic and subatomic processes.

Today, physicists are still actively researching quantum phenomena and new applications are likely in the areas of computing and cryptography.

We have no right to assume that any physical laws exist, or if they have existed up to now, that they will continue to exist in a similar manner in the future.

Max Planck, German physicist, *The Universe in the Light of Modern Physics*

1900 German physicists Johann Elster and Hans F Geitel invent the first practical photoelectric cell.

1901 French physicist Henri Becquerel reports the first radiation burn. It is caused by a sample of radium carried in his waistcoat pocket and leads to the use of radium for medical purposes.

Radiotherapy is now an accepted treatment for many tumours.

1901 German physicist Wilhelm Röntgen receives the first Nobel Prize for Physics for his discovery of X-rays, also known as Röntgen rays.

Röntgen's citation reads, 'In recognition of the extraordinary services he has rendered by the discovery of the remarkable rays subsequently named after him.'

1902 Canadian-born US physicist Reginald Fessenden discovers the heterodyne principle whereby high-frequency radio signals are converted to lower frequency signals that are easier to control and amplify. It leads to the superheterodyne principle essential in modern radio and television.

Fessenden further develops radio transmission with his invention of the 'modulated carrier wave', which leads to the first radio transmission of sound on Christmas Eve, 1906.

1902 Dutch physicists Hendrik Lorentz and Pieter Zeeman share the Nobel Prize for Physics for their researches into the influence of magnetism upon radiation phenomena.

1902 English physicist Oliver Heaviside and US electrical engineer Arthur Kennelly independently predict the existence of a conducting layer in the atmosphere that reflects radio waves.

This later becomes known as the 'Heaviside layer'. Heaviside uses the notion of a conducting layer to explain why Marconi is able to transmit radio waves across the Atlantic.

1902 Italian physicist Guglielmo Marconi discovers that radio waves are transmitted further at night than during the day because they are affected by changes in the atmosphere (actually by a layer of ionized gas in the ionosphere).

1902 New Zealand-born British physicist Ernest Rutherford and English physical chemist Frederick Soddy discover thorium X and publish The Cause and Nature of Radioactivity, which outlines the theory that radioactivity involves the disintegration of atoms of one element to form atoms of another. Amongst other things, the discovery lays the foundation for radiometric dating of natural materials.

All science is either physics or stamp collecting.

Ernest Rutherford, New Zealand physicist, quoted in J B Birks *Rutherford at Manchester*

1903 French physicists Henri Becquerel and Pierre and Marie Curie share the Nobel Prize for Physics for their discovery of radioactivity. Marie Curie is the first woman to win a Nobel prize.

*It is impossible. It would be contrary to scientific spirit . . .
Physicists always publish their researches completely. If our discovery
has a commercial future, that is an accident by which we must not profit.
And radium is going to be of use in treating disease . . .
It seems to me impossible to take advantage of that.*

Marie Curie, Polish-born French scientist, on the patenting of radium.
Discussion with her husband, Pierre, quoted in Eve Curie 'The Discovery of Radium' in *Marie Curie*
transl. V Sheean (1938)

1903 New Zealand-born British physicist Ernest Rutherford discovers that a beam of alpha particles is deflected by electric and magnetic fields. From the direction of deflection he is able to prove that they have a positive charge and from their velocity he determines the ratio of their charge to their mass. He also names the high-frequency electromagnetic radiation escaping from the nuclei of atoms, first discovered in 1900, as gamma rays.

1903 Scottish chemist William Ramsay shows that helium is produced during the radioactive decay of radium – an important discovery for the understanding of nuclear reactions.

This leads to an understanding of the alpha particle as the nucleus of the helium atom.

1904 English electrical engineer John Fleming patents the diode valve, which allows electricity to flow in only one direction.

Diodes become an essential part of the electronics industry and are used to 'convert' alternating (AC) current into direct (DC) current.

1904 English physicist John Strutt (Lord Rayleigh) receives the Nobel Prize for Physics for his discovery of argon.

1904 English physicist Joseph John Thomson puts forward a model of the atom in which negatively charged electrons are 'embedded' in a sphere of positive charge. This becomes known as the 'plum pudding' model.

It is soon replaced, following work by Ernest Rutherford and Niels Bohr.

1904 Japanese physicist Hantaro Nagaoka proposes a model of the atom in which the electrons are located in an outer ring and orbit the positive charge which is located in a central nucleus. The model is ignored because it is thought the electrons would fall into the nucleus.

This model is revived by Niels Bohr in 1913 and becomes known as the Bohr model. Bohr draws on experimental evidence produced in 1911 by Geiger and Marsden working under Rutherford.

1904 New Zealand-born British physicist Ernest Rutherford publishes *Radio-activity*, summarizing his work on the subject and pointing out that radioactivity produces more heat per atom than chemical reactions do.

Rutherford announces that the enormous amounts of latent heat in all atoms explain why Irish physicist Lord Kelvin's 1862 calculation of the age of the Earth was less than appeared to be possible by looking at layer upon layer of rock.

1905 German physicist Philipp Lenard receives the Nobel Prize for Physics for his work on cathode rays.

1905 German-born US physicist Albert Einstein publishes four important scientific papers while working as a patent examiner in Switzerland. In 'On the Motion – Required by the Molecular Kinetic Theory of Heat – of Small Particles Suspended in a Stationary Liquid' he explains Brownian motion. In 'On a Heuristic Viewpoint Concerning the Production and Transformation of Light' he explains the photoelectric effect by proposing that light consists of photons and also exhibits wavelike properties. In 'On the Electrodynamics of Moving Bodies' he formulates his special theory of relativity. In 'Does the Inertia of a Body Depend on its Energy Content?' he argues that mass and energy are equivalent, which can be expressed by the formula $E = mc^2$.

Imagination is more important than knowledge.

Albert Einstein, German-born US physicist, *On Science*

1906 English physicist Charles Glover Barkla demonstrates that each element can be made to emit X-rays of a characteristic frequency.

In other words, each element can be said to have a characteristic X-ray spectrum.

1906 English physicist J J Thomson receives the Nobel Prize for Physics for his theoretical and experimental investigations on the conduction of electricity by gases.

1906 US chemist Frederick Gardner Cottrell develops the first practical electrostatic precipitator for removing impurities from the air.

This development is based on the earlier work of Oliver Lodge in 1884.

1906 US physicist Lee De Forest invents the 'audion tube', a triode vacuum tube with a third electrode, shaped like a grid, between the cathode and anode that controls the flow of electrons and permits the amplification of signals. It is an essential element in the development of radio, radar, television, and computers.

De Forest's patent leads to a courtroom battle with the English scientist John Fleming, inventor of the two-electrode tube. The US supreme court finally rules in De Forest's favour in 1943.

1907 French physicist Pierre-Ernest Weiss develops the domain theory of ferromagnetism, which suggests that in a ferromagnetic material, such as lodestone, there are regions, or domains, where the molecules are all magnetized in the same direction. His theory leads to a greater understanding of rock magnetism.

1907 US physicist Albert Michelson receives the Nobel Prize for Physics for his optical precision instruments and his spectroscopic and meteorological investigations.

Michelson is perhaps best known for his work on the velocity of light and the motion of the Earth through the 'ether' (1887).

1908 Dutch physicist Heike Kamerlingh Onnes liquefies helium at 4.2 K (−268.95°C/ −452.11°F).

1908 French physicist Gabriel Lippmann receives the Nobel Prize for Physics for his method of reproducing colours photographically based on the phenomenon of interference.

1908 French physicist Jean-Baptiste Perrin proves the existence of molecules; an achievement which earns him the Nobel Prize for Physics in 1926.

Perrin studies Brownian motion – the apparently random movement of particles in liquids – by adding a dye of known particle size to a cylinder of water and then observing the manner in which it settles. He notices that Brownian motion is opposing the downward passage of the particles. Perrin is able to calculate the force the water molecules exert to oppose gravity and uses an equation derived by German-born US physicist Albert Einstein to calculate the size of a water molecule. His results convince the sceptics that molecules exist.

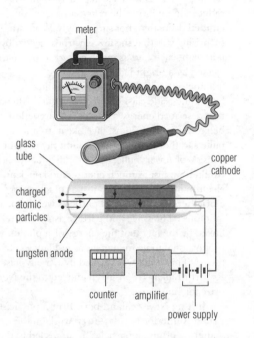

meter

glass tube

copper cathode

charged atomic particles

tungsten anode

counter   amplifier

power supply

A Geiger–Müller counter, developed in 1908, detects and measures ionizing radiation (alpha particles, beta particles, and gamma rays) emitted by radioactive materials. Any incoming radiation creates ions (charged particles) within the counter, which are attracted to the anode and cathode to create a measurable electric current.

1908 German physicist Hans Geiger and New Zealand-born British physicist Ernest Rutherford develop the Geiger counter, which counts individual alpha particles emitted by radioactive substances.

These tubes are often known as Geiger–Müller tubes.

1909 Dutch physicist Hendrik Lorentz publishes *The Theory of Electrons and its Applications to the Phenomena of Light and Radiant Heat*, in which he explains the production of electromagnetic radiation in terms of electrons.

1909 German physicist Ferdinand Braun and Italian physicist Guglielmo Marconi share the Nobel Prize for Physics for their development of wireless telegraphy.

1909 German-born US physicist Albert Einstein introduces his idea that light exhibits both wave and particle characteristics.

This becomes known as wave–particle duality and is still an active area of research today.

A Brief Chronology of Quantum Physics

by Gren Ireson

INTRODUCTION

December 2000 saw the centenary of the quantum; therefore, as you read this, the subject of quantum physics is a little more than a hundred years old. In that time quantum ideas have been successful in explaining observations from the subatomic to cosmological scales, as they provide an excellent mathematical model for subatomic particle interactions. For example, quantum ideas are now used in medical imaging and are expected to contribute to the fields of computing and cryptography.

Whilst scientists working in many diverse fields can use quantum ideas to explain and predict certain behaviour, no one can give a complete explanation as to why the basic quantum rules are followed by nature.

THE 'DISCOVERY' OF QUANTUM PHYSICS

The starting point of quantum theory can be taken as 1900 when German physicist Max Planck (1858–1947) hypothesized that electromagnetic radiation in black bodies (perfect absorbers) could not be given out or absorbed in arbitrary amounts, but only in separate packets called quanta, a process he called 'quantization'.

Planck put forward an argument that relied on classical ideas of entropy, the amount of 'disorder' in a system. This generated a formula that matched the experimental data perfectly for all wavelengths. However, Planck was unable, initially, to offer any theoretical justification for the results.

Planck's genius was that, after an intense period of work, he related the energy emitted and absorbed to the frequency of the radiation. So, low frequency infrared radiation only emitted and absorbed in low-energy quanta, while ultraviolet radiation had high frequency, and so high-energy quanta.

THE DEVELOPMENT OF QUANTUM PHYSICS

In 1905 German-born US physicist Albert Einstein (1879–1955) took the notion of

quantization further. He wanted to explain some puzzling behaviour of the newly discovered electrons. When light was shone onto certain metals, electrons were ejected. If the light was reduced in intensity, the number of electrons emitted decreased, not surprisingly. What was surprising was that the few electrons ejected by a faint light each had as much energy as the many ejected by a bright light, the so-called 'photoelectric effect'. Einstein accounted for this by putting aside classical electromagnetic theory that described energy as waves. He argued that it behaved in this circumstance like a stream of bullets, rather than a wave. A faint ray of light consists of fewer 'bullets' – but individually each would have just as much energy, and knock an electron out of metal with just as much energy, as a brighter ray. These light particles were named 'photons'. The wave and particle aspects were linked by the fact that the energy of a photon was proportional to its frequency, as in Planck's theory. This result was verified by US physicist Robert Millikan (1868–1953) and led to the Nobel prize for Einstein.

In 1913 New Zealand-born British physicist Ernest Rutherford's (1871–1937) work on the scattering of alpha particles by atoms led to the concept of an atom as a small, very hard nucleus around which orbit electrons. In considering the line spectrum of hydrogen, Danish physicist Niels Bohr (1885–1952) accepted Rutherford's nuclear atom and proposed that electrons in atoms orbit in a number of possible orbits called 'stationary states'.

Bohr considered the condition for a stable orbit to be that the electrostatic attraction between the negative electron and the positive nucleus provided the required force to keep it orbiting. However, this model could not be supported by classical theory since the accelerating electron should spiral into the nucleus. In order to obtain a set of stable orbits Bohr suggested, without any explanation, that electrons are confined to certain stationary states, with circular orbits, and that they emit radiation only when they make transitions from a higher state to a lower one. The notion of a transition from one state to another considers that the atom emits a single photon of energy equal to the difference between the orbit energies.

In accordance with Planck's quantization of energy, Bohr considered an electron infinitely far away from the nucleus falling into an allowed orbit and quantized the energy of the photon in terms of the energy of the electron in its final orbit as $E = (n/2) \times hf$, where n is a positive integer. In developing his theory, Bohr considered only circular orbits, but later work by others extended these ideas to elliptical orbits.

In order to maintain the notion of a transition from one stationary state to another, Bohr was forced to regard the transition as 'instantaneous'. It is now common to refer to such instantaneous transitions as 'quantum leaps'. Einstein demonstrated that quantum theory could not predict the timing of such a transition and neither could it predict the direction of the emitted photon of electromagnetic radiation. Quantum theory could, and still can, only predict the probability of such a transition taking place. Einstein did not accept such a probabilistic theory as final, leading to his comment in a letter to German physicist Max Born (1882–1970): 'You believe in a God who plays dice and I in complete law and order.'

The early studies in quantum mechanics relied on Newtonian ideas and sought to supplement Isaac Newton's laws with quantization conditions. This allowed for the selection of the preferred stationary states in the Bohr model of the atom.

The first move away from the Newtonian influence was French physicist Louis de Broglie's notion that 'particles' have 'wave' properties. De Broglie (1892–1987) suggested that the frequency of the wave associated with a particle is related to the energy of that particle in the same way as for the energy of a photon. By building on the relativistic connection between energy and momentum and frequency and

wavelength, de Broglie was able to work out that the wavelength of a wave associated with a particle is related to the momentum of the particle.

While this work was in progress, Austrian physicist Erwin Schrödinger (1887–1961) formulated what is now commonly called the Schrödinger wave equation. It described the motion of any particle as a wave motion, in accordance with de Broglie's ideas. However, this would have allowed for the possibility of an electron being cut in two by an obstacle. The wave function idea was, in some ways, rescued by Max Born who proposed that it predicted the probability of an electron's position.

It is this formulation of the 'new' quantum mechanics that was to become the most popular among physicists. However, even before the Schrödinger wave equation had been published an alternative formulation had been developed by German physicist Werner Heisenberg. Heisenberg took the view that classical quantities such as position and momentum had no meaning in quantum mechanics since they could not be measured. He replaced all such classical quantities with ones directly related to the quantum mechanical stationary states. Both approaches generated the same results and soon Schrödinger was able to generate a mathematical proof that any formula constructed using one approach could be translated into the other.

MODERN THINKING

It could be argued that mathematically quantum physics is no longer a problem. The problem is now one of interpretation. Many interpretations exist, but there is still ongoing debate about the issues.

Perhaps the next development will come from experimental physics. Could an experiment be carried out that produces results that quantum theory cannot predict? If results show that something is wrong with the predictions made by the theory then perhaps we would be closer to understanding it.

1910–1919

Astronomy

19 May 1910 Halley's comet – which comes near the Earth roughly every 75 years – returns, with the Earth passing through the comet's tail.

1912 US astronomer Edward Barnard discovers the second nearest star to the Sun, now known as Barnard's star. It displays the greatest proper movement of any star relative to others.

The star is a red dwarf, invisible to the naked eye, located at a distance of 5.98 light years in the constellation Ophiuchus. The star's large proper motion, 10.28 arc seconds per year, is equivalent to half the Moon's apparent diameter in a century.

1912 US astronomer Henrietta Leavitt establishes a relationship between the period and luminosity of Cepheid variable stars (stars which pulsate and vary regularly in brightness). The relationship is later used to calculate interstellar and intergalactic distances.

During her career, Leavitt discovers more than 2,400 variable brightness stars, about half of the known total in her day. The Cepheids are a class of variable star, in which Leavitt finds a direct correlation between the brightness of the star and the period with which its brightness varies. With this relationship, other astronomers, such as Edwin Hubble, are able to investigate the scale of the universe.

1912–1925 US astronomer Vesto Slipher measures the radial velocities of spiral nebulae by examining small changes in the Doppler effect which suggest that they must be external to our galaxy.

Slipher's work is later built upon by US astronomer Edwin Hubble to show that the universe is expanding.

1913 Danish astronomer Ejnar Hertzsprung introduces a luminosity scale of Cepheid variable stars to measure their distance.

1913 US astronomer Henry Russell refines the correlation between a star's brightness and its spectrum, further improving the determination of stellar distances.

1914 British astronomer John Franklin publishes the Franklin–Adams charts, the first photographic star charts of the entire sky.

1914 English astrophysicist Arthur Eddington publishes *Stellar Movement and the Structure of the Universe*, in which he theorizes that spiral nebulae are galaxies similar to the Milky Way.

Science is one thing, wisdom is another. Science is an edged tool, with which men play like children, and cut their own fingers.

Arthur Stanley Eddington, British astronomer

1915 Proxima Centauri is discovered. The closest star to the Sun, it is 4.2 light years away.

This is so close to our Solar System that its motion (proper motion as opposed to parallax) can be measured as it moves against the background of more distant stars.

1916 German astronomer Karl Schwarzschild offers a solution to Einstein's gravitational field equations, which predicts the existence of black holes, collapsed stellar bodies.

A black hole forms when an object collapses to a small size (perhaps to a singularity) and the escape velocity in its neighbourhood is so great that light cannot escape. The boundary of this region is called the event horizon because any event that occurs inside is invisible to outside observers; its radius is called the Schwarzschild radius.

1917 Dutch astronomer Willem de Sitter shows that Einstein's theory of general relativity implies that the universe must be expanding.

Einstein had introduced a quantity he called the cosmological constant to solve a problem which had troubled Newton before him: why does the universe not collapse under gravitational attraction? Einstein introduced this constant of integration admitting it was not justified by our actual knowledge of gravitation.

1917 The 2.5 m/100 in Hooker reflecting telescope is installed at Mount Wilson Observatory, California. It is the world's largest reflecting telescope to date.

1917 US astronomer Harlow Shapley calculates that the Sun is situated about 30,000 light years from the central plane of the Galaxy.

1918 US astronomer Harlow Shapley estimates the size of the Milky Way as 45,000 light years.

1919 The International Astronomical Union (IAU) is founded to promote cooperation in astronomy.

1919 US astronomer William Pickering predicts the existence and location of the planet Pluto.

It is not discovered until 18 February 1930, by US astronomer Clyde Tombaugh, at the Lowell Observatory, Arizona.

Biology

1910 US geneticist Thomas Hunt Morgan discovers that certain inherited characteristics of the fruit fly *Drosophila melanogaster* are sex linked. He later argues that because all sex-related characteristics are inherited together they are linearly arranged on the X-chromosome.

In 1916, Morgan and his co-workers construct the first chromosome map for fruit flies in which genes are set in linear arrangement on chromosomes, showing the relative positions of some of the sex-linked genes in the fruit fly. Their map shows the relative positions of more than 2,000 genes on the four chromosomes found in the cell nucleus of the fruit fly.

Easy to keep, fast-breeding, and prolific (the female lays hundreds of eggs in a few days) *Drosophila melanogaster*, the humble fruit fly, was the workhorse of the early geneticists. Thirty generations of fruit flies could be tracked over the course of a year. *Corbis*

221

1913 German biochemists Leonor Michaelis and Maude Menten develop a mathematical equation describing the rate of enzyme-catalysed reactions.

Michaelis proposes a constant – the Michaelis constant – which summarizes the effect of substrate concentration on the rate of an enzyme-controlled reaction. In the same year Menten develops the work. Using the constant it is possible to predict that the reaction between substrate and enzyme is preceded by the formation of a substrate–enzyme complex, a prediction confirmed 50 years later by direct experiment. The constant is sometimes called the Michaelis–Menten constant.

1914 German-born US biochemist Fritz Lipmann explains the role of adenosine triphosphate (ATP) as the carrier of chemical energy from the oxidation of food to the energy consumption processes in the cells.

1915 English bacteriologist Frederick Twort demonstrates the existence of bacteriophages (viruses that infect bacteria).

This is independently demonstrated in 1917 by French bacteriologist Félix d'Hérelle. Bacteriophages are later used to investigate the genetics, replication, and structure of viruses.

1915 US geneticist Hermann Muller begins experiments using X-rays as a means of inducing mutations.

Muller's work is of great importance, providing a ready source of mutations, which are a key to the study of genetics. He is awarded the Nobel Prize for Physiology or Medicine in 1946.

1915 US geneticists Thomas Hunt Morgan, Alfred Sturtevant, Calvin Bridges, and Hermann Muller publish *The Mechanism of Mendelian Heredity*, which outlines their work on the fruit fly *Drosophila*, demonstrating that genes can be mapped on chromosomes.

This becomes the definitive statement of the view.

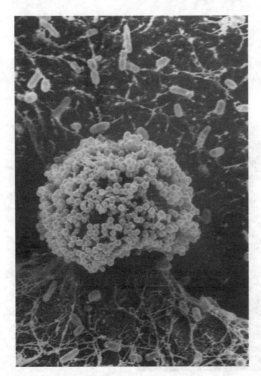

Even bacteria have enemies, not only other bacteria competing for resources, but also they face attack from bacteriophages, viruses that have evolved to invade bacterial cells. In the years following their discovery, bacteriophages were unsuccessfully considered as treatments for bacterial diseases. *Corbis*

1919 Austrian zoologist Karl von Frisch proposes that bees communicate the distance and direction of nectar to each other by two types of rhythmic movements, which he calls wagging and circle dances.

The circle dance indicates food within 75 m/300 ft of the hive, and the wagging dance indicates food at a greater distance. In 1949 von Frisch demonstrates that bees use their perception of polarized light to make the Sun function as a compass.

Chemistry

1910 Polish-born French physicist Marie Curie and French chemist André-Louis Debierne complete the task of isolating pure metallic radium, which they began in 1898.

1910 German chemist Otto Wallach receives the Nobel Prize for Chemistry for his work on alicyclic compounds.

1911 Russian-born US chemist Phoebus Levene discovers D-ribose, the sugar that forms the basis of RNA.

Levene isolates a five-carbon sugar from a sample of nucleic acid, the substance from which the genetic material of a cell is made up. He calls the sugar 'ribose' and, because of this discovery, every nucleic acid containing this sugar becomes known as 'ribose nucleic acid' or, more commonly, by the abbreviation RNA.

1911 Polish-born French chemist Marie Curie receives the Nobel Prize for Chemistry for her discovery of radium and polonium, and the isolation and study of radium.

1912 Dutch physical chemist Peter Debye proposes his theory of a polar molecule. This work earns him the Nobel Prize for Chemistry in 1936.

The distribution of electrons in a molecule can be symmetrical or non-symmetrical. In the non-symmetrical case, there is a localized surplus or deficiency of charge, which causes those parts of the molecule to have slight negative and positive charges respectively. This is called a 'polar molecule' or 'molecular dipole' and the charge distribution across it is known as a 'dipole moment'. Debye works out a mathematical representation to predict chemical behaviour of polar molecules.

1912 German chemist Alfred Stock is the first to synthesize boranes; his synthesized compounds are eventually used to make high-energy additives for rocket fuel.

Also known as boron hydrides, boranes are structurally equivalent to alkanes or saturated hydrocarbons, except that their molecules are composed of B–B bonds instead of C–C bonds. The lighter members of this class are highly reactive (being sensitive to air and moisture) volatile, and toxic. Stock develops high-vacuum apparatus and techniques in order to study boranes.

1912 Polish-born US biochemist Casimir Funk isolates vitamin B_1 (thiamine) and coins the name 'vitamines'. This proves a vital discovery in the treatment of the disease beriberi, which is caused by a deficiency of the vitamin.

Funk is searching for the 'accessory factors' in food that were proposed by Frederick Hopkins, when he isolates an amine compound from a yeast that cures beriberi. The substance Funk has discovered is part of what becomes known as the vitamin B complex. He mistakenly thinks that all essential substances contain amine groups and so renames 'accessory factors' as 'vitamines' from the Latin for 'life amine'. This name becomes shortened to 'vitamin' when it is discovered that not all of these compounds contain amine groups.

1912 French chemists Victor Grignard and Paul Sabatier share the Nobel Prize for Chemistry: Grignard for his discovery of Grignard reagent in 1900, and Sabatier for the catalytic hydrogenation of organic compounds in 1897.

1913 English physicists William and Lawrence Bragg develop X-ray crystallography further by establishing laws that govern the orderly arrangement of atoms in crystal interference and diffraction patterns. They also demonstrate the wave nature of X-rays.

The Braggs show that X-ray diffraction patterns in crystalline materials are caused by interference between the atoms of a crystal and the X-rays. This leads to the conclusion that crystals are composed of regular, repeated arrays of atoms. They develop a mathematical system to relate diffraction patterns to crystal structure, which becomes known as 'Bragg's law'. This leads to the determination of the molecular structure of complex molecules such as insulin and penicillin, allowing their synthesis.

Physicists use the wave theory on Mondays, Wednesdays and Fridays, and the particle theory on Tuesdays, Thursdays and Saturdays.

William Bragg, British physicist, attributed remark

1913 German chemist Friedrich Bergius publishes *The Use of High Pressure in Chemical Actions* in which he describes his process of converting coal at high pressure to produce petrol.

Bergius's process adds hydrogen to coal or heavy oils to produce petrol and kerosene. The hydrogenation process is used to supply Germany with petrol during World War II. Bergius shares the Nobel Prize for Chemistry in 1931 for his development of high-pressure chemical processes.

1913 US biochemist Elmer McCollum isolates vitamin A (retinol) and is the first to use letters in the naming of vitamins, calling it fat-soluble A.

McCollum isolates a vitamin from the fats in butter and egg yolks and names the substance fat-soluble A to distinguish it from the water-soluble compound discovered by Polish-born US biochemist Casimir Funk in 1912, which he designates as water-soluble B. The use of letters becomes the standard way of naming vitamins.

1913 US chemist William Burton patents a thermal cracking process which doubles the yield of petrol from crude oil.

Burton's process breaks up the less volatile, heavy hydrocarbon fractions of petroleum, which are then distilled and chemically converted into petrol. This doubles the yield of petrol from crude oil. Thermal cracking methods developed from the Burton process are still used to refine crude petroleum today.

1913 Swiss chemist Alfred Werner receives the Nobel Prize for Chemistry for his work on the bonding of atoms within molecules.

October 1914 Chemical agents are used for the first time as a battlefield weapon when tear gas is used by the German forces against French soldiers at Neuve Chapelle.

The development of the chemical industry has made possible the production of chemical agents in sufficient quantities to make an effective weapon. This use of a chemical agent breaks the spirit of the Hague convention of 1907. French forces retaliate against the use of chemical weapons by using tear gas in March 1915 against German soldiers at Argonne, France.

1914 US chemist Theodore Richards is awarded the Nobel Prize for Chemistry for his accurate determination of the atomic masses of many elements.

April 1915 Poison gas (chlorine) is used in warfare for the first time by German forces at the battle of Ypres in Belgium.

1915 German chemist Richard Willstätter receives the Nobel Prize for Chemistry for research into plant pigments, especially chlorophyll.

1916 US chemist Gilbert Lewis states a new valency theory, in which electrons are shared between atoms.

Lewis suggests that the complete transfer of valence shell electrons from one element to another forms ionic bonds, but a sharing of two electrons occurs to form a covalent bond. He stresses that atoms in a compound combine in order to attain the electronic configuration closest to the valence shell of a noble gas. He publishes his theory in *Valence and the Structure of Atoms and Molecules* in 1926. This theory is still used to explain the nature of chemical bonds.

1917 German physical chemist Otto Hahn and Austrian physicist Lise Meitner independently discover the radioactive element protactinium-231 (atomic number 91).

The element is present in all uranium ores and transmutes into the element actinium by radioactive decay. Polish chemist Kasimir Fajans had identified the shorter-lived isotope protactinium-234 in 1913, but Hahn and Meitner are given credit for the discovery of the element because protactinium-231 is the isotope of the element that has the longest half-life. It is initially named 'protoactinium' meaning 'before actinium', but this name is shortened to 'protactinium' in 1949.

1918 English physicist Francis Aston invents the mass spectrometer, which allows him to separate ions or isotopes of the same element. The mass spectrometer will become an important tool in stable isotope geochemistry.

The device breaks down a compound into charged atoms and fragments, which are introduced into a vacuum chamber. Powerful magnetic and electric fields are then used to separate these particles according to their mass. The machine is capable of accurately separating different isotopes of an element. In 1922, Aston publishes 'Isotopes' in which he details the isotopic composition of over 50 elements. This study of isotopes earns Aston the Nobel Prize for Chemistry in 1922.

1918 German chemist Fritz Haber receives the Nobel Prize for Chemistry for the synthesis of ammonia from its elements, nitrogen and hydrogen.

1919 English physicist Joseph John Thomson discovers that neon exists as a number of isotopes.

Thomson uses the recently invented mass spectrograph to study the gaseous element neon. His results prove that the element is composed of at least two isotopes, Ne-20 and Ne-22. Up to this time, it had been thought that isotopes were restricted to radioactive elements and their decay products. Thomson's discovery proves that the theory of isotopes could also be applied to the stable elements.

Earth Sciences

1910 US geographer Frank Bursley Taylor publishes a paper in which he rejects the hypothesis of a contracting Earth and instead suggests that there have been lateral movements

225

of the Earth's crust. He explains the Himalayan mountain range as the result of a large-scale crustal pile-up as the continents dispersed from polar regions towards the Equator.

1910 US physicist Percy Bridgman invents a device, called the 'collar', that allows him to squeeze all kinds of materials to pressures comparable with the base of the Earth's crust, giving rise to the new fields of high-pressure physics and mineral physics.

The work is important in determining the physical and chemical state of the deep interior of the Earth.

1910 US seismologist Harry Fielding Reid publishes *The California Earthquake of April 18, 1906: The Mechanics of the Earthquake*, in which he outlines his theory of elastic rebound.

The theory states that earthquakes occur when strain builds up along a fault. Eventually the rocks break, and when they do, the strain is released as seismic energy. The theory is immediately accepted by the geological community (G K Gilbert's almost identical ideas 25 years previously had not). A major difference between Reid's theory and previous theories is that it does not involve volcanic activity.

1912 German physicist Max von Laue discovers that crystals diffract X-rays and that the pattern in which the diffraction occurs can be used to determine the position of the atoms within the crystal. The discovery revolutionizes the fields of crystallography, mineral identification, and mineral discovery.

1912 German geophysicist Alfred Wegener proposes that, 250 million years ago, a single land mass formed a supercontinent he calls 'Pangaea'. He argues that the supercontinent split into two components from which different portions broke free, forming the present continents, which occupy their current positions through continental drift.

Wegener's proposal comes in a series of lectures. In 1915, he publishes *Die Entstehung der Kontinente und Ozeane / The Origin of Continents and Oceans*, in which these ideas are fully described. Wegener seeks to account for the distribution of fossils of closely related species discovered in different parts of the world, but to begin with, his theory is largely rejected because of the absence of a mechanism to account for the movement of large land masses. In the 1960s, the theory of plate tectonics provides an explanation, and continental drift becomes a central element within the theoretical framework of geophysics.

200 million years ago   140 million years ago   today

The changing positions of the Earth's continents. Millions of years ago, there was a single large continent, Pangaea. This split 200 million years ago: the continents had started to move apart, to form Gondwanaland in the south and Laurasia in the north. By 50 million years ago the continents were almost in their present positions.

German meteorologist Alfred Wegener in 1926, when he was rector of the Handel School in Germany.
UPI/CORBIS–Bettmann

If it turns out that sense and meaning are now becoming evident in the whole history of the Earth's development, why should we hesitate to toss the old views overboard?

Alfred Wegener, German geologist, quoted in Martin Schwarzbach
Alfred Wegener, the Father of Continental Drift

1913 English geologist Arthur Holmes publishes *Age of the Earth* in which he records his use of radioactivity to date rocks, suggesting that the Earth is 4.6 billion years old.

The book is a review of the previous half-century's methods and results in calculating the age of the Earth and also includes the beginnings of a quantitative time scale – the actual age in years of the various stratigraphic units.

1913 French physicist Charles Fabry discovers the ozone layer in the upper atmosphere.

Ozone, the molecule O_3, occurs in concentrations of 1–10 parts per million at altitudes of 15–30 km/9.3–18.6 mi. Atmospheric ozone (as opposed to ground ozone) is now known to play an important role in absorbing ultraviolet wavelengths of light, thus limiting the amount of ultraviolet radiation that reaches the ground.

1914 German-born US geologist Beno Gutenberg uses seismic waves to calculate the depth of a discontinuity between the Earth's lower mantle and core at about 2,800 km/1,750 mi.

The discontinuity, consisting of an increase in the velocity of seismic waves, is interpreted as an increase in rock density. This is now known as the core/mantle boundary or the Gutenberg discontinuity.

1915 Finnish chemist and geologist Penotti Eskola develops the concept of metamorphic facies.

A metamorphic facies is defined as a particular combination of metamorphic minerals that make up a metamorphic rock. A particular facies is thought to result from a particular set of physical conditions such as pressure and temperature. The concept will thus be crucial in describing metamorphic rocks and in ultimately understanding the conditions under which they form.

1915 German meteorologist Alfred Wegener publishes *Die Entstehung der Kontinente und Ozeane/On the Origin of Continents and Oceans*, in which he discusses his theory of continental displacement, now known as continental drift.

The book includes detailed arguments outlining the climatological, palaeontological, and geophysical data support the theory of continental drift. The idea becomes extremely controversial, the main criticism being that no driving force for continental drift can be found.

1918 The Bergen School, a group of Scandinavian scientists headed by Norwegian meteorologist Vilhelm Bjerknes puts forward the theory that most weather occurs along boundaries between air masses. They refer to the boundaries as 'fronts' in analogy to the battle fronts of World War I. The theory revolutionizes the understanding of weather and weather forecasting.

Ecology

1913 US biochemist Lawrence Joseph Henderson publishes *The Fitness of the Environment*, in which he points out that the physical characteristics of the environment are a unique collection of properties that enable organisms to survive.

The concept suggests that life can only exist within a limited range of physical parameters and that these are found on Earth as a result of its geophysical and geochemical characteristics, and its position in the Solar System.

1913 US Congress passes legislation to protect migratory game and insect-eating birds.

1 September 1914 The last passenger pigeon dies in the Cincinnati zoo, Ohio.

These birds once flew in their millions over the skies of North America. It is estimated that there were about 5 billion passenger pigeons in North America in the 16th century.

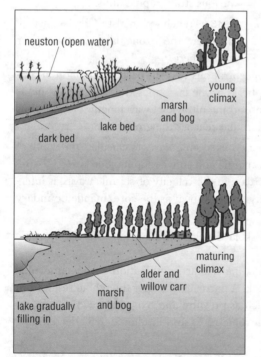

The succession of plant types along a lake. As the lake gradually fills in, a mature climax community of trees forms inland from the shore. Extending out from the shore, a series of plant communities can be discerned with small, rapidly growing species closest to the shore. The concept of succession was put forward in 1916.

By 1880 commercial harvesting, developed to supply cheap meats to the growing cities on the east coast, had seriously depleted their numbers. In 1876, for example, Oceana county in Michigan was selling over 1.5 million birds each year. The last wild bird was seen in Ohio in 1900. The death of the last specimen marks the extinction of a species and its disappearance serves as a catalyst for the US conservation movement.

1915 The US Bureau of the Biological Survey in the Department of Agriculture, under the directorship of C Hart Merriam, begins a massive campaign to eradicate pests and vermin in the National Forests and the National Parks.

Founded in 1905, the Bureau was awarded $125,000 this year for the killing of wolves on private and public land. The measure is part of a crusade against 'varmints' such as wolves and coyotes that are loathed because of the economic damage they cause. Merriam is convinced that humans have the right and even duty to manipulate the distribution of species on the planet for human welfare.

1916 US ecologist Frederic Clements publishes his major work *Plant Succession: An Analysis of the Development of Vegetation.*

Clements stresses the dynamic nature of plant communities in which one formation can give rise to another. He stresses the role of climate in determining what type of climax community will emerge in any geographical region. Clements adopts a holistic view of ecology but emphasizes that it is a practical science that will help humans better manage agricultural land.

1919 The UK's Forestry Act sets up the Forestry Commission with the aim of increasing the supply of timber. World War I had highlighted the fact that Britain's percentage of forested land is one of the lowest in Europe with only 3% of the land forested.

The Forestry Commission is charged with the responsibility of improving timber supplies on a profitable basis. In the first year, 190 sq km/73 sq mi are purchased and 7 sq km/2.7 sq mi planted. The speed of development, however, encourages poor forestry techniques, and for many the new conifer forests are aesthetically unappealing and not conducive to wildlife.

Mathematics

1913 English philosopher Bertrand Russell publishes the final volume of *Principia Mathematica/Principles of Mathematics* in collaboration with English mathematician and philosopher Alfred North Whitehead. They attempt to derive the whole of mathematics from a logical foundation.

1919 English philosopher Bertrand Russell publishes *Introduction to Mathematical Philosophy*.

> *Mathematics may be defined as the subject in which we never know what we are talking about, nor whether what we are saying is true.*
>
> Bertrand Russell, English philosopher and mathematician, *Mysticism and Logic* 1917

1919 German mathematician Felix Hausdorff introduces the idea of the Hausdorff dimension used in the study of fractals, objects containing the small-scale structure of repetitive mathematical shapes, many of which occur in nature (for example, in snowflakes).

Between 1975 and 1980, fractals will be developed more fully by the Polish-born French mathematician Benoit Mandelbrot.

Physics

1910 Dutch physicist Johannes van der Waals receives the Nobel Prize for Physics for his work on the equation of state for gases and liquids.

1910 German physicist Wolfgang Gaede develops the molecular vacuum pump, which can generate a vacuum of 0.00001 mm of mercury.

This opens the way to a wide range of research projects.

1910 French industrial chemist Georges Claude develops neon lighting when he discovers that the gas emits light when an electric current is passed through it. As it is initially only possible to produce red lighting, its potential is mainly restricted to advertising.

229

1911 Dutch physicist Heike Kamerlingh Onnes discovers superconductivity, the characteristic of a substance displaying zero electrical resistance when cooled to just above absolute zero.

Onnes discovers this phenomenon when investigating the resistance of mercury. At 4.15 K ($-269°C/-452.2°F$) he find no resistance.

1911 German physicist Albert Einstein calculates the deflection of light caused by the Sun's gravitational field.

His prediction is not tested until Arthur Eddington and others carry out measurements during a solar eclipse in May 1919. By then Einstein had revised his prediction of the deflection to be 1.75 seconds of arc; Eddington measures it as 1.60 ± 0.31 seconds of arc. Subsequent measurements have also supported general relativity.

1911 German physicist Wilhelm Wien receives the Nobel Prize for Physics for his laws regarding the radiation of heat by blackbodies.

1911 New Zealand-born British physicist Ernest Rutherford proposes the concept of the nuclear atom, in which the mass of the atom is concentrated in a nucleus occupying 1/10,000 of the diameter of the atom and which has a positive charge balanced by surrounding electrons.

Rutherford's model has most of the atom's mass located in a small, positively charged central nucleus circled by negatively charged electrons, the majority of the atom being empty space. The alpha particles in his experiment had passed through the empty space, except those few which collided with atomic nuclei.

1911 US physicist Robert Millikan measures the electric charge on a single electron in his oil-drop experiment, in which the upward force of the electric charge on an oil droplet precisely counters the known downward gravitational force acting on it.

Danish physicist Niels Bohr, one of the major figures of atomic research. Successfully combining Rutherford's classical model of the atom with Planck's quantum theory, he developed a model of the atom that revolutionized atomic theory. *AEA Technology*

His results show that the charge on an oil drop is an integer multiple of a 'unit of charge'. This leads to the modern notion of charge being 'quantized'.

1912 Swedish physicist Nils Dalén receives the Nobel Prize for Physics for his invention of automatic regulators for use in gas lights for lighthouses and buoys.

1913 Danish physicist Niels Bohr proposes that electrons orbit the atomic nucleus in fixed orbits, thus upholding Ernest Rutherford's model proposed in 1911.

Bohr suggests that electrons move in fixed circular orbits around the nucleus. Initially the theory can only be applied to the hydrogen atom, but later it is extended to other elements. The theory states that only a limited number of electrons can occupy each orbit; every element thus has a unique electronic configuration which determines its chemical properties. Bohr is awarded the Nobel Prize for Physics in 1922 for this work.

Our task is not to penetrate into the essence of things,
the meaning of which we don't know anyway,
but rather to develop concepts which allow us to talk
in a productive way about phenomena in nature.

Niels Bohr, Danish physicist, letter to H P E Hansen 20 July 1935

1913 Dutch physicist Heike Kamerlingh Onnes receives the Nobel Prize for Physics for his investigations on the properties of matter at low temperatures and the production of liquid helium.

1913 English physical chemist Frederick Soddy coins the term 'isotope' (from the Greek *isos*, 'equal', and *topos*, 'place') to describe atoms of the same chemical element but with different atomic numbers.

1913 English physicist Henry Moseley discovers the characteristic feature of an element, the atomic number. Moseley discovers that the X-ray spectra of the elements have a deviation that changes regularly through the periodic table.

Atomic number is later found to correspond to the number of protons in the nucleus of the atom. This quantity is the characteristic feature of all isotopes of an element and replaces atomic weight as the property used to place elements in the periodic table.

1913 English physicist J J Thomson develops a mass spectrometer called a parabola spectrograph. A beam of charged ions is deflected by a magnetic field to produce parabolic curves on a photographic plate.

1913 German-born US physicist Albert Einstein formulates the law of photochemical equivalence, which states that for every quantum of radiation absorbed by a substance one molecule reacts.

1914 German physicist Max von Laue receives the Nobel Prize for Physics for his discovery of the diffraction of X-rays by crystals.

1914 German physicists James Franck and Gustav Hertz provide the first experimental evidence for the existence of discrete energy states in atoms and thus verify Danish physicist Niels Bohr's atomic model.

1915 English physicists William and Lawrence Bragg win the Nobel Prize for Physics for their work showing that the atomic structure of crystals can be analysed from the diffraction patterns of X-rays.

1916 German physicist-born US Albert Einstein publishes *The Foundations of the General Theory of Relativity*, in which he postulates that space is curved locally by the presence of mass, and that this can be demonstrated by observing the deflection of starlight around the Sun during a total eclipse. This replaces previous Newtonian ideas that invoke a force of gravity.

1916 US physical chemist William David Coolidge patents an X-ray tube that can produce highly predictable amounts of radiation. It serves as the prototype of the modern X-ray tube.

This leads to advances in medical diagnostics, and improved safety for patients and operators.

1917 English physicist Charles Barkla receives the Nobel Prize for Physics for his discovery of the characteristic X-ray radiation of the elements.

1918 English astrophysicist Arthur Eddington publishes 'Report on the relativity theory of gravitation', which is the first explanation of Einstein's theory of relativity in English.

Until this time Einstein's work has only been available in German.

1918 German physicist Max Planck receives the Nobel Prize for Physics for his discovery of energy quanta.

1919 German physicist Johannes Stark receives the Nobel Prize for Physics for showing that an electric field splits the spectral lines of hydrogen and for predicting that high-velocity rays of positive ions will demonstrate the Doppler effect.

1919 New Zealand-born British physicist Ernest Rutherford splits the atom by bombarding a nitrogen nucleus with alpha particles, discovering that it ejects hydrogen nuclei (protons). It is the first artificial disintegration of an element and inaugurates the development of nuclear energy.

Rutherford recognizes that hydrogen nuclei are fundamental particles and names them 'protons'.

29 May 1919 English astrophysicist Arthur Eddington and others observe the total eclipse of the Sun on Príncipe Island (West Africa), and discover that the Sun's gravity bends the light from the stars beyond the edge of the eclipsed Sun, thus fulfilling predictions made according to Albert Einstein's theory of general relativity.

1920–1929

Astronomy

1920 German astronomer Walter Baade discovers the asteroid Hidalgo, which is unusual in that its orbit is tilted out of the plane of the Solar System by 43°.

1920 US inventor Robert Goddard publishes 'A Method of Reaching Extreme Altitudes', about the use of rockets as a means to reach the Moon.

Goddard details his search for methods of sending weather-recording instruments to higher altitudes than is possible with sounding balloons. This search leads him to develop the mathematical theory of rocket propulsion, and suggests sending a rocket to the Moon. Goddard's work largely anticipates in technical detail the later German V-2 missiles, including gyroscopic control, steering by means of vanes in the jet stream of the rocket motor, gimbal steering, power-driven fuel pumps, and other devices.

1920 US physicist Albert Michelson, using a stellar interferometer, measures the diameter of the star Betelgeuse to be 386,160,000 km/241,350,000 mi, which is about 300 times the diameter of the Sun. It is the first time an accurate measurement of the size of a star other than the Sun has been made.

The modern value for the diameter of Betelgeuse is 1,100 million km/700 million mi; that is, 800 times the diameter of the Sun.

12 May 1922 A 20.3 tonne/20 ton meteorite lands in a field near Blackstone, Virginia, leaving a 46 sq m/500 sq ft hole in the ground.

1923 German mathematician Hermann Oberth publishes *Die Rakete zu den Planetenräumen/The Rocket into Interplanetary Space*, a treatise on space-flight, in which he is the first to provide the mathematics of how to achieve escape velocity.

1923 German-born British physicist Frederick Lindemann investigates the size of meteors and the temperature of the upper atmosphere.

His assistant, Gordon Dobson, deduces that the high temperature of the stratosphere is due to the presence of a stratospheric ozone layer. Dobson goes on to devise the Dobson ozone spectrometer, which uses ultraviolet absorption lines to estimate the column abundance and vertical distribution of ozone above the observer. In due course a worldwide network of these devices is set up, and used to detect the Antarctic ozone hole in the 1960s.

1924 US astronomer Edwin Hubble demonstrates that certain Cepheid variable stars are several hundred thousand light years away and thus outside the Milky Way galaxy.

The nebulae in which they are situated are the first galaxies to be discovered outside the Milky Way.

1925 US inventor Robert Goddard conducts a static test of a liquid-propelled rocket.

Goddard's liquid-propelled rocket flies in 1929 carrying the first scientific payload, a barometer and a camera.

1925 Swedish astronomer Bertil Lindblad discovers that the Milky Way rotates around its centre. One rotation takes 210 million years.

He shows that the system of stars can be divided into several subsystems rotating about the same axis, each subsystem having its own rotation rate and consequent degree of flattening. The 'high-velocity' stars are actually moving slowly; the Sun is moving faster. Lindblad confirms Harlow Shapley's direction and approximate distance to the centre of the Galaxy (1917), and he estimates the galactic mass and the period of the Sun's orbit. His work leads directly to Jan Oort's theory of differential galactic rotation.

1926 English astrophysicist Arthur Eddington publishes *Internal Constitution of the Stars*, in which he shows that the luminosity of a star is a function of its mass.

Eddington calculates the abundance of hydrogen in a star and produces a theory to explain the pulsation of Cepheid variable stars. Eddington also leads eclipse expeditions to Brazil and Príncipe Island in West Africa. The results from the Africa expedition provide the first confirmation of Einstein's theory that gravity bends the path of light when it passes near a massive star.

1927 Belgian astronomer Georges Lemaître proposes that the universe was created by an explosion of energy and matter from a 'primaeval atom' – the beginning of the Big Bang theory.

Lemaître envisions a solitary 'space particle', the first object in existence. Radioactive elements in the particle started a chain reaction that forced the immediate and rapid expansion of the universe and also created life. Forces of repulsion ensured that expansion became dominant and from that time (about 9 billion years ago), galaxies began to separate. The important point of Lemaître's theory is not so much the expansion of the universe, but the assumption that something actually started it. It was this assumption that gives birth to the 'Big Bang' theory.

1927 The Verein für Raumschiffahrt (Society for Space Travel) is founded in Germany for rocket experimentation.

233

1927 US astronomer Edwin Hubble shows that galaxies are receding and that the further away they are, the faster they are receding.

c. 1928 English physicist James Hopwood Jeans proposes the steady-state hypothesis, which states that the universe is constantly expanding and maintaining a constant average density through the continuous creation of new matter.

The steady-state hypothesis asserts that the universe is the same at all times as well as from any point within it. Thus, as the universe expands there must be a continuous creation of matter at such a rate as to maintain constant mean density throughout the increasing volume in a universe with no beginning or end. New matter supposedly is being created continuously, forming new galaxies, so that the density of galaxies remains the same despite their mutual recession. Through computation of an average volume equivalent of an average galaxy according to the steady-state theory, a new galaxy is created once every 1,000 years. This corresponds roughly to a rate of creation of matter of one hydrogen atom per cubic kilometre per year.

1928 It is discovered that Neptune's moon Triton rotates in the retrograde direction, that is, opposite to that of Neptune's spin.

This behaviour is usually considered characteristic of a captured asteroid, rather than a natural moon. Certainly, such a capture event could account for the peculiarities of Triton's axis of rotation, tilted 157° with respect to Neptune's axis (which is in turn inclined 30° from the plane of Neptune's orbit), as well as the retrograde orbit. However, Triton (diameter 2,700 km/1,700 mi) is the only large satellite in the Solar System to behave in this way, and its density (2.06 g/cm²) is closer to the other large, icy satellites of the gas giant planets than to a rocky asteroid.

1929 *The Universe Around Us* is published by English physicist James Hopwood Jeans and helps to popularize astronomy.

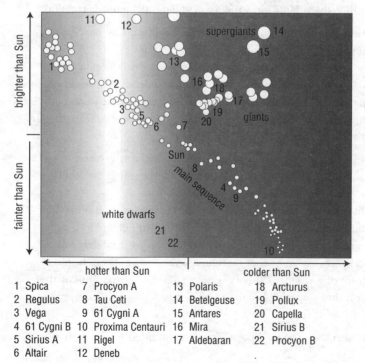

The Hertzsprung–Russell diagram relates the brightness (or luminosity) of a star to its temperature. Most stars fall within a narrow diagonal band called the main sequence. A star moves off the main sequence when it grows old. The Hertzsprung–Russell diagram is one of the most important diagrams in astrophysics.

1	Spica	7	Procyon A	13	Polaris	18	Arcturus
2	Regulus	8	Tau Ceti	14	Betelgeuse	19	Pollux
3	Vega	9	61 Cygni A	15	Antares	20	Capella
4	61 Cygni B	10	Proxima Centauri	16	Mira	21	Sirius B
5	Sirius A	11	Rigel	17	Aldebaran	22	Procyon B
6	Altair	12	Deneb				

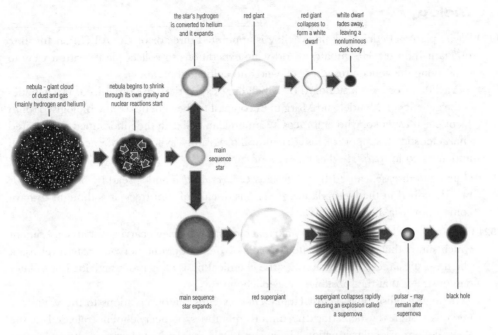

the star's hydrogen is converted to helium and it expands

red giant

red giant collapses to form a white dwarf

white dwarf fades away, leaving a nonluminous dark body

nebula - giant cloud of dust and gas (mainly hydrogen and helium)

nebula begins to shrink through its own gravity and nuclear reactions start

main sequence star

main sequence star expands

red supergiant

supergiant collapses rapidly causing an explosion called a supernova

pulsar - may remain after supernova

black hole

The life cycle of a star. New stars are being formed all the time when nebulae (giant clouds of dust and gas) contract due to the action of gravity. As the star contracts and heats up eventually nuclear reactions begin and the star becomes a main sequence star. If the star is less than 1.2 times the mass of the Sun, it eventually forms a white dwarf that finally fades to a dark body. If it is a massive star, then the main sequence star expands to become a red supergiant that eventually explodes as a supernova. It leaves part of the core as a neutron star (pulsar), or as a black hole if the mass of the collapsing supernova core is three times greater than the Sun.

1929 English philosopher Alfred North Whitehead publishes *Process and Reality: An Essay in Cosmology*.

1929 US astronomer Edwin Hubble publishes Hubble's law, which states that the ratio of the speed of a galaxy to its distance from Earth is a constant (now known as Hubble's constant).

1929 US astronomer Henry Russell publishes 'Stellar Evolution', in which he suggests that stars begin as huge cool red bodies, shrink to become hot yellow stars then hot white and blue dwarfs, then shrink to become cool red stars.

This relationship is captured in the Hertzsprung–Russell or H–R diagram, an important astronomical tool which shows the luminosity of a star as a function of its surface temperature.

1929 US physicist Robert Goddard launches his first liquid-propelled rocket, carrying the first scientific payload of a barometer and a camera.

1939 German physicists Hans Bethe and Carl von Weizsäcker propose that nuclear fusion of hydrogen is the source of a star's energy.

This theory confirms earlier predictions made by Russian-born US nuclear physicist and cosmologist George Gamow, and not only does it explain the synthesis of stellar material but also the 'hydrogen bomb'.

Biology

1920 Soviet plant geneticist Nikolai Vavilov is appointed director of the All Union Institute of Plant Industry. He initiates world-wide expeditions to collect plants with a view to conserving the genetic resources of wild and cultivated plants.

Vavilov's work sets a standard by which later expeditions are organized. His genetics is the genetics of Mendel and Morgan in contrast to the genetics of the biologist Trofim Denisovich Lysenko, who embraces a Lamarckian belief in the inheritance of acquired characteristics in support of his agricultural program. Vavilov is denounced by Lysenko and arrested in 1940. He dies in a labour camp.

1921 Dutch elm disease, caused by the fungus *Ceratocystis ulmi* and spread by bark beetles, is first described in the Netherlands. A serious disease of elm trees, it is thought to have come from Asia after World War I.

1921 German-born US physician and pharmacologist Otto Loewi carries out an experiment which shows that, when stimulated, nerves of the autonomic nervous system release a chemical substance. This establishes the chemical basis of nervous activity. Loewi later demonstrates that the substance is acetylcholine.

In 1936 English physiologist Henry Dale extends the observations to the voluntary nervous system, by detecting the minute quantities of acetylcholine released at the skeletal neuromuscular junctions. Loewi and Dale share a Nobel prize in the same year. Their work leads to an understanding of how drugs affect the nervous system.

1922 English biologist Walter Garstang publishes 'The Theory of Recapitulation: A Critical Re-statement of the Biogenetic Law', in which he attacks Haekel's biogenetic law that states 'ontogeny recapitulates phylogeny'. In other words, during its development (ontogeny), an organism repeats its evolutionary history (phylogeny).

In the paper Garstang coins the word 'paedomorphosis' to describe juvenile adaptations, which might affect adults of the same species and thereby perhaps affect the course of evolution. He summarizes his views by stating that 'ontogeny does not recapitulate phylogeny: it creates it'.

1926 US geneticist Hermann Muller uses X-rays to cause mutations in the fruit fly *Drosophila*. It permits a greater understanding of the mechanisms of variation.

1926 US geneticist Thomas Hunt Morgan publishes *The Theory of the Gene*, in which he demonstrates that the gene will form the foundation of all future genetic research.

1927 Biologist Ernst Munch proposes the mass flow hypothesis to account for the transport of sucrose and other substances (translocation) through the phloem tissue of plants.

The hypothesis is one of several put forward over the years to account for translocation. It gains support from experimental data confirming some of its predictions.

1928 Scottish bacteriologist Alexander Fleming discovers penicillin when he notices that the mould *Penicillium notatum*, which has invaded a culture of staphylococci, inhibits the growth of the bacteria.

Fleming observes that a culture of the bacterium *Staphylococcus aureus*, which he had left standing for several days, had failed to grow in areas where it had been accidentally contaminated with a green mould. He eventually identifies this as *Penicillium notatum*, a close relative of the common bread mould. His studies show that the mould is an effective antibacterial agent and kills many bacteria known to be harmful to humans. Although he publishes his findings the following year the significance of the discovery is not developed until World War II.

1928 English biologist Walter Garstang gives the presidential address 'The Origin and Evolution of Larval Forms' to the Zoology Section of the British Association, in which he establishes the idea of paedomorphosis.

Garstang states that the adaptations of juvenile forms might not only affect adults of the same species but also influence the course of evolution. In the same year he puts forward the view that the chordates are derived by paedomorphosis from the pelagic larval forms of sedentary echinoderms. Garstang couples the idea of paedomorphosis with that of neotony, whereby the sexual development of juvenile forms accelerates so that they are capable of reproduction at a pre-adult stage. Examples of neotony occurring in nature and the recent growth of developmental genetics gives impetus to the concept, which provides a probable explanation for the contention that evolution is not a process of gradual change but occurs in 'jumps'.

1929 German biochemist Adolf Butenandt and, simultaneously and independently, US biochemist Edward Doisy isolate the hormone oestrone, which is involved in the growth and development of females.

Chemistry

1920 US biochemists led by Elmer McCollum discover vitamin D (cholecalciferol), which is subsequently successfully used to treat rickets.

McCollum isolates a fat-soluble substance from cod liver oil that is found to cure the deficiency disease rickets and the eye condition xerophthalmia. It is the fourth vitamin to be discovered and is called vitamin D. In 1921, the same team discover that rats, which have been deprived of vitamin D in their diet, do not develop rickets if they are exposed to sunlight. This indicates that sunlight converts a substance in the skin to vitamin D.

1920 German chemist Walther Nernst receives the Nobel Prize for Chemistry for work on thermochemistry. He formulated the third law of thermodynamics in 1906.

1921 English physical chemist Frederick Soddy receives the Nobel Prize for Chemistry for his work on radioactive substances, especially isotopes.

1921 US chemist Thomas Midgley Jr invents an anti-knocking additive for petrol.

Knocking is a problem in automobiles that is caused by premature ignition of petrol vapours in the engine cylinders. Midgley discovers that the addition of the organometallic compound, tetraethyl lead, inhibits the combustion of fuel just enough to prevent an engine from knocking. However, lead is now discharged through the exhaust pipe of automobiles, contributing significantly to the build-up of lead in the environment and ultimately in the blood stream of city dwellers.

1922 German organic chemist Hermann Staudinger coins the word 'macromolecule'.

Staudinger uses the word to describe a chain of isoprene units that make up a molecule of natural rubber. He suggests that the isoprene chain is composed of tens of thousands of atoms held together by ordinary bonds. This is not well received by the scientific community, who favour the prevailing view that polymers are composed of disorderly conglomerates of small molecules. Staudinger is proven to be correct the following year. 'Macromolecule' becomes used to describe any molecule that has a very high molecular mass (>10,000 atoms) including proteins.

1922 US anatomist Herbert McLean Evans discovers vitamin E.

Evans finds that rats become sterile when limitations are placed on their diet. This condition is not relieved even when the four known vitamins, A, B, C, and D are introduced. However, if the animals are fed wheat germ, lettuce, or alfalfa, their fertility returns. He deduces that the introduced foodstuffs must contain an unknown vitamin, which he designates as vitamin E.

1922 English physicist Francis Aston receives the Nobel Prize for Chemistry for pioneering the mass spectrometry of isotopes of radioactive elements and enunciating the whole-number rule.

1923 Danish chemist Johannes Nicolaus Brønsted and British chemist Thomas Martin Lowry simultaneously and independently propose the concept of acid–base pairs.

The theory of ionic dissociation (see 1884) states that an acid is a substance that breaks up in solution to liberate hydrogen ions. Brønsted and Lowry point out that hydrogen ions are protons, which cannot exist freely in solution and must therefore combine with other molecules immediately after dissociation. They call these groupings acid–base pairs, the acid being the substance that gives up a proton and the base being the one that accepts it. This broader concept clarifies and enhances the original theory.

1923 Dutch physicist Dirk Coster and the Hungarian-born Swedish physicist Georg von Hevesy discover the element hafnium (atomic number 72).

They identify the new element in a sample of the mineral zircon using an X-ray technique developed by Coster. They make their discovery in Copenhagen, Denmark, and name the new element 'hafnium' after the Latin name for Copenhagen, *Hafnia*.

1923 Swedish chemist Theodor Svedberg develops the ultracentrifuge, which he uses to show that polymers are composed of very large molecules comprising hundreds of thousands of atoms.

A conventional centrifuge does not exert the force necessary to separate colloidal particles smaller than red blood corpuscles. Svedberg develops a machine that can exert the equivalent separating force of several hundreds of thousands of times the normal force of gravity. This allows the separation by mass of normal protein molecules. Svedberg is awarded the Nobel Prize for Chemistry in 1926 for this invention and his contributions to colloidal chemistry.

1923 Austrian chemist Fritz Pregl receives the Nobel Prize for Chemistry for the micro-analysis of organic substances.

1924 Czech chemist Jaroslav Heyrovský develops the polarograph, used for electrochemical analysis.

This device uses a mercury electrode that is configured to drop mercury through a solution into a mercury pool below. An electric current is passed through the solution and the potential is increased until the current reaches a stable plateau. The concentration of ions in the solution can be calculated from the height of the plateau. Heyrovský is able to use his device to determine the ion concentrations in solutions of unknown compositions. He calls his technique 'polarimetry' and is awarded the Nobel Prize for Chemistry in 1959 for this work.

1925 Austrian chemist Richard Zsigmondy receives the Nobel Prize for Chemistry for his work on colloids.

1925 German chemists Ida Tacke, Walter Noddack, and Otto Berg discover the rare metallic element rhenium (atomic number 75).

They isolate an unknown substance from a platinum ore, which they recognize as being a new element. They also detect the element in molybdenite and spend the next three years isolating a single gram of the pure metal from over 600 kg/1,320 lb of the mineral. They name their discovery 'rhenium' after 'Rhenus', the Latin name for the River Rhine.

1926 German chemist Hermann Staudinger proves that polymers are formed by chemical interaction between small monomer units.

Staudinger disproves the previously held belief that polymers are composed of random conglomerates of molecules held together by physical attractions. He shows that small molecules, called 'monomers', chemically combine to form long-chain straight polymer structures. In 1930, he develops a method for determining the molecular weight of a polymer molecule based on the polymer's viscosity, which becomes known as Staudinger's law. His pioneering work in the field of polymer chemistry earns Staudinger the Nobel Prize for Chemistry in 1953.

1926 US biochemist James Sumner crystallizes the enzyme urease. It is the first enzyme to be crystallized. Sumner's achievement demonstrates that enzymes are proteins.

Sumner is extracting the enzyme 'urease' from jack beans, when he notices that tiny crystals have formed in one of his sample fractions. He assumes that the crystals are a mixture of the enzyme and some other organic component. The crystals show strong enzyme activity, but every attempt at separating this activity from the crystals fails. He correctly concludes that he has crystallized a highly pure form of the enzyme. Sumner is awarded a share in the Nobel Prize for Chemistry in 1946 for this discovery.

1926 Swedish chemist Theodor Svedberg receives the Nobel Prize for Chemistry for his investigation of dispersed systems.

1927 German biochemist Heinrich Wieland receives the Nobel Prize for Chemistry for his research into the constitution of bile acids and related substances.

1928 German chemists Kurt Alder and Otto Diels develop a process to synthesize cyclic carbon compounds – this proves to be an important breakthrough in the development of synthetic rubber and plastics.

Alder and Diels discover a method that joins together compounds known as 'dienes', which are hydrocarbons that contain two unsaturated C=C double bonds, with an activated alkene or alkane, to form cyclic carbon compounds. The process is correctly called 'diene synthesis', but becomes commonly known as the 'Diels–Alder Reaction'. Many natural compounds containing such groups can now be artificially synthesized for the first time. It also proves to be an important breakthrough in the development of synthetic rubber and plastics.

1928 Hungarian-born US biochemist Albert Szent-Györgyi is first to isolate crystals of vitamin C (ascorbic acid).

Szent-Györgyi isolates a crystalline substance form adrenal glands and cabbages, which he finds has the ability to lose or gain pairs of hydrogen atoms easily. He determines that the substance is composed of six carbon atoms and has the properties of a sugar, so he calls his discovery 'hexuronic acid' after the Greek for 'six' and the suffix 'uronic', which denotes a sugar. He does not realize that hexuronic acid is actually vitamin C, which is first identified in 1932.

Discovery consists of seeing what everybody has seen,
thinking what nobody has thought.

Albert Szent-Györgyi, Hungarian-born US biochemist, quoted in
I G Good (ed) *The Scientist Speculates* (1962)

1928 German biochemist Adolf Windaus receives the Nobel Prize for Chemistry for his work on the constitution of sterols and related vitamins.

He also discovers histamine, an important compound in allergic responses.

1929 French chemist Eugene Houdry develops the fixed-bed catalytic process of cracking crude oil to obtain petrol, now known as the Houdry process.

Houdry's process uses a series of heat exchanger reactors, which incorporate a bed containing a clay catalyst over which the crude petroleum is passed. This is able to break down heavy hydrocarbon fractions into a much wider range of lighter fractions than is possible using the earlier thermal cracking technology. In particular, a higher proportion of unsaturated hydrocarbon gases are produced. These are important raw materials for the growing plastics industry.

1929 Russian-born US chemist Phoebus Levene discovers 2-deoxyribose, the five-carbon sugar that forms the basis of deoxyribonucleic acid, DNA.

Levene is studying nucleic acids that do not contain ribose, when he isolates the sugar, deoxyribose. This substance is identical to ribose with the exception that it has one less oxygen atom in its molecular structure. This proves that there are two types of nucleic acid, ribose nucleic acid, or RNA, and DNA. The latter is later found to be present in chromosomes.

1929 US chemist William Giauque discovers that natural oxygen consists of three isotopes of masses 16, 17, and 18. Since oxygen is used as the standard against which relative atomic mass is determined, this means that all atomic weights are inaccurate.

The problem of choosing a new standard is not solved until 1961, when carbon-12 is selected by international agreement as the new standard weight.

1929 English biochemist Arthur Harden and German chemist Hans von Euler-Chelpin share the Nobel Prize for Chemistry for their investigations of the fermentation of sugars and fermentative enzymes.

Earth Sciences

1920 Using new echosounding technology to make bathymetric maps of the seafloor, the German *Meteor* expedition discovers a rift, or valley, running down the middle of the mid-Atlantic Ridge. Its importance to understanding seafloor genesis is discovered in the 1950s.

1920 Yugoslavian meteorologist and mathematician Milutin Milankovitch shows that the amount of energy, or heat, received by the Earth from the Sun varies with long-term changes in Earth's orbit. Decades later, scientists will correlate fluctuations in global temperature to his 'Milankovitch cycles'.

1921 Norwegian meteorologist Vilhelm Bjerknes publishes *On the Dynamics of the Circular Vortex with Applications to the Atmosphere and to Atmospheric Vortex and Wave Motion*, in which he summarizes his work on the movement of air masses and weather forecasting.

1922 British mathematician Lewis Fry Richardson publishes *Weather Prediction by Numerical Process*. Richardson suggests that it might be possible to predict the weather through complex numerical calculations. He admits that to predict the weather regularly, one would require 64,000 mathematicians working full-time. With the advent of computers, his idea becomes possible.

1922 A nest of fossilized dinosaur eggs is discovered in the Gobi Desert by US palaeontologist Roy Chapman Andrews. It is the first such nest to be found.

1922 The torsion balance, a device for measuring gravitational field strength, is used for the first time to locate salt domes, and possibly oil, along the coast of the Gulf of Mexico.

Variations in the Earth's gravitational field are a result of variations in density of the rocks of the crust. Salt domes can trap oil and because salt is of particularly low density and salt domes have a characteristic shape, they are relatively easy to identify with gravity measurements.

1924 English physicist Edward Appleton discovers that radio emissions are reflected by an ionized layer of the atmosphere. He estimates the height of the Kennelly–Heaviside atmospheric layer at 80 km/50 mi.

In 1902 the British-born US electrical engineer Arthur Kennelly and the English engineer Oliver Heaviside had independently advanced the idea that the upper atmosphere contains a layer of electrical charges or ions that serve to reflect radio signals. This predicted layer is called the Kennelly–Heaviside layer. Appleton finds another layer of ions at 241 km/150 mi high, which are subsequently called Appleton layers. The layer of air above the stratosphere now becomes called the ionosphere.

1925 The US Navy develops a pulse modulation technique to measure the distance above the Earth of the ionizing layer in the atmosphere.

1926 At a time when very few people believe German meteorologist Alfred Wegener's theory of continental drift, US geologist Reginald Daly advocates the theory and publishes *Our Mobile Earth*.

1927 English geologist Arthur Holmes publishes a booklet which includes a table of all radiometric ages calculated for rocks so far. He also includes his estimate of the age of the Earth at 1.6–3.0 billion years.

1927 South African geologist Alexander du Toit publishes *A Geological Comparison of South America and South Africa*, in which he outlines the similarities in geology and fossil flora and fauna between the two land masses. Du Toit is a firm supporter of Alfred Wegener, whose continental drift theory explains the observations he makes.

1928 Canadian geochemist Norman Levi Bowen publishes *The Evolution of the Igneous Rocks* in which he suggests that Earth's crust is the product of melting of parts of the mantle, a process known as differentiation. His work firmly establishes the potential of the physical chemical approach to geology.

1928 Japanese geophysicist Kiyoo Wadati shows that earthquake foci near the Japan trench descend as they move away from the trench towards Asia. His observation is rediscovered 20 years later by Hugo Benioff who makes similar observations.

1929 By studying the magnetism of rocks, the Japanese geologist Motonori Matuyama shows that the Earth's magnetic field periodically reverses direction.

This discovery, first made by French physicist Bernard Brunhes in 1906, will be crucial evidence for the seafloor spreading hypothesis in the 1960s.

1929 English geologist Arthur Holmes describes a mantle convection mechanism for continental drift.

With mantle convection, Holmes provides the much-needed mechanism for continental drift. The idea re-emerged with Harry Hess in the 1960s.

1929 French engineer and chemist Georges Claude shows that the temperature difference between the upper and lower depths of the ocean can be used to generate electricity.

1929 Norwegian chemist Victor Goldschmidt produces the first table of ionic radii which is useful for predicting crystal structures.

Ecology

1920 British economist Arthur Cecil Pigou publishes *The Economics of Welfare*.

Until this time neo-classical economics has failed to tackle the problem of externalities, conceiving of economic welfare as a mixture of profits to the producer and the satisfaction received from the product by the consumer. Pigou adds a missing part of the framework: some means of accounting for market-external occurrences such as pollution or damage to scenery which also affects human welfare. With these ideas, he lays the foundations of environmental economics.

1921 English philosopher and conservationist Henry Salt publishes his autobiography, *Seventy Years Among Savages*.

In this work he describes how he gave up his life of a schoolmaster in 1884 and moved into the countryside in Surrey, England, to follow a vegetarian diet, advocate socialism, and promote a 'broad democratic sentiment of universal sympathy'. Salt argues against capitalism, meat-eating, vivisection, and economic inequality. He suggests that humans and nature should be reunited by extending the circle of human ethics to include other animals.

1924 Austrian philosopher, occultist, and educationalist Rudolf Steiner, founder of the mystical philosophy anthroposophy, gives a series of lectures at a meeting of anthroposophical farmers on the estate of Count Keyserling in Germany.

Steiner calls for an organic system of farming that works with the life forces inherent in the cosmos. Steiner's ideas on biodynamic farming are influential and accord well with German ecological ideas of the period.

1924 US Congress passes the Oil Pollution Act, prohibiting oil producers from polluting the environment.

1927 English zoologist Charles Elton publishes *Animal Ecology*. The work is hailed as a foundation stone of the 'New Ecology'. Elton describes his approach as 'the sociology and economics of animals'.

In his work Elton draws upon economic ideas and applies them to the science of ecology. He enunciates four principles to describe the economy of nature: the idea of a food chain, food supply, population pyramid, and an ecological niche. In the food chain, Elton shows how each animal performs a certain role in relation to the supply of nutrients. Animals can be producers, consumers (primary or secondary), or decomposers. The higher a species is on the food chain, the lower the population of individual

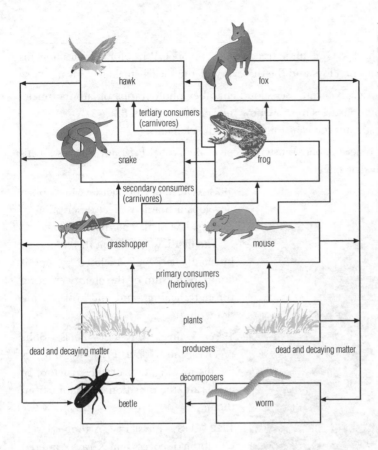

The complex interrelationships between animals and plants in a food web. A food web shows how different food chains are linked in an ecosystem. Note that the arrows indicate movement of energy through the web. For example, an arrow shows that energy moves from plants to the grasshopper, which eats the plants. The food chain was described in 1927.

animals. Hence, the plants at the base of the food chain form most of the biomass on Earth, while a much lower mass is accorded to the carnivores at the top. Elton also defines a niche according to food supply, and he formulates the principle of exclusion, the idea that no two species can occupy the same niche.

Mathematics

1921 English economist John Maynard Keynes publishes his *Treatise on Probability*, which is to have a profound effect on statistics as well as economics.

1922 Polish mathematician Stefan Banach begins his work on a development of normed vector spaces, an important tool in general analysis.

With the Polish-born US mathematician Alfred Tarski, he goes on to make important contributions to set theory.

Physics

1920 French physicist Charles Guillaume receives the Nobel Prize for Physics for his discovery of anomalies in alloys and the discovery of the nickel-steel alloy invar.

1921 English physicist Patrick Blackett uses cloud-chamber photographs of atomic nuclei bombarded with alpha particles to show how they are disintegrated.

Blackett later investigates the transmutation of the elements.

1921 German-born US physicist Albert Einstein receives the Nobel Prize for Physics for his work on the photoelectric effect.

Max Born, who left Germany for Britain when the Nazis came to power, was awarded the Nobel Prize for Physics in 1954 for his work on quantum mechanics. *Corbis*

1921 German physicist Max Born develops a mathematical description of the first law of thermodynamics.

The first law states that the change in internal energy of a closed system is equal to the sum of the energy entering through heating and the work being done on it.

1922 Danish physicist Niels Bohr receives the Nobel Prize for Physics for his investigation of the structure of atoms and the radiation emanating from them.

1922 US physicist Arthur Holly Compton discovers that X-rays scattered by an atom undergo a shift in frequency. He explains the phenomenon, known as the Compton effect, by treating the X-rays as a stream of particles, thus confirming the wave–particle idea of light.

1923 German-born US physicist Albert Michelson measures the speed of light, obtaining a value close to the modern value of 299,792 km per sec/186,282 mi per sec.

1923 US physicist Robert Millikan receives the Nobel Prize for Physics for his work on the elementary charge of electricity and on the photoelectric effect.

1923 French physicist Louis de Broglie argues that particles can also behave as waves, laying the foundations for wave mechanics. He demonstrates that a beam of electrons has a wave motion with a short wavelength. The discovery permits the development of the electron microscope.

De Broglie formulated what is known as the de Broglie wavelength:

$$\lambda = h/\rho$$

where h is the Planck constant and ρ is the momentum.

1924 Swedish physicist Karl Siegbahn receives the Nobel Prize for Physics for his research in the field of X-ray spectroscopy.

1925 German physicists James Franck and Gustav Hertz receive the Nobel Prize for Physics for their discovery of the laws governing the impact of an electron upon an atom.

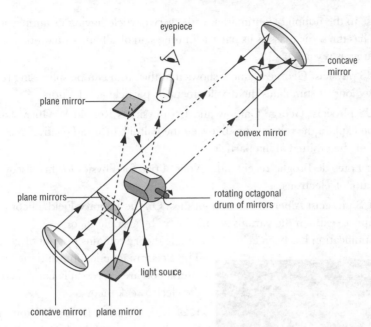

eyepiece

plane mirror

concave mirror

convex mirror

plane mirrors

rotating octagonal drum of mirrors

light source

concave mirror   plane mirror

German-born US physicist Albert Michelson determined the velocity of light by using a rotating drum faced with eight mirrors to time a light beam over a distance of more than 70 km/43 mi.

1926 French physicist Jean Perrin receives the Nobel Prize for Physics for his work on the discontinuous structure of matter, and especially for his discovery of sedimentation equilibrium.

1926 German-born US physicist Albert Michelson uses an eight-sided rotating prism to determine the speed of light. The value determined by Michelson is 2.99774×10^8 m s^{-1}.

1927 German physicist Werner Heisenberg propounds the 'uncertainty principle' in quantum physics, which states that it is impossible to simultaneously determine the position and momentum of a particle without some degree of error or uncertainty – the better one is determined, the greater the error in the other.

It explains why Newtonian mechanics is inapplicable at the atomic level.

An expert is someone who knows some of the worst mistakes that can be made in his subject and how to avoid them.

Werner Heisenberg, German physicist, *The Part and the Whole*

1927 US physicist Arthur Holly Compton and Scottish physicist Charles Wilson share the Nobel Prize for Physics: Compton for his discovery of the change in wavelength in scattered X-rays and Wilson for his method of making the paths of electrically charged particles visible by condensation of vapour.

1927 US physicist Clinton Davisson and, independently, English physicist George Paget Thomson show that electrons can be diffracted.

This shows that 'matter' can exhibit 'wavelike' properties.

1928 English physicist Owen Richardson receives the Nobel Prize for Physics for his work on the thermionic phenomenon and especially for the discovery of Richardson's law relating the current emitted by a heated metal to its temperature.

1928 English physicist Paul Dirac describes the electron by four wave equations. The equations imply that negative energy states of matter must exist.

This gives rise to the notion of antimatter, with every particle having its antiparticle. An antiparticle has the same mass as its particle but the sign of all other quantities, such as spin and charge, is reversed.

1928 Russian-born US physicist George Gamow shows that the atom can be split using relatively low-energy ions. It stimulates the development of particle accelerators.

1928 Russian-born US physicist George Gamow and US physicists Ronald W Gurney and Edward Condon explain the relationship between the half-life of a radioactive element and the energy of the emitted alpha particle.

1929 French physicist Louis de Broglie receives the Nobel Prize for Physics for his discovery of the wave nature of electrons.

1929 German-born US physicist Albert Einstein publishes 'On the Unitary Field Theory' in which he attempts to explain the various atomic forces by a single theory.

Whilst some unification has been achieved, a single theory remains just out of reach. The electromagnetic force and the weak interaction are successfully combined in the electroweak theory.

English nuclear physicist John Cockcroft. In the 1920s he played a leading role in the development of the particle accelerator, and in 1951 shared a Nobel prize with Ernest Walton for splitting the nucleus of an atom with artificially accelerated particles. *AEA Technology*

1929 Irish physicist Ernest Walton and English physicist John Cockcroft develop the first particle accelerator.

Particle accelerators soon grow into vital tools for physicists working in many fields. Small accelerators are used to produce radioisotopes for use in medicine, while the largest accelerators produce particle collisions with energy so high that new fundamental particles can be created. In this way, evidence for the existence of the six 'quarks' has been observed.

1929 Russian-born US physicist George Gamow, English astrophysicist Robert Atkinson, and German physicist Fritz Houtermans suggest that thermonuclear processes are the source of solar energy.

Gamow is able to explain that as the Sun 'burns' hydrogen to build helium and heavier elements, energy is released and this causes the Sun to heat up. Gamow also develops a theory of alpha particle decay based on 'quantum mechanical tunnelling', which explains how alpha particles could be emitted with less energy than classical physics would suggest is possible.

1930–1939

Astronomy

1930 Swiss-born US astronomer Robert Trumpler discovers the existence of interstellar material that reduces the apparent brightness of distant stars.

18 February 1930 US astronomer Clyde Tombaugh, at the Lowell Observatory, Arizona, discovers the ninth planet, Pluto.

1932 German-born US physicist Albert Einstein and Dutch astronomer Willem de Sitter publish a joint paper in which they proposed the Einstein–de Sitter model of the universe. This is a particularly simple solution of the field equations of general relativity for an expanding universe. They argue that there might be large amounts of matter which do not emit light and have not been detected. This matter, now called 'dark matter', remains a mystery in that its nature is still unknown, although its existence is generally accepted and it is today the subject of major research efforts.

1932 US engineer Karl Jansky discovers that the interference in telephone communications is caused by radio emissions from the Milky Way. He thus begins the development of radio astronomy.

1932 US scientist Carl Anderson, while analysing cosmic rays, discovers positive electrons ('positrons'), the first form of antimatter to be discovered.

1933 English physicist Arthur Eddington publishes *The Expanding Universe*, in which he lays out his theory that the universe is constantly increasing in size.

1933 US astronomer Walter Baade identifies two different types of supernova, dubbed types I and II.

In type I supernovae there is a massive outburst of energy. In this type of supernova, the star can literally blow itself to bits, leaving nothing behind. In type II supernovae the star swells into a red supergiant, while the core yields to gravity and begins shrinking. As it shrinks, it grows progressively hotter and denser and eventually explodes. If the mass of the core remnant is less than 1.44 solar masses it becomes a white dwarf. If the core remnant measures between 1.44 and 3 solar masses, the stellar remnant will become a neutron star (rapidly rotating neutron stars are known as pulsars). If the core remnant of a supernova exceeds about 3 solar masses, it continues to contract and collapses to a black hole.

1937 Austrian astronomer Marietta Blau examines cosmic radiation using a photographic plate.

Biology

1930 English geneticist and statistician Ronald Fisher publishes *The Genetical Theory of Natural Selection*, in which he synthesizes Mendelian genetics and Darwinian evolution.

1930 Russian-born Swiss biochemist Paul Karrer formulates the structure of beta-carotene, the precursor to vitamin A (retinol).

Karrer shows that vitamin A is related to a family of plant pigments known as the 'carotenoids'. These substances give a distinctive orange colouring to plants, the best-known of which is 'carotene', the colouring agent in carrots. Karrer shows that the structure of vitamin A is similar to that of carotene. He proves his theory by synthesizing beta-carotene, which can then be converted into vitamin A. Karrer shares the Nobel Prize for Chemistry in 1937 for his study of carotenoids and vitamin A.

1930 US biochemist Edward Doisy crystallizes the hormone oestriol, the first oestrogen hormone to be crystallized.

In 1939 he isolates vitamin K, for which he wins the Nobel Prize for Physiology or Medicine in 1943.

247

1930 US biochemist John Northrop crystallizes pepsin and trypsin, demonstrating that they are proteins.

1930 US biochemist John Northrop develops a technique for crystallizing pepsin, a digestive enzyme present in stomach juices, and demonstrates it to be a protein.

Northrop later isolates a bacterial virus using the same technique, which he shows to be a nucleoprotein. He shares the Nobel Prize for Chemistry in 1946 for this work.

1932 British zoologist and mathematician J B S Haldane publishes *The Causes of Evolution*, in which he revitalizes interest in studying evolutionary causes, and refines the mathematical basis of population genetics and the genetical basis of natural selection.

Haldane's work is one of three simultaneous projects by mathematical population geneticists that provide a rigorous formal basis for studies of evolution at the population level and give a boost to biologists interested in evolutionary mechanisms, especially natural selection. Other projects include Ronald Fisher's *The Genetical Theory of Natural Selection* (1930) and Sewall Wright's *Evolution in Mendelian Populations* (1931).

1933 Canadian biologist Ludwig von Bertalanffy writes *Theoretical Biology* in which he attempts to develop a common methodological approach to all sciences based on the tenets of organismic biology.

1933 German electrical engineer Ernst Ruska makes an electron microscope capable of a magnification of 12,000 times.

The instrument follows on from an earlier design of 1928, and his later improvements are capable of magnification up to 1 million times. The electron microscope revolutionizes the study of cell structure. In 1937 Ruska joins Siemens-Reiniger-Werke AG as a research engineer and in 1939 the company produces the first commercial electron microscope. Ruska's later improvements produce magnifications up to 1 million times. Ruska shares the Nobel Prize for Physics with Gerd Binnig and Heinrich Rohrer in 1986.

The invention of the electron microscope in 1933 brought new insights into the world of the very small that lay beyond the resolving power of the light microscope, revealing never-before-seen details of cell structure and showing viruses for the first time. *Corbis*

1935 Hungarian-born US biochemist Albert Szent-Györgi works on the biochemistry of muscle. He isolates the muscle proteins actin and myosin and shows that they combine to form actomyosin. When ATP is added to fibres of actomyosin, they contract.

1935 US biochemist Edward Kendall isolates the steroid hormone cortisone from the adrenal cortex.

Cortisone and other related corticosteroids are used as anti-inflammatories and to treat allergic and rheumatic diseases. Kendall shares the Nobel Prize for Medicine or Physiology in 1950 with Philip Hench and Tadeus Reichenstein.

1935 US biochemist Wendell Stanley crystallizes tobacco mosaic virus, thereby showing that viruses are not submicroscopic organisms but are proteinaceous in nature.

Stanley shares the Nobel Prize for Chemistry in 1946 with James Sumner and John Northrop.

1937 Austrian zoologist Konrad Lorenz coins the term 'imprinting' to describe the process by which visual and auditory stimuli from animals around them cause young ducklings to identify with their own species. Lorenz suggests this is evolution's mechanism for locating the biologically 'right' object species for their upbringing.

This discovery becomes the basis of the discipline of ethology.

December 1938 A coelacanth, an ancient fish assumed to be extinct, is captured live off the southern coast of Africa in the Indian Ocean. The discovery is publicized by South African zoologists Marjorie Courtney Latimer and J L B Smith.

Though considerable effort is expended to locate other specimens, it is not until 1952 that another living coelacanth is caught, far to the north in the Indian Ocean, off the Comoros Islands.

1939 English plant biologist Robin Hill isolates chloroplasts and shows that, when illuminated, they produce oxygen.

The oxygen comes from the photolysis of water. The reaction is light-dependent and is the first stage of photosynthesis. It is called the Hill reaction.

Chemistry

1930 Swedish biochemist Arne Tiselius invents electrophoresis. This process, also known as 'cataphoresis', is defined as the movement of electrically charged particles through a liquid under the influence of an applied electric field.

Tiselius develops the technique to separate proteins in suspension, based on their electrical charge. His invention of the technique and his subsequent studies earn Tiselius the Nobel Prize for Chemistry in 1948.

1930 US chemist Thomas Midgley Jr discovers the chlorofluorocarbon (CFC) Freon-12.

The gas replaces ammonia and sulphur dioxide as a refrigerant in household refrigerators and air conditioning systems and is eventually used as an aerosol propellant. Unfortunately, Freon-12 is later found to be an efficient ozone depleter and is banned in the USA and Europe to limit damage to the Earth's ozone layer.

1930 German biochemist Hans Fischer receives the Nobel Prize for Chemistry for his analysis and synthesis of haemin (the iron-bearing group in haemoglobin) and for his chlorophyll research.

Later, he discovers the structure of the bile pigment bilirubin, which is related to haemin. Bilirubin is synthesized in 1944.

1931 German biochemist Adolf Butenandt isolates androsterone, a male sex hormone, from male human urine.

His research does much to develop understanding of the steroid group of compounds.

1931 US chemists of the DuPont company led by Wallace Hume Carothers develop the first commercially successful synthetic rubber, neoprene.

They base their process on work carried out by Belgian-born US chemist Julius Nieuwland in 1929. Polymerization is a term used to describe the chemical joining of

249

identical small molecules to form a large chain molecule, known as a polymer. The DuPont process involves the polymerization of chlorobutadiene to produce a synthetic material that is more resistant than natural rubber to organic solvents such as petrol. It is marketed as 'Duprene', but is renamed 'neoprene' in 1937.

1931 German chemists Carl Bosch and Friedrich Bergius share the Nobel Prize for Chemistry for their development of chemical high-pressure methods in 1913.

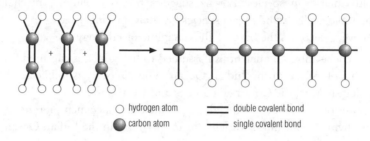

○ hydrogen atom   ══ double covalent bond

● carbon atom   ── single covalent bond

In polymerization, small molecules (monomers) join together to make large molecules (polymers). In the polymerization of ethene to polyethene, electrons are transferred from the carbon–carbon double bond of the ethene molecule, allowing the molecules to join together as a long chain of carbon–carbon single bonds.

1932 German chemist Gerhard Domagk discovers that the red azo dye Prontosil can control streptococcal infections in mice. This is the first antibacterial sulphonamide drug ('sulfa drug').

In 1935, Domagk uses the dye to cure blood poisoning in his youngest daughter after all other treatments have failed. This is the first use of a synthetic antibacterial drug on a human. The active agent in the dye is found to be 'sulfanilimide'. It forms the basis of a series of similar compounds.

1932 German-born British biochemist Hans Adolf Krebs discovers the urea cycle, in which ammonia is turned into urea in mammals.

Krebs shows that amino acids degrade in mammals to form the soluble compound urea, which is then excreted from the system. This is an important step in the understanding of metabolism.

1932 US chemist Irving Langmuir receives the Nobel Prize for Chemistry for his work on surface chemistry.

1933 Austrian-born German organic chemist Richard Kuhn isolates vitamin B_2 (riboflavin).

Kuhn announces the determination of the molecular structure of vitamin B_2 at the same time as Swiss chemist Paul Karrer, but is first to isolate the vitamin. He isolates a yellow, water-soluble compound that he calls 'riboflavin' after the Latin word meaning 'yellow'. He receives the Nobel Prize for Chemistry in 1938 for his study of the carotenoids and vitamins. Karrer synthesizes vitamin B_2 in 1935.

1933 English organic chemist Walter Haworth and Polish-born Swiss biochemist Tadeus Reichstein independently synthesize vitamin C (ascorbic acid).

The ability to synthesize vitamin C allows it to be mass-produced and it soon becomes a common food additive and dietary supplement. Haworth is awarded a share in the Nobel Prize for Chemistry in 1937 for his study of carbohydrates and vitamin C.

1933 US biochemist Charles Glen King determines the structure of vitamin C (ascorbic acid).

He isolates the vitamin in crystalline form from cabbages and determines its molecular structure. He shows that the vitamin C molecule is made up of six carbon atoms and is similar in structure to sugars. King suggests that the isolated vitamin be called 'ascorbic acid' from the Greek phrase meaning 'no scurvy'. Szent-Györgyi, who in 1928 had isolated vitamin C but had not identified it, publishes his work within weeks of King, causing a fierce debate over who will get credit for the discovery.

1934 US chemist Wallace Carothers develops the polyamide-based synthetic fibre nylon, the first synthetic polymer fibre to be commercially produced (1938).

Carothers succeeds in polymerizing the dicarboxylic acid, adipic acid, with the diamine compound hexamethylene diamine. Carothers finds that the resultant synthetic fibre is stronger than silk. The fibre is given the trade name 'Nylon' and is commercially produced by US company DuPont in 1938.

1934 US chemist Harold Urey receives the Nobel Prize for Chemistry for discovering deuterium (heavy hydrogen) in 1931.

1935 British chemist Michael Perrin and his group working for Imperial Chemical Industries (ICI) polymerize ethylene to make polyethylene, the first true plastic.

The first patent for the process is issued in 1936 under the brand name 'Alkthene' and commercial production starts in 1939. This form of the plastic is low-density polyethylene (LDPE) and is used as an insulating material.

1935 French physicists Frédéric and Irène Joliot-Curie share the Nobel Prize for Chemistry for their synthesis of new radioactive elements in 1934.

1936 US chemist Robert Runnels Williams synthesizes vitamin B_1 (thiamine).

In 1933, Williams had isolated the crystalline form of vitamin B_1 and called it 'thiamine' after the Greek word for sulphur. He succeeded two years later in developing a process to synthesize the vitamin, which goes into commercial production in 1936. Williams is instrumental in the decision in the USA to use thiamine as a food additive to flour, cereal grains, and cornmeal. This effectively eradicates the deficiency disease beriberi in the USA.

1936 Dutch chemist Peter Debye receives the Nobel Prize for Chemistry for his work on 'polar' molecules of 1912, and on X-ray diffraction of gases.

1937 Italian-born US physicist Emilio Segrè and Italian mineralogist Carlo Perrier identify technetium (atomic number 43).

US physicist Ernest Lawrence creates an unknown radioactive substance by bombarding molybdenum (atomic number 42) with the atomic nuclei of deuterium atoms (atomic number 1) using a cyclotron. He sends the material to Perrier who discover that the cyclotron bombardment has transmuted molybdenum into element 43. They name the element 'technetium' after the Greek word meaning 'artificial'. It is the first element to be created in a laboratory.

1937 German-born British biochemist Hans Krebs describes the citric acid cycle in cells, which converts sugars, fats, and proteins into carbon dioxide, water, and energy – the 'Krebs cycle'.

Krebs studies the mechanisms involved in the metabolism of carbohydrates. He discovers a cyclic metabolic process that begins and ends with citric acid, and shows that sugar molecules enter the system to produce carbon dioxide and hydrogen atoms. Hydrogen goes through a series of reactions with enzymes before eventually combining

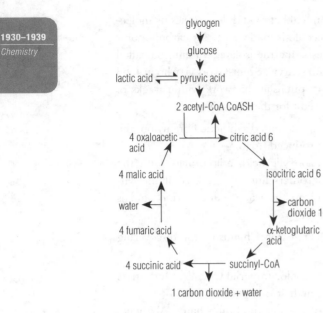

glycogen

glucose

lactic acid ⇌ pyruvic acid

2 acetyl-CoA CoASH

4 oxaloacetic acid → citric acid 6

isocitric acid 6

4 malic acid

carbon dioxide 1

water

α-ketoglutaric acid

4 fumaric acid

4 succinic acid ← succinyl-CoA

1 carbon dioxide + water

The purpose of the Krebs (or tricarboxylic acid) cycle is to complete the biochemical breakdown of food to produce energy-rich molecules, which the organism can use to fuel work. Acetyl coenzyme A (acetyl CoA) – produced by the breakdown of sugars, fatty acids, and some amino acids – reacts with oxaloacetic acid to produce citric acid, which is then converted in a series of enzyme-catalysed steps back to oxaloacetic acid. In the process, molecules of carbon dioxide and water are given off, and the precursors of the energy-rich molecules ATP are formed. (The numbers in the diagram indicate the number of carbon atoms in the principal compounds.)

with oxygen to form water. This process is shown to provide the body with a source of usable energy. The process becomes known as the 'citric acid cycle' as well as the 'Krebs cycle'.

1937 Russian-born Swiss biochemist Paul Karrer and English biochemist Walter Haworth share the Nobel Prize for Chemistry, Karrer for his study of carotenoids and vitamin A, Howarth for his work on carbohydrates and vitamin C.

1938 German chemist Richard Kuhn is awarded the Nobel Prize for Chemistry for his work on vitamins. He declines.

6 April 1938 US chemist Roy Plunkett discovers the stable and slippery substance polytetrafluoroethylene (PTFE) (a synthetic resin), marketed by DuPont as Teflon. The most slippery substance known, it becomes commercially available in 1947–48 and is used for electrical insulation and to produce nonstick coatings.

Plunkett accidentally discovers a white waxy residue while searching for a new refrigerant based on the gas tetrafluoroethylene. He discovers that the conditions used in his test cause the gas to polymerize to form PTFE. This substance is resistant to solvents and heat and it is found that fats will not stick to a surface coated in the material.

1939 A team of US chemical engineers led by Warren K Lewis and Edward R Gilliand at the Massachusetts Institute of Technology develop fluidized bed catalytic cracking of petroleum. It proves to be the most efficient method of cracking ever to be developed and is still the primary method for refining petroleum in use today.

This method is similar to the Houdry process developed in 1929, with the exception that oil is introduced through a bed of catalyst instead of over it. The fluidized bed ensures that petroleum is in contact with the catalyst for the maximum amount of time.

1939 British pathologist Howard Florey and German-born chemist Ernst Chain succeed in isolating the antibiotic penicillin from the bread mould *Penicillium notatum*, publishing their results the following year.

The process is delicate and time-consuming and it is not until 1941 that they obtain enough penicillin to perform the first clinical trials on rats. It is a complete success. Florey and Chain do not have the resources in Britain to mass-produce the drug and so take their research to the USA. In 1943, penicillin is purified and produced in commercial quantities. Large-scale clinical trials on human patients are a complete success. Penicillin is still in common use as an antibiotic drug.

1939 French chemist Marguérite Perey discovers the radioactive element francium (atomic number 87).

The element exists naturally as a decay product of actinium, which is itself a decay product of uranium. Perey is studying the radioactive properties of actinium, when she detects an unusual radiation coming from her sample. She deduces that this radiation is being emitted by a decay product of actinium and eventually proves this to be element 87. She names the element 'francium' after the country of her birth.

1939 Swiss chemist Paul Müller synthesizes dichlorodiphenyltrichloroethane (DDT) and discovers its insecticidal properties.

DDT had been synthesized in the 19th century; however, Müller discovers that it is an effective agent against insects. It is cheap and easy to manufacture and soon becomes the most widely used insecticide in the world. DDT is used in a World Health Organization programme to eradicate the mosquito in an attempt to rid the world of the disease malaria. Unfortunately, the durability of the chemical leads to a progressive contamination of the environment and the food chain, with corresponding detrimental effects to human health. DDT is banned in the USA in 1972 and in Britain in 1984.

1939 US biochemist Edward Albert Doisy determines the structure of vitamin K (phytomenadione) and synthesizes it.

The lack of vitamin K lowers the ability of blood to clot efficiently and leads to spontaneous haemorrhaging. This was discovered in 1929 by Danish biochemist Carl Dam, who named the vitamin K after 'koagulation', the German 'coagulation'. Doisy determines the molecular structure of the vitamin and then successfully develops a process to synthesize it. The two researchers share the Nobel Prize for Physiology or Medicine for their study of vitamin K.

1939 German biochemist Adolf Butenandt shares the Nobel Prize for Chemistry with Swiss chemist Leopold Ruzicka, Butenandt for his work on sexual hormones and Ruzicka for his work on polymethelenes. Butenandt declines the prize.

1939 US chemist Linus Pauling consolidates his theory of the chemical bond in *The Nature of the Chemical Bond and the Structure of Molecules and Crystals*. Pauling applies quantum mechanics to the study of chemical bonds.

Pauling applies experimental techniques ranging from X-ray diffraction to magnetic effects and heat measurements to provide data for the accurate calculation of angles between bonds and the interatomic distance of their association. He also proposes the concept of hybrid orbitals and explains the properties of covalent bonds. He is awarded the Nobel Prize for Chemistry in 1954 for his study of chemical bonding.

253

US chemist Linus Pauling holds up a sign in 1962 as he pickets the White House during a mass protest against resumption of US atmospheric nuclear testing. Pauling carries out his Nobel prizewinning work on the nature of chemical bonds, in the 1930s.
Bettmann/CORBIS

Earth Sciences

c. 1930 The radio meteorograph (radiosonde) is developed by the US Weather Service; it transmits information on temperature, humidity, and barometric pressure from uncrewed balloons. A network of stations is also inaugurated. The radiosonde allows scientists to collect data over a wide area and to better understand the physics of the atmosphere.

1930 The Woods Hole Oceanographic Institution is established in Massachusetts, USA.

1931 In a submarine gravity survey over the Caribbean trench, Dutch geophysicist Feliz Vening-Meinesz and his crew discover that the gravitational pull is much weaker than expected. This suggests that the trench is not a simple valley, and perhaps that the entire crust is down-warped in the area.

1933 US geologist Walter Bucher explains deep-sea trenches as cracks formed as the Earth contracts.

1935 US seismologist Charles Richter introduces the Richter scale for measuring the magnitude of earthquakes at their epicentre.

Unlike the Mercalli scale, the Richter scale is a quantitative scale that describes the energy released by an earthquake rather than the effects of earthquakes on human beings and human-made structures. This allows earthquakes around the world, regardless of location with respect to civilization, to be analysed.

1936 Danish seismologist Inge Lehmann postulates the existence of a solid inner core of the Earth from the study of seismic waves.

Lehmann notices that some seismic waves are arriving in places and at times that are unexpected given the known structure of the Earth. She shows that the anomalies could be explained by the existence of a solid (or very dense) inner core.

1937 English physicist William Bragg's book *Atomic Structure of Minerals* is published.

1937 Finnish chemist and geologist Victor Goldschmidt tabulates the absolute abundances of chemical elements in Earth from solar and meteorite chemical data.

1937 South African geologist Alexander du Toit publishes *Our Wandering Continents*, advocating German meteorologist Alfred Wegener's theory of continental drift.

1938 German seismologist Beno Gutenberg and US seismologist Charles Richter report the deepest earthquake shock on record; it occurred in 1934 at a depth of 720 km/450 mi beneath the floor of the Flores Sea, southern Indonesia.

1939 US geophysicist Walter Maurice Elsasser formulates the 'dynamo model' of the Earth, which proposes that eddy currents in the Earth's molten iron core cause its magnetism.

Ecology

1934 Soviet biologist Georgii F Gause suggests that an evolutionary niche is occupied by only one species' population at any one time.

This is called the Gause hypothesis and shows that the separation of niches reduces competition between different species. It builds on previous work by US zoologist Joseph Grinnell.

May 1934 As the drought in the US dust bowl enters its second year, about 270 billion kg/300 million tons of topsoil from 40 million hectares/100 million acres in Kansas, Texas, Colorado, and Oklahoma is blown into the Atlantic, causing large-scale human migration to California and other states.

1935 A 'Green Belt' scheme is put into operation around London, England, to prevent excessive development.

1935 English botanist Arthur Tansley publishes his essay 'The Use and Abuse of Vegetational Concepts and Terms'. Tansley rejects the notion of plant communities as quasi-organisms not reducible to smaller parts, and advocates a more reductionist approach.

Tansley advocates the avoidance of the word 'community' because of its anthropomorphic overtones. He suggests that ecology should be purged of all ideas not subject to quantification and analysis; he wishes to distance the science from the vague and moralizing ethos that it has attracted since the Romantic period. Instead of the notion of a community, Tansley suggests that the concept of an ecosystem should be used to investigate the mechanisms by which plants and animals exchange matter and energy with each other. Tansley places particular value on energy flows and energy accounting in ecosystems.

1936 For the third time in six years, there is a drought in the US wheat belt and the dust bowl phenomenon worsens. Thirteen states are affected. The Resettlement Administration and the Works Progress Administration, set up in the depression years, work hard to cope with the number of farmers driven off their smallholdings by drought and dust storms.

Wind blows topsoil away with loss of agricultural land. In Nebraska, 9 million acres of corn are believed lost. It is estimated that cultivable land the size of Germany has been lost since the first European settlers arrived in America. Forests were cleared and the grassland ploughed to sustain constant yields of wheat, which, it was argued, spoilt the structure of the soil and allowed it to be washed away by water and wind. The Roosevelt Administration instigates counter-measures such as the planting of a forest belt 160 km/100 mi wide and 1,600 km/1,000 mi long across the plains, and the construction of dams to contain water. In this year, the Great Plains Committee presents a report titled 'The Future of the Great Plains' to Roosevelt arguing that the dust storms are caused by human interference.

1936 US philosopher and historian of ideas Arthur Lovejoy publishes *The Great Chain of Being*.

In this work, Lovejoy charts the evolution of the concept of nature as an unbroken series of entities from the inorganic to the organic. Lovejoy shows how this notion of an ascending ladder was both a taxonomic system and an indication of ideas of ecological relations that influenced Western thought for centuries.

Mathematics

1930 US electrical engineer Vannevar Bush builds a differential analyser. An analogue computer, it is used to solve differential equations and is the forerunner of modern computers.

1931 Austrian mathematician Kurt Gödel publishes 'Gödel's proof' (*Über formal unentscheidbare Sätze der Principia Mathematica und verwandter Systeme/On Formally Undecidable Propositions of Principia Mathematica and Related Systems*). His proof refutes the possibility of establishing dependable axioms in mathematics, showing that any axiomatic theory strong enough to verify the consistency of the laws of arithmetic must have its own consistency in doubt. It also shows that there will be statements that cannot be proved true or false.

1933 Russian mathematician Andrei Nikolaevich Kolmogorov presents the first full formal axiomatic treatment of probability in *Foundations of the Theory of Probability*.

1935 US mathematician Alonzo Church invents lambda calculus, a mathematical method for representing mechanical computations.

This becomes particularly important to the development of computer programming languages.

1936 English mathematician Alan Turing supplies the theoretical basis for digital computers by describing a machine, now known as the Turing machine, capable of universal rather than special-purpose problem solving.

A Turing machine can be used to decide whether a given problem is capable of solution by a computer – a test of computability.

We do not need to have an infinity of different machines doing different jobs. A single one will suffice. The engineering problem of producing various machines for various jobs is replaced by the office work of 'programming' the universal machine to do these jobs.

Alan Mathison Turing, English mathematician, quoted in
A Hodges *Alan Turing: The Enigma of Intelligence* 1985

1937 US mathematician Georges Stibitz builds the first binary circuit that can add two binary numbers based on Boolean algebra. Consisting of batteries, lights, and wires, it is instrumental in the development of subsequent electromechanical computers.

1938 German inventor Konrad Zuse constructs the first binary calculator using a binary code (Boolean algebra); it is the first working computer.

Physics

1930 Indian physicist Chandrasekhara Raman receives the Nobel Prize for Physics for his work on the scattering of light and for the Raman effect, which concerns the loss or gain of energy by the scattered light.

1931 US chemist Harold C Urey and atomic physicist Edward Washburn discover that electrolysed water is denser than ordinary water, leading to the discovery of deuterium ('heavy hydrogen'). The discovery ushers in the modern field of stable isotope geochemistry.

Urey evaporates liquid hydrogen to concentrate any heavy isotope of hydrogen that may be present and examines the residue spectroscopically. He identifies a set of spectral lines that are slightly different from those produced by normal hydrogen. He names this new isotope 'deuterium' for the Greek word *deuteros*, which means 'second'. Deuterium has a nucleus composed of one proton and one neutron, making it double the mass of hydrogen. Urey is awarded the Nobel Prize for Chemistry in 1934 for this discovery.

1932 English physicist James Chadwick discovers the neutron, an important discovery in the development of nuclear reactors.

The discovery of the neutron helps physicists gain a deeper understanding of the atom.

1932 German physicist Werner Heisenberg receives the Nobel Prize for Physics for his formulation of the indeterminacy principle in quantum mechanics (1927).

1933 English physicist Paul Dirac and Austrian physicist Erwin Schrödinger share the Nobel Prize for Physics for their work on wave mechanics in quantum mechanics.

A theory with mathematical beauty is more likely to be correct than an ugly one that fits some experimental data. God is a mathematician of a very high order, and He used very advanced mathematics in constructing the universe.

Paul Dirac, British physicist, *Scientific American*, May 1963

1933 German physicists Walter Meissner and Robert Ochensfeld discover that superconducting materials expel their magnetic fields when cooled to superconducting temperatures – the Meissner effect.

1934 Italian physicist Enrico Fermi and his team in Rome systematically use neutrons to bombard nuclei of various chemical elements. They produce many new radioactive substances and also discover the 'moderator effect', in which collisions with hydrogen and other light nuclei slow down the neutrons and thus heighten their ability to induce nuclear reactions.

If I could remember the names of all these particles I'd be a botanist.

Enrico Fermi, Italian-born US physicist, quoted in R L Weber *More Random Walks in Science*

1934 Russian physicist Pavel Cherenkov discovers that light is emitted when particles pass through liquids or transparent solids faster than the speed of light in the same medium. The phenomenon becomes known as 'Cherenkov radiation'.

The characteristic 'blue glow' seen in cooling pools at nuclear power plants is Cherenkov radiation.

1935 English physicist James Chadwick receives the Nobel Prize for Physics for his discovery of the neutron (1932).

1935 Japanese physicist Hideki Yukawa proposes the existence of a new particle, an exchange particle, to explain the strong nuclear force.

These particles, now called π mesons or pions, are not to be detected for another 12 years.

1936 Austrian physicist Victor Hess shares the Nobel Prize for Physics with US physicist Carl Anderson; Hess for discovering cosmic radiation and Anderson for discovering the positron.

1936 US physicists Carl D Anderson and Seth Neddermeyer discover the muon, an electron-like particle over 200 times more massive than an electron in cloud chamber experiments.

1937 US physicist Clinton Davisson shares the Nobel Prize for Physics with English physicist George Paget Thomson for their experiments on electron interference in crystals.

1938 Italian physicist Enrico Fermi receives the Nobel Prize for Physics for the artificial production of elements through neutron irradiation.

1938 German chemists Otto Hahn and Fritz Strassmann discover that uranium nuclei sometimes split when struck by neutrons. Austrian physicists Lise Meitner and Otto Robert Frisch explain the physical cause of the phenomenon, which is soon dubbed 'fission'. It is the key to nuclear reactors and weapons.

1939 French physicists Frédéric Joliot and Irène Curie-Joliot demonstrate the possibility of a chain reaction when they split uranium nuclei and show extra neutrons are released.

Uncontrolled chain reactions may lead to explosions in nuclear weapons, but when controlled can be used to generate electricity in nuclear power stations.

1939 US physicist Ernest Lawrence receives the Nobel Prize for Physics for inventing the cyclotron, a particle accelerator which he built in 1931 with his student M Stanley Livingston.

1940–1949

Astronomy

1940 Belgian astronomer Marcel Minnaert publishes *Photometric Atlas of the Solar Spectrum*, a standard reference text providing measurements of the absorption lines from 3,332 angstroms to 8,771 angstroms.

1942 Radar operators in the British army detect radio emissions from the Sun for the first time.

1942 US radio engineer Grote Reber makes the first radio maps of the sky, locating individual radio sources.

1944 Dutch astronomer Hendrik van de Hulst predicts that cosmic hydrogen will emit line radiation at 21 cm/8.3 in.

1945 Hungarian scientist Lajos Jánossy investigates cosmic radiation in Manchester and later in Dublin.

1945 Hungarian astronomer Zoltán Bay and the US Army Signal Corps Laboratory at Fort Monmouth, New Jersey, receive radar echoes from the Moon.

1946 Cygnus A, the first radio galaxy (a galaxy that is a strong source of electromagnetic waves of radio wavelength), and the most powerful cosmic source of radio waves, is discovered by the English physicist James Hey.

1946 English physicists Edward Appleton and Donald Hay discover that sunspots emit radio waves.

1948 US astronomer Gerard Kuiper discovers and photographs Miranda, the fifth moon of Uranus.

1949 US astronomer Walter Baade discovers the close approach asteroid Icarus; except for comets, it has the most eccentric orbit of any body in the Solar System, and passes closest to the Sun (28 million km/18 million mi).

Biology

1940 Austrian-born US immunologist Karl Landsteiner and US physician and immuno-haematologist Alexander Wiener discover the rhesus (Rh) factor in blood.

1940 US geneticist George Beadle and US microbiologist Edward Tatum establish the one-gene–one-enzyme hypothesis, showing that one gene is responsible for the production of one specific enzyme.

1943 Russian-born US biologist Selman Waksman discovers the antibiotic streptomycin, which is used as a treatment for tuberculosis; he coins the term 'antibiotic' to describe the range of antibacterial drugs developed since the discovery of penicillin.

Waksman performs a study of the antibacterial properties of microscopic fungi found in soils. He finds a member of the *Streptomycetes* class of fungi that shows promising antibiotic properties. He calls his drug 'streptomycin' and it proves to be the first effective treatment against tuberculosis. His discovery of streptomycin earns Waksman the Nobel Prize for Physiology or Medicine in 1952.

1944 The role of deoxyribonucleic acid (DNA) in genetic inheritance is first demonstrated by US bacteriologist Oswald Theodore Avery, US biologist Colin M MacLeod, and US biologist Maclyn McCarthy.

They show that *Pneumococcus* bacteria that lack the ability to produce capsules can be given this ability by the transference of DNA obtained from capsule-producing bacteria. They correctly deduce that the genetic material for the bacteria is contained in the DNA molecule. This discovery opens the door to the elucidation of the genetic code and marks the beginning of the science of molecular genetics.

1946 US biochemist Melvin Calvin begins his work focusing on the sequence of reactions in the stroma of chloroplasts that results in the biosynthesis of carbohydrate.

US biochemist Melvin Calvin, holding a sucrose molecule in 1961, the year in which he was awarded the Nobel prize for his work on photosynthesis, began in 1946.
Bettmann/CORBIS

The research is able to progress because the sensitive analytical methods of chromatography and radioisotope labelling are now available. The pathway of reactions is called the Calvin cycle. The reactions are the light-independent components of photosynthesis, in contrast to the light-dependent reactions discovered earlier by Robin Hill (1939). Calvin is awarded the 1961 Nobel Prize for Chemistry.

1946 US geneticists Joshua Lederberg and Edward Lawrie Tatum pioneer the field of bacterial genetics with their discovery that sexual reproduction occurs in the bacterium *Escherichia coli*.

Lederberg finds that sexual reproduction (conjugation) occurs in a number of species of bacterium, and that it can be used to map bacterial genes.

1947 English ornithologist David Lack publishes *Darwin's Finches: An Essay on the General Biological Theory of Evolution*.

The work provides detailed case studies of the polytypic species concept, that species should be understood to be reproductive communities, and may have substantial regional variation. His research demonstrates Darwinian principles of adaptation by natural selection and competitive exclusion.

1947 Dutch ethologist Nikolaas Tinbergen moves to Oxford, England, where he develops his work on animal behaviour.

Among various studies which seek to establish patterns of animal behaviour that, are those showing that the mating of sticklebacks follows a stereotypical pattern. The key components of the behaviour follow a fixed sequence which determines the outcome of territorial encounters between males. Tinbergen shares the Nobel Prize for Physiology in 1973. Recent work seems to challenge Tinbergen's account.

1948 Soviet biologist Trofim D Lysenko outlaws Mendelian orthodox genetics in favour of 'Michurin' genetics in the USSR. Purges of geneticists (and assertions that, for example, wheat plants can be easily and economically treated to become more productive) obstruct agricultural development.

1948 Scottish anthropologist Arthur Keith publishes *A New Theory of Human Evolution*, in which he emphasizes competition as a major factor in human evolution and argues that racial prejudice is inborn.

Both cooperation and competition are deeply rooted in the evolutionary history of all social species.

1948 US biologist Alfred Mirsky discovers ribonucleic acid (RNA) in chromosomes.

Chemistry

1940 Italian-born US physicist Emilio Segrè and US physicists Dale Corson and K R Mackenzie synthesize astatine (atomic number 85).

Segrè, Corson, and Mackenzie bombard bismuth with alpha particles using the cyclotron at Berkeley, California, in a deliberate attempt to create element 85. They achieve their aim, but find that the new element is highly unstable and only has a half-life of a few hours. Accordingly they name their discovery astatine after the Greek word *astatos*, meaning 'unstable'.

1940 Canadian biochemist Martin Kamen discovers carbon-14. It becomes a vital tool in dating archaeological and geological samples.

Carbon-14 is a naturally occurring radioactive isotope of carbon. It is continually being formed in the Earth's atmosphere by the bombardment of nitrogen-14 with neutrons produced by cosmic rays. It is highly stable and enters the biological carbon cycle by absorption of air by green plants. Kamen discovers the radioisotope while searching for radioactive isotopes of light elements. Carbon-14 has a radioactive half-life of over 5,000 years and decays to produce nitrogen.

1940 US physicist Edwin McMillan and US physical chemist Philip Abelson synthesize neptunium (atomic number 93), the first element found with an atomic number higher than that of uranium.

They use the cyclotron at Berkeley, California, to bombard uranium with neutrons and detect a source of radiation in the sample that has a half-life of only a few days. They discover that some of the uranium has been transmuted to create element 93. They name the element neptunium after the planet Neptune. Neptunium is placed higher that uranium in the periodic table and so becomes the first 'transuranic' element. McMillan shares the Nobel Prize for Chemistry in 1951 for this discovery.

1940 US nuclear chemist Glenn Seaborg and US physicists Edwin McMillan, Joseph W Kennedy, and Arthur C Wahl discover plutonium (atomic number 94).

They bombard uranium with deuteron particles using the cyclotron at Berkeley, California, and discover a new element in the radioactive decay products. They name the element 'plutonium' after the planet Pluto. Plutonium is the second of the transuranic elements. Seaborg shares the Nobel Prize for Chemistry in 1951 for this discovery. The element eventually becomes used in the manufacture of nuclear weapons and as a fuel in nuclear power reactors.

1941 US chemist Richard O Roblin discovers sulphadiazine; it becomes the most widely used sulfa drug.

Sulphadiazine is a synthetic compound and a member of the sulphonamide class of antibacterial drugs. It is effective against a wide spectrum of bacterial infections and has very few side effects. Sulphadiazine becomes the most widely used antibacterial drug until the development of penicillin. The drug is still used to treat infections in patients who are allergic to penicillin-based antibiotics.

1943 Hungarian-born Swedish chemist Georg von Hevesy receives the Nobel Prize for Chemistry for the use of isotopes as tracers.

1944 English biochemists Archer Martin and Richard Synge invent paper chromatography.

In 1942 they had improved the technique of partition chromatography by replacing starch with silica gel in the separating column (see 1906). They continue to develop the technique and use a column of absorbent filter paper instead of silica gel. However, they introduce the liquid from the base of the column, forcing it to creep up through the separation medium. Unlike previous methods, the paper column can be dipped into a series of solvents, which greatly increases the efficiency of the separation process. This work earns them the Nobel Prize for Chemistry in 1952.

1944 German chemist Otto Hahn receives the Nobel Prize for Chemistry for his discovery of nuclear fission.

1944 US chemists Robert Burns Woodward and William Doering synthesize quinine.

Their process builds up the complex molecule from simple compounds using a series of chemical reactions. This is an example of a totally artificial chemical synthesis, as it does not use any compounds that have been produced by a living organism. Woodward is awarded the Nobel Prize for Chemistry in 1965 for outstanding achievements in organic synthesis.

1944 US nuclear chemist Glenn Seaborg and his associates discover americium (atomic number 95) and curium (atomic number 96).

They bombard plutonium with neutron and alpha particles using the cyclotron at Berkeley, California, to transmute it into the two new radioactive elements. They name them 'americium' after the USA and 'curium' after the founders of nuclear chemistry, Pierre and Marie Curie (see 1898).

1945 Finnish chemist Artturi Virtanen receives the Nobel Prize for Chemistry for his invention of a method of preserving fodder.

1945 German-born US biochemist Fritz Lipmann discovers coenzyme A.

Lipmann explains the role of adenosine triphosphate (ATP) as the carrier of chemical energy from the oxidation of food to the energy consumption processes in the cells and shows that coenzyme A is an important catalyst in this process. He isolates coenzyme A in 1947 and determines its molecular structure in 1953. He shares the Nobel Prize for Physiology or Medicine in 1953 for this study.

1945 US chemists Charles DuBois Coryell, Jacob Marinsky, and Lawrence Glendenin, discover the radioactive element promethium (atomic number 61).

They discover the element among the waste deposits from a uranium fission nuclear reactor at the research site at Oak Ridge, Tennessee. In recognition that the element was recovered from a nuclear fire, they name it 'promethium' after Prometheus, the mythological Greek hero who stole fire from the gods.

1946 US physicists Edward Mills Purcell and Felix-Bloch independently discover nuclear magnetic resonance, which is used to study the structure of pure metals and composites.

1946 US biochemists James Sumner, John Northrop, and Wendell Stanley share the Nobel Prize for Chemistry, Sumner for his discover of the crystallization of enzymes in 1926, and Northrop and Stanley for the crystallization of pure enzymes and virus proteins in 1930 and 1935 respectively.

1947 US chemist and physicist Willard Libby develops carbon-14 dating.

Libby deduces that carbon-14 will be incorporated into plants during photosynthesis by the absorption of carbon dioxide. A living plant will continue to replenish any carbon-14 lost by radioactive decay, but this will stop when it dies. The amount of radioisotope present at that point will then diminish at a constant rate proportional to the half-life of carbon-14, which allows the determination of age by the measurement of how much remains. The age of any plant product, such as cloth, paper, and wood, can be determined in this way. Libby is awarded the Nobel Prize for Chemistry in 1960 for this work.

1947 Scottish chemist Robert Robinson receives the Nobel Prize for Chemistry for his work on alkaloids.

1948 Swedish chemist Arne Tiselius receives the Nobel Prize for Chemistry for his work on electrophoresis and serum proteins.

1949 US chemist Glenn Seaborg and his team discover berkelium (atomic number 97).

They bombard americium with helium particles using the cyclotron at Berkeley, California, to transmute it into the new radioactive element. Berkelium is highly unstable and has a half-life of less than six hours. They name it after Berkeley, the home of the University of California.

1949 US chemist William Giauque receives the Nobel Prize for Chemistry for his work on the behaviour of substances at extremely low temperatures.

Earth Sciences

1940 English astronomer and physicist Harold Jeffreys and New Zealand seismologist Keith E Bullen publish *Seismological Tables*, in which they estimate travel times of seismic body waves through the Earth. This allows geophysicists to calculate earthquake epicentres using arrival times of seismic waves at a number of observing stations.

1943–1945 The USS *Cape Johnson*, led by US geophysicist Harry Hess, surveys the Pacific ocean floor with echosounders. Among other features, they discover flat-topped seamounts which Hess calls 'guyots' after the founder of Princeton University's department of geology.

1945 Single-stage sounding rockets, reaching speeds of 4,800–8,000 kph/3,000–5,000 mph, and a maximum altitude of 160 km/100 mi, are launched carrying instrumentation to gather information about the upper atmosphere.

1946 US geochemist Alfred Nier improves the mass spectrometer. He builds many machines so that geochemists are able to perform radiometric age dating.

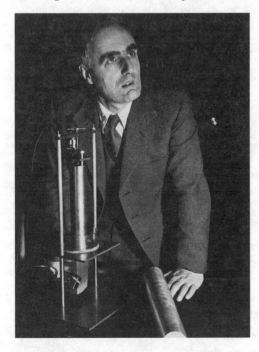

English geophysicist Edward C Bullard, director of the National Physical Laboratory. In 1941, at the age of 34, he was one of the youngest Fellows ever elected to the Royal Society. *Hulton-Deutsch/CORBIS*

1948 US chemist Harold C Urey suggests the use of stable isotopes as 'geothermometers' and chemical tracers in the earth sciences.

1948 US chemists Lyman Aldrich and Alfred Nier find argon from decay of potassium in four geologically old minerals, confirming predictions by German physicist Carl von Weizsacker made in 1937. This provides a basis for potassium–argon dating.

1949 English geophysicist Edward C Bullard and colleagues design a probe for measuring Earth's heat flow through the ocean floor (published in 1952).

Bullard and his colleagues find that heat flow beneath the oceans is about the same as that below the continents. It is surprisingly high, considering the lack of radioactive minerals in the ocean crust. The observation will later be interpreted as evidence for convection in the mantle (bringing hot material up from depth).

263

1949 Japanese metamorphic petrologist Akiho Miyashiro estimates the pressure and temperatures for the aluminosilicate polymorphs (sillimanite, kyanite, and andalusite).

The estimate allows geologists to estimate the pressure and temperature conditions (and thus the depth and tectonic environment) of formation of rocks that contain these minerals.

1949 US seismologist Hugo Benioff identifies planes of earthquake foci, extending from deep ocean trenches to beneath adjacent continents or ocean crust at approximately 45°, as fault zones.

Although not accepted today, Benioff explains the zones as a result of instability along the boundaries of continents and oceans, which causes surface flow towards the ocean basin. The observation will be key to the theory of plate tectonics. These fault zones, later called Benioff zones or Wadati–Benioff zones, will later be identified as the upper portion of the descending lithospheric slab.

Ecology

1940 The US Fish and Wildlife Service is established to conserve, protect, and enhance fish and wildlife and their habitats.

1942 US biologist Raymond Lindeman publishes 'The Trophic-Dynamic Aspect of Ecology'. The work fuses British and US approaches to ecological energetics and signals the consolidation of the 'New Ecology' in the USA.

Lindeman shows how an ecosystem can be split into a number of trophic levels, each one passing energy to the levels above. He uses the term 'autotrophs' for plants that generate their own food by photosynthesis and 'heterotrophs' for organisms such as animals that must feed on organic matter created by the autotrophs or other heterotrophs.

1944 Smog damages vegetation in Los Angeles basin, California. It continues to be a problem for the next 50 years.

Smog is thought to be a petrochemical fog caused by the action of sunlight on the oxides of nitrogen and on hydrocarbons emitted from car exhausts. Ironically, the cleaner air above cities enables sunlight to penetrate and act upon nitrogen dioxide to produce oxygen free radicals and thence ozone. Ozone is now the most economically damaging air pollutant in the USA.

1948 Sulphur from the Donora steel mill in Pennsylvania combines with moisture in the air to form a sulphuric acid fog that affects over 5,000 nearby residents and kills 22 people.

1949 A book of essays by the US forester and academic Aldo Leopold is published posthumously under the title *A Sand County Almanac and Sketches Here and There*. The most influential essay in the volume is 'The Land Ethic', in which Leopold expounds his philosophy that the good is that which promotes and preserves 'the integrity, stability, and beauty of the biotic community'.

Leopold advocates a position of ecocentrism stressing the intrinsic value of the natural world and not just its instrumental or utilitarian value for humans. In Leopold's view, land is more than just property, and has its own moral standing. Leopold stood at the transition from a utilitarian to an ecological approach to conservation. Leopold argues that traditional ethics are only concerned with the relationship between humans, and he calls for a new biocentric ethic that extends the bounds of the moral community to encompass other living things.

1949 An ecology group at the University of Chicago, headed by Warder Allee and including Alfred Emerson, Thomas Park, Orlando Park (from Northwestern University), and Karl Schmidt, publishes the massive and authoritative *Principles of Animal Ecology*.

The book firmly establishes the concept of the 'organismic community'. Allee and his colleagues are adamant that ecology can teach social and moral lessons about human society. The work stresses the interdependence of all creatures. In the other outpourings of the group at this time, it is stressed that selfish individualism should be transcended and that ecology provides the foundation for a holistic value system. The book becomes called the 'Great AEPPS Book' after the initials of the authors.

March 1949 The National Parks and Access to Countryside Act is passed in England and Wales, creating national parks and long-distance footpaths.

The Act grants official status to the Nature Conservancy (NC) established one year earlier and chaired by the English botanist Arthur Tansley. The initiative is in response to the work of the Wild Life Conservation Special Committee set up after World War II to identify sites of special interest for conservation and research. The NC is charged with the task of providing scientific advice on conservation of flora and fauna, the maintenance of nature reserves and the organization of scientific research. It has the powers of compulsory purchase of land and has the duty to notify to local authorities Sites of Special Scientific Interest (SSSI). The NC is the world's first statutory, non-voluntary conservation body.

Mathematics

1943 Colossus, the first electronic computer and code-breaker, is developed at Bletchley Park, England, to break German codes. Designed by Thomas Flowers, Max Newman, and English mathematician Alan Turing, it has 1,500 vacuum tubes and is the first all-electronic calculating device.

1944 The US firm IBM construct the second all-electronic computer.

Called the Mark I or Automatic Sequence Controlled Calculator, it uses valves (for calculations) and punched paper tape (for programming). Frequent failure of the valves causes major problems with the machine.

May 1944 Hungarian-born US mathematician John von Neumann and the German-born US economist Oscar Morgenstern publish *Theory of Games and Economic Behavior*. Games theory is important in the study of economics, biology, and sociology.

Von Neumann goes on to study hydrodynamics and becomes involved in the design of computers for solving the complex computations involved.

In mathematics you don't understand things. You just get used to them.

John von Neumann, Hungarian-born US mathematician, attributed remark

---

1946 ENIAC (Electronic Numerical Integrator, Analyser, and Calculator), the first general purpose, fully electronic digital computer, is completed at the University of Pennsylvania for use in military research. It uses 18,000 vacuum tubes instead of mechanical relays, and can make 4,500 calculations a second. It is 24 m/80 ft long and is built by electrical engineers John Presper Eckert and John Mauchly.

1948 US mathematician Claude Elwood Shannon invents information theory which has important applications in computer science and communications. He summarizes his theories in *A Mathematical Theory of Communication*, which he writes with Warren Weaver.

Physics

1940 US physicist John R Dunning leads a research team that uses a gaseous diffusion technique to isolate uranium-235 from uranium-238. Because uranium-235 readily undergoes fission into two nuclei, and in doing so releases large amounts of energy, it is used for fuelling nuclear reactors.

Large quantities of uranium-235 are needed for development of the atomic bomb.

1943 German-born US physicist Otto Stern receives a Nobel Prize for Physics for his development of the molecular ray method and his discovery of the magnetic moment of the proton.

1944 US physicist Isidor Rabi receives a Nobel Prize for Physics for his discovery of the resonance method for registering the magnetic properties of atomic nuclei.

1945 The first atomic bomb, a plutonium implosion device, is detonated in the 'Trinity test' in New Mexico, USA, on 16 July. It releases as much energy as the explosion of about 18,000 tons of TNT.

1945 Austrian physicist Wolfgang Pauli receives a Nobel Prize for Physics for the discovery of the exclusion principle, also called the Pauli principle.

I don't mind your thinking slowly: I mind your publishing faster than you think.

Wolfgang Pauli, Austrian-born Swiss physicist, attributed (from H Coblaus)

1945 US physicist Edwin M McMillan and the Soviet physicist Vladimir I Veksler (1943) independently describe the principle of phase stability and so make it possible to increase greatly the energy of particle accelerators.

1946 English physicist Edward Appleton discovers what is now known as the 'Appleton' layer in the ionosphere, one of the layers that reflect radio waves and make long-range radio communication possible. He is awarded the Nobel Prize for Physics in 1947 for his discovery.

1946 US physicist Percy Bridgman receives the Nobel Prize for Physics for his work on high pressure physics.

1948 US physicists John Bardeen, William Shockley, and Walter Brattain develop the transistor in research at Bell Telephone Laboratories, USA. A solid-state mechanism for generating, amplifying, and controlling electrical impulses, it revolutionizes the electronic industry by enabling the miniaturization of computers, radios, and televisions, as well as the development of guided missiles.

1948 English physicist Patrick Blackett receives the Nobel Prize for Physics for his work in nuclear physics and cosmic radiation.

1948 Russian-born US physicist George Gamow and US physicist Ralph Alpher develop the 'Big Bang' theory of the origin of the universe, which says that a primeval explosion led to the universe expanding rapidly from a highly compressed original state.

1949 Japanese physicist Yukawa Hideki receives the Nobel Prize for Physics for his prediction of the existence of mesons on the basis of theoretical work on nuclear forces.

1950–1959

Astronomy

1950 Dutch astronomer Jan Oort proposes that comets originate in a vast cloud of bodies (the 'Oort cloud') that orbits the Sun at a distance of about one light year.

The Oort cloud is difficult to observe but may account for a significant fraction of the mass of the Solar System, perhaps as much as or even more than Jupiter.

1951 US astronomer Gerard Kuiper proposes the existence of a ring of small, icy bodies orbiting the Sun beyond Pluto, thought to be the source of short-period comets. It is discovered in the 1990s and is named the Kuiper belt.

It is estimated that there are at least 35,000 Kuiper belt objects greater than 100 km/ 62 mi in diameter, which is several hundred times the number (and mass) of similarly sized objects in the main asteroid belt. Triton, Pluto, and its moon Charon, are considered by some to be the largest examples of Kuiper belt objects.

14 September 1951 The 'close-approach' asteroid Geographos is discovered by astronomers at the Mount Palomar Observatory in California.

20 September 1951 The US Air Force makes the first successful recovery of animals from a rocket flight when a monkey and 11 mice are recovered from a flight to an altitude of 72,000 m/236,000 ft.

21 May 1954 The Central Observatory of the Soviet Academy of Sciences, near Leningrad, USSR, opens.

1955 French astronomer Audouin Dolfus ascends 7.2 km/4.5 mi above the Earth in a balloon to make photoelectric observations of Mars.

5 May 1955 US astronomers Bernard Burke and Kenneth Franklin announce their discovery that Jupiter emits radio waves at a frequency of 22.2 MHz.

The emission occurs in episodes or 'storms' lasting from a few minutes to several hours. Two distinctive types of burst can be received: long bursts that last from a few seconds to several tens of seconds and have instantaneous bandwidth of a few MHz, and short bursts that arrive at a rate of a few to several hundred per second.

19 August 1957 US astronomers, using a 33 cm/12 in telescope on board the balloon-telescope *STRATOSCOPE I*, take the first clear photographs of the Sun from 24,384 m/80,000 ft.

4 October 1957 The USSR launches the first artificial satellite, *Sputnik 1*, to study the cosmosphere. It weighs 84 kg/184 lb and circles the Earth in 95 minutes, inaugurating the space age.

3 November 1957–13 April 1958 The Soviet spacecraft *Sputnik 2* is placed in orbit carrying a dog, Laika. It is the first vehicle to carry a living organism into orbit. Laika dies in space.

1958 Radio astronomers receive the reflection of radio waves from Venus. Similar echoes from other objects in the Solar System allow an accurate measurement of their distance.

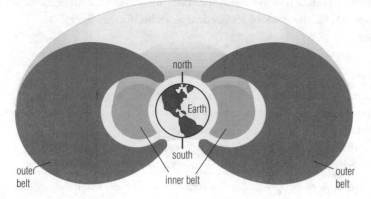

The Van Allen belts of trapped charged particles are a hazard to spacecraft, affecting on-board electronics and computer systems. They were discovered in 1958. Similar belts have been discovered around the planets Mercury, Jupiter, Saturn, Uranus, and Neptune.

31 January 1958 The US Army launches *Explorer 1* into Earth orbit. The first US satellite, it is used to study cosmic rays.

10 February 1958 The aurora borealis (northern lights) reach as far south as latitude 40° in the USA.

May 1958 Using data from the *Explorer* rockets, US physicist James Van Allen discovers a belt of charged particles from the Sun trapped by the Earth's magnetic field. It is further investigated in 1959 by the US space probe *Explorer 6*.

15 May 1958 The USSR places *Sputnik 3* in orbit. It contains the first multipurpose space laboratory and transmits data about cosmic rays, the composition of the Earth's atmosphere, and ion concentrations.

29 July 1958 The US National Aeronautics and Space Administration (NASA) is created for the research and development of vehicles and activities involved in space exploration.

11 October 1958 The USA launches the space probe *Pioneer 1* into orbit; this is the first spacecraft launched by NASA.

 Pioneers 0, 1, and *2* are the first US spacecraft to attempt to leave Earth orbit. Each of the three vehicles is designed to go into orbit around the Moon and photograph the Moon's surface. None of the vehicles accomplish their intended mission, although some useful data is returned.

1959 English astronomer Martin Ryle and colleagues publish the *Third Cambridge Catalogue*, a catalogue of radio sources that leads to the discovery of the first quasar.

1959 US astronomer Harold Babcock discovers that the Sun periodically reverses its magnetic polarity.

2 January 1959 The USSR launches *Luna 1*. The first spacecraft to escape Earth's gravity, it passes within 6,400 km/4,000 mi of the Moon.

April 1959 NASA selects a pool of nine military test pilots to compete to be the first US astronaut to orbit the Earth.

14 September 1959 The Soviet spacecraft *Luna 2* (launched on 12 September) becomes the first spacecraft to strike the Moon.

7 October 1959 The Soviet *Luna 3* (launched on 4 October) takes the first photographs of the far side of the Moon.

Biology

1951 Czech-born US biochemist Erwin Chargaff builds on work undertaken by Philip Levine in the 1920s. He notes that the quantities of purines within DNA samples are equal to quantities of pyrimidines: that is, the amount of the purine adenine is similar to that of the pyrimidine thymine, and the amount of the purine guanine is similar to that of the pyrimidine cytosine. This pattern holds true for all organisms.

In other words the number of the respective bases in DNA follows the rule that adenine = thymine and guanine = cytosine – the so called Chargaff's rule. The relationship is important to Crick and Watson as a clue to the structure of DNA. The two helical strands are held together by hydrogen bonds between the complimentary bases.

1951 The director of the UK's Natural History Museum, Gavin de Beer, recognizes the importance of Garstang's concept of paedomorphosis (neoteny) in his book *Embryos and Ancestors*.

De Beer analyses the respective contributions of paedomorphosis and the changes through time brought about by modification of adult structures on the evolutionary process. He suggests that the former is responsible for the larger steps in the course of evolutionary change and the latter for the smaller ones. In so doing, de Beer raises the issue of the mechanisms of macro- and micro-evolution.

1951 US biochemist Robert Woodward synthesizes cortisone and cholesterol. He is awarded the Nobel Prize for Chemistry in 1965.

1952 English biophysicist Rosalind Franklin uses X-ray diffraction to study the structure of DNA. She suggests that its sugar-phosphate backbone is on the outside – an important clue that leads to the elucidation of the structure of DNA the following year.

1952 English physiologists Andrew Huxley and Alan Hodgkin publish the results of their work on the nerve impulse. They show that when a nerve impulse passes along an axon, the action potential reverses the polarity of the interior of the axon with respect to the surface of the surrounding membrane.

The nerve impulse is described as a wave of depolarization related to the change in distribution of sodium ions and potassium ions across the membrane.

1952 US biologists Alfred Day Hershey and Martha Chase use radioactive tracers to show that bacteriophages infect bacteria with DNA and not protein.

They tag the DNA core of a phage with radioactive phosphorous and the protein coat with radioactive sulphur. The work shows that when the labelled phage attacks bacteria, the phage injects its DNA into the bacterial cell. The protein coat is left on the outside. The injected DNA brings about replication of a new phage complete with the protein coat. Hershey and Chase conclude that the DNA carries the information for the replication of new phage particles.

1953 British immunologist Peter Medawar and Australian physician Macfarlane Burnet share the Nobel Prize for Physiology or Medicine. The award reflects Medawar's pioneering research, which supports Burnet's idea that an animal's rejection of transplanted tissue is not inherited, but develops in the fetus.

The work fosters the concept of 'immunological tolerance' and leads to transplantation surgery based on tissue-typing to achieve as close a match as possible between the organ donor's and recipient's tissues.

269

The Discovery of the Structure of DNA

by Julian Rowe

THE FIRST ANNOUNCEMENT

'We wish to suggest a structure for the salt of deoxyribose nucleic acid (DNA). This structure has novel features which are of considerable biological interest.'

So began a 900-word article that was published in the journal *Nature* in April 1953. Its authors were English molecular biologist Francis Crick (1916–) and US biochemist James Watson (1928–). The article described the correct structure of DNA, a discovery that many scientists have called the most important since Austrian botanist and monk Gregor Mendel (1822–1884) laid the foundations of the science of genetics. DNA is the molecule of heredity, and by knowing its structure, scientists can see exactly how forms of life are transmitted from one generation to the next.

THE PROBLEM OF INHERITANCE

The story of DNA really begins with British naturalist Charles Darwin (1809–1882). When, in November 1859, he published *On the Origin of Species by Means of Natural Selection* outlining his theory of evolution, he was unable to explain exactly how inheritance came about. At that time it was believed that offspring inherited an average of the features of their parents. If this were so, as Darwin's critics pointed out, any remarkable features produced in a living organism by evolutionary processes would, in the natural course of events, soon be swamped by more average features.

The work of Gregor Mendel, only rediscovered 18 years after Darwin's death, provided a clear demonstration that inheritance was not a 'blending' process at all. Mendel's mathematical alalysis showed. He concluded that each of the features he studied, such as colour or stem length, was determined by two 'factors' of inheritance, one coming from each parent. Each egg or pollen cell contained only one factor of each pair.

In this way a particular factor, say for the colour red, would be preserved through subsequent generations.

GENES

Today, we call Mendel's factors genes. Through the work of many scientists, it came to be realized that genes are part of the chromosomes located in the nucleus of living cells and that DNA is the hereditary material.

THE DOUBLE HELIX

In the early 1950s, scientists realized that X-ray crystallography, a method of using X-rays to obtain an exact picture of the atoms in a molecule, could be successfully applied to the large and complex molecules found in living cells.

It had been known since 1946 that genes consist of DNA. At King's College, London, New Zealand-born British biophysicist Maurice Wilkins (1916–) had been using X-ray crystallography to examine the structure of DNA, together with his colleague, English X-ray crystallographer Rosalind Franklin (1920–1958), and had made considerable progress.

While in Copenhagen, James Watson had realized that one of the major unresolved problems of biology was the precise structure of DNA. In 1952, he came as a young postdoctoral student to join the Medical Research Council Unit at the Cavendish Laboratory, Cambridge, where Francis Crick was already working. Convinced that a gene must be some kind of molecule, the two scientists set to work on DNA.

Helped by the work of Wilkins and Franklin, they were able to build an accurate model of DNA. They showed that DNA had a double helical structure, rather like a spiral staircase. Because the molecule of DNA was made from two strands, they envisaged that as a cell divides, the strands unravel, and each could serve as a template as new DNA was formed in the resulting daughter cells. Their model also explained how genetic information might be coded in the sequence of the simpler molecules that DNA comprises. Here for the first time was a complete insight into the basis of heredity. James Watson commented that this result was 'too pretty not to be true!'

CRACKING THE CODE

Later, working with South African-born British molecular biologist Sidney Brenner (1927–), Crick went on to work out the genetic code, and determine exactly what DNA sequence was a code for each protein. These triumphant results created a tremendous flurry of scientific activity around the world. The pioneering work of Crick, Wilkins, and Watson was recognized in the award of the Nobel Prize for Physiology or Medicine in 1962.

The unravelling of the structure of DNA was a key early accomplishment of a nascent discipline, molecular biology, and laid the foundation for genetic engineering.

For some years scientists had suspected that DNA was involved in passing on characteristics from one generation to the next. Watson and Crick's piecing together of the evidence to reveal the structure of this remarkable molecule showed just how it was able to do so. *Corbis*

1953 US biochemist Stanley Lloyd Miller shows that amino acids can be formed when simulated lightning is passed through containers of water, methane, ammonia, and hydrogen – conditions under which life may have arisen.

Other intense energy sources such as gamma radiation produce more complex organic molecules, including the nucleotide base adenine. It is not clear how mixtures of organic molecules evolve into living systems.

25 April 1953 English molecular biologist Francis Crick and US biologist James Watson announce the discovery of the double helix structure of DNA, the basic material of heredity. They also theorize that if the strands are separated then each can form the template for the synthesis of an identical DNA molecule. It is one of the most important discoveries in biology.

They suggest that the DNA molecule consists of two chains of amino acids arranged as a double helix, with phosphate groups positioned on the outside of the molecule and purine and pyrimidine groups positioned inside. A crucial part of their analysis is based on X-ray photographs of DNA obtained by English biophysicist Rosalind Franklin. Crick and Watson share the Nobel Prize for Physiology or Medicine in 1962 for this discovery.

1954 Russian-born US cosmologist George Gamow proposes that there is a genetic code written in nucleotide triplets in the DNA molecule.

1955 Spanish-born US molecular biologist Severo Ochoa discovers polynucleotide phosphorylase, the enzyme responsible for the synthesis of RNA (ribonucleic acid), which allows him to synthesize RNA.

1955 Swedish biochemist Axel Theorell receives the Nobel Prize for Physiology or Medicine for his discoveries concerning the nature and mode of action of oxidation enzymes.

1955 US geneticists Joshua Lederberg and Norton Zinder discover that some viruses carry part of the chromosome of one bacterium to another; called transduction it becomes an important tool in genetics research.

1956 Romanian-born US biologist George Palade discovers ribosomes, which contain RNA (ribonucleic acid).

He shows that they are the sites of protein synthesis in the cell.

1956 US biologists Maklon Hoagland and Paul Zamecnik discover transfer RNA (ribonucleic acid), which transfers amino acids, the building blocks of proteins, to the correct site on the messenger RNA.

1957 US molecular biologists Matthew Meselson and Franklin Stahl demonstrate that when the two strands of parent DNA unzip, new nucleotides attach to each strand according to the rule of complementary base pairs.

Each daughter strand of DNA therefore consists of one parental strand and one new strand. The work shows that DNA replication is semi-conservative and verifies Crick and Watson's earlier (1953) ideas on the role of DNA.

1959 The US geneticist Harry Harris publicizes and expands on the ideas, long neglected, of the British physician Archibald Garrod, whose early 20th century work on heritable disorders of metabolism became, in the 1960s, a critical resource for the rise of human biochemical genetics.

1959 Austrian-born British biochemist Max Perutz determines the structure of haemoglobin using X-ray diffraction methods.

The work began in 1937 and continues after he and his student, John Kendrew, share the Nobel Prize for Chemistry in 1962.

17 July 1959 Kenyan-British anthropologists Louis and Mary Leakey discover the first fossilized fragment of *Homo habilis* ('Handy Man') at Olduvai Gorge, Tanganyika (modern Tanzania).

At the time it is realized that our species *Homo sapiens* was preceded by at least two earlier hominid species: the Australopithecines that appeared about 4 million years ago, and *Homo erectus* that appeared about 1.5 million years ago. The find by the Leakeys is thought to be the earliest representative of our genus. It acquires the name *Homo habilis* since it is associated with the first use of deliberately fashioned tools. *Homo habilis* is dated to around 2 million years ago.

Chemistry

1950 US chemist Glenn Seaborg and his colleagues at the University of California discover californium (atomic number 98).

They bombard curium with helium particles using the cyclotron at Berkeley, California, to transmute it into the new radioactive element. Californium has a half-life of 55 days, and is named after the state and the University of California.

1950 German chemists Otto Diels and Kurt Alder share the Nobel Prize for Chemistry for their discovery of diene synthesis in 1928.

1951 US chemists Linus Pauling and Robert Corey establish the helical or spiral structure of proteins.

Their study of the hydrogen bond shows that proteins form two types of molecular structure: either a pleated sheet or a helix (or a combination of the two). Their work helps in the later determination of the molecular structure of DNA in 1953.

1951 US chemists Edwin McMillan and Glenn Seaborg receive the Nobel Prize for Chemistry for their discovery of and research on transuranic elements from 1940.

1952 British biochemists Archer Martin and Richard Synge are awarded the Nobel Prize for Chemistry for their development of gas chromatography, a technique for separating the elements of a gaseous compound.

Mixtures of gaseous compounds are passed through a liquid solvent or over an absorbent solid using an inert carrier gas, such as nitrogen. The individual components of the gas are carried at different speeds and so separate out. This technique is extremely sensitive and can be used to detect trace quantities of impurities in a sample.

1952 US chemist Albert Ghiorso and colleagues at the University of California, Berkeley, discover the radioactive elements einsteinium (atomic number 99) and later fermium (atomic number 100) in the radioactive debris collected by drone aeroplanes flown through the radioactive cloud from a hydrogen bomb explosion in the Pacific. They are named after German-born US physicist Albert Einstein and US physicist Enrico Fermi.

1952 US chemist and physicist Robert S Mulliken develops a quantum mechanical theory to explain his molecular orbital theory of chemical bonds and structure.

Mulliken develops the theory of molecular orbitals and proposes that the electron configurations of atoms merge and conform to a molecular configuration when part of a molecule. He develops a quantum mechanical theory to describe this behaviour, for which he receives the Nobel Prize for Chemistry in 1966.

1953 English biochemist Frederick Sanger determines the structure of the insulin molecule. The largest protein molecule to have its chemical structure determined to date, it is essential in the laboratory synthesis of insulin.

Sanger spends ten years studying the structure of the bovine insulin molecule before he correctly determines the exact order of all the amino acids in the peptide chain. He shares the Nobel Prize for Chemistry in 1958 for this work.

1953 German chemist Karl Ziegler discovers a chemical catalyst that permits polyethylene plastics to be produced at atmospheric pressure. Previous methods required pressures of 30,000 lb per sq in.

Ziegler discovers that resins containing aluminium or titanium help to catalyse the polymerization of ethylene. This new form of polyethylene is called high-density poly-ethylene and is stronger and more resistant to heat than the previous type. He shares the Nobel Prize for Chemistry in 1963 for this discovery.

1953 German chemist Hermann Staudinger receives the Nobel Prize for Chemistry for his discoveries in macromolecular chemistry.

1953 US biochemist Vincent du Vigneaud synthesizes oxytoxin, the first protein hormone to be synthesized in a laboratory.

Vigneaud's studies of the hormone oxytocin show him that the protein is composed

273

of only eight different amino acids, where most proteins are composed of hundreds of acid units. He then systematically builds up the peptide chain corresponding to oxytoxin using a series of chemical reactions. This shows that any protein can be synthetically manufactured, and earns Vigneaud the Nobel Prize for Chemistry in 1955.

1954 German chemist Georg Wittig develops a method of synthesizing chemical compounds containing unsaturated carbon bonds, a process that proves useful in the manufacture of vitamins A and D, prostaglandins and sterols.

Wittig discovers that a class of sulphur compounds known as ylids can be used in reactions with aldehydes and ketones to synthesize compounds containing C=C double bonds. The reaction becomes known as the Wittig synthesis, and he is awarded a share of the Nobel Prize for Chemistry in 1979 for this work.

1954 Italian chemist Giulio Natta develops isotactic polymers; he polymerizes propylene to obtain polypropylene.

Natta develops catalysts discovered by Ziegler in 1953 to produce polymers in which all the side groupings attached to the monomer units are pointing in the same direction. This discovery highly influences future polymer development. This work earns Natta a share of the Nobel Prize for Chemistry in 1963.

1954 US chemist Linus Pauling receives the Nobel Prize for Chemistry for his discoveries concerning the nature of chemical bonds.

1954 US chemist Robert Burns Woodward synthesizes the poison strychnine, and the basis of LSD, lysergic acid.

The alkaloid strychnine is a highly complex molecule that is built up of seven interrelated rings of atoms. By synthesizing this compound, Woodward establishes himself as one of the greatest synthetic chemists in history. He synthesizes lysergic acid, the basis of the hallucinogenic drug lysergic acid diethylamide, LSD.

1955 English biochemist Dorothy Hodgkin determines the complex structure of vitamin B_{12} (cyanocobalamin) using X-ray photographs which she obtains from crystals of the vitamin.

The molecular structure of vitamin B_{12} is one of the most complex non-protein compounds and the task has taken Hodgkin over seven years to complete. She receives the Nobel Prize for Chemistry in 1964.

1955 Russian-born Belgian physical chemist Ilya Prigogine describes the thermodynamics of irreversible processes.

Prigognine applies the second law of thermodynamics to complex systems. He suggests that the behaviour of complex systems cannot be predicted by theory if energy and matter are continually entering the system. This work is highly influential in the development of chaos theory. He is awarded the Nobel Prize for Chemistry in 1977.

1955 US chemist Albert Ghiorso and colleagues at the University of California, Berkeley, discover the radioactive element mendelevium (atomic number 101). It is named after Russian chemist Dmitry Mendeleyev, the founder of the periodic table.

They create the element by bombarding an isotope of einsteinium with helium ions using the Berkeley 152 cm/60 in cyclotron. The radioactive element has a half-life of only 76 minutes and is the first element to be discovered one atom at a time.

1955 US biochemist Vincent Du Vigneaud receives the Nobel Prize for Chemistry for his investigations into sulphur compounds, and the first synthesis of a polypeptide hormone.

1956 US biochemist and physician Arthur Kornberg discovers how DNA (deoxyribonucleic acid) molecules replicate, allowing him to synthesize DNA in a test tube.

Kornberg uses radioactively tagged nucleotides to discover that the bacteria *Escherichia coli* uses an enzyme, now known as DNA polymerase, to replicate DNA. He is awarded the Nobel Prize for Physiology and Medicine in 1959 for this work.

1956 English chemist Cyril Hinshelwood and the Russian chemist Nikolai Semenov receive the Nobel Prize for Chemistry for their work on the kinetics of chemical reactions.

1957 Scottish chemist Alexander Todd receives the Nobel Prize for Chemistry for his work on nucleotides and nucleotide coenzymes.

1958 US company DuPont develops the elastic polymer Lycra, subsequently commonly used in the manufacture of clothes, especially sportswear.

This class of synthetic fibres is known as the elastomers, as they have highly elastic properties and retain their original shape even after stretching and deformation. Lycra is a polymeric material based on polyurethane and is usually covered in a protective layer of nylon.

1958 US chemist Albert Ghiorso and colleagues at the University of California, Berkeley, discover nobelium (atomic number 102).

They use the heavy ion linear accelerator to bombard curium with carbon-12 ions to create the element. However, groups in Sweden and the USSR had already created a number of different isotopes of the element in 1957. Eventually the Swedish group are given permission to name the element, which they do so after the Swedish chemist Alfred Nobel.

1958 English biochemist Frederick Sanger receives the Nobel Prize for Chemistry for his determination of the structure of insulin in 1953.

His research turns to the problem of the structure and sequence of nucleic acids.

1959 Czech chemist Jaroslav Heyrovský receives the Nobel Prize for Chemistry for his polarographic methods of chemical analysis.

Earth Sciences

1950 Hungarian-born US mathematician John von Neumann and his colleagues make the first 24-hour weather forecast using a computer: the Electronic Numerical Integrator and Computer (ENIAC).

By 1953, numerical predictions will be made on a regular basis.

1950 The Mid-Pacific Expedition, led by US oceanographer Roger Revelle, embarks to study the Pacific seafloor. Using echosounders, seismic arrays, corers, and dredges, they discover that the ocean floor is not composed of the most ancient crust on Earth, but is, compared to continental crust, very young, less than 200 million years old.

1950–1960 A Worldwide Standarized Seismographic Network of 120 stations in 60 countries is established.

The network allows geophysicists to calculate precise earthquake hypocentres, magnitudes, and source mechanisms. Its results have been instrumental in the development of the theory of plate tectonics.

1951 Raimond Castaing publishes *Microanalysis by Means of Electron Probes*. The discovery sparks research into creating electron probes and by the late 1950s they are available commercially.

Electron microprobes are now used routinely to make elemental analyses of rocks and minerals. They have revolutionized the field of mineralogy and petrology, allowing geologists to analyse extremely small samples with incredible precision.

1952 US geophysicist Francis Birch predicts fundamental changes in the mineralogy of the Earth's mantle with depth.

1953 US chemist Loring Coes invents an apparatus for obtaining high temperatures at high pressures in the laboratory. He uses this apparatus to grow high-pressure minerals, including a new form of dense silica that bears his name, coesite.

1953 US geochemist Clair Patterson makes lead isotope measurement of the Canyon Diablo meteorite. By 1956, he has enough data to suggest that the Earth is 4.55 billion years old.

Because meteorites are thought to have formed at the same time as the Earth but have not undergone a significant amount of geological processing since, they might be used to date the age of the Earth, and of the Solar System. Using Patterson's measurements, the German-born Dutch physicist Friedrich Houtermans calculates an age of 4.5 billion years, which Patterson later refines.

1953 US geophysicist W Maurice Ewing announces that there is a crack, or rift, running along the middle of the Mid-Atlantic Ridge.

1954 Researchers at the US electrical company General Electric produce the first synthetic diamonds.

Synthetic diamonds are now used in things such as industrial abrasives and glass-cutting technology.

1954 The existence of pre-metazoan life becomes widely accepted when diverse fossil microscopic organisms are discovered in the Gunflint rocks along the north shore of Lake Superior. The Gunflint biota suggest significant evolutionary activity at least 2 billion years before present.

1956 US geochemist Clair Patterson uses the isotopic composition of lead to determine that three stony meteorites and two iron meteorites have a common age of 4.55 billion years. Patterson uses lead isotopes to show that Earth is the same age.

1956 US geologists Bruce Heezen and W Maurice Ewing discover a global network of oceanic ridges and rifts 60,000 km/37,000 mi long that divide the Earth's surface into 'plates'.

1957 US geophysicist W Maurice Ewing notices that mid-ocean ridges have rifts running down their axes. This suggests that ridges are not compressional mountain chains, but rather sites of crustal extension.

1958 Australian structural geologist S Warren Carey publishes *A Tectonic Approach to Continental Drift*.

Carey suggests that the continents were as one during the late Palaeozoic and that they split apart as a result of the expansion of the Earth from three-quarters of its current diameter to its present size.

1958 The Equatorial Undercurrent is discovered in the equatorial Pacific Ocean. It has a width of 320–480 km/200–300 mi, a height of 200–300 m/650–1,000 ft, and flows 50–150 m/165–500 ft below the surface.

1959 US geologists Marion Hubbert and William Rubey suggest that the overthrusting of large horizontal planes of rock that produces folded mountains is due to the reduction of friction caused by fluids in the rocks.

1959 US geophysicist Harry Hess publicizes the theory of seafloor spreading, in which molten material wells up along the mid-oceanic ridges as the seafloor moves away from the ridges. The flow is thought to be the cause of continental drift.

Ecology

1 November 1952 The USA explodes the first thermonuclear fusion device, or hydrogen bomb, at Eniwetok Island in the Marshall Islands, although this is not revealed until February 1954.

The bomb uses a fission trigger to fuse hydrogen-2 and hydrogen-3, releasing an amount of energy equivalent to several million tonnes of TNT. The atoll is completely destroyed as the bomb yields energy equivalent to 500 times that of the bomb the USA dropped on Hiroshima.

5–13 December 1952 Smog in London, England, claims the lives of at least 4,000 people. The cause is thought to be a synergistic reaction between smoke particles and sulphur dioxide.

By the week ending on 13 December, three times more deaths than expected are registered amongst infants aged four weeks to a year and adults aged over 55. The causes of the increased deaths are mostly related to disorders of the circulatory and respiratory systems. One of the main sources of smoke and sulphur dioxide is the burning of coal in domestic fires. This incident leads to the passage of the Clean Air Acts in 1956, which set up, among other things, smokeless zones.

1953 US biologists Eugene and Howard Odum publish *Fundamentals of Ecology*, which provides a general framework for systems ecology and ecological energetics.

The Odums develop and extend the energy flow models of Raymond Lindeman (1942) and introduce the idea of a mature ecosystem. They urge the study of the flow of materials and energy in ecosystems, much as social scientists study economies to help humans plan and regulate their activities for a healthier planet. They advocate a holistic rather than reductionist approach.

Large coal-fired power stations such as this one at Didcot, southern England, require huge amounts of coal to operate – Didcot consumes up to 1,000 truckfuls per day. The electricity produced per tonne of coal may be as high as 2,000 kWh. Concern about acid rain and global warming, however, has led to a move away from coal-fired power generation. *AEA Technology*

1953–1956 Mercury poisoning affects thousands of people living around Minamata Bay in Japan; at least 43 die as a direct result.

The mercury is discharged into the bay in the form of mercury sulphate (a waste product of the manufacture of acetaldehyde) by the Chisso company. Investigating scientists discover that bacterial action on the mercury sulphate results in the formation of methyl mercury, which accumulates in the marine food chain and results in fish with high levels of mercury. By 1970, 1,000 cases of mercury poisoning will have been recorded, along with 3,000 suspected cases.

1954 Australian ecologists Herbert Andrewartha and Charles Birch argue that climatic factors are largely responsible for the natural control of populations.

They believe that biotic factors such as predator–prey relationships, which depend on the density of the respective populations for their effect, do not significantly regulate numbers. Their observations focus on insect populations, the numbers of which are susceptible to climatic variation.

1 March 1954 The US government tests an H-bomb at Bikini Atoll in the Marshall Islands. A Japanese fishing vessel, the *Fukurya Maru/Lucky Dragon* carrying 23 fishermen sailing 129 km/80 mi west of the island, is caught in a shower of radioactive ash. The fishermen suffer radiation burns and sickness.

Due to unpredicted shifts in wind velocity at high altitudes, 300 Marshall Islanders and task force personnel are also affected and 84 are subjected to severe exposure. Fish in the area show signs of radioactive contamination and are declared unfit for consumption.

1954 A study in the USA finds that cars contribute 80% of the pollution in the greater Los Angeles community, in California.

1955 A British government research report entitled 'Precautionary Measures against Toxic Chemicals used in Agriculture' highlights the problems caused over the previous few years by the widespread use of dinitro- and organophosphorus compounds such as Parathion. These substances are being sprayed on vegetable crops to combat aphids but have caused the death of a large number of birds.

The Ministry of Transport decides to limit the spraying of chemicals on roadside verges to major roads only. This encourages a switch to organochlorine compounds such as aldrin, dieldrin, endrin, and DDT. These are more persistent, however, and concentrations in the food chain begin to rise.

1956 The Clean Air Act is passed by the UK Parliament. It prohibits the burning of untreated coal in London, England, and successfully reduces the emission of sulphur dioxide pollution.

1957 English ecologist James Lovelock develops the electron detector. The instrument is able to detect low levels of pollutants and leads to the discovery that pesticide residues and other contaminants at low concentrations are widespread in the environment.

10 October 1957 A fire in a military reactor producing plutonium at the English nuclear facility Windscale (now Sellafield) releases large amounts of radioactivity into the surrounding area, news of which is suppressed by the UK government.

Milk from the affected area, eventually extending to 518 sq km/200 sq mi and containing over 100 farms, is destroyed for a period of some weeks as the radioactive iodine-131 content rises to over six times the permitted safe levels. Iodine-131 has a half-

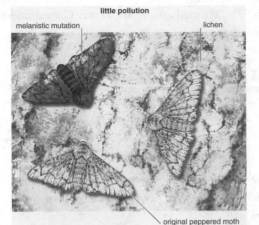

little pollution

melanistic mutation

lichen

original peppered moth

high pollution

Industrial melanism in the peppered moth, first noticed by English geneticist Henry Kettlewell. Kettlewell observed that whereas in rural areas peppered moths were light in colour to camouflage them against the lichens, in industrial areas where the tree trunks were dirtied with soot, peppered moths were darker. Natural selection favoured the darker mutation in industrial areas because it offered better camouflage there, so it had become widespread, whereas in rural areas the darker mutant was highly visible against the lighter tree trunks and so was easy prey to insect feeders.

life of about eight days and by the end of the month most milk-producing areas are declared safe once again.

1958 English biologist and animal ecologist Charles Elton, publishes *The Ecology of Invasions by Animals and Plants*, in which he discusses numerous examples of population explosions that have occurred comparatively recently as a result of human agency.

Pandemics of disease as a result of the multiplication and dispersal of infectious pathogens, and the harmful expansion of populations of species like the prickly pear cactus or the rabbit, are among the examples cited.

1959 English biologist Henry Kettlewell announces experimental evidence to support the theory of industrial melanism.

By using dark- and light-coloured moth variants, he suggests that birds are better able to spot and eat light-coloured moths against the background of polluted, darkened trees, and thus favourably select the dark-coloured individuals.

Mathematics

June 1951 US engineers John Mauchly and John Presper Eckert build UNIVAC 1 (Universal Automated Computer), the first commercially available electronic digital computer, in Philadelphia, Pennsylvania. Built for the US Bureau of the Census by the Remington Rand corporation, it uses vacuum tubes, is the first to handle both numeric and alphabetic information easily, has a memory of 1.5 kilobytes, and is the first to store data on magnetic tape.

1955 Homological algebra is developed by the French mathematicians Henri Cartan and Samuel Eilenberg, thus helping to unify mathematics by combining algebraic topology and abstract algebra.

1956 US computer programmer Jack Backus at IBM invents FORTRAN (formula translation), the first computer programming language. It is used primarily by scientists and mathematicians.

Physics

1950 English physicist Cecil Powell receives the Nobel Prize for Physics for his development of the photographic method of studying nuclear processes and his discoveries regarding mesons made with this method.

1951 English physicist John Cockcroft and the Irish physicist Ernest Walton share the Nobel Prize for Physics for their work on the transmutation of atomic nuclei by artificially accelerated atomic particles.

1951 US physicist Edward Purcell discovers line radiation of wavelength 21 cm/8.3 in emitted by hydrogen in space.

   Line radiation is radiation emitted at only one specific wavelength. This discovery allows the distribution of hydrogen clouds in galaxies and the speed of the Milky Way's rotation to be determined.

1952 US physicists Felix Bloch and Edward Purcell are jointly awarded the Nobel Prize for Physics for their development of nuclear magnetic resonance in solids.

1952 US nuclear physicist Donald Glaser develops the bubble chamber to observe the behaviour of subatomic particles. It uses a superheated liquid instead of a vapour to track particles. He is awarded the Nobel Prize for Physics in 1960 for the invention.

   Unlike cloud chambers, bubble chambers do not rely on expansion to provide visible tracks. The absence of the expansion allows photographic records to be made with greatly reduced distortion.

1953 Dutch physicist Frits Zernike receives the Nobel Prize for Physics for his development of phase contrast microscopy and for his invention of the phase contrast microscope.

1953 US physicist Murray Gell-Mann introduces the concept of 'strangeness', a property of certain subatomic particles, to explain their behaviour.

   Later work will demonstrate the existence of the strange quark, the carrier of strangeness.

1954 German-born British physicist Max Born and the German physicist Walther Bothe share the Nobel Prize for Physics: Born for his research in quantum mechanics, especially his statistical interpretation of the wave function, and Bothe for his invention of the coincidence method.

1955 US physicists Willis Lamb and Polykarp Kusch share the Nobel Prize for Physics, Lamb for his discoveries concerning the fine structure of the hydrogen spectrum, and Kusch for his precision determination of the magnetic moment of the electron.

1956 US physicists William Shockley, John Bardeen, and Walter Brattain share the Nobel Prize for Physics for their work on semiconductors and their discovery of the transistor effect.

1956 US physicists Clyde Cowan and Fred Reines detect the existence of the neutrino, a particle with no electric charge and at the time believed to have no mass, at the Los Alamos Laboratory.

Debate still continues regarding the mass of the neutrino and the effect that this mass, if any, could have on the evolution of the universe.

1957 Experiments at Columbia University, New York, confirm the theoretical predictions of Chinese-born US physicists Chen Ning Yang and Tsung-Dao Lee that parity is not conserved in weak interactions.

In the same year Yang and Lee receive the Nobel Prize for Physics for their penetrating investigation of the so-called parity laws, which has led to important discoveries regarding the elementary particles.

1957 Japanese physicist Leo Esaki makes important discoveries about tunnelling, the ability of electrons to penetrate barriers in semiconductors by acting as waves.

1957 US physicists John Bardeen, Leon Cooper, and John Schrieffer formulate a theory to explain superconductivity, the characteristic of a solid material of losing its resistance to electric current when cooled below a certain temperature.

The theory centres on electrons forming so-called 'Cooper pairs'.

1958 Soviet physicists Pavel Cherenkov, Ilya Frank, and Igor Tamm are jointly awarded the Nobel Prize for Physics for their discovery and interpretation of the Cherenkov effect.

1959 US physicists Emilio Segrè and Owen Chamberlain are jointly awarded the Nobel Prize for Physics for their discovery of the antiproton.

1960–1969

Astronomy

11 March 1960 The USA launches *Pioneer 5*, which relays the first measurements of deep space.

12 August 1960 NASA launches *Echo I*, a 30-m/100-ft aluminium-coated balloon used as a passive communications satellite to reflect radio-waves. It remains in orbit for eight years and is a conspicuous object in the night sky.

1961 The Soviet space probe *Venera 1* passes within 99,000 km/62,000 mi of Venus but fails to transmit data due to a telemetry failure.

12 April 1961 Soviet cosmonaut Yuri Gagarin, in *Vostok 1*, is the first person to enter space. His flight lasts 108 minutes.

5 May 1961 US astronaut Alan Shepard in the Mercury capsule *Freedom 7* makes a 14.8-minute single suborbital flight. He is the first US astronaut in space.

21 May 1961 US President John F Kennedy commits the country to 'landing a man on the Moon and returning him safely to Earth before this decade is out'.

1962 Italian astronomers Riccardo Giacconi, Herbert Gursky, Frank Paolini, and Bruno Rossi discover the first astronomical X-ray source – in Scorpio.

20 February 1962 US astronaut John Glenn, in the Mercury capsule *Friendship 7*, becomes the first US astronaut to orbit the Earth. He makes three orbits.

26 April 1962 The USA and UK launch the Earth satellite *Ariel*. Designed to study the ionosphere, it is the first international cooperative launch.

26 August 1962 US space probe *Mariner 2*, the world's first successful interplanetary spacecraft, is launched. It makes a flyby of Venus (14 December), passing within 34,000 km/ 21,000 mi of the planet's surface and taking measurements of temperature and atmospheric density.

Mariner 2 records the temperature at Venus for the first time, revealing the planet's very hot (about 500°C/900°F) lower atmosphere. The spacecraft's solar wind experiment measures for the first time the density, velocity, composition, and variation over time of the solar wind.

1963 An international team of astronomers discovers the first quasar (3C 273), an extraordinarily distant object brighter than the largest known galaxy yet with a star-like image.

The word quasar is derived from 'quasi-stellar radio source'. This name is retained today, even though it is now known that most quasars are only faint radio emitters. Most quasars are larger than our Solar System and are thought to be the most distant objects yet detected in the universe. Despite their brightness, no quasars can be seen with the unaided eye due to their great distance from Earth. Their light takes billions of years to reach the Earth's atmosphere, providing information about the early stages of the universe.

16 June 1963 Soviet cosmonaut Valentina Tereshkova, the first woman in space, is launched into a three-day orbital flight aboard *Vostok 6*, to study the problems of weightlessness.

1964 English astronomer Fred Hoyle and Indian astronomer Jayant Narlikar propound a new theory of gravitation that solves the problem of inertia.

Space isn't remote at all. It's only an hour's drive away if your car could go straight upwards.

Fred Hoyle, English astronomer and writer, *The Observer* September 1979

---

28 November 1964 The USA launches *Mariner 4* to Mars. It will pass within 9,800 km/6,118 mi of the planet on 14 July 1965, and relay the first close-up photographs of the surface, as well as information on the Martian atmosphere.

1965 US astronomers Arno Penzias and Robert Wilson detect microwave background radiation in the universe and suggest that it is the residual radiation from the Big Bang.

18 March 1965 Soviet cosmonaut Alexei Leonov leaves spacecraft *Voskhod 2* and demonstrates the ability of humans to function in outer space when he floats in space for 20 minutes – the first space walk.

3–7 June 1965 US astronaut Edward White, during the *Gemini 4* space mission, makes a 22-minute space walk, the first by a US astronaut. He is also the first to use a personal propulsion pack during the walk.

1966 Astronomers at the US Naval Research Laboratory, Washington, DC, discover powerful X-rays emitted from within the constellation Cygnus.

3 February 1966 Soviet spacecraft *Luna 9* (launched 31 January) makes the first soft landing on the Moon and transmits photographs and soil data for three days.

2 June 1966 NASA spacecraft *Surveyor I* (launched 30 May) makes the first US soft landing on the Moon, and transmits over 10,000 photographs of the lunar surface.

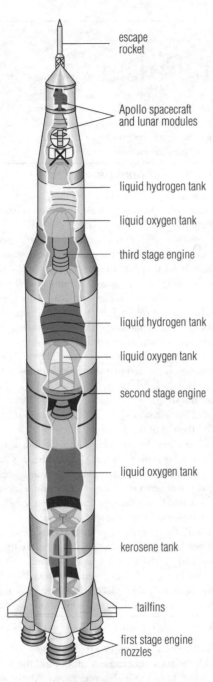

escape rocket

Apollo spacecraft and lunar modules

liquid hydrogen tank

liquid oxygen tank

third stage engine

liquid hydrogen tank

liquid oxygen tank

second stage engine

liquid oxygen tank

kerosene tank

tailfins

first stage engine nozzles

The three-stage *Saturn V* rocket used in the Apollo moonshots of the 1960s and 1970s. It stood 111 m/365 ft high, as tall as a 30-storey skyscraper, weighed 2,700 tonnes/3,000 tons when loaded with fuel, and developed a power equivalent to 50 Boeing 747 jumbo jets.

7 July 1967 English astronomers Jocelyn Bell and Anthony Hewish, at the Mullard Radio Astronomy Observatory, Cambridge, England, discover the first pulsar (announced in 1968).

Pulsar radio emission consists of a uniform series of pulses, spaced with great precision at periods of between a few milliseconds and several seconds. Over 700 radio pulsars are known. The regularity of the pulses is such that observers can predict the arrival times of pulses a year ahead with an accuracy better than a millisecond. Pulsars are rapidly rotating, very strongly magnetized neutron stars.

18 October 1967 The Soviet spacecraft *Venera 4* (launched 12 June) lands on Venus. The first soft landing on another planet, its instrument-laden capsule transmits information about Venus's atmosphere.

12 June 1968 The first radar observations of an asteroid are made when Icarus approaches within 6 million km/3.7 million mi of Earth.

14–21 September 1968 The Soviet spacecraft *Zond 5* flies around the Moon and returns to Earth – the first spacecraft to do so.

1969 A multichannel spectrometer is installed at the Mount Palomar Observatory, California. It permits the rapid and accurate collection of data through the simultaneous observation of 32 wavelength bands.

1969 The Sacramento Peak Observatory at Sunspot, New Mexico, becomes operational. All the air from its 76 cm/30 in diameter, 54.9 m/180 ft long solar telescope has been evacuated to prevent it from overheating.

16 January 1969 Two cosmonauts aboard Soviet spacecraft *Soyuz 5* (launched 15 January) dock with and transfer to *Soyuz 4* (launched 14 January). Locked together for four hours, the craft form the first experimental space station.

February–March 1969 The US space probe *Mariner 6* (launched on 24 February) passes within 3,410 km/2,131 mi of the surface of Mars. *Mariner 7* (launched on 27 March) photographs the Martian landscape, makes thermal maps of the planet, and analyses its atmosphere.

The Discovery of Pulsars

by Peter Lafferty

LITTLE GREEN MEN?

Making a major scientific discovery is not as many people imagine; often, there is no 'eureka moment' or single instant of discovery. Discovery is more often a process of checking and rechecking, of gradually eliminating spurious effects, until the truth is apparent. The discovery of the first pulsar shows this process in action.

Luck played a part in the discovery. In 1967 Antony Hewish at the Mullard Radio Astronomy Laboratory, Cambridge, England, constructed a new type of radio telescope – a large array of 2,048 aerials covering an area of 1.8 hectares/4.4 acres – to study the 'twinkling' or scintillation of radio galaxies. This is caused by clouds of ionized gas ejected from the Sun. It is most noticeable at metre wavelengths, so the new telescope had to be sensitive to radiation of this wavelength. Most radio telescopes achieve high sensitivity by averaging incoming signals for several seconds, and so are unsuitable for studying rapidly varying signals. They collect radiation with wavelengths of around a centimetre. The new telescope's ability to detect rapidly varying signals of metre wavelength was just what was needed to detect a pulsar.

TWINKLE, TWINKLE, LITTLE GALAXY

The telescope began work in July 1967. Its first task was to locate all radio galaxies twinkling in the area of sky accessible to it. Each day, as the Earth rotated, the telescope swept its radio eye across a band of sky. A complete scan took four days. Initially a graduate student, Jocelyn Bell, ran the survey and analysed the results, output on about 30 m/100 ft of chart paper each day.

After a few weeks, Bell noticed an unusual signal: not a single blip, but an untidy bunch of squiggles on the chart. It was nothing like the signals she was looking for, so she marked a query on the chart and did not investigate further. Later, when she saw the same signal on another chart, she realized it merited closer attention. The signal came from a part of the sky where scintillations were normally weak. It occurred at night, and scintillations are strongest during the day. A faster chart recorder was installed for a

more detailed look at the signal. It would stretch out the signal over a longer chart, like a photographic enlargement.

WHAT WAS GOING ON?

For a while, Bell and Hewish's efforts were frustrated – the signal weakened and vanished. For a month, there was no sign of it. The researchers feared that they had seen a one-off event: possibly a star flaring brightly for a short time. If so, they had missed the chance to study it in detail. However, on 28 November, it returned. This time, the new recorder revealed the true nature of the signal. It was a series of short pulses about 1.3 seconds apart. Timing the pulses more accurately showed that they were in step to within one-millionth of a second. Their short duration indicated that they were coming from a very small object. Something peculiar was going on, but what?

The task now was to rule out spurious effects. Did the signal result from a machine malfunction? Was it caused by a satellite signal? Or a radar echo from the Moon? The fact that the signal appeared at regular intervals hinted that it was not a machine malfunction. Having calculated when the signal would next appear, at the appointed time the research team stood around the recorder. Nothing! Hewish and the others began to wander away. But before they reached the door, they were called back. 'Here it is!' said a student. They had miscalculated when the signal would be picked up. They knew now that it came from outside the laboratory.

IS THERE ANYBODY THERE?

Further study established that the signal rotated with the stars, and came from beyond the Solar System, but from within our galaxy. At one stage, Hewish thought that the signal might be a message from an extraterrestrial civilization. It was given the name 'LGM', for 'little green men'. However, this possibility was ruled out. Other beings must live on a planet circling round a star. The planetary motion would show up as a slight variation in the pulse rate. After carefully timing the pulses for several weeks, the idea of a planetary origin was given up.

On 21 December 1967 Bell discovered a second signal elsewhere in the sky. This clinched the reality of the phenomenon, and the name 'pulsar' was quickly coined to describe the object radiating the pulsation. The results were published in February 1968; within a year it was generally accepted that the signals came from rapidly spinning neutron stars. Since then more than 500 pulsars have been discovered. Antony Hewish shared the 1974 Nobel Prize for Physics with English radio astronomer Martin Ryle for their work in radio astronomy and, in particular, the discovery of pulsars.

20 July 1969 Having been launched from Cape Kennedy (now Canaveral) on 16 July, *Apollo 11* lands in the Sea of Tranquility on the Moon. Neil Armstrong and Buzz Aldrin become the first humans to step onto another world.

They spend only a few hours on the Moon, with little time to set up scientific experiments, but the Early Apollo Scientific Experiments Package is deployed, including a Passive Seismic Experiments Package and a Laser Ranging Retro-Reflector. More extensive scientific studies are carried on later Apollo missions.

22 November 1969 The lunar module from the US spacecraft *Apollo 12* (launched on 14 November) lands on the Moon in the Ocean of Storms. The crew collects 34 kg/75 lb of Moon rocks, inspects *Surveyor 3*, which landed nearby 2.5 years earlier, and uses a radioisotope-fuelled generator to power experiments.

Biology

1960 German chemist K H Hofman synthesizes pituitary hormone.

1960 US biochemist Robert Woodward and German biochemist Martin Strell independently synthesize chlorophyll.

After ten years' work, Woodward and others synthesize vitamin B_{12} in 1971.

1961 English molecular biologist Francis Crick and South African chemist Sydney Brenner discover that each base triplet on the DNA strand codes for a specific amino acid in a protein molecule.

In the same year, Brenner shows that ribosomes receive mRNA as a code which instructs them to carry out protein synthesis.

Progress in science depends on new techniques, new discoveries, and new ideas, probably in that order.

Sydney Brenner, South African-born British molecular biologist, *Nature* 1980

1961 US biochemist Melvin Calvin receives the Nobel Prize for Chemistry for his discoveries concerning the chemical processes of photosynthesis.

1962 Soviet scientist K Chudinov claims to have revived fossil algae approximately 250 million years old.

1962 Austrian-born British biochemist Max Perutz and the English biochemist John Kendrew receive the Nobel Prize for Chemistry for determining the structure of globular proteins.

285

1963 Austrian zoologist Konrad Lorenz, in *On Aggression*, states that only humans intentionally kill their own species and that aggressive behaviour is inborn but can be modified by the environment.

1963 English biochemist Peter Mitchell moves to Bodmin in Cornwall, England, and sets up his own laboratory and research institute.

Here he proposes a mechanism for the generation of energy in mitochondria and chloroplasts. Called the chemiosmotic theory, it is at first opposed by other scientists but further work convinces sceptics. He is awarded the Nobel Prize for Chemistry in 1978.

1963 US biochemist Robert Woodward synthesizes the plant chemical colchicine.

The compound proves useful in research on the division of the nucleus.

1964 British biologist Henry Harris and British virologist John Watkins fuse human cells together with the cells of a mouse. The work throws light on the mechanisms that regulate gene expression, and facilitates gene mapping.

It makes possible the later achievement of Milstein and Kohler (1975) who create hybridomas – a fusion of malignant and antibody-generating cells.

1965 French biochemists François Jacob and Jacques Monod discover messenger ribonucleic acid (mRNA), which transfers genetic information to the ribosomes, where proteins are synthesized.

The work particularly focuses on how gene action is switched 'on' and 'off', and introduces the idea of operons. These are gene clusters activated by an end sequence called an operator. Another sequence called a repressor combines with and switches off the operator. mRNA is a part of the scheme, carrying genetic information from the DNA of the chromosomes to the ribosomes where protein synthesis takes place.

1965 US biochemist Robert Woodward synthesizes the antibiotic cephalosporin C.

Woodward develops what becomes known as the Woodward–Hoffman rules for a large category of addition reactions.

1965 US biochemist Robert Woodward receives the Nobel Prize for Chemistry for synthesis of organic substances, especially chlorophyll.

1966 Molecular biologists discover that DNA is not confined to chromosomes but is also contained within the mitochondria.

1966 English molecular biologist Francis Crick publishes *Of Molecules and Men*, in which he discusses the revolution occurring in molecular biology and its implications.

1966 US geneticists Mark Ptashne and Walter Gilbert separately identify the first repressor genes.

They identify the repressor substance, *lac*-repressor, as a protein. It represses the gene that expresses an enzyme that acts on lactose.

1967 English ethologist Jane Goodall publishes *In the Shadow of Man*, a study of chimpanzee behaviour in the Gombe Stream National Park in Tanzania.

Through extensive observations of chimpanzee social interactions, Goodall argues that chimpanzees are capable of complex behaviours and the full suite of human emotional relationships. Later, she champions the cause of conservation.

1967 US biochemist Marshall Nirenberg establishes that mammals, amphibians, and bacteria all share a common genetic code.

1967 US scientist Charles Caskey and associates demonstrate that identical forms of messenger RNA produce the same amino acids in a variety of living beings, showing that the genetic code is common to all life forms.

1968 US scientist Elso Sterrenberg Barghorn and associates report the discovery of the remains of amino acids in rocks 3 billion years old.

1969 US geneticist Jonathan Beckwith and associates at the Harvard Medical School isolate a single gene for the first time.

Chemistry

1960 English biochemist John Kendrew, using X-ray diffraction techniques, elucidates the three-dimensional molecular structure of the muscle protein myoglobin.

Myoglobin is used in the body to store oxygen in muscle tissue and then release it when oxygen is needed. Kendrew shares the Nobel Prize for Chemistry in 1962 for his study of globular proteins.

1960 US chemist Willard Libby receives the Nobel Prize for Chemistry for his development of radiocarbon dating in 1947.

1961 Carbon-12 becomes the internationally recognized standard by which all atomic weights are determined.

The previous standard, oxygen, had in 1929 been discovered to exist in a number of stable isotopes, and it takes scientists over 30 years to agree on a replacement; this isotope of carbon is given the atomic weight of 12.0000.

1961 US chemist Albert Ghiorso and colleagues at the University of California, Berkeley, discover the radioactive element lawrencium (atomic number 103).

They create the element by bombarding a californium target consisting of four different isotopes with two types of boron particles. This process produces a tiny amount of the new element, which they name after the inventor of the cyclotron, US physicist Ernest Lawrence.

1961 US biochemist Melvin Calvin receives the Nobel Prize for Chemistry for his discoveries concerning the chemical processes of photosynthesis.

1962 British-born US chemist Herbert Brown publishes *Hydroboration*, in which he outlines his discovery of organoboranes.

Brown synthesizes organoboranes by reacting sulphur-containing boron hydride compounds with unsaturated organic chemicals, such as alkenes, in an ether solvent at room temperature. These compounds are highly versatile reagents used in organic synthesis. He shares the Nobel Prize for Chemistry in 1979 for his discovery of these reagents.

1962 Canadian-born British chemist Neil Bartlett prepares the first compound of a noble gas.

Noble gases are known as the 'inert gases' at this time because no one has succeeded in producing one of their compounds. Barlett correctly deduces that the heavier gases will be the most reactive. He immerses the very reactive compound platinum fluoride in xenon gas, which produces the compound xenon hexafluoroplatinate. Soon after, compounds of radon and krypton are also reported. The elements become known as the 'noble gases' from this time onwards.

287

1962 Austrian-born British biochemist Max Perutz and the English biochemist John Kendrew receive the Nobel Prize for Chemistry for determining the structure of globular proteins.

At the same ceremony, Francis Crick, James Watson, and Maurice Wilkins share the Nobel Prize for Physiology or Medicine for their work elucidating the structure of DNA.

1963 German chemist Karl Ziegler shares the Nobel Prize for Chemistry with the Italian chemist Giulio Natta for their work on the chemistry and technology of high polymers.

1964 US chemist Albert Ghiorso and colleagues at the University of California, Berkeley, and researchers at the Joint Nuclear Research Institute at Dubna, USSR, independently discover rutherfordium (atomic number 104).

The element is created by bombardment of plutonium with accelerated neon nuclei. Both groups claim credit for the discovery. The US group name the element 'rutherfordium' after the New Zealand-born British physicist Ernest Rutherford, while the Soviet group name it 'Kurchatovium' after the Soviet nuclear physicist Igor Kurchatov. The element is temporarily known as 'unnilquadium', the Latin for '104'.

1964 English chemist Dorothy Hodgkin receives the Nobel Prize for Chemistry for her determination of the structures of biochemical compounds, notably penicillin and vitamin B_{12}, using X-ray crystallography (she is the third woman to win the prize).

1965 US biochemist Robert Woodward receives the Nobel Prize for Chemistry for his synthesis of organic substances, especially chlorophyll.

1966 US chemist Robert Mulliken receives the Nobel Prize for Chemistry for his molecular orbital theory of chemical bonds and structures, developed in 1952.

1967 Korean-born US chemist Charles Pedersen discovers crown ethers.

These are two-dimensional cyclic polyethers that are composed of twelve carbon atoms and six oxygen atoms arranged in a distinctive 'crown' shape with a hollow centre. He shares the Nobel Prize for Chemistry in 1987 for his discovery.

1967 German chemist Manfred Eigen and English chemists Ronald Norrish and George Porter receive the Nobel Prize for Chemistry for their investigation of rapid chemical reactions by means of very short pulses of energy.

1968 US chemist Lars Onsager receives the Nobel Prize for Chemistry for his discovery of the thermodynamics of irreversible processes.

1969 English biochemist Dorothy Hodgkin determines the molecular structure of insulin.

This important compound is used to control sugar levels in the body and is an essential drug to sufferers of diabetes. Insulin is a highly complex molecule and it has taken Hodgkin 34 years to complete her structural determination of the insulin molecule. Her work allows insulin to be artificially synthesized for the first time.

1969 French researcher Jean-Marie Lehn invents 'cryptands', synthetic crystalline materials that can capture metal ions which, once captured, act with unique chemical and electrical properties; they are used in medicine as tracers and in the production of polymers in industry.

Lehn uses the recently discovered crown ethers to design a molecule capable of accepting a metal ion. He shows that two crown ether molecules can be linked together to form a three-dimensional structure that has a large empty cavity at its centre. This cavity will accept metal ions, and this discovery leads to the development of the 'host–guest' branch of organic chemistry, where the crown ether is the 'host' and a foreign species is the 'guest'. This work earns Lehn a share in the Nobel Prize for Chemistry in 1987.

1969 US chemist Donald Cram discovers 'cryptates'. They make salts soluble in organic solvents such as chloroform.

Cram's study culminates in dissolving an inorganic salt into an organic solvent, a process considered to be impossible at this time. This procedure has broad applications in organic synthesis and biochemistry and earns Cram a share in the Nobel Prize for Chemistry in 1987.

1969 English chemist Derek Barton shares the Nobel Prize for Chemistry with Norwegian chemist Odd Hassel for the concept and applications of conformation.

Earth Sciences

1 April 1960 The USA launches the first polar-orbiting satellite, *TIROS 1* (Television and Infra-Red Observation Satellite). A weather satellite, it is equipped with television cameras, infrared detectors, and videotape recorders.

Over 78 days it will take 23,000 photographs, giving meteorologists a view of the future of weather detection and prediction.

1961 L O Nicolaysen invents the 'isochron' method of rubidium–strontium and uranium–lead dating of geological materials.

1961 Satellite imagery of Cyclone Carla is used to plan the evacuation of thousands of inhabitants along the coast of the Gulf of Mexico. It is the first use of a satellite in weather hazard mitigation.

1961 US geophysicist Francis Birch relates seismic velocities to the density and the average atomic mass of rocks in Earth's mantle.

1961 US geophysicist Robert Dietz publishes a short paper in the British journal *Nature*, in which he puts forward his theory of seafloor spreading.

Dietz's theory is almost exactly the same as that put forward by the US geophysicist Harry Hess the following year. Both play a pivotal role in the acceptance of plate tectonics.

1962 English geophysicist Stanley Runcorn publishes 'Paleomagnetic Evidence for Continental Drift' in *Continental Drift*, a book of papers on the subject edited by himself.

In the paper Runcorn summarizes the observations and interpretations that have led to his complete acceptance of the theory since the mid-1950s. These include the observation that the north pole appears to wander over time, but that its wandering curve can be best described by movement of the continents relative to the pole rather than by movement of the pole itself.

1962 US geophysicist Harry Hess publishes *History of Ocean Basins*, in which he formally proposes seafloor spreading, the idea that the ocean crust is like a giant conveyor belt produced by volcanism at the mid-ocean ridges, that moves away from the ridge axis, and is eventually destroyed by plunging down into the mantle at deep-sea trenches. The hypothesis, influenced by the earlier work of Arthur Holmes, is embraced by the scientific community.

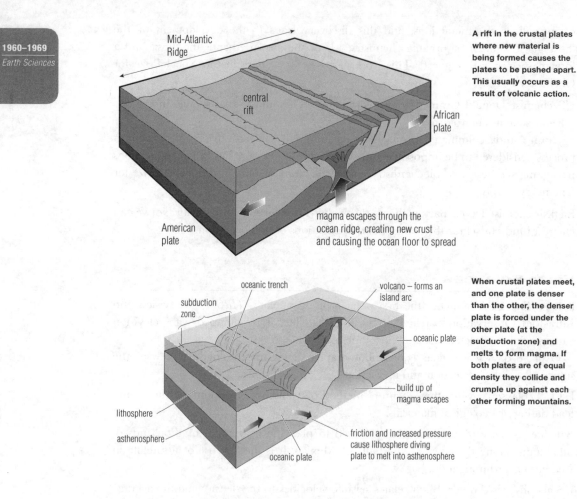

Mid-Atlantic
Ridge

central
rift

African
plate

American
plate

A rift in the crustal plates
where new material is
being formed causes the
plates to be pushed apart.
This usually occurs as a
result of volcanic action.

magma escapes through the
ocean ridge, creating new crust
and causing the ocean floor to spread

oceanic trench

volcano – forms an
island arc

subduction
zone

oceanic plate

build up of
magma escapes

lithosphere

asthenosphere

oceanic plate

friction and increased pressure
cause lithosphere diving
plate to melt into asthenosphere

When crustal plates meet,
and one plate is denser
than the other, the denser
plate is forced under the
other plate (at the
subduction zone) and
melts to form magma. If
both plates are of equal
density they collide and
crumple up against each
other forming mountains.

1963 English geophysicists Fred Vine and Drummond Matthews analyse the magnetism of
rocks on the Indian Ocean floor, which assume a magnetization aligned with the
Earth's magnetic field at the time of their creation. It provides further evidence of
seafloor spreading.

1963 Canadian geophysicist Laurence Morley interprets the 'striped' pattern of magnetism
in rocks of the ocean floor as being a result of seafloor spreading (published in 1964).

Whereas Vine and Matthews are influenced by Hess, Morley bases his hypothesis on
Dietz's article (1961).

1963 Geologist Ian G Gass suggests that the Troodos Massif, Cyprus, is a fragment of
Mesozoic ocean floor. It is a significant new interpretation of ophiolites.

Ophiolites are considered to be fragments of ocean floor that have been thrust up (or
obducted) onto a continent during seafloor subduction. They are used as a guide to the
structure of the oceanic crust.

1964 The US National Science Foundation establishes a consortium of four leading institutions, called the Joint Oceanographic Institutions for Deep Earth Sampling (JOIDES),
for the purpose of drilling into the Earth's ocean floor. Results from the project over the
next few years will eventually help confirm the theory of seafloor spreading by establishing that the oceanic crust everywhere is less then 200 million years old.

1965 Canadian geologist John Tuzo Wilson publishes 'A New Class of Faults and Their Bearing on Continental Drift', in which he formulates the theory of plate tectonics to explain continental drift and seafloor spreading.

1965 NASA launches *GEOS 1* (Geodynamics Experimental Ocean Satellite). Its aim is to provide a three-dimensional map of the world accurate to within 10 m/30 ft.

1966 Australian geochemists Alfred Ringwood and Alan Major relate the seismic discontinuity at a depth of 400 km/250 mi to the change of the mineral olivine to the structure of the mineral spinel.

1966 The US National Science Foundation puts Scripps Institute of Oceanography in charge of the JOIDES project and establishes the Deep Sea Drilling Project (DSDP).

1967 British geophysicist D McKenzie (published with R L Parker in 1967) and US geophysicist Jason Morgan (published in 1968) describe the motions of plates across the Earth's surface. Morgan calls the plates 'tectosphere'. They are later referred to as plates of lithosphere.

1967 US geophysicist Lynn R Sykes uses first-motion seismic studies to show that mid-ocean ridges form with offsets (sections of the ridges are displaced relative to one another along faults so that the crest of the ridge has a stepped appearance) rather than being offset later. This is a major advance in the understanding of the formation of ocean basins.

1968 English geochemist Ernest R Oxburgh and US geodynamicist Donald L Turcotte calculate the thermal consequences of subduction.

1968 French geophysicist Xavier Le Pichon, working at the Lamont Observatory in New York, describes the motions of Earth's six largest plates using poles of rotation derived from the patterns of magnetic anomalies and fracture zones about mid-ocean ridges.

Ecology

6 April 1960 The California legislature approves the USA's first state-sponsored antismog bill.

1961 The World Wildlife Fund (WWF; now the World Wide Fund for Nature) is formed.

The Fund is set up by the International Union for the Protection of Nature which itself was founded in 1948. After its formation the panda logo of the WWF becomes familiar throughout the world and WWF becomes the world's largest conservation body.

1962 US marine biologist and writer Rachel Carson publishes *Silent Spring*. Carson shows how, since 1940, over 200 new chemical substances have been created to kill insects, weeds, rodents, and other pests.

Carson attacks the mentality behind the idea that humans have dominion over nature. She reports how the chemicals synthesized to control pests are also killing friendly organisms, and ultimately threatening human health. The book has a major impact on ecological consciousness for years to come.

As cruel a weapon as the cave man's club, the chemical barrage has been hurled against the fabric of life.

Rachel Carson, US marine biologist and writer, *Silent Spring* (1962)

US biologist, writer, and conservationist Rachel Carson working at the US Bureau of Fisheries where she worked 1936–49. She wrote in her spare time before becoming a full-time writer in the 1950s.
Underwood & Underwood/CORBIS

3–7 December 1962 A thick smog settles over London, England. Of the 340 excess deaths attributed to this smog, two-thirds are over 70 years of age.

Some of the smokeless zones established after a previous smog incident in 1952 have yet to come into effect. The biggest proportionate rise in the cause of mortality is bronchitis. Figures show levels of sulphur dioxide to be about the same as in 1952 but levels of smoke to be lower, reflecting the impact of the 1956 Clean Air Act in reducing sulphur emissions above the capital.

1963 The US Office of Science and Technology issues a report titled *Use of Pesticides*. The report endorses the message carried in Rachel Carson's *Silent Spring* (1962), which condemned the use of synthetic pesticides, and industry and government agencies are criticized for their lax handling of the substances.

In June Rachel Carson appears before committees of the US Senate and House of Representatives and argues the case for more restrictive legislation.

1963 US economists Harold Barnett and Chandler Morse publish *Scarcity and Growth: The Economics of Natural Resource Availability*. Barnett and Morse were hired by Resources for the Future, a non-profit corporation for research and education set up to investigate the likelihood of resource shortage or increasing resource costs.

Barnett and Morse suggest that while resource scarcity might not appear to be a problem for the first world, the results of technological progress in the form of pollution might threaten the quality of life. Barnett and Morse stimulate a debate on the economics of resource scarcity.

1965 The authorities in Iceland declare Surtsey, an island that erupted from the sea in 1963, to be a nature reserve category one.

This grants recognition to the new island as a site of scientific importance where natural processes are allowed to take place in the absence of any direct human interference and where public access is prohibited.

1966 US economist Kenneth Boulding publishes *The Economics of the Coming Spaceship Earth*. Boulding expresses neo-Malthusian concerns about the future of the Earth's resources, especially energy resources.

Boulding advocates that in the spaceship economy humans must cut their resource consumption since the world is a closed system with a finite capacity to absorb waste and deliver resources.

1966 The US Department of the Interior issues the first rare and endangered species list; 78 species are included. Congress passes the Rare and Endangered Species Act to protect them.

1967 The International Whaling Commission prohibits hunting blue, right, grey, and humpback whales.

1967 US historian Lynn White publishes her paper 'The Historical Roots of Our Ecologic Crisis'. White argues that the anthropocentrism at the heart of Western civilization has brought about destruction of the natural world.

White argues that Western Christianity, with its emphasis on the rights of humans to dominate and exploit nature for their own ends, has led to an unhealthy environmental philosophy of technological imperialism in which the value of the natural world is merely to satisfy human desires and in which humans and nature are separated.

1967 US scientists Syukuvo Manabe and Richard T Wetherald suggest that the increase in carbon dioxide in the atmosphere is produced by human activities and is causing a 'greenhouse effect', which might raise atmospheric temperatures and cause a rise in sea levels.

18 March 1967 The Liberian-registered 120,000 tonne tanker *Torrey Canyon* strikes the Seven Stones, a submerged reef, and runs aground off the Isles of Scilly, off the southwest coast of Britain, and spills 860,000 barrels (around 119,000 tonnes) of crude oil into the sea. It is the biggest oil spill to date.

The oil forms a slick measuring 20 km/12 mi by 10 km/6 mi, polluting the French and British coastlines. Around £1.3 million is spent on unsuitable detergents that cause further harm to wildlife.

1968 US entomologist Paul Erlich publishes *The Population Bomb*, which becomes an instant and influential best-seller.

Erlich highlights the problem of a growing world population proceeding at an exponential rate and he predicts that, by 1977, one in seven people will die due to food shortages. The book is widely criticized and its many flaws as an academic work are exposed, but, nevertheless, it draws a large following.

British Petroleum workers practising methods for controlling oil pollution. Oil spills became a major environmental problem in the 1960s and 1970s as a result of oil exploration along the continental shelf and several major oil spills involving supertankers. *British Petroleum*

293

Shades of Green: Ethics and Value in Ecology

by John Cartwright

THE INTELLECTUAL ROOTS OF ATTITUDES TO NATURE

The 1960s was a decade of rising ecological consciousness in the West. The World Wildlife Fund (now the World Wide Fund for Nature) was founded in 1961, and in 1962 Rachel Carson (1907–1964) published her influential work *Silent Spring*. It was becoming clear that humans, despite their incredible technical achievements, were transforming whole areas of the planet for the worse for themselves and many other species. This was clearly a problem in need of explanation and remedy and economists, ecologists, and historians were not slow in supplying answers. In 1967 the historian Lynn White Jr published an influential paper entitled 'The Historical Roots of Our Ecologic Crisis'. White laid the blame for the environmental problems of the West firmly at the door of orthodox Christian attitudes to nature that had shaped science, technology, and people's value systems. White thought that the answer to the crisis did not lie in more or better science: 'our present science and our present technology are so tinctured with orthodox Christian arrogance toward nature that no solution for our ecologic crisis can be expected from them alone'.

White's thesis provoked much controversy. Theologians, understandably, were annoyed. Conceding the point that the Bible did enjoin humans to 'fill the earth and subdue it' and to 'rule over the fish of the sea and . . . over all creatures that move along the ground', they were keen to point out that elsewhere in the Bible the relationship suggested is one of stewardship, where humans look after God's creation as would a tenant after an owner's property. Philosophers such as John Passmore, however, in his *Man's Responsibility for Nature* (1974), although disagreeing with White's emphasis on Christianity as the sole source of anthropocentrism, nevertheless was in accord that 'Western metaphysics and Western ethics . . . have done a great deal to encourage the ruthless exploitation of nature.'

The sense that science had somehow taken a wrong turn was not a new one. As early as 1926 the philosopher Alfred North Whitehead was stating that the future task of philosophy should be to 'end the divorce of science from the affirmations of our aesthetic and ethical experiences'. But it was in the last quarter of the 20th century that the feeling that humans were steadily destroying the ecological base to their own existence became acute. For many, reductionism, materialism, and consumerism were to blame. Some seemed to revel in the new responsibility that the scientific world view laid on the shoulders of its adherents. The French Nobel prizewinner Jacques Monod (1910–1976), for example, noted in his work *Chance and Necessity* (1970) how 'The ancient covenant is in pieces; man at last knows that he is alone in the unfeeling immensity of the universe, out of which he emerged by chance. Neither his destiny nor his duty have been written down'. Others turned to the science of ecology for ethical and practical guidance on how to manage both the planet and human affairs. With its holistic emphasis on the interconnectedness of things, and the important role played by even the smallest of organisms in the great economy of nature, ecology seemed a likely source of value in a world deprived of ultimate meaning by the reductionism of physics, chemistry, and biochemistry.

THE DEEP AND THE SHALLOW ECOLOGY MOVEMENT

A distinction useful to understanding the multifarious ways in which ecology and ethics became entangled was introduced by the Norwegian philosopher Arne Naess (1912–) in his paper of 1973 entitled 'The Shallow and the Deep, Long-Range Ecology Movement'. Naess distinguished between 'shallow' and 'deep' environmental thought. Shallow ecology expresses concern for the environment on the basis that a deteriorating environment would worsen the conditions for human existence. Shallow ecology is therefore anthropocentric and takes the view that the environment must be properly managed to foster human welfare. Deep ecology, in contrast, shows respect for all organisms irrespective of their value to humans. Here we meet a distinction that is fundamental to understanding environmental debates.

Shallow greens subscribe to a belief in instrumental value, the view that the environment as a whole and other life forms in it are worthy of respect in so far as they are instrumental in satisfying human needs and desires. This need not be so callous as it sounds. We should conserve forests and fish stocks and manage them properly, for example, because they are a source of timber and food, respectively. But this same ethic would lead to the conservation of wilderness and endangered species in so far as humans take delight in these things. Within this mindset, the values ascribed to other living things are hierarchical. Hence, we would probably place more value on, say, pandas than on parasites that cause disease. Deep greens, however, argue that the world has intrinsic value, a value that is irrespective of its use for humans. From this position, we should make efforts to conserve whales not because their sustainable hunting provides useful products such as sushi for expensive restaurants, nor solely because humans take pleasure in seeing them or knowing of their existence, but because whales have their own life interests and hence a right to exist, a right that brings into question, therefore, whether they should be hunted at all. This division between intrinsic and instrumental value, between deep and shallow green positions, and between ecocentric and anthropocentric philosophies, is now a fundamental component of environmental debates.

CATASTROPHISTS AND CORNUCOPIANS

As the deep greens debated with the shallow greens, another polarization of opinion was taking place. In the 1970s a group, sometimes called the 'catastrophists' or neo-Malthusians, were predicting dire shortages of resources, and widespread famine on a massive scale by the end of the 20th century as the effects of the human population explosion were felt. Typical of this view was the report *The Limits to Growth* issued by Dennis Meadows and his co-workers in the USA in 1972, which painted a bleak future for the human race unless drastic measures were taken. The fact that some of the more depressing predictions of the catastrophists failed to materialize lent credence to a group (usually economists) who came to be called the 'cornucopians'. True to their name, these thinkers, exemplified in the USA by Julian Simon and Herman Kahn and in the UK by Wilfred Beckerman, argued that the world's resources are not running out, that there is no environmental crisis, and that industrial capitalism and free enterprise will ultimately deliver solutions to any short-term problems.

For many politicians, alarmed at the radical and almost certainly unpopular measures proposed by the catastrophists to avert an ecological crisis, yet worried that environmental issues were not entirely safe in the hands of the free market, the concept of sustainable development, highlighted in the Brundtland Report of 1987, became a sort of intellectual life raft. It seemed as if humans could have their cake and eat it. One of the most widely quoted definitions of sustainability comes from the Brundtland Report itself: 'Development that meets the needs of the present without compromising the ability of future generations to meet their own needs.' Some on the 'New Right' questioned whether there was any ethical obligation to future generations, and if so how far into the future this should extend. Ecocentrics pointed out that the whole concept rested upon the old anthropocentric notion of meeting human needs. They were also concerned that economic growth got the noun (development) while ecology had to make do with the adjective (sustainable).

Following the publication of the Brundtland Report, the UN General Assembly resolved in 1989 to convene a conference on environment and development. That conference was held in Rio de Janeiro, Brazil, in June 1992 and became known as the Earth Summit. States participating in the summit agreed four documents:

- Forest Principles – a non-binding statement of principles for the sustainable management of forests
- Biodiversity Convention – an agreement on how to protect the Earth's wide range of species and habitats
- Climate Change Convention – an agreement on a framework to proceed with the reduction of greenhouse gas emissions
- Agenda 21 – a plan of action for the 21st century intended to promote sustainable development.

Shades of Green: Ethics and Value in Ecology *continued*

It can be seen that sustainability is a concept that infused the thinking of the summit. There are, however, many problems, both theoretical and political, that remain with the whole concept of sustainability. At the theoretical level, much effort is expended in refining the notion of sustainability and in developing indicators that enable it to be measured. At the political level, there is the danger that sustainability becomes a concept that no one disagrees with and, consequently, is not taken seriously as a guide to action. More than ever before, politicians, scientists, economists, and the consuming public need to talk and listen.

December 1968 US biologist Garrett Hardin publishes his essay 'The Tragedy of the Commons'.

Hardin expresses pessimism about the idea that population growth can continue at present rates. He points to the tragedy of the commons – a situation where environmental destruction is brought about by individuals maximizing their own utility without regard for the whole. The metaphor is then applied to the global commons: air, land, and water. It pays any individual to pollute the oceans, for example, because the benefit to the polluter in not having to treat the waste returns entirely to the individual, whereas all the community of which the polluter is only one small part shares the social cost of poisoned oceans. Hardin's essay is criticized but it is widely recognized that he has revived a potent metaphor for many of the environmental problems that the world faces.

1969 After resigning from the Sierra Club in frustration at its slow reaction to environmental issues, US conservationist David Brower sets up a new organization called Friends of the Earth.

Mathematics

1961 US meteorologist Edward Lorenz discovers a mathematical system with chaotic behaviour, leading to a new branch of mathematics known as chaos theory.

It soon finds applications in a number of branches of science, such as fluid dynamics, electronics, and meteorology. Chaos theory attempts to predict the *probable* behaviour of such systems, based on a rapid calculation of the impact of as wide a range of elements as possible.

Physics

1960 US physicist Theodore Maiman constructs the first laser (light amplification by stimulated emission of radiation), a device producing an intense beam of coherent light.

The principle of laser action was put forward by Arthur L Schawlow and Charles H Townes in 1958. However, Maiman uses a ruby to construct his laser, whereas they had suggested that the greatest chance of success lay in the use of gases. It is some months after Maiman's demonstration that a helium/neon gas laser is produced.

1960 US nuclear physicist Donald Glaser receives the Nobel Prize for Physics for his invention of the bubble chamber in 1952.

1961 US physicist Robert Hofstadter and German physicist Rudolf Mössbauer share the Nobel Prize for Physics: Hofstadter for his pioneering studies of electron scattering in atomic nuclei and for his consequent discoveries concerning the structure of the nucleons, and Mössbauer for his research on the resonance absorption of gamma radiation and his discovery of the Mössbauer effect.

1961 US physicist Murray Gell-Mann and Israeli physicist Yuval Ne'eman independently propose a classification scheme for subatomic particles that comes to be known as the Eightfold Way.

This classification leads to the prediction of the Ω^- particle. It is discovered in bubble chamber photographs at the Brookhaven National Laboratory in 1963.

1962 Soviet physicist Lev Landau receives the Nobel Prize for Physics for his theories concerning condensed matter, especially liquid helium.

1963 US physicists Eugene Wigner and Maria Goeppert-Mayer and German physicist Hans Jensen receive the Nobel Prize for Physics: Wigner for his development of the shell theory of the atomic nucleus, and Goeppert-Mayer and Jensen for their contributions to the theory of the atomic nucleus and elementary particles, particularly through the discovery and application of fundamental symmetry principles.

1964 US physicist Charles Townes shares the Nobel Prize for Physics with Soviet physicists Nicolay Basov and Alexandr Prokhorov.

Townes is honoured for fundamental work in the field of quantum electronics, which has led to the construction of oscillators and amplifiers based on the maser-laser principle. Basov and Prokhorov are honoured for basic researches in the field of experimental physics, which led to the discovery of the maser and laser.

1964 US physicists Murray Gell-Mann and George Zweig suggest the existence of quarks, the building blocks of hadrons, which are subatomic particles that experience the strong nuclear force.

Using the three quarks, 'up', 'down', and 'strange', Gell-Mann and Zweig are able to explain the classification of the then known particles. Gell-Mann receives the Nobel Prize for Physics for this work in 1969.

1965 Japanese physicist Sin-Itiro Tomonaga and US physicists Julian Schwinger and Richard Feynman are jointly awarded the Nobel Prize for Physics for their work in quantum electrodynamics.

One does not, by knowing all the physical laws as we know them today, immediately obtain an understanding of anything much.

Richard Feynman, US physicist, *The Character of Physical Law* (1965)

1966 French physicist Alfred Kastler receives the Nobel Prize for Physics for his discovery of optical methods of studying Hertzian resonance in atoms.

1967 US physicist Hans Bethe receives the Nobel Prize for Physics for his contributions to the theory of nuclear reactions, especially his discoveries concerning energy production in stars.

1967 US nuclear physicists Sheldon Lee Glashow and Steven Weinberg and Pakistani nuclear physicist Abdus Salam separately develop the electroweak unification theory, which provides a single explanation for electromagnetic and 'weak' nuclear interactions.

The quest to include the two other fundamental forces, the strong force and gravitation, continues.

> *The whole history of particle physics, or of physics, is one of getting down the number of concepts to as few as possible.*
>
> Abdus Salam, Pakistani physicist, quoted in L Wolpert and A Richards *A Passion for Science* (1988)

1968 US physicist Luis Alvarez receives the Nobel Prize for Physics for his work in elementary particle physics.

In particular he is honoured for the discovery of a large number of resonance states, made possible through his development of the hydrogen bubble chamber technique.

1969 US physicist Murray Gell-Mann receives the Nobel Prize for Physics for his work on the classification of elementary particles and their interactions.

1969 The University of Utah, Stanford Research Institute, and the Santa Barbara and Los Angeles campuses of the University of California are linked by an experimental computer network.

Funded by the US Advanced Research Projects Agency, it is known as the Arpanet. From this simple beginning will grow the Internet we now know.

1970–1979

Astronomy

1970 NASA's Orbiting Astronomical Observatory (launched December 1968) and Orbiting Geophysical Observatory detect hydrogen in the tail of a comet.

1970 The Small Astronomy Satellite (SAS) is launched by the USA. It catalogues X-ray sources and leads to the development of the High Energy Astronomy Observatory (HEAO).

1970 The Soviet spacecraft *Venera 7* transmits information from the surface of Venus.

19 April 1971 The USSR launches *Salyut 1*, the first space station. It is visited by cosmonauts in June, and re-enters the Earth's atmosphere in October.

26 July–7 August 1971 NASA spaceship *Apollo 15*, commanded by the US astronaut David R Scott, is sent to the Moon. It contains a Lunar Roving Vehicle that enables Scott and his fellow astronaut James B Irwin to explore 27 km/17 mi of the lunar surface.

24 November 1971 The US space probe *Mariner 9* (launched in May) becomes the first artificial object to orbit another planet (Mars); it transmits 7,329 photographs of the planet and its two moons, Deimos and Phobos.

5 January 1972 US President Richard Nixon authorizes a $5.5 billion six-year programme to develop plans for a spaceship capable of undertaking multiple missions, thereby launching the space shuttle programme.

2 March 1972 NASA uncrewed spacecraft *Pioneer 10* takes off towards Jupiter.

Pioneer 10 is the first spacecraft to travel through the asteroid belt, and the first spacecraft to make direct observations and obtain close-up images of Jupiter. The spacecraft makes valuable scientific investigations in the outer regions of the Solar System until the end of its mission on 31 March 1997.

1973 Comet Kohoutek is first observed by Czech astronomer Lubos Kohoutek.

May 1973 NASA launches *Skylab*, the first US space station. It contains a workshop for carrying out experiments in weightlessness, an observatory for monitoring the Sun, and cameras for photographing the Earth's surface. *Skylab* is subsequently visited by three three-person crews, and astronauts make observations of the Sun, manufacture superconductors, and conduct other scientific and medical experiments.

1974 English physicist Stephen Hawking suggests that 'black holes aren't black' – they 'evaporate' by emitting subatomic particles.

1974 English radio astronomers Martin Ryle and Antony Hewish receive the Nobel Prize for Physics for their work in radio astronomy.

Ryle is honoured for his observations and inventions, in particular of the aperture synthesis technique, and Hewish for his decisive role in the discovery of pulsars.

1974 The US probe *Mariner 10* (launched 3 November 1973) photographs the upper atmosphere of Venus (February) and then takes the first photographs of the surface of Mercury (March and September), flying within 740 km/460 mi of the planet's surface.

September 1974 US astronomer Charles T Kowal announces the discovery and naming of Leda, the 13th moon of Jupiter.

1975 US radio astronomers Russell Hulse and Joseph Taylor identify PSR1913+16 as a binary star (a pair of stars in orbit around each other); it loses energy at a rate that Einstein's theory of general relativity predicts for the emission of gravitational waves (ripples in the structure of space-time that may occur singly or as continuous radiation).

15 March 1975 The US-German space probe *Helios 1* (launched 10 December 1974) passes the Sun at a distance of 45 million km/28 million mi and returns information about the Sun's magnetic field and solar wind.

1 August 1975 The European Space Agency is founded in Paris, France, to undertake research and develop technologies for use in space.

22–25 October 1975 The Soviet spacecraft *Venera 9* and *10*, launched on 8 June and 14 June respectively, land on Venus and transmit the first pictures from the surface of another planet.

1976 Astronomers at Harvard College Observatory, Cambridge, Massachusetts, discover bursts of X-rays coming from a star cluster 30,000 light years from Earth.

July 1976 The US spacecraft *Viking 1* and *Viking 2* (launched in 1975) soft-land on Mars (20 July and 7 August). They make meteorological readings of the Martian atmosphere and search for traces of bacterial life, which prove inconclusive.

March 1977 US astronomer James Elliot, with several groups of other US astronomers, discovers rings around Uranus when the planet occludes a relatively bright star.

1 November 1977 The interplanetary body Chiron is discovered by US astronomer Charles T Kowal; at least 200 km/120 mi in diameter, it is initially thought to be an asteroid but is later identified as a giant cometary nucleus. It is subsequently classified by the International Astronomical Union (IAU) in 1995 as a 'centaur', the first of a new category of interplanetary object, named after the centaur (half man, half horse) in Greek mythology.

May–9 December 1978 NASA launches *Pioneer Venus Orbiter*, which enters Venus's orbit on 4 December.

The *Orbiter* carries 17 experiments (with a total mass of 45 kg/99 lb), including a cloud photopolarimeter to measure the vertical distribution of the clouds, a surface radar mapper to determine topography and surface characteristics, and an infrared radiometer to measure IR emissions from the Venus atmosphere.

22 June 1978 US astronomer James W Christy discovers Charon, a moon orbiting Pluto.

The similarity of Pluto's size to Charon's, and the fact that their common centre of gravity, about which they orbit, is outside either solid body, makes Pluto–Charon a double, or binary, planet, rather than a planet–moon system.

1979 The IRAM array telescope begins operation at Plateau de Bruce, France; its four 15.0 m/19.2 ft dishes make it the largest millimetre telescope in the world.

1979 The Multiple Mirror Telescope begins operation on Mount Hopkins, Arizona; it focuses the light from six 180 cm/70 in telescopes to form one image, giving the light-gathering power of a single 4.5 m/15.7 ft telescope. It becomes the prototype for larger optical telescopes.

March 1979 NASA spacecraft *Voyager 1* comes within 275,000 km/172,000 mi of Jupiter, in what proves to be its closest approach.

Images returned by the spacecraft show the complex, swirling turbulence of Jupiter's atmosphere in exquisite detail. The Great Red Spot is revealed as a giant storm, three times the size of Earth, raging in Jupiter's upper atmosphere, surrounded by rippling currents that rotate around it. *Voyager 1* finds nine active volcanoes erupting on Io, the innermost of Jupiter's four major moons. A thin, dusty ring is also discovered around Jupiter, forcing revision of theories about the origins and mechanics of planetary ring systems.

July 1979 The US space station *Skylab* falls back to Earth after travelling 140 million km/87 million mi in orbit since 1973.

1 September 1979 The US space probe *Pioneer 11* (launched 6 April 1973) travels through the rings of Saturn to within 20,900 km/13,000 mi of the planet.

The rings are found to be made of dirty water ice. Additional rings and high-energy particles within Saturn's magnetosphere are also discovered.

Biology

1970 Argentine biochemist Luis Leloir receives the Nobel Prize for Chemistry for his discovery of sugar nucleotides and their role in carbohydrate biosynthesis.

1970 US geneticist Hamilton Smith discovers the Hind II restriction enzyme that breaks the DNA strand at predictable places, making it an invaluable tool in recombinant DNA technology.

The work confirms and extends research by Werner Arber in the 1960s on an early example of such an enzyme. Smith shares the Nobel Prize for Physiology or Medicine in 1978 with Arber and Daniel Nathans.

September 1970 Indian-born US biochemist Har Gobind Khorana assembles an artificial yeast gene from its chemical components.

Earlier, he synthesizes all 64 codons, each formed from a triplet of nucleotide bases, thereby helping to establish the 'dictionary' of the genetic code for amino acids.

1971 US dendrochronologist Charles Ferguson of the University of Arizona establishes a tree-ring chronology dating back to *c.* 6000 BC.

1972 US microbiologist Daniel Nathans uses a restriction enzyme that splits DNA (deoxyribonucleic acid) molecules to produce a genetic map of the monkey virus (SV40), the simplest virus known to produce cancer; it is the first application of these enzymes to an understanding of the molecular basis of cancer.

Nathans shares the Nobel Prize for Physiology or Medicine in 1978 for this research.

1973 The Nobel Prize for Physiology or Medicine is awarded jointly to Austrian zoologists Karl von Frisch and Konrad Lorenz, and Dutch-born British zoologist Nikolaas Tinbergen for their discoveries concerning individual and social behaviour patterns in animals.

1973 US biochemists Stanley Cohen and Herbert Boyer develop the technique of recombinant DNA (deoxyribonucleic acid). Strands of DNA are cut by restriction enzymes from one species and then inserted into the DNA of another; this marks the beginning of genetic engineering.

Cohen and Boyer remove DNA from the *E. coli* bacteria and then cut it into fragments. They join different fragments to the original pieces and reinsert them into the bacteria, which then reproduce normally.

1975 Argentine immunologist César Milstein and German immunologist Georges Köhler develop the first monoclonal antibodies – lymphocyte and myeloma tumour cell hybrids that are cloned to secrete unlimited amounts of specific antibodies – in Cambridge, England.

They inject experimental animals with a particular antigen in order to stimulate the formation of corresponding antibodies by the animals' B-lymphocytes. Antibody-rich B-lymphocytes are then harvested from the spleen tissue and fused with malignant cells forming hybridomas. The hybridomas divide to produce a potentially immortal cell line that produces large quantities of pure antibody (hence 'monoclonal'). Monoclonal antibodies are invaluable tools for disease testing and diagnosis.

1975 British scientist Derek Brownhall produces the first clone of a rabbit, in Oxford, England.

1975 The gel-transfer hybridization technique for the detection of specific DNA sequences is developed; it is a key development in genetic engineering.

1976 Japanese molecular biologist Susumu Tonegawa demonstrates that antibodies are produced by large numbers of genes working in combination.

28 August 1976 Indian-born US biochemist Har Gobind Khorana and his colleagues announce the construction of the first artificial gene to function naturally when inserted into a bacterial cell.

In 1970, Khorana had synthesized an artificial gene from individual nucleotides but had no evidence that his compound could replace a natural gene. This later work provides the evidence he needs. He has successfully developed a process to synthesize fully functional artificial genes. This is a major breakthrough in genetic engineering.

1977 English biochemist Frederick Sanger describes the full sequence of 5,386 bases in the DNA of virus *phi*X174 in Cambridge, England; the first sequencing of an entire genome.

This virus is a bacteriophage, a virus that infects bacteria, and is the first organism to have its complete genome determined. Sanger shares the Nobel Prize for Chemistry in 1980 for this work; he is only the fourth person in history to be awarded a second Nobel prize.

1977 Scientists from the project FAMOUS (French-American Mid-Ocean Undersea Study) in their deep-sea submersible vehicle *ALVIN* discover a host of strange life forms, such as large red and white tube worms, near undersea hot springs heated by ocean-ridge volcanism. The discovery shows that life can exist in extreme conditions.

1977 Swedish neurologist Tomas Hökfelt discovers that most neurons contain not one but several neurotransmitters.

Chemistry

1970 US chemist Albert Ghiorso and colleagues at the University of California, Berkeley, synthesize the radioactive element dubnium (atomic number 105).

They create it by bombarding californium with a beam of nitrogen nuclei using a linear accelerator. A Soviet research group at the Joint Nuclear Research Institute at Dubna, USSR, again disputes the discovery (see 1964), claiming they had created the element in 1967. The US group propose the name 'hahnium' after the discoverer of nuclear fission, Otto Hahn, while the Soviet group suggest 'neilsbohrium' in honour of Danish physicist Niels Bohr. The International Union of Pure and Applied Chemistry (IUPAC) allots the temporary name unnilpentium. In 1997, the name dubnium is adopted, at the recommendation of IUPAC.

1971 Canadian chemist Gerhard Herzberg receives the Nobel Prize for Chemistry for his work on electron structure and the geometry of molecules, particularly free radicals.

1971 US chemist Robert Burns Woodward and Swiss chemist Albert Eschenmoser synthesize vitamin B_{12} (cyanocobalamin).

The molecular structure of coenzyme B_{12} is so complex that it takes the researchers over a decade to complete the synthesis of the compound. The process involves over 100 separate chemical reactions and is too complex and costly to be used in commercial production of the vitamin. However, the chemical route and the procedures used to attain it prove to be of great importance for the synthesis of many other complex compounds, such as vitamins and antibiotics.

1972 US biochemists Christian Anfinsen, Stanford Moore, and William Stein receive the Nobel Prize for Chemistry for their work on amino-acid structure and biological activity of the enzyme ribonuclease.

1973 German chemist Ernst Fischer and English chemist Geoffrey Wilkinson receive the Nobel Prize for Chemistry for their work on organometallic sandwich compounds.

1974 US chemist Paul Flory receives the Nobel Prize for Chemistry for his work on the physical chemistry of macromolecules.

1974 US chemist Albert Ghiorso and colleagues at the university of California, Berkeley, and Soviet scientist Georgii Flerov and colleagues at the Jomt Institute for Nuclear Research, Dubna, both claim to have synthesized the element seaborgium (atomic number 106). The element is, in the end, named after the US nuclear chemist Glenn Seaborg.

Having used different methods, the two groups had produced different isotopes of the element. Both isotopes are synthesized by researchers in Dubna in 1993.

1975 Australian chemist John Cornforth and Swiss chemist Vladimir Prelog receive the Nobel Prize for Chemistry for their work in stereochemistry.

1976 Soviet scientists at the Joint Institute for Nuclear Research in Dubna, USSR, announce the synthesis of bohrium (atomic number 107).

They bombard a rotating cylinder coated with bismuth with charged chromium particles and detect the new element, which only lasts a few thousandths of a second. German physicist Peter Armbruster and collegues at Darmstadt, Germany, substantiate the work in 1981. They name the new element 'bohrium' after the Danish physicist Niels Bohr.

1976 US chemist William Lipscomb receives the Nobel Prize for Chemistry for his work on the structure and chemical bonding of boranes, compounds of boron and hydrogen.

1977 Japanese researcher Hideki Shirakawa and US researchers Alan MacDiarmid and Alan Heeger add iodine to one of the new electrically conductive plastics, vastly improving its conductive characteristics. They receive the Nobel Prize for Chemistry in 2000.

1977 Soviet-born Belgian chemist Ilya Prigogine receives the Nobel Prize for Chemistry for his contributions to the thermodynamics of irreversible processes, particularly to the theory of dissipative structures.

1977 English biochemists Frederick Sanger and Alan Coulson, and US molecular biologists Walter Gilbert and Allan Maxam, independently develop a rapid gene-sequencing technique that uses gel electrophoresis.

This technique greatly increases the speed at which nucleotides can be added together to form synthetic proteins. It is of particular importance to the rapidly growing genetics industry.

1978 English chemist Peter Mitchell receives the Nobel Prize for Chemistry for his studies of biological energy transfer and chemiosmotic theory.

1979 US chemist Herbert Brown and German chemist Georg Wittig receive the Nobel Prize for Chemistry for use of boron and phosphorus compounds, respectively, in organic syntheses.

Earth Sciences

1970 British geologist John F Dewey with J M Bird relate the positions of the Earth's mountain belts to the motions of lithospheric plates. Mountain belts are shown to form at the boundaries between two plates that are colliding.

23 July 1972 The USA launches *Landsat 1*, the first of a series of satellites for surveying the Earth's resources from space.

1974 Earth scientist John Liu discovers that the lower mantle is likely to be composed of silicate perovskite, a mineral with a structure wildly different from the minerals found in the Earth's upper mantle and crust.

1974 The Global Atmospheric Research Program (GARP) is launched. An international project, its aim is to provide a greater understanding of the mechanisms of the world's weather by using satellites and by developing a mathematical model of the atmosphere.

1975 Five nations, the USSR, West Germany, France, Japan, and the UK join the Deep Sea Drilling Project (DSDP) to form the International Phase of Ocean Drilling (IPOD).

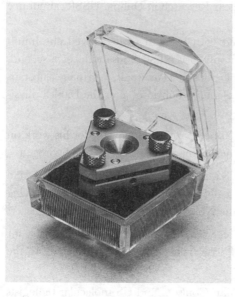

Diamond anvil cell used in laboratory experiments to simulate the high pressure and temperature found inside planets. *High Pressure Diamond Optics, Inc*

1975 Mineral physicists David Mao and Peter Bell of the Geophysical Laboratory in Washington, DC, use a diamond-anvil cell to produce pressures exceeding a million atmospheres.

The instrument is used to simulate the high pressures and temperatures thought to exist deep within the Earth and even towards the centre of large planets such as Jupiter.

Ecology

1970 The year is designated as European Conservation Year. The aims are to seek agreement on policies to conserve the environment and educate citizens in the importance of conservation. Conferences are held throughout the year in the UK.

1970 US population biologists Paul and Anne Ehrlich publish *Population, Resources, and Experimental Issues in Human Ecology*, in which they predict famine in the 1970s through depletion of resources.

February 1970 US president Richard Nixon sends Congress the blueprint for an environmental clean-up bill. Among other initiatives, the president proposes earmarking $10 billion for waste treatment.

22 April 1970 Millions of Americans participate in the first 'Earth Day' as a series of mass meetings, speeches, and events take place across the USA.

Various theories are promulgated about causes of environmental degradation. The idea of an Earth day originates with Senator Gaylord Nelson of Wisconsin but the main organizer is Sam Brown, an anti-war activist. Numerous lectures are given on this day at campuses all over the USA, including talks by Barry Commoner, Paul Erlich, René Dubos, and Ralph Nader.

December 1970 The Environmental Protection Agency (EPA) comes into being in the USA. It becomes the most powerful regulatory agency in the federal government and reports directly to the president. It is given wide-ranging environmental jurisdiction, including the regulation of air pollution, water pollution, drinking water quality, hazardous waste disposal, and the regulation of toxic substances.

The EPA also creates the Council on Environmental Quality with a remit to make annual reports to the US president on the state of the environment. The EPA requires an 'environmental impact statement' for any federally funded project that might damage the environment.

December 1970 US president Richard Nixon signs the Clean Air Act. The act imposes stricter air pollution standards and requires US car manufacturers to reduce emissions of nitrogen dioxide, carbon monoxide, and hydrocarbons by 90% by the mid-1970s.

1971 Greenpeace, the environmental campaign organization, is founded. It grows from a group called the 'Don't Make a Wave Committee' set up to protest against nuclear testing on Amichitka Island, Alaska.

1971 US biologist Barry Commoner publishes *The Closing Circle: Man, Nature and Technology*. It is a book that links scientific with social and economic analyses of environmental problems. He suggests three principal causes of damage to the environment: population growth, increasing affluence, and new technologies, and concludes that it is largely the nature of modern technology and industrial capitalism that is to blame for pollution.

16 January 1971 The official opening of the Aswan High Dam built across the Nile takes place. Its reservoir, Lake Nasser, is 300 km/186 mi long and necessitated the relocation of the Abu Simbel temple complex. The dam allows Egypt to control the annual flooding of the Nile but increases incidence of the disease schistosomiasis.

1972 As a result of pressure from conservationists, Britain's Ministry of Agriculture, Fisheries, and Foods (MAFF) finally withdraws subsidies to farmers for hedgerow removal.

In the mid-1940s, Britain had about 980,000 km/609,000 mi of hedgerows. By 1972, around 230,000 km/143,000 mi have been lost with about 85% grubbed up by farmers receiving government grants to improve farming efficiency. Conservationists argue that hedgerows are fundamentally important in supporting populations of birds, mammals, insects, and plants. The removal of subsidies slows down the loss, but hedgerow removal continues.

1972 *The Limits to Growth* by Donella Meadows, Dennis Meadows, and Jørgen Randers is published by the Massachusetts Institute of Technology, USA. Based on a Club of Rome report and computer simulation, it predicts environmental catastrophe if the depletion of the Earth's resources, overpopulation, and pollution are not acted upon immediately.

1972 In the UK, the *Ecologist* magazine publishes a short book titled *Blueprint for Survival*.

The book calls for radical social change such as decentralization of power and the formation of self-sufficient communities. The book is criticized for simply interpreting current trends without taking account of the complexities of the free market and the reality of technological innovation.

1972 The United Nations Environment Programme (UNEP) is established; its aim is to advise and coordinate environmental activities within the United Nations.

1972 The US Environmental Protection Agency announces a ban on virtually all uses of the pesticide DDT on the grounds that it harms fish and bird life and that there is a possibility it could cause cancer in humans. The USA still exports DDT to other countries, a trade worth £26 million each year.

The ban follows research indicating that the populations of several fish-eating birds, such as the brown pelican, osprey, and the national symbol of the USA, the bald eagle, have fallen to dangerously low levels. Following the ban, the bird populations improve and the bald eagle is eventually taken off the endangered species list.

June 1972 The United Nations Conference on the Human Environment is held in Stockholm, Sweden; the first international conference on the state of the environment, its aim is to improve the world's environment through monitoring, resource management, and education.

305

Residents of Seoul, Korea, being sprayed with DDT in 1951 to prevent the spread of typhus. DDT is banned in the USA in 1972 because of its accumulation in the food chain. *Bettmann/CORBIS*

The Swedish participants suggest that environmental damage is caused by acid rain. Partly as a result of this pressure, the Organization for Economic Cooperation and Development (OECD) instigates an important international research programme on cross-border pollution in Europe. As a working text to accompany the Conference, English economist Barbara Jackson and French-born US microbiologist René Dubos write *Only One Earth: The Care and Maintenance of a Small Planet.*

1973 Australian physicist Robert May publishes *Stability and Complexity in Model Ecosystems.* In this work, May overturns the accepted notion that the greater the species diversity in an ecosystem, the more stable it will be.

May shows that simple and accepted mathematical models do not capture the chaotic way in which populations of organisms rise and fall. May finds that systems with many species are often fragile and species numbers fluctuate in an unpredictable way.

1973 Herman E Daly, professor of economics at Louisiana State University, USA, publishes *The Steady State Economy: Toward a Political Economy of Biophysical Equilibrium and Moral Growth.* In this work, Daly revives the ideas of 19th century philosopher John Stuart Mill that economic growth for its own sake is futile and society should aim for a steady-state economy.

1973 Representatives from 80 nations sign the Convention on International Trade in Endangered Species (CITES) that prohibits trade in 375 endangered species of plants and animals and the products derived from them, such as ivory; the USA does not sign.

1973 Norwegian philosopher Arne Naess, a professor at the University of Oslo, publishes his paper 'The Shallow and the Deep, Long-Range Ecology Movement'.

The paper is influential in clarifying a division that is becoming increasingly apparent between different types of environmentalists: the 'shallow' and the 'deep'.

According to Naess, the shallow environmentalists show concern for the environment because the planet is the life support system for humans. Deep ecologists show respect for all life forms irrespective of their value to humans. Followers of Naess see this paper as a landmark in deep green thought.

1974 Mexican chemist Mario Molina and US chemist F Sherwood Rowland suggest that the chlorofluorocarbons (CFCs) used in fridges and as aerosol propellants may be damaging the atmosphere's ozone layer that filters out much of the Sun's ultraviolet radiation, and that they could persist in the stratosphere for decades.

Rowland and Molina show that, in the stratosphere, ultraviolet radiation could initiate the release of monoatomic chlorine from CFCs and this chlorine would then react with ozone to produce chlorine monoxide. They calculate that a single CFC molecule could destroy thousands of ozone molecules. They are awarded the Nobel Prize for Chemistry in 1995.

1975 The Ramsar Convention on Wetlands of International Importance Especially as Waterfowl Habitat, formulated in 1971, now comes into force. It is the first international convention dealing solely with habitat, and there are 52 contracting parties.

The convention establishes a number of requirements of its signatories. There is a general duty to promote the conservation of wetlands, a requirement that sites must be designated, that each signatory must designate at least one site in its territory and that a site can only be deleted or reduced in size on the grounds of urgent national interests. The convention also provides for a Ramsar bureau based in Switzerland.

1975 US biologist and entomologist Edward Wilson publishes his controversial *Sociobiology: The New Synthesis.*

Wilson captures and synthesizes the rediscovery of Darwin that has taken place in ecology over the previous decade. He shows how an evolutionary approach illuminates the behaviour of animals and how they interact with each other in ecosystems. Most controversially, he applies these ideas to human behaviour.

1976 US scientists experiment with algae and micro-organisms that consume crude oil as a means of clearing up oilspills.

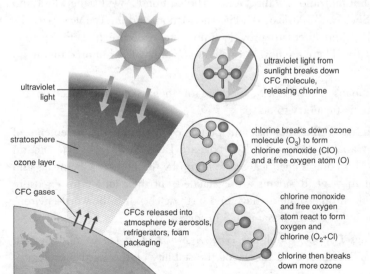

ultraviolet light

stratosphere

ozone layer

CFC gases

CFCs released into atmosphere by aerosols, refrigerators, foam packaging

ultraviolet light from sunlight breaks down CFC molecule, releasing chlorine

chlorine breaks down ozone molecule (O_3) to form chlorine monoxide (ClO) and a free oxygen atom (O)

chlorine monoxide and free oxygen atom react to form oxygen and chlorine (O_2+Cl)

chlorine then breaks down more ozone

The destruction of the ozone layer by chloro-fluorocarbons (CFCs). CFCs discharged into the atmosphere break down in sunlight releasing chlorine, which breaks down the ozone to form chlorine monoxide and a free oxygen atom. These products react together to form oxygen and chlorine, leaving the chlorine to break down another ozone molecule, and so on. This process was first described in 1974.

April 1976 Indian president Indira Gandhi, operating under emergency powers, issues a National Population Policy Statement, making it clear that there would be no objection to a programme of forced male sterilization to enable local regions to reach their sterilization targets.

Regional states use a variety of methods to coerce men into having vasectomies. These include force, withholding of salary, or refusal to consider for employment. The measures evoke widespread revulsion in India and the rest of the world.

26 July 1976 At a factory owned by a subsidiary of the Swiss company Hoffmann–La Roche in the village of Seveso near Milan, a leak of between 2 kg/4.4 lb and 130 kg/287 lb of tetrachlorodibenzo-*p*-dioxin (TCDD) occurs. Over 700 people are evacuated from the area and 500 show symptoms of poisoning.

The factory has been manufacturing 2,4,5-trichlorphenoxyacetic acid (2,4,5-T) used as a herbicide. Domestic animals in the area are slaughtered.

1977 Dutch scientists discover that the wastes from some incinerators are contaminated by dioxins – chemicals thought to cause cancer.

1977 In the UK the Geological Conservation Review (GCR) begins. The UK is divided into 97 blocks and submissions are invited to identify key sites for geological conservation in each block.

The GCR is the most thorough and costly review of sites of geological interest for conservation purposes ever carried out. By 1999 some 3,000 sites have been identified, resulting in the designation of 2,200 Sites of Special Scientific Interest (SSSIs).

21 October 1977 At a five-day meeting sponsored by the United Nations Environment programme, 13 Mediterranean countries and the EEC agree the principles of a treaty to fight pollution of the Mediterranean. Experts advise that the first targets should be the discharge of industrial wastes, municipal sewage, and agricultural chemicals.

1978 A male silverback mountain gorilla (*Gorilla gorilla beringei*) called Digit, belonging to a group studied by US primatologist Dian Fossey, is killed by poachers. Another male, a female, and an infant are also killed. A massive publicity campaign in Britain and the USA leads to the formation of the Mountain Gorilla Project.

Four international organizations – the World Wildlife Fund (WWF), the Flora and Fauna Preservation Society, the African Wildlife Foundation, and the People's Trust for Endangered Species – join forces to provide funds and expertise to help Rwanda protect its National Park. There are now estimated to be only a few hundred mountain gorillas left.

23 January 1978 Sweden bans aerosol sprays because of their damaging effect on the environment. It is the first country to do so.

August 1978 Toxic chemicals (PCBs, dioxins, and pesticides) leak into the basements of houses in the Love Canal neighbourhood of Niagara Falls, New York. The site, an abandoned canal, was used as a chemical waste dump by the Hooker Chemicals and Plastics Corporation 1947–53. Residents are evacuated but their long-term exposure results in high rates of chromosomal damage and birth defects. It is the worst environmental disaster involving chemical waste in US history.

1979 English scientist James Lovelock publishes *Gaia: A New Look at Life on Earth*, which advances a radical and controversial theory of the stability of the Earth's ecological systems.

English scientist James Lovelock, who developed the concept of the Earth as a single organism, in what we now know as the Gaia hypothesis. *George W Wright/CORBIS*

Lovelock suggests that the species of organisms on Earth act cooperatively and symbiotically to mould conditions to support their own existence. Lovelock suggests a collective name for this living system, 'Gaia', from the Greek goddess of the Earth.

> *When I first introduced Gaia, I had vague hopes that it might be denounced from the pulpit and thus made acceptable to my scientific colleagues. As it was, Gaia was embraced by theologians and by a wide range of New Age writers and thinkers but denounced by biologists.*

James Ephraim Lovelock, English scientist, *Earthwatch* (1992)

1979 China begins plans to implement a policy of one child per family. The growth rate of the Chinese population is 14% per annum. A target is set to reduce this to 0.5% by 1985 and zero population growth by 2000.

Laws are prepared to set up financial and social incentives for couples to limit family size. One is that permission is required for marriage to take place. Any couple cohabiting and producing a child out of wedlock will be fined two months wages and the child will be confiscated. Once wed, couples with only one child will be treated favourably.

Mathematics

1972 French mathematician René Thom formulates catastrophe theory, an attempt to describe abruptly changing phenomena mathematically. It has important applications in biology and optics.

1975 US mathematician Mitchell Feigenbaum discovers a new fundamental constant (approximately 4.6692016), which plays an important part in chaos theory.

1976 US mathematicians Kenneth Appel and Wolfgang Haken use a computer to prove the four-colour problem – that the minimum number of colours needed to colour a flat map such that no two adjacent sections have the same colour is four. The proof takes 1,000 hours of computer time and hundreds of pages.

1977 Public-key codes are introduced by US scientists Leonard M Adleman, Ronald L Rivest, and Adi Shamir.

They employ a system of cryptography that makes use of the difficulty of factorizing large numbers. The encoding method can be revealed to anyone, but there is at the same time no reduction in security. Public-key cryptography had been invented earlier by UK workers at the secret government establishment of GCHQ, but for reasons of security they did not publish at the time.

Physics

1970 Swedish physicist Hannes Alfvén shares the Nobel Prize for Physics with the French physicist Louis Néel: Alfvén for his work in magneto-hydrodynamics and plasma physics, and Néel for his work in antiferromagnetism and ferrimagnetism and solid-state physics.

1970 US physicist Sheldon Glashow and associates postulate the existence of a fourth quark, which they name 'charm'.

The charm quark becomes the fourth member of the six in the quark family. The six quarks are now considered to form the 'complete set'.

1971 English theoretical physicist Stephen Hawking suggests that after the Big Bang, miniature black holes no bigger than protons but containing more than a billion tonnes of mass were formed, and that they were governed by the laws of both relativity and quantum mechanics.

Black holes are states of matter so dense that the escape velocity, the velocity with which something must travel to leave the surface, is greater than the speed of light.

1971 The Nobel Prize for Physics is awarded to Hungarian-born British physicist Dennis Gabor for the invention of holography.

1972 The central research laboratories of EMI in the UK develop the first X-ray scanner.

The technique, called computerized axial tomography, becomes a major clinical breakthrough for imaging the human brain and other parts of the body.

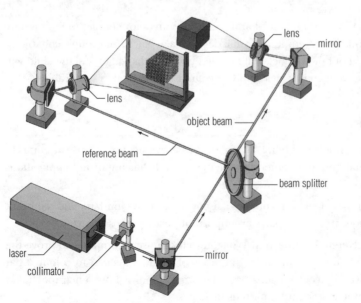

Recording a transmission hologram. Light from a laser is divided into two beams. One beam goes directly to the photographic plate. The other beam reflects off the object before hitting the photographic plate. The two beams combine to produce a pattern on the plate, which contains information about the 3-D shape of the object. If the exposed and developed plate is illuminated by laser light, the pattern can be seen as a 3-D picture of the object. Holography is developed in 1971.

1972 US physicist Murray Gell-Mann presents the theory of quantum chromodynamics (QCD), which explains how quarks interact. Strongly interacting particles consist of quarks, which are bound together by gluons.

1972 US physicists John Bardeen, Leon Cooper, and John Schrieffer receive the Nobel Prize for Physics for their theory of superconductivity, usually called the BCS theory.

1973 Japanese physicist Leo Esaki, Norwegian-born US physicist Ivar Giaever, and Welsh physicist Brian Josephson receive the Nobel Prize for Physics.

    Giaever and Esaki are honoured for their experimental discoveries regarding tunnelling phenomena in semiconductors and superconductors respectively, and Josephson is honoured for his theoretical predictions of the properties of a supercurrent through a tunnel barrier.

1974 US theoretical physicist Sheldon Glashow proposes the first grand unified theory – one that envisages the strong, weak, and electromagnetic forces as variants of a single superforce.

    At the present time no successful theory can accommodate the gravitational force. The 'carrier' of the gravitational force is also still unobserved.

16 November 1974 US physicists Burton Richter and Samuel Chao Chung Ting announce that they have separately discovered the J or ψ particle, which consists of a charm quark combined with its antiparticle. It confirms the existence of the charm quark.

1975 Danish physicist Aage Bohr, US-born Danish physicist Ben Mottelson, and US physicist James Rainwater receive the Nobel Prize for Physics.

    They are honoured for their discovery of the connection between collective motion and particle motion in atomic nuclei, and the development of the theory of the structure of the atomic nucleus based on this connection.

1976 US physicists Burton Richter and Samuel Chao Chung Ting receive the Nobel Prize for Physics for their discovery of a new particle, the ψ, or J particle.

1977 German-born US physicist Leon Lederman discovers the fifth ('beauty' or 'bottom') quark in proton–nucleon collisions combined with its antiquark in a particle called the upsilon meson.

    Having now observed five of the quarks, physicists race to find the sixth and final one.

1977 US physicists Philip Anderson, John van Vleck, and English physicist Nevill Mott receive the Nobel Prize for Physics for their fundamental theoretical investigations of the electronic structure of magnetic and disordered systems.

1978 Soviet physicist Peter Kapitza shares the Nobel Prize for Physics with US physicists Arno Penzias and Robert Wilson: Kapitza for his basic inventions and discoveries in the area of low temperature physics, and Penzias and Wilson for their discovery of cosmic microwave background radiation.

1979 US physicists Sheldon Glashow and Steven Weinberg and Pakistani physicist Abdus Salam receive the Nobel Prize for Physics for their contributions to the theory of the unified weak electromagnetic interaction between elementary particles, including the prediction of the weak neutral current.

1980–1989

Astronomy

1980 A thin layer of iridium-rich clay, about 65 million years old, is found around the world. US physicist Luis Alvarez suggests that it was caused by the impact of a large asteroid or comet, which threw enough dust into the sky to obscure the Sun and cause the extinction of the dinosaurs.

1980 US astronomer Uwe Fink and associates report the discovery of a thin atmosphere on Pluto.

Pluto has a very thin atmosphere of nitrogen and methane, but there is no confirmed evidence for an atmosphere on Pluto's moon Charon.

1980 US astrophysicist Alan Guth proposes the theory of the inflationary universe – that the universe expanded very rapidly for a short time after the Big Bang.

12 November 1980 The US space probe *Voyager 1* flies past Saturn within 124,000 km/77,000 mi; it discovers the planet's 13th, 14th, and 15th moons and transmits information about the planet, its moons, and its rings.

1981 The most massive star in the universe, R136, is discovered; it is 2,500 times more massive than our Sun and ten times as bright.

1981 The Very Large Array (VLA) radio telescope at Socorro, New Mexico, enters service. Its 27 25-m/82-ft diameter dishes can be steered and moved on railway tracks, and are equivalent to one dish 27 km/17 mi in diameter; together they provide high-resolution radio images.

1982 Astronomers at Villanova University in Pennsylvania, USA, announce the discovery of rings around Neptune.

1982 Astrophysicists at Groningen University in the Netherlands postulate the existence of a black hole at the centre of the Milky Way.

1983 Studies from the US *Lageos* satellite (launched 4 May 1976 to monitor slight crustal movements to help predict earthquakes) indicate that the Earth's gravitational field is changing.

1983 The Search for Extraterrestrial Intelligence (SETI) programme is established at NASA's Ames Research Centre, Mountain View, California.

Radio telescopes, like this one (the world's largest) at Arecibo, Puerto Rico, allow astronomers to analyse a broad range of low-frequency electro-magnetic waves – visible light is only a small part of the electromagnetic spectrum. Pulsars and quasars were first discovered by radio telescopes. *National Aeronautics and Space Agency Administration*

4 April 1983 The first US space shuttle, *Challenger*, is launched for the first time. It carries the first five-person crew on 18 June.

13 June 1983 The US space probe *Pioneer 10*, launched 3 March 1972, becomes the first artificial object to leave the Solar System.

25 June 1983 The Infrared Astronomical Satellite (IRAS), an orbiting observatory, is launched; it is designed to detect infrared radiation from objects in space and surveys almost the entire infrared sky. It also finds the first evidence of planetary material around the star Vega outside our Solar System.

1984 Astronomers at Cornell University, New York, report the discovery of eight infrared galaxies – thought to resemble primeval galaxies – located by the Infrared Astronomical Satellite (IRAS).

1984 The USA launches the Earth resource satellite *Landsat 4*. It carries a new kind of technology, the Thematic Rapper, used to provide information on the Earth's natural resources.

1984 US astronomers working in Chile photograph a partial ring system around Neptune.

1985 European, Japanese, and Soviet probes are launched to rendezvous with Halley's comet in 1986.

1986 Scientists at Arizona State University, USA, conduct computer simulations which suggest that a Mars-sized object struck the Earth a glancing blow about 4.6 billion years ago, releasing material that eventually coalesced to form the Moon.

This 'giant impactor' theory is shown to have fewer disadvantages than earlier theories of the origin of the Moon, based on capture by, or fission from, the early Earth, and by the end of the year the impact theory is the leading hypothesis about the Moon's origin.

January 1986 The US space probe *Voyager 2* passes within 81,000 km/50,600 mi of Uranus; photographs taken by the probe reveal ten unknown satellites and two new rings.

Voyager data show that the planet's rate of rotation is 17 hours, 14 minutes. The spacecraft also finds a Uranian magnetic field that is both large and unusual. In addition, the temperature of the equatorial region, which receives less sunlight over a Uranian year, is nevertheless about the same as that at the poles.

20 February 1986 The USSR launches the core unit of the *Mir* space station; it is permanently occupied until 1999, and deliberately crashed into the Pacific Ocean in March 2001.

7–17 March 1986 The prime period of the approach to the Sun of Halley's comet, which occurs roughly every 75 years.

The European space probe *Giotto* (launched 2 July 1985) encounters Halley on 13 March 1986. Scientific data are received showing that the nucleus of Halley's comet measures 24 km/15 mi by 16 km/10 mi and consists of ices (chiefly water), various gases, and dust particles, with a very dark coating, possibly of organic material.

1987 Harvey Butcher, director of the Westerbrook Synthesis Telescope in Groningen, the Netherlands, estimates that the universe is younger than 10 billion years.

1987 Objects the size of planets are found orbiting the stars Gamma Cephei and Epsilon Eridani.

1987 Radio waves are observed from 3C326 – believed to be a galaxy in the process of formation.

1987 The James Clerk Maxwell Telescope, operated by the Royal Observatory, based in Edinburgh, Scotland, begins operation on Mauna Kea, Hawaii; its 15 m/49 ft dish makes it the largest submillimetre telescope in the world.

23 February 1987 Astronomers around the world observe a spectacular supernova in the Large Magellanic Cloud, the galaxy closest to ours, when a star (SN1987A) suddenly becomes a thousand times brighter than our own Sun. It is the first supernova visible to the naked eye since 1604.

1988 Simon Lilly of the University of Hawaii, USA, reports the location of a galaxy about 12 billion light years from Earth, adding to evidence about the date of the universe's formation.

1989 Astronomers discover a river of gas at the centre of the Milky Way, providing further evidence that a black hole, 4 million times as massive as the Sun, exists at the centre of our galaxy.

1989 Star HD 114762 is discovered to have a planet-like body circling it.

1989 The Australia Telescope is completed; it has seven 22 m/72 ft dishes and one 64 m/210 ft dish, which are spread throughout New South Wales, Australia, making it one of the largest radio telescopes in the world.

1989 US astronomers John Huchra and Margaret Geller discover a large, thin sheet of galaxies, which they name the 'Great Wall'; no current astronomical theory can explain its distinctive form. It is the largest known structure in the universe and is 500 million light years long and 250 million light years wide.

25 August 1989 The US space probe *Voyager 2* (launched 20 August 1977) reaches Neptune and transmits pictures and other scientific data.

The probe *Voyager* finds six new moons, adding to the two already known, Triton and Nereid. The craft finds Triton to be completely encircled by rings, like the other gas giants.

18 November 1989 The US Cosmic Background Explorer (COBE) satellite is launched to study microwave background radiation, thought to be a vestige of the Big Bang.

COBE measures fluctuations in the microwave background radiation with higher coverage and precision than has been possible from Earth, allowing astrophysicists to develop new models of the early stages of galaxy formation.

Biology

1980 A gene is transferred from one mouse to another by US geneticist Martin Cline and colleagues.

1981 A new family of deep-water stingray is named when a specimen of *Hexatrygon bickelli* is washed up on a South African beach.

1981 Chinese scientists make the first clone of a fish (a golden carp).

1981 US geneticists J W Gordon and F H Ruddle of the University of Ohio inject genes from one animal into the fertilized eggs of a mouse, which develop into mice with the foreign gene in many of the cells; the gene is then passed on to their offspring creating permanently altered (transgenic) animals; it is the first transfer of a gene from one animal species to another.

1981 US geneticists Robert Weinberg, Geoffrey Cooper, and Michael Wigler discover that oncogenes (genes that cause cancer) are integrated into the genome of normal cells.

1982 Dolphins are discovered to possess magnetized tissues that aid in navigation; they are the first mammals discovered to have such tissues.

1982 The Nobel Prize for Physiology or Medicine is awarded jointly to Swedish biochemists Sune Bergström and Bengt Samuelson, and British pharmacologist John Vane for their discovery of prostaglandins and related biologically active substances.

1983 The Nobel Prize for Physiology or Medicine is awarded to US geneticist Barbara McClintock for her discovery of mobile genetic elements in the 1940s.

Then, experiments on the genetics of maize led her to conclude that some genes move on the chromosome and control a number of other genes. At first her idea of 'jumping genes' was ignored, but it is now accepted though not fully understood.

1983 The skull of a creature called *Pakicetus* is discovered in Pakistan; estimated to be 50 million years old, it is intermediate in evolution between whales and land animals.

1983 US biologist Lynn Margulis discovers that cells with nuclei form by the synthesis of non-nucleated cells.

1983 US biologists Andrew Murray and Jack Szostak create the first artifical chromosome; it is inserted into a yeast cell.

1984 Allan Wilson and Russell Higuchi of the University of California at Berkeley, USA, clone large DNA segments from an extinct animal, the quagga.

This cloning allows for the comparison of zebra subspecies across Africa and helps to unravel their evolutionary relationships. It also suggests the potential for DNA hybridization in extinction prevention.

1984 English geneticist Alec Jeffreys discovers that between 20 and 30% of human DNA is made up of short, highly repetitive sequences called simple-sequence DNA. Each sequence is typically between five and ten base pairs in length. The number of repeated units in each region of simple-sequence DNA varies from person to person and is as unique to the individual as his or her fingerprints. This uniqueness is the basis for 'genetic fingerprinting', which is used in criminal investigations and to establish family relationships.

1984 From studies of DNA, US scientists Charles Sibley and Jon Ahlquist argue that humans are more closely related to chimpanzees than to other great apes, differing in their DNA by only 1%, and that humans and apes diverged approximately 5–6 million years ago.

However their conclusions are subject to serious debate, which is still on-going.

1984 Robert Sinsheimer, of the University of California, proposes that all human genes be mapped; the proposal eventually leads to the development of the Human Genome Project which gets underway in Europe and the USA in 1989 and 1990.

1987 A nest of fossilized dinosaur eggs is discovered in Alberta, Canada; it is the second such nest to be found.

1987 Foxes in Belgium are immunized against rabies by using bait containing a genetically engineered vaccine, dropped from helicopters. The success of the experiment leads to a large-scale vaccination programme.

April 1987 The US Patent and Trademark Office announces its intention to allow the patenting of animals produced by genetic engineering.

315

1988 At the Michigan Institute of Technology, German-born US biomedical researcher Rudolf Jaenisch and associates implant a human gene, connected with a hereditary disorder, into a mouse.

April 1988 The US Patent and Trademark Office grants Harvard University a patent for an 'oncomouse', so called because an oncogene has been inserted into its genetic material. The mouse is cancer-prone and a model for oncological research. It is the first transgenic animal to be patented in the USA.

1989 The first visual image of a DNA molecule is obtained by US scientists.

The scanning tunnelling electron microscope creates a three-dimensional view of the surface atoms of a molecule. Its probe, with a single atom at the tip, moves over the specimen molecule, mapping its contours in great detail.

1989 Transgenic plants with genes coding for antibodies against human diseases are developed for the first time.

Chemistry

1980 US molecular biologist Paul Berg receives the Nobel Prize for Chemistry for his studies on the biochemistry of nucleic acids, especially recombinant DNA (deoxyribonucleic acid); US molecular biologist Walter Gilbert and English biochemist Frederick Sanger receive the Nobel Prize for Chemistry for determining base sequences in nucleic acids. Sanger is only the fourth person in history to receive a Nobel prize twice.

1981 Japanese industrial chemist Kenichi Fukui and Polish-born US chemist Roald Hoffmann receive the Nobel Prize for Chemistry for their progress on theories concerning chemical reactions.

1982 South African-born molecular biologist Aaron Klug receives the Nobel Prize for Chemistry for his development of crystallographic methods for determining the structure of biologically important nucleic acid protein complexes.

1982 German physicist Peter Armbruster and colleagues at the Heavy Ion Research Institute (GSI), Darmstadt, Germany, synthesize the element meitnerium (atomic number 109).

They fuse an iron atom with a bismuth atom in a reaction producing an atom of meitnerium and an additional neutron. They name the element after Austrian physicist Lise Meitner.

April 1983 US biochemist Kary Mullis invents the polymerase chain reaction (PCR). This is a method of copying genes or known sections of a DNA molecule a million times without the need for a living cell.

The process makes previous techniques of DNA production obsolete. Mullis was awarded a share in the Nobel Prize for Chemistry in 1993 for his development of the technique.

1983 Canadian-born US chemist Henry Taube receives the Nobel Prize for Chemistry for his work on electron-transfer reactions in inorganic chemical reactions.

1984 German physicist Peter Armbruster and colleagues at the Heavy Ion Research Institute (GSI), Darmstadt, Germany, synthesize the element hassium (atomic number 108).

The team irradiate lead with charged iron particles to create element 108, but it only last for a few milliseconds. They name the element after Hassius, the Latin for the German state of 'Hess'.

1984 US chemist Bruce Merrifield receives the Nobel Prize for Chemistry for chemical syntheses of peptides and proteins on a solid matrix.

1985 English chemist Harold Kroto and US chemists Robert Curl and Richard Smalley discover fullerenes.

They use a laser supersonic cluster beam apparatus to vaporize graphite in an inert atmosphere. Their analysis shows that they have created a previously unknown allotrope of carbon. The new material is mainly composed of 60 carbon atoms linked together in 12 pentagons and 20 hexagons that fit together to form a hollow sphere. They name the material 'buckminsterfullerene' after the inventor of the geodesic dome, Richard Buckminster Fuller. They are awarded the Nobel Prize for Chemistry in 1996 for this discovery.

1985 US chemists Herbert Hauptman and Jerome Karle receive the Nobel Prize for Chemistry for their methods of determining crystal structures.

1986 US chemist Dudley Herschbach, Taiwanese-born US chemist Yuan Lee, and German-born Canadian chemist John Polanyi share the Nobel Prize for Chemistry for their work on the dynamics of chemical elementary processes.

1987 US chemists Donald Cram and Charles Pedersen and French chemist Jean-Marie Lehn receive the Nobel Prize for Chemistry for the development of molecules with highly selective structure-specific interactions.

1988 Researchers at IBM's Almaden Research Center in San José, California, using a scanning tunnelling microscope, produce the first image of the ring structure of benzene, the simplest aromatic hydrocarbon. The image confirms the structure of the molecule envisioned by Frederick Kekulé in 1865.

1988 US chemist J Wayne Rabelais and Indian chemist Srinandan Kasi were among the first to develop diamond film. It is used as an insulator and for industrial grinding.

Rabelais and Kasi develop a process to coat various substrates with a layer of micro-crystalline diamond – pure carbon that adheres tightly to surfaces. The layer retains the properties of normal diamond, such as hardness, but can be produced at a fraction of the cost.

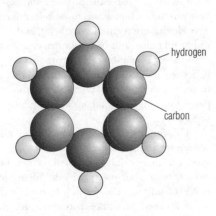

hydrogen

carbon

The molecule of benzene consists of six carbon atoms arranged in a ring, with six hydrogen atoms attached. The benzene ring structure is found in many naturally occurring organic compounds. The first image of the structure of benzene was produced using scanning tunnelling electron microscope in 1988.

The Discovery of Carbon Fullerenes

by Julian Rowe

GREAT BALLS OF CARBON

Chemists regard a diamond as just another form of carbon, an element familiar to everyone as soot, which is practically pure carbon. Equally everyday is the graphite in a 'lead' pencil, another form of carbon. The chemical difference between graphite and diamond is that the carbon atoms in each substance are arranged differently. The carbon atoms of graphite are arranged in flat, hexagonal patterns, rather like the cells of a honeycomb. Because graphite molecules are flat, they slide over one another easily. In contrast, the carbon atoms in a diamond are interlinked three-dimensionally, giving the substance its extraordinary hardness. And there until recently the matter rested: carbon was an element that came in two forms – diamond and graphite. Now chemists are excited about a third form of carbon, in which the atoms are linked together in a molecule that looks very like a soccer ball. The new form of carbon is a cagelike molecule consisting of 60 carbon atoms that make a perfect sphere. It has been named 'buckminster-fullerene' in honour of the US architect Buck-minster Fuller (1895–1983), whose work included spherical domes.

SOOT, SPACE, AND LASERS

The story of the discovery of these exotic new molecules involves soot, outer space, and lasers. It starts when two scientists, Donald Huffman from the University of Arizona, Tucson, and Wolfgang Kratschmer, were working at the Max Planck Institute for Nuclear Physics in Heidelberg, Germany. They were heating graphite rods under special conditions and examining the soot made in the process: they speculated that a similar process might take place in outer space, contributing to clouds of interstellar dust.

Meanwhile the team of Harold Kroto and David Walton at the University of Sussex had been on the trail of interstellar molecules made up of long chains of carbon atoms that might have originated in the atmosphere that surrounds red giant stars. Enlisting the help of researchers at Rice University in Houston, Texas, who were using a giant laser to blast atoms from the surface of different target substances, Kroto and his team soon found the long-chained carbon molecules. But they were struck by a surprising discovery of a very stable molecule that contained exactly 60 carbon atoms.

MODELLING THE NEW MOLECULE

Now Richard Smalley of the Rice team set out to make a model of the new molecule, using scissors, sticky tape, and paper. He soon found that a hexagonal arrangement of carbon atoms was impossible, but that a perfect sphere could be formed from 20 hexagons and 12 pentagons. Such a sphere has 60 vertices. Chemists attribute the stability of the new form of carbon to this closed cage structure. However, chemists like to prove the structures of the molecules they make: the evidence so far was merely speculative.

THE BUCKYBALL'S FINGERPRINT

Ordinary soot absorbs ultraviolet light in a characteristic way, and the new molecule also showed a characteristic ultraviolet fingerprint. The problem was to make enough of the new carbon so that exact measurements could be made. If it could be crystallized, then an X-ray analysis would enable the precise distances between the carbon atoms to be determined. The Heidelberg team forged ahead, and produced milligrams of red-brown crystals by evaporating a solution of their product in benzene. The X-ray results confirmed that the molecules were indeed spherical, and that the paper model of 20 hexagons and 12 pentagons was correct.

This result was clinched when Kroto and his colleagues, using nuclear magnetic resonance spectroscopy, not only confirmed the new 60-atom structure but also provided evidence for a family of fullerenes, as the new forms of carbon are now called. Structures containing 28, 32, 50, 60, and 70 carbon atoms are known. Chemists affectionately term such molecules 'buckyballs'.

In many ways, these new discoveries in carbon chemistry are as important as the key discovery more than a century ago of the structure of benzene. When in 1865 German chemist Friederich Kekulé (1829–1896) proposed a ring structure for this important organic molecule, the whole field of aromatic chemistry opened up, leading to dyestuffs in the first instance, and millions of new substances since. The fullerene family holds similar promise.

Chemists at Exxon's laboratories in New Jersey have already played a part in the fullerene story, and are interested in the lubricating properties of the new materials. Sumio Iijima, a Japanese scientist, has synthesized tubelike structures based on the fullerene idea, which are, naturally, called 'buckytubes'.

Other teams have now done work that suggests that such molecules may have interesting electrical properties: they may have semiconducting abilities. Cagelike molecules can contain other atoms, such as metals: a group of researchers from the University of California at Los Angeles have produced a 'doped' fullerene that behaves as a superconductor. No evidence has been found that buckminsterfullerene exists in space, but some is almost certainly produced every time you light a candle.

1988 German chemists Johann Deisenhofer, Robert Huber, and Hartmut Michel share the Nobel Prize for Chemistry for their determination of the three-dimensional structure of the reaction centre of photosynthesis.

1989 US biochemists Sidney Altman and Thomas Cech receive the Nobel Prize for Chemistry for their discovery of the catalytic properties of RNA (ribonucleic acid).

Earth Sciences

1980 North American geologists P Coney, D L Jones, and J W H Monger describe the North American Cordilleran orogen (in the Western USA and Canada) as a composite of 'suspect terranes', resulting in a new perspective on the construction of orogenic belts (in which mountain belts eventually form).

1980 The US *Magsat* satellite completes its mapping of the Earth's magnetic field.

1983 The Ocean Drilling Program (ODP) is established as the successor to the Deep Sea Drilling Project.

1984 Australian geologists Bob Pidgeon and Simon Wilde discover zircon crystals in the Jack Hills north of Perth, Australia, that are estimated to be 4.276 million years old – the oldest terrestrial mineral crystals ever discovered.

1985 The research drilling ship *SEDCO/BP 471* begins service as part of the new Ocean Drilling Program (ODP), replacing the retired *Glomar Challenger*. The ship is later renamed the *JOIDES Resolution*.

Ecology

1980 The United Nations launches its campaign for the decade entitled International Drinking Water Supply and Sanitation Decade 1981–90 – clean water and adequate sanitation for all by 1990.

Even enthusiastic supporters of the project agree that it has little chance of meeting its targets. At present, £1,000 million is spent annually on international aid for water projects; implementation of the plan would cost over £3,000 million to be spent each year.

1981 The UK government passes the Wildlife and Countryside Act. The approach is to establish a blanket of criminal offences for interfering with habitats and wildlife but with a list of exceptions such as sections dealing with pests and quarry animals. The

Act extends protection to a whole range of birds and mammals, all bats, reptiles, and amphibians.

The Act makes it an offence to bring into Great Britain any wild animal that is not a native or any species of plant or animal listed. This measure is designed to avoid the ecological havoc wrought by previous introductions of alien species such as the grey squirrel and giant hogweed.

1981 The US Committee on the Atmosphere and Biosphere reports evidence linking acid rain to sulphur emissions from power plants.

1981 The US government-commissioned *Global 2000 Report to the President of the US* is published; it predicts global environmental catastrophe if pollution, industrial expansion, and population are not brought under control.

1982 On the tenth anniversary of the Stockholm-based United Nations Conference on the Human Environment, Sweden convenes a Conference on Acidification. By now, evidence that aquatic systems in Scandinavia are being devastated by pollution from acid rain is strong. Germany also becomes alarmed at the decline of its forests and joins Norway, Sweden, Austria, Switzerland, and the Netherlands in calling for control over sulphur emissions.

1984 US biologist and conservationist Edward Wilson publishes his book *Biophilia*. In this work, Wilson argues passionately for the need to conserve biodiversity.

Wilson suggests that a feeling for life in all its richness ('biophilia') is a deep-seated human emotion and that a new conservation ethic could be constructed on our instinctive regard for other life forms.

1984 US economist Julian Simon and US physicist and futurist Herman Kahn publish *The Resourceful Earth: A Response to Global 2000*. The authors question the assumptions on the earlier *Global 2000 Report to the President of the US* was based.

The authors are confident that science, technology, and the free market will solve environmental problems and increase the quality of life.

3 December 1984 An explosion occurs at a pesticide factory owned by Union Carbide in Bhopal, India. Between 2,000 and 5,000 people die and about 250,000 people suffer ill health effects from damage to lungs and eyes.

The chemical responsible is methyl isocyanate. A report issued a year later shows that management mistakes, badly designed equipment, and poor maintenance are to blame for the tragedy. The terrible incident highlights the risks inherent in the chemical industry and, more importantly, the double standards of safety that divides the first and third worlds.

March 1985 Scientists at the British Antarctic Survey announce the discovery of thinning of the ozone layer over Antarctica, which worsens each year in the spring.

The implications of ozone thinning cause much concern. If the intensity of short wavelength ultraviolet radiation increases, then this can cause damage to plants and increase the incidence of human skin cancer.

26 April 1986 A reactor at the Chernobyl nuclear power station in the Ukraine explodes, releasing radioactive material into the surrounding area and causing a radioactive cloud to cross Europe. This is the world's worst nuclear accident.

Radiation expert Robert Gale estimates that about an extra 2,500 to 75,000 cancer deaths in Europe can be expected over the next 20 years from the radiation released.

The catastrophe at the Chernobyl nuclear power station was the result of a safety test that went badly wrong. The power rose to a hundred times its maximum design level within three to four seconds, melting the reactor core. As the molten fuel reacted with the cooling water a massive explosion destroyed the reactor building and lifted the 2,000 tonne reactor cap off. Radioactive materials from the core were lifted high into the atmosphere by the hot gas and continued to stream out for ten days before the fire was brought under control. *Corbis*

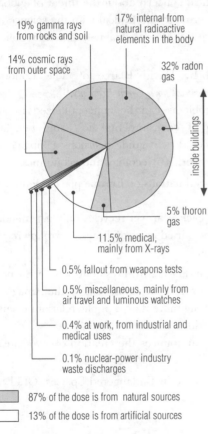

19% gamma rays from rocks and soil

17% internal from natural radioactive elements in the body

14% cosmic rays from outer space

32% radon gas

inside buildings

5% thoron gas

11.5% medical, mainly from X-rays

0.5% fallout from weapons tests

0.5% miscellaneous, mainly from air travel and luminous watches

0.4% at work, from industrial and medical uses

0.1% nuclear-power industry waste discharges

87% of the dose is from natural sources

13% of the dose is from artificial sources

Pie chart showing the various sources of radiation in the environment. Most radiation is from natural sources, such as radioactive minerals, but 13% comes from the by-products of human activities.

News of the accident only reaches Europe when abnormally high levels of radiation are reported in Sweden, Denmark, and Finland.

1987 At a conference in Montreal, Canada, an international agreement, the Montreal Protocol, is reached to limit the use of ozone-depleting chlorofluorocarbons (CFCs) by 50% by the end of the century; the agreement is later condemned by environmentalists as 'too little, too late'.

However, the Protocol is a ground-breaking piece of legislation and it is the world's first global pollution agreement. The European Community countries manufacture about half the world's CFCs.

1987 The Soviet Union announces its intention to end commercial whaling. The last minke whale is carried into the port of Odessa by the Soviet whaling ship *Soviet Ukraine* in May.

Conservationists rejoice but at the annual meeting of the International Whaling Commission held in Bournemouth, UK, the Japanese insist on their right to continue hunting whales for 'scientific' purposes. In addition to Japan, Korea, Norway, and Iceland continue to hunt whales. Norway and Iceland export most of their meat to Japan.

1987 US researchers suggest that thunderstorms can propel pollutants into the lower stratosphere when they observe high levels of carbon monoxide and nitric acid at high altitude during a thunderstorm.

April 1987 The World Commission on Environment and Development issues its first report, the 'Brundtland Report', named after the prime minister of Norway and the commission's chair, Gro Harlem Brundtland. It becomes famous for its definition of sustainable development: 'meeting the needs of the present without compromising the ability of future generations to meet their own needs'.

The report also stresses the importance of intra- and inter-generational equity, the latter being conceived as our obligation to future generations, and the former as the right of all human beings living now 'to an environment adequate for their heath and well being'.

1988 The first meeting of the Intergovernmental Panel on Climate Change (IPCC) takes place in Geneva.

The organization is created by two bodies of the United Nations, the World Meteorological Organization and the United Nations Environment Programme, at a time when there is growing scientific and political concern about the threat of global warming. The panel establishes three working groups: one to deal with the science of climate change, one to look at impacts of change, and one to deal with potential policy responses.

1989 A group of environmental economists, led by David Pearce of The London Environmental Economics Centre, publishes *Blueprint for a Green Economy*.

The authors show how environmental problems can be conceptualized in an economic framework and how the weighing of environmental costs and benefits (a procedure known as cost–benefit analysis) could help guide decision-making. It is hailed as a significant milestone in bridging the worlds of ecology and economics.

1989 A survey of recycling activities in the industrialized nations of Europe places Denmark at the top and the UK at the bottom.

The UK produces 18 million tons of waste each year, yet recycles only 2.7 million tons. In Denmark, only 12% of waste needs to be buried in landfill sites. Germany recycles 50% of its annual 30 million tonnes of waste.

1989 In the face of mounting international pressure and criticism, President Senhor Sarney of Brazil unveils a plan called 'Our Nature' to protect the rainforests of the Amazon. Western commentators criticize the plan as nothing more than a public relations event, since the funds required to implement it are lacking.

Among 50 measures signed into law, Sarney announces the creation of two forest preserves and three new national parks.

16 October 1989 The Convention on International Trade in Endangered Species (CITES) agrees to a total ban on trading in ivory.

Mathematics

1980 Mathematicians worldwide complete the classification of all finite and simple groups, a task that has taken over 100 mathematicians more than 35 years to complete. The results take up more than 14,000 pages in mathematical journals.

1980 Polish-born French mathematician Benoit Mandelbrot develops the theory of fractals. The Mandelbrot set is a spectacular shape with a fractal boundary (a boundary of infinite length enclosing a finite area), as in the shape of a snowflake.

Physics

1980 US physicists James Cronin and Val Fitch receive the Nobel Prize for Physics for their discovery of violations of fundamental symmetry principles in the decay of neutral K-mesons.

Indian-born US astrophysicist Subrahmanyan Chandrasekhar shortly after receiving the 1983 Nobel Prize for Physics. *Bettmann/CORBIS*

1981 Dutch-born US physicist Nicolaas Bloembergen, US physicist Arthur Schawlow, and Swedish physicist Kai Siegbahn receive the Nobel Prize for Physics.

Bloemberg and Schawlow are honoured for their contribution to the development of laser spectroscopy, and Siegbahn for his contribution to the development of high-resolution electron spectroscopy.

1982 US physicist Kenneth Wilson receives the Nobel Prize for Physics for his theory of phase transitions and critical phenomena.

1983 The Nobel Prize for Physics is awarded jointly to Indian-born US astrophysicist Subrahmanyan Chandrasekhar and US astrophysicist William Fowler.

Chandrasekhar is honoured for his theoretical studies of the structure and evolution of stars, and Fowler for his studies of the nuclear reactions involved in the formation of the chemical elements.

A certain modesty toward understanding nature is a precondition to the continued pursuit of science.

Subrahmanyan Chandrasekhar, Indian physicist, interview 1984

June 1983 The W and Z subatomic particles are detected in experiments at the European Laboratory for Particle Physics (CERN), Switzerland, by Italian physicist Carlo Rubbia and Dutch physicist Simon van der Meer; the existence of these particles had been predicted as carriers of the weak nuclear force.

1984 A team of international physicists at the European Laboratory for Particle Physics (CERN) in Geneva, Switzerland, discovers the sixth (top) quark; its discovery completes the theoretical scheme of subatomic building blocks.

Since everyday matter is built from neutrons and protons which in turn are built from up and down quarks, the quest is now to discover why nature produces the strange and charm, and top and bottom pairs.

1984 The Nobel Prize for Physics is awarded jointly to Italian physicist Carlo Rubbia and Dutch physicist Simon van der Meer for their basic studies on particle physics.

1985 The Nobel Prize for Physics is awarded to German physicist Klaus von Klitzing for his discovery of the quantum Hall effect.

1986 IBM researchers develop the first scanning tunnelling electron microscope. It is built on a single silicon chip only 0.2 mm/200 μm in across.

1986 The Nobel Prize for Physics is awarded jointly to German physicist Ernst Ruska for developing the electron microscope, and to German physicist Gerd Binnig and Swiss physicist Heinrich Rohrer for the development of the scanning tunnelling microscope.

1987 The Nobel Prize for Physics is awarded jointly to German physicist Georg Bednorz and Swiss physicist Alex Müller for their important breakthrough in the discovery of super-conductivity in ceramics.

1988 The Nobel Prize for Physics is awarded to US physicists Leon Lederman, Melvin Schwartz, and Jack Steinberger for creating a stream of neutrinos in a particle accelerator (the neutrino-beam method), and for demonstrating, by their discovery of the muon neutrino, that there are two types of neutrino.

1989 The Nobel Prize for Physics is awarded jointly to German nuclear physicist Wolfgang Paul and German-born US physicist Hans Dehmelt for their ion-trap method for isolating single atoms, and to US physicist Norman Ramsey for his invention of the separated oscillatory fields method and its use in the hydrogen maser and other atomic clocks.

March 1989 US physicist Stanley Pons and English chemist Martin Fleischmann announce that they have achieved nuclear fusion at room temperature (cold fusion); other scientists fail to replicate their results.

This comes to be considered one of the greatest-ever 'non-discoveries' in science and is now considered to be the best reason to ensure replication before publication.

1990–2000

Astronomy

1990 US astronomer Mark Showalter discovers an 18th moon of Saturn when analysing pictures transmitted by *Voyager 2*.

24 January–21 April 1990 Japan launches *Muses-A*, the first probe to be sent to the Moon since 1976; it places a small satellite in lunar orbit (19 March 1990).

February 1990 The US space probe *Voyager 1*, now near the edge of the Solar System, turns and takes the first photograph of the entire Solar System from space.

24 April 1990 The space shuttle *Discovery* places the Hubble Space Telescope in Earth orbit; the main mirror proves to be defective.

1 June 1990 Joint US-German-British Earth-orbiting X-ray observatory *Röntgensatellite* (ROSAT) is launched; its mission is to study X-rays given off by the coronas of stars.

ROSAT is turned off on 12 February 1999 after eight years of operation and many discoveries of outstanding scientific relevance. ROSAT performed the first All-Sky Survey with an imaging telescope of soft X-rays. The number of pointed observations reached a total of 9,000 with more than 80,000 new X-ray sources detected. The total number of sources, including those found by the survey, is about 150,000.

July 1990 US astronomers Juan Uson, Stephen Boughin, and Jeffrey Kuhn announce the discovery of the largest known galaxy; over 1 billion light years away, it has a diameter

The Hubble Space Telescope (HST) in December 1999, just after its release from the robot arm of the space shuttle *Discovery*, following a service mission. *NASA*

of 5.6 million light years, almost 80 times that of the Milky Way, and contains about 2 trillion stars.

1991 The Jodrell Bank radio astronomy centre, Cheshire, England, reports the possible discovery of a planet orbiting pulsar star PSR 1829–10.

January 1991 An asteroid 16 km/10 mi in diameter passes between the Moon and the Earth.

1992 The Cosmic Background Explorer (COBE) satellite detects ripples in the microwave background radiation, thought to originate from the formation of galaxies.

1992 The first comet-sized objects in the Kuiper belt are discovered by US astronomers David Jewitt and Jane Luu, working at Mauna Kea in Hawaii.

1992 US astronomers Jeffrey McClintock, Ronald Remillard, and Charles Bailyn identify Nova Muscae as a black hole approximately 18,000 light years from Earth.

1992 The US space probe *Magellan* has mapped 99% of the surface of Venus to a resolution of 100 m/330 ft.

The crater distribution and other features indicate that the surface of Venus is relatively young – resurfaced about 500 million years ago by widespread volcanic eruptions. The planet's present harsh environment has persisted at least since then.

5 May 1992 NASA launches the new shuttle craft *Endeavour*, named after the 18th-century vessel captained by the English explorer James Cook.

10 July 1992 The European Space Agency's *Giotto* space probe is diverted to encounter the comet Grigg-Skjellerup.

1993 US astronomers Russell Hulse and Joseph Taylor Jr are jointly awarded the Nobel Prize for Physics for their discovery of a new type of pulsar.

1993 US astronomers David Jewitt and Jane Luu discover four large ice objects in the Kuiper belt, a ring of small, icy bodies orbiting the Sun beyond the planets and thought to be the source of comets. The first comet-sized objects in the Kuiper belt were discovered in 1992.

1993 US astronomers identify part of the dark matter in the universe as stray planets and brown dwarfs. Known as MACHOs (massive astrophysical compact halo objects), they may constitute approximately half of the dark matter in the Milky Way's halo.

March 1993 A star in the galaxy M81, about 11 million light years away from Earth, erupts into a supernova. Archival photographs allow astronomers to study the behaviour of the star before it exploded.

28 August 1993 The US spacecraft *Galileo* discovers the first asteroid moon. About 1.5 km/ 0.95 mi across and named Dactyl (in 1994), it orbits the asteroid Ida.

7 December 1993 The Hubble Space Telescope (placed in Earth orbit in 1990) is repaired and reboosted into a nearly circular orbit by five US astronauts operating from the US space shuttle *Endeavour* – at a cost of $360 million.

1994 The closest pulsar to the Earth (PSR J0108–1431) is discovered; it is 280 light years away.

8 January 1994 The Russians launch space mission *Soyuz-TM 18* with Russian cosmonaut Valeri Polyakov aboard, to their *Mir* space station. Polyakov plans to spend 14 months at the space station to study the effect on the human body of being in space for the time required to travel to Mars.

25 January–10 May 1994 The US spacecraft *Clementine* is launched. The objective of the mission is to make scientific observations of the Moon and the near-Earth asteroid 1620 Geographos.

    Clementine discovers an enormous crater on the far side of the Moon, 2,500 km/1,563 mi across and 13 km/8 mi deep, making it the largest crater in the Solar System discovered so far. It also reveals the possibility of a permanently frozen water-ice deposit in a crater near the south pole of the Moon.

16–22 July 1994 Fragments of the comet Shoemaker-Levy 9 collide with Jupiter. The comet is first sighted on 24 March 1993, by US astronomers Carolyn and Eugene Shoemaker and David Levy. The impact sites on Jupiter are visible from Earth and analysis shows that most of the pieces were solid bodies about 1 km/0.6 mi diameter.

December 1994 The Apollo asteroid (an asteroid with an orbit crossing that of Earth) 1994 XM1 passes within 100,000 km/60,000 mi of Earth, the closest observed approach of any asteroid.

6 December 1994 Pictures taken by the Hubble Space Telescope of galaxies in their infancy are published.

1995 US astronomers discover the first brown dwarf, an object larger than a planet but not massive enough to ignite into a star, in the constellation Lepus. It is about 20–40 times as massive as Jupiter.

    Four other brown dwarfs are discovered in 1996.

23 July 1995 US astronomers Alan Hale and Thomas Bopp discover the Hale–Bopp comet. The brightest periodic comet, its icy core is estimated to be 40 km/25 mi wide.

1996 Astronomers announce the discovery of a galaxy in the constellation Virgo, estimated to be 14 billion light years away – the most distant galaxy ever detected.

The *Galileo* spacecraft about to be detached from the Earth-orbiting space shuttle *Atlantis* at the beginning of its six-year journey to Jupiter. *National Aeronautics and Space Administration*

1996 Astronomers from the Leiden Observatory in the Netherlands and from Johns Hopkins University in Baltimore, Maryland, USA, using data from the Hubble Space Telescope, discover a black hole in the galaxy in the constellation Virgo.

1996 The US spacecraft *Galileo* (launched 18 October 1989) shows less helium on the planet Jupiter than expected. The ratio of helium to hydrogen is similar to that of the Sun, suggesting that the composition of Jupiter has remained unchanged since its formation. The probe also records 700 kph/435 mph winds below the uppermost cloud layers, suggesting internal heating.

1996 Based on data received from the spacecraft *Galileo*, US astronomers conclude that Jupiter's moon Io has a metallic core. A 10-megawatt beam of electrons flowing between Jupiter and Io is also detected.

January 1996 US astronomers announce the discovery of three new planets orbiting stars, all within 50 light years of Earth. By July 1996 the total number of new planets discovered since October 1995 has risen to ten.

30 January 1996 The comet Hyakutake is discovered by Japanese amateur astronomer Yuji Hyakutake.

2 July 1996 The US aerospace company Lockheed Martin unveils plans for the X-33, a $1 billion wedge-shaped rocket ship. Called the *Venture Star*, it will be built and operated by Lockheed Martin and will replace the US space shuttle fleet by the year 2012.

6 August 1996 US scientist Daniel Goldwin's report on a meteorite discovered in Antarctica in 1984 claims that it was ejected from Mars millions of years ago, hitting Antarctica 12,000 years ago. The meteorite is found to contain objects resembling tiny fossils, suggesting to some that life once existed on Mars, although the fossils are much smaller than the smallest terrestrial bacteria.

327

13 August 1996 NASA scientists report that new images taken by the spacecraft *Galileo* of Europa, one of Jupiter's moons, show that icy floes on its surface may contain evidence of life.

3 December 1996 US astronomer Anthony Cook, using data from the satellite *Clementine*, announces the discovery of a frozen lake at the bottom of a crater on the dark side of the Moon. It would be important for a future Moon colony.

1997 The Solar Heliospheric Observatory satellite (SOHO) reveals that Venus's ion-packed tail is 45 million km/28 million mi in length. Discovered in the late 1970s, it stretches away from the Sun and is caused by the bombardment of the ions in Venus's upper atmosphere by the solar wind.

27 February 1997 Canadian astronomer David Gray reports that the star 51 Pegasi, thought to have a planet orbiting it, pulsates in precisely the way needed to mimic the signature of a planet in orbit around it. It casts doubt over the presence of other extra-solar planets discovered in the past 18 months.

28 February 1997 The Italian-Dutch satellite *BeppoSAX* (launched 30 April 1996) observes the first visible-light image of a cosmic gamma ray burst (GRB) – powerful flashes of gamma rays that occur daily, and randomly, and which outshine all other gamma rays combined.

The bursts release more energy in ten seconds than the Sun will emit in its entire ten billion-year lifetime, yet no source has ever been observed. The Dutch astronomer Jan van Paradijs and his Italian-Dutch team observe a light source in a distant galaxy that quickly fades after the burst. The bursts were previously thought to be relatively nearby in space.

23 March 1997 The comet Hale–Bopp comes to within 190 million km/120 million mi of Earth, the closest since 2000 BC. NASA launches rockets to study the comet. Its icy nucleus is estimated to be 40 km/25 mi wide, making it at least ten times larger than that of Comet Hyakutake and twice the size of Halley's comet.

28 April 1997 US astronomer William Purcell announces the discovery of a huge stream of antimatter at the heart of the Milky Way galaxy. The jet, the source of which is a mystery, extends for 3,000 light years above the centre of the galaxy.

5 June 1997 US astronomer Jane Luu and colleagues report the discovery of a new type of object within the Solar System – a 'worldlet', known by its catalogue number 1996TL66, which has a diameter of 500 km/300 mi. They suggest it represents a new class of object belonging to a population of possibly several thousand orbiting between the Kuiper belt and the Oort cloud.

4 July 1997 The US spacecraft *Mars Pathfinder* lands on Mars. Two days later the probe's rover *Sojourner*, a six-wheeled vehicle that is controlled by an Earth-based operator, begins to explore the area around the spacecraft.

Scientific results include an analysis of Martian dust, found to include magnetic, composite particles, with a mean size of one micrometre. The rock chemistry at the landing site is different from Martian meteorites found on Earth, appearing to be of basaltic andesite composition. However, the soil chemistry of Ares Vallis appears to be similar to that of the *Viking 1* and *2* landing sites. The possible identification of rounded pebbles and cobbles on the ground, and sockets and pebbles in some rocks, suggests conglomerates that formed in running water, during a warmer past in which liquid water was stable.

10 July 1997 Japanese astronomer Makoto Hattori and colleagues report the discovery of a knot of mass, which they call a 'dark cluster'. It has the chemical and gravitational properties of a cluster of galaxies, but is optically invisible. A new type of cosmic entity, it helps explain how light from a particular quasar has been distorted, and challenges the theories of galaxy formation.

1 August 1997 A study presented at the Division of Planetary Sciences of the American Astronomical Society in Cambridge, Massachusetts, says that the Moon was created early in Earth's history by debris thrown off from Earth after a collision with a planet three times as large as Mars.

October 1997 NASA and the European Space Agency (ESA) launch the space probe *Cassini-Huygens* mission to the planet Saturn.

    The *Cassini* spacecraft, including the orbiter and the *Huygens* probe, is one of the largest, heaviest, and most complex interplanetary spacecraft ever built. The complexity of the spacecraft is necessitated both by its trajectory or flight path to Saturn and by the ambitious programme of scientific observations to be undertaken once the spacecraft reaches its destination.

8 January 1998 At a meeting of the American Astronomical Society, US astronomers present evidence that the universe will never stop expanding and that it is about 15 billion years old, much older than previous estimates.

26 January 1998 Analysis of high-resolution images from the *Galileo* spacecraft suggests that the icy crust of Europa, Jupiter's fourth-largest moon, may hide a vast ocean that might be warm enough to support life.

27 January 1998 Al Schultz of the Space Science Institute in Baltimore, Maryland, using the Hubble Space Telescope, announces the discovery of a giant planet, larger than the Sun, orbiting Proxima Centauri, the closest star to Earth. It is the first planet outside the Solar System to be directly observed.

March 1998 The US *Mars Global Surveyor* commences a detailed photographic survey of the planet. The most significant findings include evidence for remnant magnetization of the Martian crust consistent with an early, vigorous molten interior core dynamo. These remnant anomalies in the ancient Martian crust indicate that in its early evolution Mars did have a global magnetic field, although it no longer does.

5 March 1998 US scientists announce that the *Lunar Prospector* satellite has detected hydrogen in the polar regions of the Moon, probably in the form of water, frozen in craters which never see the Sun. Scientists estimate that as much as 11 million tonnes of water may be present.

12 March 1998 Astronomers in Mauna Kea, Hawaii, announce the sighting of a new galaxy, named 0140+326RD1, which is around 90 million light years farther away than the previously known furthest galaxy from Earth.

4 July 1998 Astronomers from the University of Hawaii discover the first asteroid entirely within the Earth's orbit; it is 40 m/130 ft in diameter.

7 July 1998 A Russian nuclear submarine in the Barents Sea launches a commercial satellite into space. The first launch of its kind, it shows that launches can be made from any latitude and that an increased range of orbit can be achieved.

11 July 1998 British astronomers in Hawaii discover what they believe to be a Solar System forming around the star Epsilon Eridani, ten light years away.

The Quest for the Red Planet

by Tim Furniss

INTRODUCTION

If there is one major goal left in space explora-tion that really whets the appetite, it is the quest to find life on the Red Planet, Mars. The loss of NASA's *Mars Climate Orbiter* and the *Mars Polar Lander* in 1999 was a devastating blow to everyone who was hoping to learn more about the Red Planet, especially the scientists who had spent years devising and preparing the mission and its experiments.

MARS

Mars, named after the Roman god of war, is called the Red Planet because of its distinctive ruddy colour when seen in the night sky. This is the result of much iron oxide or rust in the soil and rocks. Mars has long been seen as a tantalizing world, especially since early observations by telescopes revealed fascinating features. Parts of Mars seemed to change shape and colour and lines were seen on its surface. Some scientists became over-excited and, based on a mistranslation of the Italian word *canali*, claimed that the linear channels were canals and that these were used for irrigation to grow vegetation using water from the melting poles of the planet. This led to speculation that the planet was inhabited by intelligent beings.

COULD THERE BE LIFE ON MARS?

The 'life on Mars' idea still intrigues people today, although at least the fear of Martian aliens has been quashed, largely owing to the first explorations of the planet by various spacecraft that flew past, orbited, and landed on the Red Planet. Some spacecraft, however, sent back images that suggested that water had indeed once flowed over the Martian surface. Like those early telescope observations, this has led to wild assumptions by scientists and the media alike that, simply if there is water, there is life. There is absolutely no evidence that life exists anywhere other than the Earth. But the quest to find life elsewhere seems unstoppable and appears to have almost become the *raison d'être* for space exploration. There is no water on Mars now but there are still many who, despite the extreme conditions that exist on the surface, believe that there may be primitive Martian lifeforms, or that these existed in the planet's early history when conditions, including lots of running water, may have been more conducive.

EXPLORATIONS OF MARS

The first astronauts could possibly travel to Mars in about 2015, if funds allow. Their voyage will have been made possible by the uncrewed precursor explorations of Mars. These began in earnest very early in the space age, which was born on 4 October 1957 with the launch of the first Earth satellite, *Sputnik 1*, by the former Soviet Union. At first Mars proved to be an elusive target. Six attempts to fly a craft to Mars – five by the Soviet Union starting in November 1960 and one by the USA – failed between 1960 and 1964 before the US *Mariner 4* was launched. It flew past Mars in July 1965, taking a series of grainy pictures, the best of which showed that Mars was covered in craters.

ORBITING MARS

Mariner 9 became the first craft to orbit Mars in 1971 but the Soviet Union continued to have a miserable time trying to explore the Red Planet. It had suffered a total of 13 failures before its first relatively successful *Mars 5* orbiter flight in 1973. One spacecraft, *Mars 6*, appears to have landed safely in 1974 but may have been covered by its parachute before it had a chance to send back data and pictures to Earth. Subsequent Russian Mars attempts have also ended in failure. Two Russian spacecraft were launched in 1988 especially to explore Phobos, one of Mars's two moons. Both failed just as they were approaching the planet and the *Mars 96* probe never left Earth orbit and crashed in South America.

US ATTEMPTS TO EXPLORE MARS

The USA has been a little more successful. Three US spacecraft, *Viking 1* and *2* and the *Mars Pathfinder* and its little rover, the *Sojourner*, made landfall on the Red Planet in 1976 and 1997 respectively. These craft found what seems to be a barren, sterile world. The *Viking 1* and *2* spacecraft consisted of orbiters and landers. The

Viking landers touched down safely in July and September 1976. Two *Viking* orbiters also conducted a systematic survey of the planet, including its polar caps, and plotted giant dust storms, which regularly blow across the Martian surface. The *Viking* landers were equipped to analyse soil using a robot arm with a scoop. Two *Viking* landers scooped up soil but found no indications of life. The *Viking* craft were also equipped with several instruments to study the Martian environment. Temperatures ranged from about $-86°C/-122.8°F$ before dawn to $-33°C/-27.4°F$ in the mid-afternoon. Pictures showed a pink sky, caused by airborne reddish dust scattering sunlight, and spectacular sunsets, which created a number of haloes. Gusts of wind reached 50 km/31 mi per hr. Carbon dioxide ice-frost was also photographed on rocks.

The *Mars Pathfinder*'s landing area in Arres Valis in 1997 created much worldwide excitement with millions of people logging on to NASA Web sites to see the images returned 'live'. A new fleet of US-led uncrewed landing and orbiting expeditions was planned, with the *Mars Climate Orbiter* due to arrive in October 1999 and the *Mars Polar Lander* in December 1999. They were to have been followed every two years by more spacecraft. It was hoped that by 2007, bits of Mars rock would be brought back to Earth. Sadly, both the *Mars Climate Orbiter* and the *Mars Polar Lander* failed, much to the embarrassment of NASA. Earlier, NASA had lost the *Mars Observer* which exploded just before it entered orbit around Mars in 1990. Following the loss of the two craft in 1999, and NASA's consequent reassessment of its plans and aspirations, it is unlikely that a mission to collect and return samples from Mars will be made until 2010.

MARS EXPRESS

The European Space Agency's (ESA) *Mars Express* was given the go-ahead in November 1998. It is scheduled to be launched in 2003. It will use radar to map possible underground water sources and will also carry a lander, *Beagle 2*, for soil collection.

MAPPING MARS

The highly successful US *Mars Global Surveyor* has, however, been orbiting Mars since 1997, returning spectacular close-ups of many features – and clearing up one puzzle. What had seemed to be a huge 'Mars face' in an earlier *Viking* spacecraft image was due to the angle of the Sun and shadows. The *Surveyor* showed that the feature was a mesa, a natural rock formation, and there was no 'face'. Mars has now been almost entirely mapped. Especially tantalizing is the fact that without these spacecraft we would not have known that much of Mars seemed to have once had water flowing over its surface. This serves as the main attraction for sending people to Mars. Apart from watching the first men walking on the Moon, nothing has captured the imagination of the public more than Mars landings. However, overcoming the technological hurdles of sending people to Mars will be extremely expensive. If there is going to be a crewed Mars mission, it is unlikely to take place before 2020.

HUMANS ON MARS?

In 1994, NASA had developed a strategy for a possible expedition to Mars to begin in 2011 involving three launches of a huge new booster. Even this plan was considered too ambitious technically and financially and NASA scaled down the proposed baseline mission. The new plan involved the use of smaller boosters but double the number of launches to assemble the components in Earth orbit for despatch to Mars. An inflatable *TransHab* module would be launched to Mars with the *Cargo 1* craft to make landfall in the prime landing zone. This would serve as a living and experiment quarters for the Mars crew after their arrival – assuming a safe and accurate landing. The crew would land in a separate spacecraft which would then be connected to *TransHab*. The upper part of the crewed *Cargo* vehicle would later take off after the proposed 500-day Mars surface exploration mission and dock with return craft, which would have been placed in Mars's orbit earlier.

A DANGEROUS QUEST

There is another important issue in human Mars exploration. The *Apollo* astronauts were three days away from the Earth but a Mars crew would face hazards with no immediate return to Earth in prospect. The quest for Mars is as dangerous and uncertain as the earliest ocean voyages. Whether the public is prepared to lose costly spacecraft, and even face the loss of the lives of astronauts in order to reach the Red Planet, remains to be seen. Meanwhile, other exploration continues as both Europe's *Mars Express* and NASA's Mars *Exploration Rover* will be launched in 2003 in a new phase of more realistically-planned exploration of the Red Planet.

October 1998 Dutch researchers discover a large galaxy close to our own. The galaxy is 20 million light years away in the Local Void, an area of space that is generally considered to be nearly empty. It escaped detection due to its faintness.

Named Cepheus 1, it is so close to Earth that it appears almost as large as the full Moon, and is thus one of the ten largest spiral galaxies in the sky. Its discovery represents another step in completing the census of galaxies in the local neighbourhood, important in determining statistics of the mass and luminosity characteristics of these fundamental building blocks of the universe.

20 November 1998 The spacecraft *Zarya*, the first module of the $60 billion International Space Station set up by a group of 16 countries, is launched from Kazakhstan. The US module *Unity* is connected to *Zarya* in December. The station, which is planned for completion in 2004, will orbit the Earth and will house research laboratories and accommodation for seven scientists.

9 January 1999 Astronomers from San Francisco State University announce the discovery of three more planets orbiting around neighbouring stars, bringing the total number of known planets outside our Solar System to 17.

11–22 February 2000 The space shuttle *Endeavour* carries out a mission in which it scans the Earth's surface with radar signals to create a detailed topographical map of the world.

25 April 2000 A team of scientists using NASA's *Chandra* X-ray observatory announce that they have measured the distance to an X-ray source by observing the delay and smearing (filtering) out of X-ray signals traversing 30,000 light years of interstellar gas and dust.

30 April 2000 The spacecraft *NEAR Shoemaker*, orbiting the near-Earth asteroid 433 Eros, is manoeuvered to within 50 km/31 mi of the surface. Eros is an S-class asteroid approximately 13 × 13 × 33 km in size, the second largest near-Earth asteroid.

Data on the elemental composition of Eros's surface show that the asteroid appears to be undifferentiated; that is, the surface has a mixture of both light and heavy elements, like chondrite meteorites that have reached the Earth. This leads scientists to believe that Eros is made of materials left over from the formation of the Solar System and has not been subjected to heating that would process the asteroid, separating light elements from heavier ones.

12 July 2000 The launch of the *Zvezda* service module from Baikonur, Kazakhstan, marks an important milestone in the development of the International Space Station.

27 July 2000 NASA announces its plans for future Mars exploration. The goals are the understanding of the planet's atmosphere and climate, the history of its rocks, the role of water and, possibly, evidence of past or present life. International participation, especially from Italy and France, is part of the plan.

11 August 2000 A Russian *Progress* cargo ship is launched from the Baikonur Cosmodrome, Kazakhstan, on a *Soyuz* rocket, carrying clothes, computers, food, and other supplies for use by the first permanent crew of the International Space Station.

Biology

1990 The Human Genome Project, a massive international research project to map the genes of selected human chromosomes formally begins.

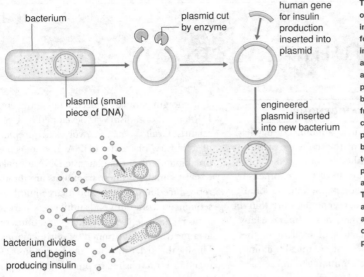

bacterium

plasmid (small piece of DNA)

plasmid cut by enzyme

human gene for insulin production inserted into plasmid

engineered plasmid inserted into new bacterium

bacterium divides and begins producing insulin

The genetic modification of a bacterium to produce insulin. The human gene for the production of insulin is collected from a donor chromosome and spliced into a vector plasmid (DNA found in bacteria but separate from the bacterial chromosomes). The plasmids and recipient bacteria are mixed together, during which process the bacteria absorb the plasmids. The plasmids replicate as the bacteria divide asexually (producing clones) and begin to produce insulin.

The two founder institutions are the US Department of Energy and the US National Institutes of Health, but many other nations join the project.

1991 British geneticists Peter Goodfellow and Robin Lovell-Badge specify the location of Tdy, the gene on the Y chromosome that determines sex.

1991 US molecular biologist Craig Venter applies for patents on more than 300 genes.

1992 An individual honey fungus (*Armallaria ostoyae*) is identified as the world's largest living thing. Discovered in Washington State and estimated to be between 500 and 1,000 years old, it has an underground network of hyphae covering 600 hectares/1,480 acres.

1992 Sperm cells are discovered to have odour receptors and may therefore reach eggs by detecting scent, by US physician David Garbers of the University of Texas.

1992 US molecular biologist Craig Venter establishes The Institute for Genomic Research, a privately funded centre for sequencing DNA.

1992 US geneticist Jack Dumbacher discovers the world's first poisonous bird, the pitohui, while on field work in New Guinea.

1993 US biochemist Kary Mullis and Canadian biochemist Michael Smith share the Nobel Prize for Chemistry: Mullis for his invention of the polymerase chain reaction (PCR) technique for amplifying DNA, and Smith for developing techniques for splicing genetic material. The Mullis technique enables small samples of DNA fragments to be replicated to yield sufficient quantities for analysis. The technique is invaluable in criminal investigations, where only trace samples of organic material are available, and in studies of trace DNA obtained from fossil remains.

1994 A previously uncategorized species of kangaroo is discovered in Papua New Guinea. Known locally as the bondegezou, it weighs 15 kg/7 lb and is 1.2 m/3.9 ft in height.

Identified by Australian zoologist Tim Flannery, the bondegezou is a black-and-white tree kangaroo formally named *Dendrolagus mbaiso*; 'mbaiso' translates as 'forbidden' in the local dialect and is used because local hunters are forbidden by custom from killing this animal.

The Human Genome Project

by Paul Wymer

DRAFTING THE HUMAN GENOME

In February 2001, scientists announced they had analysed the first draft of the sequence of the human genome, published in June 2000. While the analysis of the first draft of the genome is by no means the culmination of the international project to map the human genome, started in the USA in 1990, it is a significant landmark. Many scientists considered the task of sequencing the three billion nucleotides, the chemical building blocks that make up the human genome, impractical when it was first proposed in the mid-1980s. Large-scale sequencing efforts have been divided between the US and the UK.

As well as determining the nucleotide sequence, the Human Genome Project is designed to localize the estimated 30,000–40,000 genes within the human genome. The scientific products of the project will form a resource of detailed information about the structure, organization, and function of human DNA, information that constitutes the basic set of inherited 'instructions' for the development and functioning of a human being.

Similar efforts have since been launched to map and sequence the genomes of a variety of organisms used extensively in scientific research, for example the pathogenic bacterium *Haemophilus influenzae* and the plant *Arabidopsis thaliana*.

SEQUENCE AND CONSEQUENCE

DNA is made up of four different nucleotide bases adenine, guanine, thymine, and cytosine. The genetic code that gives us our unique characteristics is defined by the sequence of these bases along the DNA molecule. This base sequence determines the order of building blocks (amino acids) in the proteins coded for by different genes.

Progress in DNA sequencing has been very rapid. The development of advanced automation, robotics, and computer software for large-scale DNA sequencing has proceeded at a remarkable pace. The 'shotgun' strategy coupled with automated DNA sequencing machines has proved an extremely successful combination.

The shotgun strategy for sequencing the entire DNA in an organism was developed by US scientist Craig Venter. It involves the preparation of a 'library' of random DNA fragments by the breaking up of entire genomic DNA to make it easier to analyse. These fragments are then reproduced many times and a large number are sequenced from both ends until every part of the genome has been sequenced several times on average (this is determined by statistical means). Finally, the order in which the fragments fit together is established to provide the complete genome sequence.

Nonetheless, it is a huge step from elucidating the DNA sequence of the genome to knowing where within that sequence genes are located and what they do. Mere acquisition of DNA sequences conveys little more about the biology of the organisms from which they are derived than a company telephone directory can reveal about the complexities of the company's business. It is conceivable, however, that many of the major genes involved in, for example, susceptibility to heart disease and various cancers, will be identified at some point. This will make it possible to 'type' people according to the variants of these genes that they carry and hence determine the likelihood that they will develop these diseases.

MAKING SENSE OF SEQUENCE

The vast amount of DNA sequence information generated by research groups around the world is stored in computer databases, connected with each other via the Internet. In extracting meaning from sequence information, scientists are faced with a task analogous to decoding an unknown language. By themselves, the letters make no sense but their particular combination into words and sentences is crucial. As in a real language, the subtlest changes, just like changing a single letter in a word, can thoroughly alter the message. DNA sequences for microbes, plants, animals, and humans from laboratories around the world can be compared and duplications can be detected. Also, similarities that point to evolutionary relationships, new classifications, and, ultimately, better understanding of life form and function can be recognized.

GENETICS AND DISEASE

While we all share a common genome, humans as a population are not clones. The genome is what defines us as *Homo sapiens* and in DNA make us biological individuals. We are predisposed to different diseases, we respond to the environment in different ways, and we differ in the way we react to drugs – just as we are inherently different in our ability to perform particular tasks. Genetic explanations
of the mechanism of diseases will enable the development of drugs with a more precise action, producing higher response rates and lowering the risk of adverse effects. Recent research, for example, has identified some schizophrenic patients who have a mutation in a particular neuroreceptor gene. This group of patients is unlikely to benefit from the drug Clozapine, often used to treat schizophrenia. This knowledge is very valuable, as the drug is costly and can have severe side effects. However, a potential drawback for pharmaceutical companies would be the costs involved in tailoring drugs, which could be sold to fewer customers.

GENETIC SCREENING

Many genes have already been mapped to their approximate positions on the chromosomes. These positions are determined in relation to known 'markers', DNA fragments that provide landmarks throughout the genome. Once the location of a disease gene is known, it is relatively straightforward to test for its presence in different people. This is known as genetic screening. For disorders caused by a single gene, such as cystic fibrosis and Duchenne muscular dystrophy, tests can now be applied to indicate the presence or absence of the faulty gene. Such tests may be carried out antenatally on the unborn fetus, on test-tube embryos, or on adults. If the genetic conditions contributing to multi-factorial diseases are identified, it may be possible to test for increased susceptibility to predisposed illnesses, such as coronary disease.

Although genetic testing at the chromosome level has been in existence since the 1960s, for example in prenatal testing for Down's syndrome, recent progress has focused attention on the sensitive issues that this type of testing entails. For example, some genetic tests permit diagnosis of disorders in patients before they have developed any symptoms. Others can identify carriers of recessive genetic diseases. Such tests can raise very difficult choices for individuals, as illustrated by the discovery of the gene for Huntington's disease.

Huntington's disease is a degenerative condition of the central nervous system that generally develops in middle age. No cure is currently available. Sufferers gradually deteriorate to a point where they require total care before death occurs. The disease affects about one in 5,000 people in the UK. The gene responsible for Huntington's disease was identified in 1993 and mapped to the tip of chromosome 4, making direct testing possible. Since testing began, it is thought that only a small minority of those who have a 50% chance of carrying the faulty gene –that is, those with a parent who has the condition – have chosen to undergo testing; probably around 5% of such people. Those who do seek testing usually undergo extensive counselling and discussion before proceeding.

Diagnosis of a single gene disorder for which no cure is currently available is one dilemma, but as the Human Genome Project progresses, new issues with consequences for the individual and for society are regularly created. For example, much genetic information is probabalistic rather than strongly predictive. Should 'risk factors' guide therapeutic choices, for prophylactic mastectomy for example? Should the population be screened for certain genetic conditions and, if so, how do we use the information? A worst-case scenario could see the development of a genetic underclass, with people denied insurance and employment and discriminated against in society because of what is written in their genes.

OWNERSHIP AND CONTROL

The decoding of the human genome is an immense achievement and its impact on health care will be far-reaching. The major challenge that lies ahead is in ensuring appropriate management of the ownership of the new knowledge and of its potential effects on the individual.

1994 The oldest surviving fungi are found as dormant spores in the hay lining the boots of the 'iceman' Ötzi who died 5,300 years ago and whose body was preserved in a glacier in the Alps (found in 1991).

26 November 1994 Short stretches of dinosaur DNA are extracted by a team from the University of Provo, Utah, from unfossilized bone retrieved from coal deposits approximately 80 million years old.

1995 A fossil chordate *Yunnanozoon lividum* is discovered in Chengjiang, China. It is the first chordate recorded from the early Cambrian period and is 525 million years old.

1995 A transgenic sheep, containing a human gene in its genotype, is produced.

The gene expresses the protein alpha-1-antitrypsin (AAT), which protects lung tissue from the results of inflammation. Victims of the lung disease emphysema are not able to produce sufficient quantities of AAT to prevent damage. AAT is extracted from the sheep's milk. The sheep is a pioneer for other transgenic animals whose milk contains pharmacologically active compounds.

1995 The largest carnivorous dinosaur *Giganotosaurus carolinii* is discovered in Patagonia. It lived about 97 million years ago, was 12.5 m/41 ft in length, and weighed 6–8 tonnes.

1995 US molecular biologist Craig Venter announces the full DNA sequencing of the bacterium *Haemophilus influenzae*, the first time that all of the DNA of an organism has been sequenced.

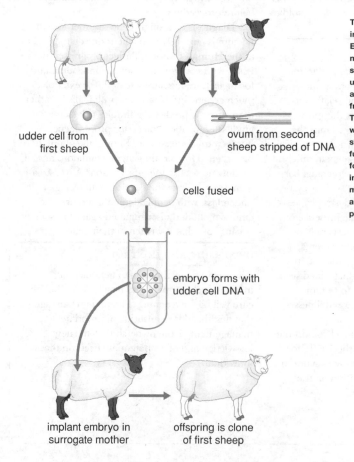

udder cell from
first sheep

ovum from second
sheep stripped of DNA

cells fused

embryo forms with
udder cell DNA

implant embryo in
surrogate mother

offspring is clone
of first sheep

The cloning of Dolly the sheep in 1997, by the Roslin Institute in Edinburgh, was a genetic milestone. It was the first successful clone produced using genetic material from an adult (udder) cell rather than from a gamete (egg or sperm). The DNA from the udder cell was fused with an ovum stripped of its own DNA. The fused cells divided in vitro to form an embryo that was then implanted into a surrogate mother. The resulting lamb was a clone of the ewe that had provided the udder cell.

1995 US scientists successfully germinate bacterial spores extracted from the gut of a bee fossilized in amber 40 million years ago.

They successfully later bring to life a previously unknown *Staphylococcus* bacterium in 1997, although their results are disputed by those who believe that the ancient specimens are actually a result of modern contamination.

1996 British palaeontologists discover the world's oldest flowering plant species *Bevhalstia pebja* in southern England. It is a wetland herb about 25 cm/10 in high and it is about 130 million years old.

1996 Publicly funded genome scientists make a concerted effort to coordinate their work in response to the challenge from privately funded genomic research. Around 30,000 human genes are located, eventually resulting in the first human gene map.

An international group of scientists issue the 'Bermuda Statement'. The statement declares that all DNA sequencing data should be immediately available to the public and not withheld for private/commercial gain.

1996 Researchers from the University of California discover anemones in the Bahamas that have been alive for 1,500–2,000 years – longer than any other known marine creature.

August 1996 US geneticists clone two rhesus monkeys from embryo cells.

February 1997 US geneticist Don Wolf announces the production of monkeys cloned from embryos. It is a step closer to cloning humans and raises acute philosophical issues.

24 February 1997 US president Bill Clinton announces a ban on using federal funds to support human cloning research, and calls for a moratorium on this type of scientific research. He also asks the National Bioethics Advisory Commission to review and issue a report on the ramifications that cloning would have on humans.

27 February 1997 Scottish researcher Ian Wilmut of the Roslin Institute in Edinburgh, Scotland, announces that British geneticists have cloned an adult sheep. A cell was taken from the udder of the mother sheep and its DNA combined with an unfertilized egg that had had its DNA removed. The fused cells were grown in the laboratory and then implanted into the uterus of a surrogate mother sheep. The resulting lamb, Dolly, came from an animal that was six years old. This is the first time cloning has been achieved using cells other than reproductive cells. The news is met with international calls to prevent the cloning of humans.

Dolly, the first ever mammal to be successfully cloned from an adult specimen.
Courtesy of the Roslin Institute

8 May 1997 English zoologists Gareth Jones and Elizabeth Barratt announce the discovery of a new British mammal – a pipistrelle bat.

Almost identical to the European pipistrelle and living in the same habitat, the two use different echolocation and courting frequencies and are genetically different, having experienced 5–10 million years of separate evolution.

16 May 1997 US geneticists identify a gene *clock* in chromosome 5 in mice that regulates the circadian rhythm.

3 June 1997 US geneticist Huntington F Wilard constructs the first artificial human chromosome. The artificial chromosome is successfully passed on to all daughter cells.

Wilard inserts telomeres (which consist of DNA and protein on the tips of chromosomes) and centromeres (specialized regions of DNA within a chromosome), removed from white blood cells, into human cancer cells, which are then assembled into chromosomes, which are about one-tenth the size of normal chromosomes.

11 July 1997 Teams of researchers from Germany and the USA use mitochondrial DNA extracted from the original remains of Neanderthal Man, discovered in the Neander Valley near Düsseldorf, Germany, in 1856, to confirm that Neanderthals and modern humans diverged evolutionarily about 600,000 years ago.

This supports the theory that modern humans arose recently in Africa as a distinct species and replaced Neanderthals with little or no interbreeding, the Neanderthals becoming extinct without evolving into modern humans.

18 July 1997 Japanese scientist Yoshinori Kuwabara announces that his team has successfully grown goat embryos.

The embryos, removed from their mother at 17 weeks into pregnancy, are placed in a tank filled with liquid to simulate amniotic fluid. The placenta is replaced by a machine to pump oxygen and nutrients into the embryo's blood. At 20 weeks' gestation the goat is 'born'. At present the procedure can only be done late in development.

August 1997 US geneticist Craig Venter and colleagues publish the genome of the bacterium *Helicobacter pylori*, a bacterium that infects half the world's population and which is the leading cause of stomach ulcers. Complete genomes are increasingly being published as gene-sequencing techniques improve.

It is the sixth bacterium to have its genome published. It has 1,603 putative genes, encoded in a single circular chromosome that is 1,667,867 nucleotide base-pairs of DNA long.

22 August 1997 Scientists from the World Wide Fund for Nature (WWF) announce the discovery of a new species of muntjac deer in Vietnam.

A dwarf species weighing only about 16 kg/35 lb, it has antlers the length of a thumbnail and lives at altitudes of 457–914 m/1,500–3,000 ft.

7 January 1998 Doctors meeting at the World Medical Association's conference in Hamburg, Germany, call for a worldwide ban on human cloning. US president Bill Clinton calls for legislation banning cloning.

30 January 1998 US scientist Angela Christiano of Columbia University, New York, publishes a study that identifies a 'hairless' gene that causes severe hair loss.

20 February 1998 Researchers at the University of Texas, in conjunction with the British company SmithKline Beecham, publish a study in which they identify a hormone that triggers hunger in humans. Scientists hope the discovery will lead to potential treatments of appetite disorders.

July 1998 China's Academy of Sciences announces a project to clone the panda. This involves transferring the nucleus of a panda cell into that of a bear species, with the same species being used as a surrogate mother.

4 July 1998 US ornithologists announce the important discovery of a new species of bird in Ecuador belonging to the genus *Antpitta*.

23 July 1998 Scientists succeed in cloning more than 50 mice at the University of Hawaii, in the first successful cloning of an adult mammal since the creation of the cloned sheep Dolly in 1997.

October 1998 US biologist French Anderson announces a technique that could cure inherited diseases by inserting a healthy gene to replace a damaged one. He calls for a full debate on the issue of gene therapy, which brings with it the dilemma of whether it is ethical to enable the choice of physical attributes such as eye colour and height.

November 1998 US researchers successfully induce the growth of new hair follicles in adult mice. It is the first time that new hair follicles have been grown in adult skin, leading to the possibility of improved treatment for baldness.

9 December 1998 A team of scientists from Kinki University, Japan, clone eight calves from an adult cow, in the third instance ever of cloning an adult mammal.

10 December 1998 The first genetic blueprint for a whole multicellular animal – a nematode worm – is completed. The 97 million-letter code, which is published on the Internet, is for a tiny worm called *Caenorhabditis elegans*. The study began 15 years previously and cost £30 million.

16 December 1998 Scientists at Kyunghee University Hospital in Seoul, South Korea, announce that they have created a human embryo cloned from a female adult, but that they have destroyed it because of the ethical controversy surrounding cloning of humans.

1999 The sequencing of DNA dramatically accelerates, following the introduction of new sequencing technology that includes coupling supercomputers with automated sequencing techniques.

The first billion bases, forming a 'working draft' of the human genome, are mapped. Chromosome 22 is fully sequenced at the Sanger Centre, England; it is the first human chromosome to be so described.

24 January 1999 US scientist Craig Venter of the Institute for Genomic Research in Maryland, USA, announces the possibility of creating a living, replicating organism from an artificial set of genes. An experiment to test this is put on hold until the moral question is discussed by religious leaders and ethicists at the University of Pennsylvania.

10 May 1999 A team of French scientists unearth what is believed to be the world's largest fossil of a land mammal. The fossil, which is 5.5 m/18 ft high and 7 m/23 ft long, is from the *Baluchitherium*, a 15–20 tonne mammal that looked like a rhinoceros and lived 30 million years ago.

24 June 1999 British minister for public health Tessa Jowell announces a temporary ban on human embryo cloning in the UK, even for medical research. Scientists warn that the ban could impede research into cures for diseases.

The ban is in response to a joint Human Fertilisation and Embryology Authority/ Human Genetics Advisory Commission report into cloning.

October 1999 A Russian expedition led by scientist Bernard Buigues discovers the carcass of a 23,000-year-old woolly mammoth frozen in Siberia. The carcass is flown to Khatanga, Russia, where scientists plan to clone the mammoth using an elephant as a surrogate mother.

9 December 1999 US scientists at the Institute for Genomic Research (TIGR) in Maryland, USA, continue with their research to create a new life form in the laboratory. They will use a microbe called *Mycoplasma genitalium*, which contains the smallest number of genes of any known living creature.

5 March 2000 Five cloned pigs are born in Blacksburg, Virginia. Scientists hope to use cloned pigs to genetically engineer their organs to be compatible with humans.

Concern about the transmission of viruses across species barriers blocks implementation.

26 June 2000 Scientists working on the Human Genome Project in London, England, and Washington, DC, announce that they have completed the first draft of the entire structure of human DNA.

Chemistry

1990 The British company Imperial Chemical Industries (ICI) develops the first practical biodegradable plastic, Biopal.

None of the plastics used in commodity applications, such as bottles and carrier bags, will quickly degrade in the environment. Increasing awareness of environmental pollution has led to the development of photodegradable polymers. They are similar in appearance and properties to LDPE polyethylene. However, these polymers are still not in common use due to higher production costs and instability during initial processing.

1990 US chemist Elias James Corey receives the Nobel Prize for Chemistry for developing novel methods for the synthesis of complex natural compounds.

1991 Swiss chemist Richard Ernst receives the Nobel Prize for Chemistry for improvements in the technology of nuclear magnetic resonance.

1992 Researchers in the US produce the first solid compound of the noble gas helium.

They mix helium and nitrogen together and apply a pressure 77,000 times greater than atmospheric pressure. The compound is thought to be held together only by van der Waals forces. If this theory is correct, then the helium/nitrogen compound is the first member in a completely new class of compounds.

1992 US chemist Rudolph Marcus receives the Nobel Prize for Chemistry for his theoretical discoveries relating to reduction and oxidation reactions.

1993 US biochemist Kary Mullis and Canadian biochemist Michael Smith share the Nobel Prize for Chemistry, Mullis for his invention of the polymerase chain reaction technique for amplifying DNA, and Smith for developing techniques for splicing genetic material.

The Mullis technique, developed in 1983, enables small samples of DNA fragments to be replicated to yield sufficient quantities for analysis. The technique is invaluable in criminal investigations, where only trace samples of organic material are available, and in studies of trace DNA obtained from fossil remains.

1994 US chemist George Olah receives the Nobel Prize for Chemistry for developing a technique for examining hydrocarbon molecules.

October 1994 Scientists at the Heavy Ion Research Institute (GSI) based in Darmstadt, Germany, discover ununnilium (atomic number 110).

They use a linear accelerator to fuse a beam of nickel ions with a lead target. Only one atom of the new element is created. The element has not been assigned a name and is known by the temporary designation 'ununnilium', the Latin word for '110'.

December 1995 Scientists at the Heavy Ion Research Institute (GSI), Darmstadt, Germany, synthesize unununium (atomic number 111).

They use a linear accelerator to fuse a beam of nickel ions with a bismuth target. Only three atoms of the new element are created. The element has not been assigned a name and is known by the temporary designation, unununium, from the Latin word for '111'.

1996 English chemist Harold Kroto and US chemists Robert Curl and Richard Smalley are awarded the Nobel Prize for Chemistry for their discovery of fullerenes in 1985.

February 1996 Scientists at the Heavy Ion Research Institute (GSI), Darmstadt, Germany, synthesize ununbium (atomic number 112). A single atom is created, which lasts for a third of a millisecond.

They use a linear accelerator to fuse a beam of zinc ions with a lead target. The element has not been assigned a name yet and is known by the temporary designation, ununbium, the Latin word for '112'.

1997 US molecular biologist Paul Boyer, English molecular biologist John Walker, and Danish biologist Jens Skou share the Nobel Prize for Chemistry, Boyer and Walker for their elucidation of the enzymatic mechanism underlying the synthesis of adenosine triphosphate (ATP), and Skou for his discovery of an ion-transporting enzyme, Na+, K+-ATPase.

1998 English scientist John Pope and US scientist Walter Kohn share the Nobel Prize for Chemistry for their work in the field of quantum chemistry.

1999 Egyptian-born US chemist Ahmed Zewail receives the Nobel Prize for Chemistry for his work on transition states of chemical reactions using femtosecond spectroscopy.

He uses laser flashes that take place on the same time scale as chemical reactions (a femtosecond is 1×10^{-15} second) to investigate what happens when chemical bonds are broken or formed.

March 1999 Russian scientists at the Joint Institute for Nuclear Research at Dubna create ununquadium (atomic number 114).

They bombard plutonium with calcium to create only a single atom of the new element. The element has not been designated a name yet and is known by the temporary assignment 'ununquadium', the Latin for '114'.

June 1999 US physicists at the Lawerence Berkeley National Laboratory, USA, create ununhexium (atomic number 116) and ununoctium (atomic number 118). They use a linear accelerator to accelerate a beam of krypton ions into a lead target.

In a three-day period, they create only three atoms of ununoctium, but this element decays to produce ununhexium. The elements have not yet been designated names and are known temporarily by the Latin words for '116' and '118'.

2000 Japanese chemist Hideki Shirakawa and US chemists Alan MacDiarmid and Alan Heeger receive the Nobel Prize for Chemistry for their work on developing electrically conductive plastics.

Earth Sciences

1991 A circular impact structure of Cretaceous–Tertiary (K–T) age is found buried beneath Mexico's Yucatan peninsula. Called the Chicxulub crater, it is the best candidate for the K–T impact site envisioned by Luis Alvarez and others (1980).

The impact structure, about 90 km/56 mi across, is found using gravity data.

1991 The World Ocean Experiment (WOCE) programme is set up to monitor ocean temperatures, circulation, and other parameters.

1994 US geochemist Sam Bowring publishes the radiometric age of the Acasta Gneiss from Canada at almost 4 billion years. It is the oldest terrestrial rock dated thus far.

4 July 1994 Electrical flashes known as 'sprites' – upper atmosphere optical phenomena associated with thunderstorms – are first examined by aircraft, by a team from the University of Alaska Statewide System, Fairbanks, Alaska, USA.

1996 US geophysicists calculate that the Earth's inner core spins slightly faster than the rest of the planet.

Using seismic data over the century, they suggest that the core, which appears to be (seismically speaking) a single crystal, has been rotating with respect to the rest of the earth since the beginning of the 20th century. They calculate that every 400 years or so, it rotates once with respect to the rest of the Earth.

24 July 1997 Canadian researcher Richard Bottomley and colleagues date the 100 km/62 mi wide Popigai impact crater in Siberia, thought to be the fifth-largest impact crater on Earth, to 35.7 million years old. They suggest that the meteorite that created it may be responsible for the mass extinction that occurred at the end of the Eocene and the start of the Oligocene geological periods.

Ecology

1990 A report given to the Royal Geographical Society in London refers to the shrinking Aral Sea in Russia as an ecological disaster.

Rivers that once reached the sea, itself once the world's fourth largest area of fresh water, have been diverted to support the growth of cotton. The report highlights how the level of water has fallen 14 m/46 ft over the last 30 years, how 80 km/50 mi of sea floor has been exposed, and how a once flourishing fishing industry has been destroyed. This picture is confirmed by accounts given by writers in the Soviet magazine *Novy Mir/New World*.

1990 A strategy for earth science conservation published by the Nature Conservancy Council in Britain outlines the future for geoconservation. The report sets up the concept of Regionally Important Geological Sites (RIGS).

These are non-statutory designated sites, notified to local authorities by county RIGS Committees as significant by virtue of their research potential, educational value, historical importance, and aesthetic appeal.

1990 A United Nations report, *The State of the World Population, 1990*, paints a catastrophic picture of the impact of population growth. The world's population is estimated at 5.3 billion with a growth rate of about a quarter of a million per day.

The report calls for the adoption of two strategies: one aimed at reducing poverty and population growth in the southern hemisphere, the other aimed at the northern hemisphere, where it is stressed that a shift towards cleaner technologies, greater energy efficiency, and resource conservation is needed.

1990 The Environmental Protection Act comes into force in Britain; it includes new powers for local authorities and public access to information as well as encouraging recycling and placing stricter control over waste disposal.

In this piece of legislation, a concept of integrated pollution control is adopted that insists that the impacts of all policy options be considered in terms of the effect on the environment as a whole and not just on one medium such as land, air, or water.

1990 The Intergovernmental Panel on Climate Change issues its first report, containing work by 170 scientists from 35 countries. The report suggests that if nothing is done to control population growth or carbon dioxide emissions, it is possible that carbon dioxide levels will double by the year 2100 and that the Earth will rise in temperature by about 2.5°C.

Rainfall is predicted to decrease in summer in both southern Europe and central North America by 2030, leading to a possible failure of crops. The sea level is also predicted to rise by about 6 cm/2.4 in per decade.

15 November 1990 The USA sees the passage of the Clean Air Act, which endorses the notion of tradable permits for the discharge of pollution.

This is seen as an acceptance of the merits of environmental economics. Within this approach, an acceptable level of pollution is prescribed for a given area and companies are then allocated permits to discharge a portion of the pollutant. The notion of an optimal or acceptable level of pollution described in the act, however, is met with disapproval by many environmental groups.

1991 'BioSphere 2', an experiment that attempts to reproduce the world's biosphere in miniature within a sealed glasshouse, is launched in Arizona, USA. Eight people remain sealed inside for two years.

1991 A European Community Directive is issued to control levels of nitrates released into soils and water. Member states must declare vulnerable zones where nitrate concentration in water sources reaches 50 parts per million.

1991 A tropical storm called 'Thelma' hits the Philippines and causes a loss of over 6,500 lives.

The worst affected area is Leyte island; officials suggest that illegal logging has contributed to the scale of the disaster. The bulldozers used to drag away the felled timber leave areas of bare soil that fail to absorb storm water.

1991 The Convention on Trade in Endangered Species (CITES) meets and considers what response to make to a proposal from South Africa that the outlawed trade in ivory should be allowed to resume.

The traditional means of controlling elephant numbers has been by culling.

343

However, this does not benefit local people, and the environmental economist David Pearce suggests in a book *Elephants, Economics, and Ivory* (1991) that the ivory trade could be harnessed to provide revenue for local populations. Such proposals are vehemently opposed by Richard Leakey, director of Kenya's wildlife service. The problem of the African elephant now seems a vivid case study of the problems facing conservationists.

1991 The Organization for Economic Cooperation and Development (OECD) produces its report *State of the Nation*, which gives a pessimistic survey of the environmental costs of development in the organization's 24 member states.

The report lists the positive achievements over the period 1970–90, which include the elimination of disease-contaminated drinking water, an increase in the number of national parks, and better collection, disposal, and recycling of municipal waste. Sulphur emissions have also declined. Waste per person has increased, however, by over 20%. It argues that environmental concerns should be made central to economic decision-making.

4 April 1991 The US Environmental Protection Agency announces ozone layer depletion at twice the speed previously predicted.

1992 As the black rhinoceros faces extinction, Zimbabwe's Department of National Parks and Wildlife Management takes drastic action to save the 400 remaining individuals in the country.

The animals are pursued by helicopters and tranquillized by darts, then their horns are sawn off. This action is designed to thwart poachers who kill the animals for their horns, which in Hong Kong can fetch £3,350 per kilo as an aphrodisiac.

1992 On the 20th anniversary of the publication of *Limits to Growth*, Dennis Meadows, Donnella Meadows, and Jørgan Randers publish a sequel called *Beyond the Limits*.

Computers are used once more, with revised data and more sophisticated models. The predicted outcomes are again pessimistic. The authors argue that pollution has already surpassed rates that are physically sustainable, and that the world must accept a reduction in material and energy flows.

1992 The first Earth Summit takes place in Rio de Janeiro in Brazil. Officials from over 150 countries attend, together with 9,000 journalists and about 20,000 environmentalists.

More world leaders assemble in one place than at any time before. Two new international treaties are signed, one on biodiversity and one on climate change. Agenda 21 – a non-binding framework of recommendations designed to protect the environment and achieve sustainable development – is agreed and the Rio declaration – a statement of 27 principles related to sustainable development – is issued.

1992 The World Conservation Monitoring Centre based in Cambridge, England, publishes *Global Biodiversity* – an encyclopedia of existing species. The group estimates the total number of species at anything between 10 and 80 million.

The encyclopedia names about 1.4 million species. The remainder are mostly beetles and other insects not yet classified. The work also lists the 464 species of mammals, birds, insects, fish, and other fauna known to have become extinct since 1600. These include the great auk, last recorded in Iceland in 1844, the Falkland Islands wolf, the quagga (a South African relative of the zebra), and the Caribbean monk seal.

11 February 1992 US President George Bush announces that the USA will phase out the ozone-depleting chemicals chlorofluorocarbons (CFCs) by January 1996, four years earlier than originally planned. In Britain, ICI, one of the world's largest manufacturers of CFCs, announces that it intends to phase out CFCs two years ahead of its previous deadline.

1993 A report issued by the European Commission shows that the full impact of the release of oil by Iraq into the Gulf at the end of the Gulf war is worse than had been feared.

The intentional destruction by Iraqi troops of 700 oil wells and other installations is thought to be the world's biggest oil spill. Saudi Arabia now embarks on the fourth phase of a £70 million clean-up operation along 644 km/400 mi of coastline. Catches of shrimp fall to one tenth of their former value. Shortage of fish forces white-cheeked terns to abandon their eggs before hatching. Swift terns resort to eating the chicks of other species.

1993 The North Sea Task Force, a committee of scientists from European nations, presents a report to a meeting of ministers claiming that over-fishing of the North Sea is causing more damage than pollution, and is driving some species such as cod to the edge of extinction.

The report calls for the setting aside of 10% of the North Sea as a fishing-free zone to allow stocks to recover. The report also highlights the problem of the discharge of nitrates into the North Sea and the damage done to other marine organisms such as porpoises, dolphins, and sea birds through over-fishing by trawlers.

1993 The USA's Central Intelligence Agency takes satellite photographs which show that the Iraqi president Saddam Hussein's campaign against the marsh Arabs is transforming the ancient marshes of southern Iraq into a dry wasteland.

The marshes are formed by a triangle of land where the Tigris and Euphrates meet, and were once inhabited by 200,000 marsh Arabs. Like their Sumerian ancestors, they live in reed huts, spear fish, and milk water buffalo. Over recent months 5,000 marsh Arabs have fled Iraq and others have moved into cities.

February 1993 The level of ozone in the air in Mexico City reaches emergency concentrations.

The poor air quality triggers the first stage of the city's emergency plan, which shuts down some 30% of industry and restricts the use of some vehicles. The population of Mexico City is now about 20 million. The city lies at an altitude of 2,290 m/7,500 ft in a bowl that traps pollution from 4 million cars.

March–April 1993 About 400,000 people in the city of Milwaukee, Wisconsin, are affected by the world's largest cryptosporidiosis outbreak.

Oospores of the protozoan *Cryptosporidium* enter the water supply from animal manures spread onto the water catchment area and from sewage entering the water intake point. The water treatment plant taking water from Lake Michigan has difficulty in maintaining low turbidity levels, and faulty monitoring equipment means that spores enter the drinking water supply of the city, which has a population of about 800,000.

1994 An 11-member jury awards $5 billion punitive damages against Exxon Corporation for the 1989 Exxon *Valdez* oil spill in Alaska.

The damages are awarded on behalf of 12,000 people harmed by the oil spill that polluted the fishing grounds of Alaska and around 2,400 km/1,500 mi of coastline. The fine amounts to about a year's profit for the company. Exxon plans to appeal, claiming that it has already spent $3 billion on cleaning operations.

1994 At a meeting of the UN Basel Convention in Geneva, a ban on exports of hazardous waste from Organization for Economic Cooperation and Development (OECD) to non-OECD countries is agreed. The group hails the ban as a 'striking victory for global environmental justice'.

The thinking behind the measure is that a transfer of waste exposes the risk that it will not be dealt with in an environmentally sound manner, and that it is ethically dubious to use cash-hungry countries as a dumping ground for the waste of the affluent industrialized nations.

1994 The UK government announces plans to set up the Environment Agency to police emissions to the air, rivers, land, and coastal waters.

1994 The UK's Quality of Urban Air Group (part of the Department of the Environment) warns of the health dangers of expanding the number of diesel cars.

The warning is based on health studies carried out by US academic Douglas Dockery in six US cities. The research shows a correlation between exposure to particulates in diesel exhausts and increased mortality.

1994 The United Nations International Conference on Population and Development takes place in Cairo, Egypt. Some 20,000 representatives from 180 UN member nations attend. The conference is heated and controversial.

Despite the controversy, the conference ends with considerable agreement on the need to take urgent steps to educate women, improve access to family planning, bolster development in Third World countries, and stabilize the world's growing population. The 20-year programme of action is not binding on any of the signatories but is expected to have some moral force.

October 1994 The UK's Royal Commission on Environmental Pollution issues a report on transport and the environment, which observes that 'the growth in traffic which has been forecast cannot be accommodated in a sustainable transport policy'.

The report makes over 100 recommendations including a doubling in the price of petrol by 2005, improved fuel efficiency for new vehicles, a halving of expenditure on trunk roads, greater support for public transport, and targets to cut carbon dioxide emissions.

1995 A series of articles collected by the International Whaling Commission in Cambridge, England, shows that many whale species are recovering from the edge of extinction.

Blue whales are increasing by 5% each year in the northeast Atlantic, though world population is estimated at 5,000 compared to 228,000 in 1910. In general, 77% of the whale populations monitored are believed to be increasing slightly. Norway and Japan continue to hunt the small minke whale, although a moratorium on commercial hunting came into force in 1986. Conservationists fear that the new evidence for population increase may spur the industry to call for renewed hunting of other species.

1995 An iceberg the size of Majorca breaks away from the Antarctic, fuelling fears that global warming might be a root cause.

Environmentalists point out that eight out of the last ten years have been the warmest on record around the world. Tests on the tree rings of the world's oldest tree, a huon pine found in Tasmania, indicate that the last 30 years have been the warmest for 2,000 years.

1995 At the international climate conference held in Melbourne, Australia, it is reported that periodic disruptions of surface currents (which may cause climate changes) have been discovered in the Atlantic and Indian Oceans.

1995 Delegates from more than 100 nations meet in Berlin, Germany, for a summit to review progress since the United Nations Conference on Environment and Development in Rio de Janeiro, Brazil, in 1992. The Paris-based research organization DRI predicts that 14 European nations will meet targets agreed at Rio for the stabilization of greenhouse gas emissions.

The DRI suggests that, although European industries are becoming more energy-efficient, the switch from nuclear power to fossil fuels as a means of generating electricity will cause greenhouse gas emissions to increase early in the next century.

1995 Groups including the World Wide Fund for Nature (WWF) (formerly World Wildlife Fund) and the Royal Society for the Protection of Birds estimate that Britain has lost 150 species of animals and plants in the 20th century.

Losses include Blair's wainscot, a moth last seen in 1966 before its only known breeding site on the Isle of Wight was destroyed by draining and burning for agriculture, and the large tortoiseshell butterfly declared extinct in Britain in 1993. The last male of the mouse-eared bat died in 1990, females having vanished in 1974. The burbot, an eel-like fish, was declared extinct in the late 1970s, a victim of over-fishing and pollution.

1995 The European Community grants marketing approval for the first genetically engineered major crop to reach commercialization.

The Belgian firm Plant Genetic Systems intends to market its oilseed rape initially in the UK. The UK government's Department of the Environment grants approval despite objections from environmentalists. Applications of oilseed rape include margarine, vegetable oil, animal feed, industrial feedstock, and biofuel. The rape seed contains genes inserted from four species of bacteria and two species of plant. The transplanted genes will ensure that the crop remains sterile, thereby avoiding cross-pollination. Another gene makes the plant resistant to selected herbicides. This enables competitor plants to be eradicated shortly after growth. The environmentalist group Greenpeace objects that the use of herbicide-resistant genes could lead to an increased use of herbicides and to the spread of resistance by unwanted gene transfer.

1995 US chemist F Sherwood Roland, the Mexican chemist Mario Molina, and the Dutch chemist Paul Crutzen receive the Nobel Prize for Chemistry for explaining the chemical process of the ozone layer.

1996 The World Heritage Convention, drawn up by the United Nations Educational, Scientific, and Cultural Organization (UNESCO) in 1972, has now been signed by 144 countries.

UNESCO was created in 1945 to preserve the world's cultural heritage and coordinate scientific and educational programmes. In 1972 the World Heritage Convention aimed to identify places of 'outstanding universal value from the point of view of art, history, science, or natural beauty' in each country, whose protection should be of concern to the international community.

2 June 1996 US scientists at the National Oceanic and Atmospheric Administration in Washington, DC, announce the first decline in levels of ozone-depleting chemicals in the air.

347

4 October 1996 The World Conservation Union (International Union for the Conservation of Nature (IUCN)) publishes the latest Red List of endangered species. Over 1,000 mammals are listed, far more than on previous lists. The organization believes it has underestimated the risks to habitats from pollution and that the number of endangered species is greater than previously thought.

February 1997 US zoologists Bill Detrich and Kirk Malloy show that the increased ultraviolet radiation caused by the hole in the ozone layer above Antarctica kills large numbers of fish in the Southern Ocean. Because their transparent eggs and larvae stay near the surface for up to a year, they are exposed to the full force of the ultraviolet rays. It is the first time ozone depletion in the Antarctic has been shown to harm organisms larger than one-celled marine plants.

26 March 1997 German ecologist Venugopalan Ittekkot shows that dams on the River Danube keep back silicate sediments and thus starve the Black Sea of food for many marine plants, and create ideal conditions for the growth of toxic competitors, altering the sea's ecosystem. Silicate-loving sea grasses have been replaced by nitrate-loving species such as dinoflagellates, which cause poisonous red tides. It raises concerns that the world's thousands of dams may be slowly killing the world's seas.

7 September 1997 Australian researcher William de la Mare, using old whaling records that record data on every whale caught since the 1930s, including the ship's latitude, announces the discovery that Antarctic sea-ice could have decreased by up to a quarter between the mid-1950s and the 1970s. The finding has major implications, both for global climate conditions as well as for whaling.

December 1997 One of the most complex and far-reaching environmental agreements is reached at Kyoto in Japan. A protocol to limit greenhouse gas emissions is signed. Developed nations must cut their emission by 5% of 1990 levels and 30% below projected levels by 2008–12.

The protocol allows for emission trading whereby countries which are likely to meet their targets can trade emissions for dollars with other countries which are unlikely to meet their targets.

8 April 1998 The World Conservation Union (or International Union for the Conservation of Nature (IUCN)), based in Switzerland, publishes a survey which suggests that one in every eight known plant species in the world is in danger of becoming extinct.

October 1998 Despite measures taken to reduce chlorofluorocarbon (CFC) emissions, it is found that the ozone hole over Antarctica has increased in size to three times the area of the USA, the largest it has ever been recorded to date.

The expected size difference due to seasonal fluctuations also indicates that the hole is not regenerating as fast as it has in the past. The problem may have been exacerbated by the effects of the climatic phenomenon El Niño.

2 November 1998 Britain's Meteorological Office predicts that large areas of the Amazon rainforest will begin to die in around 2050, releasing carbon into the atmosphere and causing global warming to accelerate faster than previously forecasted.

1999 After ten years of debate, the European Union (EU) passes its landfill directive. The directive states that member states should reduce municipal biodegradable waste to 75% of 1995 levels by 2006 and to 35% by 2016.

The directive bans the co-disposal of hazardous and non-hazardous wastes and restricts landfilling with liquid wastes and tyres. Member states now have just over two years to incorporate the directive into national laws. It is expected that the effect of the directive will be to reduce groundwater contamination and encourage recycling in EU countries.

March 1999 Scientists at an international conference held in Switzerland report progress in understanding the extent and mechanism of endocrine disruption. John Sumpter of Brunel University in Britain reports that a high proportion of male wild fish (such as roach and gudgeon) downstream of sewage discharge points have become intersexual.

In the early 1990s, US researchers showed that certain chemicals have the ability to mimic the effects of sex hormones. Researchers also report that the synthetic oestrogens used in contraceptive pills is relatively resistant to environmental degradation.

17 November 1999 Scientists from the University of Washington in Seattle, USA, publish data showing that the Arctic icecap has shrunk by around 40% over the past 50 years. The Canadian Wildlife Service reports that polar bears in the area are in danger of starving to death because of their shortened hunting season.

2000 In the UK, the government sets out plans to introduce mandatory targets for recycling, in an effort to improve the poor performance of local authorities.

A non-statutory target of recycling 25% of household waste was set in 1990 for the year 2000, but by 1999 levels had only reached 9.5%. The government plans to introduce a number of targets including a recycling rate of at least 30% for households by 2010 and increased composting rates for biodegradable material.

January 2000 The World Bank sets up a Prototype Carbon Fund (PCF). The fund will pay for greenhouse gas abatement in developing countries.

So far, government contributions have come from Finland, the Netherlands, Norway, and Sweden. Companies from Japan and Belgium have also invested in the scheme. The PCF sponsors hope that the emission credits they gain from projects can be used to offset their emission targets as prescribed in the Kyoto Protocol.

October 2000 US anthropologist John Oates and co-workers announce the likely extinction of Miss Waldron's red colobus monkey (*Procolobus badius waldroni*). This is the first recorded extinction of a primate in over 300 years.

Despite its inclusion in the World Conservation Union's list of endangered species, it seems that hunting by humans for meat and the destruction of habitat for farming are the primary causes of the extinction of this monkey. The range of this species was restricted to Ghana and Côte d'Ivoire.

Mathematics

1994 English mathematician Andrew Wiles proves Fermat's last theorem, a problem that had remained unsolved since 1647. The most notorious problem in mathematics, the last theorem was created in the 17th century by the French judge Pierre de Fermat, who studied mathematics in his spare time.

26 June 1997 English mathematician Andrew Wiles is awarded the Wolfskehl Prize for solving Fermat's last theorem. In 1908 the German industrialist Paul Wolfskehl bequeathed DM100,000 (£1 million/$1.4 million by today's value) to be given to the first person to prove the theorem.

18 August 1998 Cambridge University professors Richard Borchers and Tim Gowers win Fields Medals in Mathematics for work in vertex algebra and probabilistic number theory.

1 June 1999 The 38th Mersenne prime, comprising about two million digits, is found by the Great Internet Mersenne Prime Search project.

A Mersenne prime is a prime of the form $2^q - 1$, where q is also a prime.

Physics

1990 Researchers at IBM's Almaden Research Center in California, USA, are the first to manipulate individual atoms on a surface; they use a scanning tunnelling microscope and spell out the initials 'IBM'.

1990 The Nobel Prize for Physics is awarded to US physicists Jerome Friedman and Henry Kendall and Canadian physicist Richard Taylor.

They are honoured for their pioneering investigations into deep inelastic scattering of electrons on protons and bound neutrons, which have been of essential importance for the development of the quark model in particle physics.

1991 Experiments at the CERN research centre in Switzerland demonstrate the existence of three 'families' of elementary particles, each with two quarks and two leptons.

1991 French physicist Pierre-Gilles de Gennes receives the Nobel Prize for Physics for discovering that methods developed for studying order phenomena in simple systems can be generalized to more complex forms of matter, in particular to liquid crystals and polymers.

9 November 1991 The Joint European Torus (JET) at Culham, near Oxford, England, produces a 1.7 megawatt pulse of power in an experiment that lasts 2 seconds.

It is the first time that a substantial amount of fusion power has been produced in a controlled experiment, as opposed to a thermonuclear bomb. If successful, fusion holds the key to sustainable energy on a global scale.

1992 French physicist Georges Charpak receives the Nobel Prize for Physics for his invention of the multiwire proportional chamber particle detector.

1992 Two Japanese researchers develop a material that becomes superconducting at the relatively high temperature of $-103°C / -153°F$.

High-temperature superconductors have many applications, especially in medical imaging.

1994 Canadian physicist Bertram Brockhouse and US physicist Clifford Shull share the Nobel Prize for Physics.

Brockhouse is honoured for his pioneering contributions to the development of neutron scattering techniques for studies of condensed matter, specifically the development of neutron spectroscopy, and Shull for his pioneering contributions to the development of neutron scattering techniques for studies of condensed matter, specifically for the development of the neutron diffraction technique.

1994 US scientists at Fermilab, near Chicago, Illinois, announce the discovery of the sixth and final 'top' quark.

1995 US physicists Martin Perl and Frederick Reines share the Nobel Prize for Physics: Perl for the discovery of the tau lepton and Reines for the detection of the neutrino.

June 1995 US physicists announce the discovery of a new form of matter, called a Bose–Einstein condensate (its existence had been predicted by Albert Einstein and Indian physicist Satyendra Bose), created by cooling rubidium atoms to just above absolute zero.

1996 Construction of the world's largest neutrino detector, the Antarctic Muon and Neutrino Detector Array (AMANDA), begins at the South Pole.

1996 US physicists David Lee, Douglas Osheroff, and Robert Richardson are jointly awarded the Nobel Prize for Physics for their discovery of superfluidity in helium-3.

4 January 1996 A team of European physicists at the CERN research centre in Switzerland create the first atoms of antimatter: nine atoms of antihydrogen survive for 40 nano-seconds.

Despite the very short lifetime of these atoms, it is a major breakthrough. The ability to create antimatter atoms is important for many areas of research including the understanding of the matter–antimatter asymmetry in the universe.

1997 The Nobel Prize for Physics is awarded to the US physicists Steven Chu and William D Phillips and the French physicist Claude Cohen-Tannoudji for their development of methods to cool and trap atoms with laser light.

11 September 1997 Israeli physicist Rafi de-Picciotto claims to show the formation, within semiconductor materials, of 'quasiparticles', which have a charge one-third that of an electron. They challenge the idea that charge always comes in discrete units based on the charge of a single electron.

1998 US astronomers Saul Perlmutter and Alex Filippenko suggest that not only is the universe expanding but that the rate of expansion is increasing.

Some scientists believe that this acceleration is caused by quantum effects, which would give a non-zero cosmological constant. Einstein introduced the cosomological constant into his theory of general relativity and once called it his 'biggest blunder'.

1998 The Nobel Prize for Physics is awarded jointly to US physicists Robert Laughlin and Daniel Tsui, and German physicist Horst Störmer, for their discovery of the fractional quantum Hall effect.

11 July 1998 Researchers at the Fermi National Accelerator laboratory in Batavia, Illinois, announce the discovery of the tau neutrino, the least stable elementary particle of the lepton class.

This completes the search for the particles that make up the three families of particles, each containing two quarks and two leptons.

1999 The Nobel Prize for Physics is awarded to Gerardus 't Hooft and Martinus Veltman of the Netherlands for elucidating the quantum structure of electroweak interactions in physics.

February 1999 Scientists succeed in slowing down light from its normal speed of 299,792 km/186,282 mi per second to 61 km/38 mi per hour, opening up potential for the development of high-precision computer and telecommunications technologies, as well as for the advanced study of quantum mechanics.

2000 Physicists and astronomers start the search for 'quintessence'. Quintessence has the property of being able to cause the rate of expansion of the universe to increase. It is believed to achieve this feat by having 'repulsive gravitational force'.

2000 The Nobel Prize for Physics is jointly awarded to Russian physicist Zhores Alferov and German physicist Herbert Kroemer for developing semiconductor heterostructures used in high-speed and opto-electronics; and to US physicist Jack Kilby for his part in the invention of the integrated circuit.

Biographies

Biographies
Anderson,
Philip

°Abbe, Ernst 1840–1905
German physicist who, working with German optician Carl Zeiss 1816–1888, greatly improved the design and quality of optical instruments, particularly the compound microscope.

Abel, Neils Henrik 1802–1829
Norwegian mathematician who, in a very brief career, became the first to demonstrate that an algebraic solution of the general equation of the fifth degree is impossible.

Abetti, Giorgio 1882–1982
Italian astrophysicist best known for his studies of the Sun.

Adams, John Couch 1819–1892
English astronomer who was particularly skilled mathematically and deduced the existence of Neptune in 1845.

Adams, Walter Sydney 1876–1956
US astronomer who was particularly interested in stellar motion and luminosity. He developed the use of spectroscopy and found that luminosity and the relative intensities of spectral tints can distinguish giant stars from dwarf stars.

Agassiz, (Jean) Louis Rodolphe 1807–1873
Swiss palaeontologist who developed the idea of the ice age.

Agnesi, Maria Gaetana 1718–1799
Italian mathematician who is noted for her work in differential calculus.

Agricola, Georgius 1490–1555
born Georg Bauer, German mineralogist who wrote pioneering works on mining, technology, and metallurgy.

Airy, George Biddell 1801–1892
English astronomer who, as Astronomer Royal for 46 years, was responsible for greatly simplifying the systematization of astronomical observations and for expanding and improving the Royal Observatory at Greenwich.

Aitken, Robert Grant 1864–1951
US astronomer whose primary contribution to astronomy was the discovery and observation of thousands of double stars.

Alder, Kurt 1902–1958
German organic chemist who, with Otto Diels, developed diene synthesis, a fundamental process that has become known as the Diels–Alder reaction.

Aleksandrov, Pavel Sergeevich 1896–1982
Russian mathematician who was a leading expert in the field of topology and one of the founders of the theory of compact and bicompact spaces.

Alfvén, Hannes Olof Gösta 1908–1995
Swedish astrophysicist who made fundamental contributions to plasma physics, particularly in the field of magnetohydrodynamics (MHD) – the study of plasmas in magnetic fields.

al-Haytham, Abu Alī al-Hassan ibn
(Latin **Alhazen**) c. 965–c. 1040
Arabian scientist who made significant advances in the theory and practice of optics.

Alpher, Ralph Asher 1921–
US cosmologist who carried out the first quantitative work on nucleosynthesis and was the first to predict the existence of primordial background radiation.

Alvarez, Luis Walter 1911–1988
US physicist who won the 1968 Nobel Prize for Physics for developing the liquid-hydrogen bubble chamber and detecting new resonant states in particle physics.

Ampère, André-Marie 1775–1836
French physicist, mathematician, chemist, and philosopher who founded the science of electromagnetics (which he named electrodynamics) and gave his name to the unit of electric current.

Anaximander the Elder c. 611–547 BC
Greek philosopher who formulated some basic natural philosophical views, often in opposition to his teacher, Thales of Miletus.

Anderson, Carl David 1905–1991
US physicist who did pioneering work in particle physics, notably discovering the positron – the first antimatter particle to be found – and the muon (or mu-meson).

Anderson, Philip Warren 1923–
US physicist who shared the 1977 Nobel Prize for Physics with Nevill Mott and John Van

353

Vleck for his theoretical work on the behaviour of electrons in magnetic, non-crystalline solids.

Andrews, Thomas 1813–1885
Irish physical chemist, best known for postulating the idea of critical temperature and pressure from his experimental work on the liquefaction of gases, which demonstrated the continuity of the liquid and gaseous states.

Ångström, Anders Jonas 1814–1874
Swedish physicist and astronomer, one of the early pioneers in the development of spectroscopy.

Antoniadi, Eugène Marie 1870–1944
Turkish-born French astronomer who had a particular interest in the planet Mars and later became an expert also on the scientific achievements of ancient civilizations.

Apollonius of Perga c. 245–c.190 BC
Greek mathematician whose treatise on conic sections represents the final flowering of Greek mathematics.

Appleton, Edward Victor 1892–1965
English physicist famous for his discovery of the Appleton layer of the ionosphere, which reflects radio waves and is therefore important in communications.

Arago, (Dominique) François Jean 1786–1853
French scientist who made contributions to the development of many areas of physics and astronomy, particularly in optics and magnetism.

Archimedes c. 287–212 BC
Greek mathematician and physicist, generally considered to be the greatest in the ancient world.

Argand, Jean Robert 1768–1822
Swiss mathematician who invented a method of geometrically representing complex numbers and their operations.

Argelander, Friedrich Wilhelm August 1799–1875
Prussian astronomer whose approach to the subject was one of great resourcefulness and thoroughness.

Aristarchus of Samos c. 320–c. 250 BC
Greek astronomer who was the first to argue that the Earth moves around the Sun.

Aristotle 384–322 BC
Greek polymath, one of the most imaginative and systematic thinkers in history, whose writings embraced virtually every aspect of contemporary thought, including cosmology.

Arrhenius, Svante August 1859–1927
Swedish physical chemist who first explained that in an electrolyte (a solution of a chemical dissolved in water) the dissolved substance is dissociated into electrically charged ions.

Aston, Francis William 1877–1945
English chemist and physicist who developed the mass spectrograph, which he used to study atomic masses and to establish the existence of isotopes.

Atkinson, Robert d'Escourt 1898–1982
Welsh astronomer and inventor whose contributions were fundamental to our basic understanding of how stars evolve.

Audubon, John James Laforest 1785–1851
French-born US ornithologist who painted intricately detailed studies of birds and animals.

Avery, Oswald Theodore 1877–1955
Canadian-born US bacteriologist whose work on transformation in bacteria established that DNA is responsible for the transmission of heritable characteristics.

Avogadro, Amedeo 1776–1856
Conte de Quaregna, Italian scientist who shares with his contemporary Claud Berthollet the honour of being one of the founders of physical chemistry.

Ayrton, Hertha (Phoebe Sarah) Marks 1854–1923
English physicist and inventor who studied electricity and wrote the standard textbook *The Electric Arc*.

Ayrton, William Edward 1847–1908
English physicist and electrical engineer who invented many of the prototypes of modern electrical measuring instruments.

Baade, (Wilhelm Heinrich) Walter 1893–1960
German-born US astronomer who is known for his discovery of stellar populations and whose research proved that the observable universe is larger than had originally been believed.

Babbage, Charles 1792–1871
English mathematician, one of the greatest pioneers of mechanical computation.

Babcock, Harold Delos 1882–1968
US astronomer and physicist whose most important contributions were to spectroscopy and the study of solar magnetism.

Bacon, Francis 1561–1626
English philosopher, politician, writer, and a founder of modern scientific method, based on induction.

Bacon, Roger *c.* 1220–*c.* 1292
English philosopher and scientist who was among the first medieval scholars to realize and promote the value of experiment in reaching valid conclusions.

Baekeland, Leo Hendrik 1863–1944
Belgian-born US industrial chemist famous for his invention of Bakelite, the first commercially successful thermosetting plastic resin.

Baer, Karl Ernst Ritter von 1792–1876
Estonian embryologist famous for his discovery of the mammalian ovum, who made a significant contribution to the systematic study of the development of animals.

Baeyer, Johann Friedrich Wilhelm Adolf von 1835–1917
German organic chemist famous for developing methods of synthesis, the best known of which is his synthesis of the dye indigo.

Baily, Francis 1774–1844
English astronomer who is best known for his discovery of the phenomenon called 'Baily's beads' during an eclipse.

Bainbridge, Kenneth Tompkins 1904–1996
US physicist best known for his work on the development of the mass spectrometer.

Baker, Alan 1939–
English mathematician whose chief work has been devoted to the study of transcendental numbers.

Balmer, Johann Jakob 1825–1898
Swiss mathematics teacher who devised mathematical formulae that give the frequencies of atomic spectral lines.

Banneker, Benjamin 1731–1806
US mathematician, astronomer, and surveyor who is chiefly known for his almanacs published in the 1790s.

Banting, Frederick Grant 1891–1941
Canadian physiologist who discovered insulin, the hormone responsible for the regulation of the sugar content of the blood (an insufficiency of which results in the disease diabetes mellitus).

Bardeen, John 1908–1991
US physicist whose work on semiconductors, together with William Shockley and Walter Brattain, led to the first transistor, an achievement for which all three men shared the 1956 Nobel Prize for Physics. Bardeen was also awarded the 1972 prize for work on superconductivity.

Barkla, Charles Glover 1877–1944
English physicist who made important contributions to our knowledge of X-rays, particularly the phenomenon of X-ray scattering.

Barnard, Edward Emerson 1857–1923
US observational astronomer whose keen vision and painstaking thoroughness made him an almost legendary figure.

Barrow, Isaac 1630–1677
English mathematician, physicist, classicist, and Anglican divine, one of the intellectual luminaries of the Caroline period.

Bartlett, Neil 1932–
English chemist who achieved fame by preparing the first compound of one of the rare gases, previously thought to be totally inert and incapable of reacting with anything.

Barton, Derek Harold Richard 1918–1998
English organic chemist whose chief work concerned the stereochemistry of natural compounds.

Bates, H(enry) W(alter) 1825–1892
English naturalist and explorer whose discovery of a type of mimicry (called Batesian mimicry) lent substantial support to Charles Darwin's theory of natural selection.

Bateson, William 1861–1926
English geneticist who was one of the founders of the science of genetics (a term he introduced), and a leading proponent of Mendelian views after the rediscovery in 1900 of Gregor Mendel's work on heredity.

Beaufort, Francis 1774–1857
British hydrographer and meteorologist who developed the Beaufort scale of wind velocity.

Beaumont, William 1785–1853
US surgeon who did important early work on the physiology of the human stomach by

355

taking advantage of a bizarre surgical case that he treated early in his career.

Becquerel, (Antoine) Henri 1852–1908
French physicist who discovered radioactivity in 1896, an achievement for which he shared the 1903 Nobel Prize for Physics with Pierre and Marie Curie.

Beijerinck, Martinus Willem 1851–1931
Dutch botanist who in 1898 published his finding that an agent smaller than bacteria could cause diseases, an agent that he called a virus (the Latin word for poison).

Bell Burnell, (Susan) Jocelyn 1943–
Northern Irish astronomer who discovered pulsating radio stars (pulsars) in the 1960s.

Beltrami, Eugenio 1835–1900
Italian mathematician whose work ranged over almost the whole field of pure and applied mathematics, but whose fame derives chiefly from his investigations into theories of surfaces and space of constant curvature, and his position as the modern pioneer of non-Euclidean geometry.

Berg, Paul 1926–
US molecular biologist who shared (with Walter Gilbert and Frederick Sanger) the 1980 Nobel Prize for Chemistry for his work in genetic engineering, particularly for developing DNA recombinant techniques that enable genes from simple organisms to be inserted into the genetic material of other simple organisms.

Bergius, Friedrich Karl Rudolf 1884–1949
German industrial chemist famous for developing a process for the catalytic hydrogenation of coal to convert it into useful hydrocarbons such as petrol and lubricating oil.

Bernays, Paul Isaak 1888–1977
English-born Swiss mathematician who was chiefly interested in the connections between logic and mathematics, especially in the field of set theory.

Bernoulli, Daniel 1700–1782
Swiss natural philosopher and mathematician, whose most important work was in the field of hydrodynamics and whose chief contribution to mathematics was in the field of differential equations.

Bernoulli, Jacques Jakob 1654–1705 and **Jean Johann** 1667–1748
Swiss mathematicians, each of whom did

important work in the early development of calculus.

Bernstein, Jeremy 1929–
US mathematical physicist who is well known for his popularizing books on various topics of pure and applied science for the lay reader.

Berthelot, Pierre Eugène Marcellin 1827–1907
French chemist best known for his work on organic synthesis and in thermochemistry.

Berthollet, Claude Louis 1748–1822
Count, French chemist with a wide range of interests, the most significant of which concerned chemical reactions and the composition of the products of such reactions.

Berzelius, Jöns Jakob 1779–1848
Swedish chemist, one of the founders of the science in its modern form.

Bessel, Friedrich Wilhelm 1784–1846
German astronomer who in 1838 first observed stellar parallax, and who set new standards of accuracy for positional astronomy.

Bethe, Hans Albrecht 1906–
German-born US physicist and astronomer, famous for his work on the production of energy within stars for which he was awarded the 1967 Nobel Prize for Physics.

Betti, Enrico 1823–1892
Italian mathematician who was the first to provide a thorough exposition and development of the theory of equations formulated by French mathematician Evariste Galois.

Bhabha, Homi Jehangir 1909–1966
Indian theoretical physicist who made several important explanations of the behaviour of subatomic particles.

Birkhoff, George David 1884–1944
US mathematician who made fundamental contributions to the study of dynamics and formulated the 'weak form' of the ergodic theorem.

Bjerknes, Vilhelm Frimann Koren 1862–1951
Norwegian scientist whose theory of polar fronts formed the basis of modern meteorology.

Black, Joseph 1728–1799
Scottish physicist and chemist whose most important contribution to physics was his work on thermodynamics, notably on latent heats and specific heats.

Black, Max 1909–
Russian-born US philosopher and mathematician, one of whose concerns has been to investigate the question 'what is mathematics?'

Blackett, Patrick Maynard Stuart 1897–1974
Baron Blackett, English physicist who made the first photograph of an atomic transmutation and developed the cloud chamber into a practical instrument for studying nuclear reactions.

Blakemore, Colin Brian 1944–
English physiologist who has made advanced studies of how the brain works, especially in connection with memory and the senses.

Bloch, Konrad Emil 1912–
German-born US biochemist whose best-known work has been concerned with the biochemistry and metabolism of fats (lipids), particularly reactions involving cholesterol.

Bode, Johann Elert 1749–1826
German mathematician and astronomer who contributed greatly to the popularization of astronomy.

Boerhaave, Hermann 1668–1738
Dutch physician and chemist who dominated and greatly influenced various branches of science in Europe.

Bohr, Niels Henrik David 1885–1962
Danish physicist who established the structure of the atom and led in the development of quantum mechanics.

Bok, Bart J(an) 1906–1992
Dutch astrophysicist best known for his discovery of the small, circular dark spots in nebulae (now known as Bok's globules).

Boksenberg, Alexander 1936–
English astronomer and physicist who devised a new kind of light-detecting system that can be attached to telescopes and so vastly improve their optical powers.

Boltzmann, Ludwig Eduard 1844–1906
Austrian theoretical physicist who contributed to the development of the kinetic theory of gases, electromagnetism, and thermodynamics.

Bolzano, Bernardus Placidus Johann Nepomuk 1781–1848
Czech philosopher and mathematician who made a number of contributions to the development of several branches of mathematics.

Bond, George Phillips 1825–1865
US astronomer whose best work was on the development of astronomical photography as an important research tool.

Bond, William Cranch 1789–1859
US astronomer who, with his son, George Phillips Bond, established the Harvard College Observatory as a centre of astronomical research.

Bondi, Hermann 1919–
Austrian-born British scientist who was trained as a mathematician, but who went on to make important contributions to many disciplines both as a research scientist and as an enthusiastic administrator.

Boole, George 1815–1864
English mathematician who, by being the first to employ symbolic language and notation for purely logical processes, founded the modern science of mathematical logic.

Born, Max 1882–1970
German-born British physicist who pioneered quantum mechanics, the mathematical explanation of the behaviour of an electron in an atom.

Bosch, Karl 1874–1940
German chemist who developed the industrial synthesis of ammonia, leading to the cheap production of agricultural fertilizers and of explosives.

Boveri, Theodor Heinrich 1862–1915
German biologist who performed valuable early work on chromosomes.

Bowden, Frank Philip 1903–1968
Australian physicist and chemist who studied friction, lubricants, and surface erosion.

Bowen, Ira Sprague 1898–1973
US astrophysicist who is best known for his study of the spectra of planetary nebulae.

Bowen, Norman Levi 1887–1956
Canadian geologist whose work helped found modern petrology.

Boyle, Robert 1627–1691
Irish natural philosopher and one of the founders of modern chemistry.

Boys, Charles Vernon 1855–1944
English inventor and physicist who is probably best known for designing a very sensitive apparatus (based on Henry Cavendish's earlier experiment) to determine Newton's gravitational constant and the mean density of the Earth.

357

Bradley, James 1693–1762
English astronomer of great perception and practical skill who discovered aberration of starlight and nutation.

Bragg, William Henry 1862–1942 and **(William) Lawrence** 1890–1971
English physicists, father and son, who pioneered and perfected the technique of X-ray diffraction in the study of the structure of crystals.

Brahe, Tycho 1546–1601
Danish astronomer, sometimes known by his first name only, who is most noted for his remarkably accurate measurements of the positions of stars and the movements of the planets.

Brattain, Walter Houser 1902–1987
US physicist who was awarded the Nobel Prize for Physics in 1956, with William Shockley and John Bardeen, for development of the transistor.

Braun, Emma Lucy 1889–1971
US botanist, an early pioneer in recognizing the importance of plant ecology and conservation.

Bredig, Georg 1868–1944
German physical chemist who contributed to a wide range of subjects within his discipline but is probably best known for his work on colloids and catalysts.

Brenner, Sydney 1927–
South African-born British molecular biologist noted for his work in the field of genetics. He discovered RNA and is one of the pioneers of genetic engineering.

Brewster, David 1781–1868
Scottish physicist who investigated the polarization of light, discovering the law named after him for which he was awarded the Rumford Medal by the Royal Society in 1819.

Bridgman, Percy Williams 1882–1961
US physicist famous for his work on the behaviour of materials at high temperature and pressure, for which he won the 1946 Nobel Prize for Physics.

Briggs, Henry 1561–1630
English mathematician, one of the founders of calculation by logarithms.

Broglie, Louis Victor Pierre Raymond de 1892–1987
7th duc de Broglie, French physicist who first developed the principle that an electron or any other particle can be considered to behave as a wave as well as a particle.

Brongniart, Alexandre 1770–1847
French geologist and palaeontologist, among the first to use fossils to date rock strata and one of the first scientists to arrange geological formations of the Tertiary period in chronological order.

Brønsted, Johannes Nicolaus 1879–1947
Danish physical chemist whose work in solution chemistry, particularly electrolytes, resulted in a new theory of acids and bases, the Brønstead–Lowry theory, which described acids and bases in terms of proton exchange or the transfer of protons between compounds.

Brown, Robert 1773–1858
Scottish botanist whose discovery of the movement of suspended particles has proved fundamental in the study of physics.

Brown, Robert Hanbury 1916–
Indian-born British radio astronomer who was involved with the early development of radio astronomy techniques and who has since participated in designing a radio interferometer that permits considerably greater resolution in the results provided by radio telescopes.

Buchner, Eduard 1860–1917
German organic chemist who discovered non-cellular alcoholic fermentation of sugar – that is, that the active agent in the reaction is an enzyme contained in yeast, and not the yeast cells themselves.

Buckland, William 1784–1856
English geologist and palaeontologist, a supporter of catastrophism.

Buffon, Georges-Louis Leclerc 1707–1788
comte de Buffon, French naturalist who compiled the vast encyclopedic work *Histoire naturelle générale et particulière*.

Bullard, Edward Crisp 1907–1980
English geophysicist who, with US geologist Maurice Ewing, is generally considered to have founded the discipline of marine geophysics.

Bunsen, Robert Wilhelm 1811–1899
German chemist who pioneered the use of the spectroscope to analyse chemical compounds and discovered caesium and rubidium. He is credited with the invention of the Bunsen burner.

Burali-Forti, Cesare 1861–1931
Italian mathematician who is famous for the paradox named after him and for his work on the linear transformations of vectors.

Burbidge, Geoffrey 1925–   and **(Eleanor) Margaret** (born **Peachey**) 1919–
British husband-and-wife team of astrophysicists distinguished for their work, chiefly in the USA, on nucleosynthesis – the creation of elements in space – and on quasars and galaxies.

Burnside, William 1852–1927
English applied mathematician and mathematical physicist whose interest turned in his later years to a profound absorption in pure mathematics.

Cailletet, Louis Paul 1832–1913
French physicist and inventor who is remembered chiefly for his work on the liquefaction of the 'permanent' gases: he was the first to liquefy oxygen, hydrogen, nitrogen, and air, for example.

Callendar, Hugh Longbourne 1863–1930
English physicist and engineer who carried out fundamental investigations into the behaviour of steam.

Calvin, Melvin 1911–1997
US chemist who worked out the biosynthetic pathways involved in photosynthesis, the process by which green plants use the energy of sunlight to convert water and carbon dioxide into carbohydrates and oxygen.

Cameron, Alastair Graham Walter 1925–
Canadian-born US astrophysicist responsible for theories regarding the formation of the unstable element technetium within the core of red giant stars and of the disappearance of Earth's original atmosphere.

Campbell, William Wallace 1862–1938
US astronomer and mathematician, now particularly remembered for his research into the radial velocities of stars.

Cannizzaro, Stanislao 1826–1910
Italian chemist who, through his revival of Avogadro's hypothesis, laid the foundations of modern atomic theory.

Cannon, Annie Jump 1863–1941
US astronomer renowned for her work in stellar spectral classification, with particular reference to variable stars.

Cantor, Georg Ferdinand Ludwig Philipp 1845–1918
German mathematician and philosopher who is now chiefly remembered for his development of the theory of sets, for which he was obliged to devise a system of mathematics in which it was possible to consider infinite numbers or even transfinite ones.

Carathéodory, Constantin 1873–1950
German mathematician who made significant advances in the calculus of variations and in function theory.

Cardano, Girolamo 1501–1576
Italian physician, mathematician, philosopher, astrologer, and gambler. He is remembered for his theory of chance, use of algebra, and the first clinical description of typhus fever.

Carnot, (Nicolas Léonard) Sadi 1796–1832
French physicist who founded the science of thermodynamics.

Carothers, Wallace Hume 1896–1937
US organic chemist who did pioneering work on the development of commercial polymers, producing nylon (a polyamide) and neoprene (a polybutylene, one of the first synthetic rubbers).

Carr, Emma Perry 1880–1972
US chemist, teacher, and researcher, internationally renowned for her work in the field of spectroscopy.

Carrington, Richard Christopher 1826–1875
English astronomer who was the first to record the observation of a solar flare, and is now most remembered for his work on sunspots.

Carson, Rachel Louise 1907–1964
US biologist, conservationist, and writer. Her book *Silent Spring* (1962), attacking the indiscriminate use of pesticides, inspired the creation of the modern environmental movement.

Cassegrain, Laurent 1629–1693
French inventor of the system of mirrors within many modern reflecting telescopes – a system by transference also sometimes used in large refraction telescopes.

Cassini, Giovanni Domenico (Jean Dominique) 1625–1712
Italian-born French astronomer with a keen interest in geodesy. He discovered four moons

359

of Saturn and the gap in Saturn's rings now called the Cassini division.

Cauchy, Augustin-Louis 1789–1857
French mathematician who did important work in astronomy and mechanics, but who is chiefly famous as the founder, with Karl Gauss, of the modern subject of complex analysis.

Cavendish, Henry 1731–1810
English natural philosopher whose main interests lay in the fields of chemistry and physics.

Cayley, Arthur 1821–1895
English mathematician who was responsible for the formulation of the theory of algebraic invariants.

Celsius, Anders 1701–1744
Swedish astronomer, mathematician, and physicist, now mostly remembered for the Celsius scale of temperature.

Cesaro, Ernesto 1859–1906
Italian mathematician whose interests were wide-ranging, but who is chiefly remembered for his important contributions to intrinsic geometry.

Chadwick, James 1891–1974
English physicist who discovered the neutron in 1932.

Chain, Ernst Boris 1906–1979
German-born British biochemist who, in collaboration with Howard Florey, first isolated and purified penicillin and demonstrated its therapeutic properties.

Chamberlin, Thomas Chrowder
1843–1928
US geophysicist who developed theories of the origin of the Earth.

Chandrasekhar, Subrahmanyan
1910–1995
Indian-born US astrophysicist who was particularly concerned with the structure and evolution of stars.

Chardonnet, (Louis-Marie) Hilaire Bernigaud 1839–1924
comte de Chardonnet, French industrial chemist who invented rayon, the first type of artificial silk.

Chargaff, Erwin 1905–
Austrian-born US biochemist noted for his work on nucleic acids and for the Chargaff rules that demonstrate the mathematical relationship between the nitrogenous bases of DNA.

Charles, Jacques Alexandre César
1746–1823
French physicist and mathematician who is remembered for his work on the expansion of gases and his pioneering contribution to early ballooning.

Chase, Mary Agnes 1869–1963
born Mary Agnes Meara, US botanist and suffragist who made outstanding contributions to the study of grasses, despite a lack of higher education and any formal qualifications.

Chevreul, Michel-Eugène 1786–1889
French organic chemist who in a long lifetime devoted to scientific research studied a wide range of natural substances, including fats, sugars, and dyes.

Child, Charles Manning 1869–1954
US zoologist who tried to elucidate one of the central problems of biology – that of organization within living organisms.

Chladni, Ernst Florens Friedrich
1756–1827
German physicist who studied sound and invented musical instruments, helping to establish the science of acoustics.

Christoffel, Elwin Bruno 1829–1900
German mathematician who made a fundamental contribution to the differential geometry of surfaces, carried out some of the first investigations that later resulted in the theory of shock waves, and introduced what are now known as the Christoffel symbols into the theory of invariants.

Church, Alonzo 1903–1995
US mathematician who in 1936 published the first precise definition of a calculable function, and so contributed enormously to the systematic development of the theory of algorithms.

Clairaut, Alexis Claude 1713–1765
French mathematician who studied celestial mechanics.

Clark, Wilfrid Edward Le Gros 1895–1971
English anatomist and surgeon who carried out important research that made a major contribution to the understanding of the structural anatomy of the brain.

Clausius, Rudolf Julius Emanuel
1822–1888
German theoretical physicist who is credited with being one of the founders of thermodynamics, and with originating its second law.

Clifford, William Kingdon 1845–1879
English mathematician and scientific philosopher who developed the theory of biquaternions and proved a Riemann surface to be topologically equivalent to a box with holes in it.

Cockcroft, John Douglas 1897–1967
English physicist who, with Ernest Walton, built the first particle accelerator and achieved the first artificial nuclear transformation in 1932.

Cohn, Ferdinand Julius 1828–1898
German botanist of distinction and one of the founders of bacteriology. He showed that bacteria are not able to generate spontaneously.

Compton, Arthur H(olly) 1892–1962
US physicist who is remembered for discovering the Compton effect, a phenomenon in which electromagnetic waves such as X-rays undergo an increase in wavelength after having been scattered by electrons.

Coolidge, Julian Lowell 1873–1954
US geometrician and a prolific author of mathematical textbooks, in which he not only reported his results but also described the historical background together with contemporary developments.

Cooper, Leon Niels 1930–
US physicist who shared the 1972 Nobel Prize for Physics with John Bardeen and J Robert Schrieffer for developing the 'BCS' theory of superconductivity.

Copernicus, Nicolaus 1473–1543
Polish Mikołaj Kopernik, Polish doctor and astronomer who, against the long-held and almost universal belief tht the Earth is the centre of the universe, proposed instead that the Earth is a planet revolving around a stationary Sun.

Coriolis, Gustave Gaspard de 1792–1843
French physicist who discovered the Coriolis force that governs the movements of winds in the atmosphere and currents in the ocean.

Cornforth, John Warcup 1917–
Australian organic chemist who shared the 1975 Nobel Prize for Chemistry with the Swiss biochemist Vladimir Prelog for his work on the stereochemistry of biochemical compounds.

Correns, Carl Franz Joseph Erich
1864–1933
German botanist and geneticist who is credited with rediscovering Mendel's laws (Hugo de Vries is similarly credited).

Coulomb, Charles Augustin de 1736–1806
French physicist who established the laws governing electric charge and magnetism.

Coulson, Charles Alfred 1910–1974
English theoretical chemist whose major contribution to the science was his molecular orbital theory and the concept of partial valency.

Courant, Richard 1888–1972
German-born US mathematician who taught mathematics from a very early age, wrote several textbooks that are now standard reference works, and founded no fewer than three highly influential mathematical institutes.

Cowling, Thomas George 1906–1990
English applied mathematician and physicist who contributed significantly to modern research into stellar energy, with special reference to the Sun.

Cramer, Gabriel 1704–1752
Swiss mathematician who contributed to the solution of linear equations and to the systematic organization of mathematical knowledge.

Crick, Francis Harry Compton 1916–
English biophysicist who shared the Nobel Prize for Physiology or Medicine with James Watson and Maurice Wilkins for their work on determining the structure of DNA.

Crookes, William 1832–1919
English physicist and chemist who discovered the element thallium, invented the radiometer, and studied electric discharges in evacuated tubes, known as 'Crookes tubes'.

Crutzen, Paul 1933–
Dutch meteorologist who shared the Nobel Prize for Chemistry in 1995 with Mario Molina and F Sherwood Rowland for their work on the formation and decomposition of ozone in the atmosphere.

Curie, Marie 1867–1934, born **Maria Sklodowska** and **Pierre** 1859–1906
French scientists, husband and wife, who were early investigators of radioactivity. They discovered and studied polonium and radium.

Curtis, Heber Doust 1872–1942
US astronomer who carried out important research into the nature of spiral nebulae.

Cuvier, Georges Léopold Chrêtien Frédéric Dagobert 1769–1832
Baron Cuvier, French zoologist, one of the founders of modern palaeontology, and one of the first to arrange geological formations using characteristic fossils.

d'Alembert, Jean le Rond 1717–1783
French mathematician and theoretical physicist who was a great innovator in the field of applied mathematics, discovering and inventing several theorems and principles – notably d'Alembert's principle – in dynamics and celestial mechanics.

Dalton, John 1766–1844
English chemist who proposed the theory of atoms and produced the first list of relative atomic masses.

Daniell, John Frederic 1790–1845
English meteorologist, inventor, and chemist, famous for devising the Daniell cell, a primary cell that was the first reliable source of direct-current electricity.

Darboux, Jean Gaston 1842–1917
French geometrician who contributed immensely to the differential geometry of his time, and to the theory of surfaces.

Darwin, Charles Robert 1809–1882
English naturalist famous for his theory of evolution and natural selection as put forward in 1859 in his book *The Origin of Species*.

Davis, William Morris 1850–1934
US physical geographer who analysed landforms.

Davisson, Clinton Joseph 1881–1958
US physicist who made the first experimental observation of the wave nature of electrons.

Davy, Humphry 1778–1829
English chemist who is best known for his discovery of the elements sodium and potassium and for inventing a safety lamp for use in mines.

Dawkins, (Clinton) Richard 1941–
British zoologist and evolutionary theorist whose book *The Selfish Gene* (1976) popularized sociobiology.

De Beer, Gavin Rylands 1899–1972
English zoologist known for his important contributions to embryology and evolution, notably disproving the germ-layer theory and developing the concept of paedomorphism (the retention of juvenile characteristics of ancestors in mature adults).

Debye, Peter Joseph Willem 1884–1966
Dutch-born US physical chemist who was a pioneer in X-ray powder crystallography and in a long career made many important contributions to the science.

Dedekind, (Julius Wilhelm) Richard 1831–1916
German mathematician, a great theoretician whose work on continuity and irrational numbers – in which he devised a system known as Dedekind's cuts – led to important and fundamental studies on the theory of numbers.

Dehn, Max Wilhelm 1878–1952
German-born US mathematician who in 1907 provided one of the first systematic studies of what is now known as topology – the branch of mathematics dealing with geometric figures whose overall properties do not change despite a continuous process of deformation, by which a square is (topologically) equivalent to a circle, and a cube is (topologically) equivalent to a sphere.

De la Beche, Henry Thomas 1796–1855
born Henry Thomas Beach, English geologist who secured the founding of the Geological Survey.

de la Rue, Warren 1815–1889
British pioneer of celestial photography.

Democritus *c.* 460–*c.*370 BC
Greek philosopher who is best known for formulating an atomic theory of matter and applying it to cosmology.

De Morgan, Augustus 1806–1871
British mathematician whose work in logic led him into a bitter controversy with his contemporary William Hamilton.

Descartes, René 1596–1650
French philosopher and mathematician whose work in attempting to reduce the physical sciences to purely mathematical principles – and particularly geometry – led to a fundamental revision of the whole of mathematical thought.

Deslandres, Henri Alexandre 1853–1948
French physicist and astronomer, now remembered mostly for his work in spectroscopy and for his solar studies.

Desmarest, Nicolas 1725–1815
French geoloist who became a champion of the volcanic origin of basalt.

Désormes, Charles Bernard 1777–1862
French physicist and chemist whose principal contribution was to determine the ratio of the specific heats of gases.

De Vaucouleurs, Gerard Henri 1918–1995
French-born US astronomer who carried out important research into extragalactic nebulae.

Dewar, James 1842–1923
Scottish physicist and chemist whose great experimental skills enabled him to carry out pioneering work on cryogenics, the properties of matter at extreme low temperatures.

Dicke, Robert Henry 1916–1997
US physicist who carried out considerable research into the rates of stellar and galactic evolution.

Dickson, Leonard Eugene 1874–1957
US mathematician who gave the first extensive exposition of the theory of fields.

Diels, Otto Paul Hermann 1876–1954
German organic chemist who made many fundamental discoveries, including (with Kurt Alder) the diene synthesis or Diels–Alder reaction.

Diophantus lived *c.* AD 270–280
Greek mathematician who, in solving linear mathematical problems, developed an early form of algebra.

Dirac, Paul Adrien Maurice 1902–1984
English theoretical physicist of great international standing who shared the Nobel Prize for Physics in 1933 (with Erwin Schrödinger) for his work on the development of quantum mechanics.

Dirichlet, (Peter Gustav) Lejeune 1805–1859
German mathematician whose work in applying analytical techniques to mathematical theory resulted in the fundamental development of the theory of numbers.

Dobzhansky, Theodosius 1900–1975
Russian-born US geneticist whose synthesis of Darwinian evolution and Mendelian genetics established evolutionary genetics as an independent discipline.

Dollfus, Audouin Charles 1924–
French physicist and astronomer who was the first to detect a faint atmosphere around Mercury.

Donati, Giovanni Battista 1826–1873
Italian astronomer whose principal astronomic interests were the study of comets and cosmic meteorology.

Doppler, Christian Johann 1803–1853
Austrian physicist who discovered the Doppler effect, which relates the observed frequency of a wave to the relative motion of the source and the observer.

Draper, Henry 1837–1882
US amateur astronomer, noted for his work on stellar spectroscopy and commemorated by the Henry Draper Catalogue of stellar spectral types.

Dreyer, John Louis Emil 1852–1926
Danish-born Irish astronomer and author, best known for a biographical study of the work of the Danish astronomer Tycho Brahe, and for his meticulous compilation of catalogues of nebulae and star clusters.

Driesch, Hans Adolf Eduard 1867–1941
German embryologist and philosopher who is best known as one of the last advocates of vitalism, the theory that life is directed by a vital principle and cannot be explained solely in terms of chemical and physical processes.

Dubois, (Marie) Eugène François Thomas 1858–1940
Dutch palaeontologist who in 1891 discovered the remains of *Pithecanthropus erectus*, known as Java Man.

Du Bois-Reymond, Emil Heinrich 1818–1896
German physiologist who showed the existence of electrical currents in nerves.

Dulong, Pierre Louis 1785–1838
French chemist, best known for his work with Alexis Petit that resulted in Dulong and Petit's law, which states that, for any element, the product of its specific heat and atomic weight is a constant, a quantity they termed the atomic heat.

Dumas, Jean Baptiste André 1800–1884
French chemist who made contributions to organic analysis and synthesis, and to the determination of atomic weights (relative

atomic masses) through the measurement of vapour densities.

Du Toit, Alexander Logie 1878–1948
South African geologist who contributed to theories of continental drift.

Duve, Christian René de 1917–
English-born Belgian biochemist who discovered two organelles, the lysosome and the peroxisome.

Dyson, Frank Watson 1868–1939
English astronomer especially interested in stellar motion and time determination.

Eastwood, Alice 1859–1953
US botanist who provided critical specimens for professional botanists as well as advising travellers on methods of plant collecting and arousing popular support for saving native species.

Eddington, Arthur Stanley 1882–1944
English astronomer and writer who discovered the fundamental role of radiation pressure in the maintenance of stellar equilibrium, explained the method by which the energy of a star moves from its interior to its exterior, and finally showed that the luminosity of a star depends almost exclusively on its mass – a discovery that caused a complete revision of contemporary ideas on stellar evolution.

Edlén, Bengt 1906–1993
Swedish astrophysicist whose main achievement lay in resolving the identification of certain lines in spectra of the solar corona that had misled scientists for the previous 70 years.

Eggen, Olin Jenck 1919–
US astronomer whose work included studies of high-velocity stars, red giants, and sub-luminous stars.

Ehrenberg, Christian Gottfried 1795–1876
German naturalist who developed a scheme for the classification of the animal kingdom.

Eigen, Manfred 1927–
German physical chemist who shared the 1967 Nobel Prize for Chemistry with Ronald Norrish and George Porter for his work on the study of fast reactions in liquids.

Eilenberg, Samuel 1913–1998
Polish-born US mathematician whose research in the field of algebraic topology led to considerable development in the theory of cohomology.

Einstein, Albert 1879–1955
German-born US theoretical physicist who revolutionized our understanding of matter, space, and time with his theories of relativity.

Eisenhart, Luther Pfahler 1876–1965
US theoretical geometrist whose early work was concerned with the properties of surfaces and their deformation; later he became interested in Riemann geometry from which he attempted to develop his own geometry theory.

Elsasser, Walter Maurice 1904–1991
German-born US geophysicist who pioneered analysis of the Earth's former magnetic fields.

Emeléus, Harry Julius 1903–1993
English chemist who made wide-ranging investigations in inorganic chemistry, studying particularly nonmetallic elements and their compounds.

Encke, Johann Franz 1791–1865
German astronomer whose work on star charts during the 1840s contributed to the discovery of the planet Neptune in 1846.

Eratosthenes *c.* 276–194 BC
Greek scholar and polymath, many of whose writings have been lost, although it is known that they included papers on geography, mathematics, philosophy, chronology, and literature.

Esaki, Leo 1925–
Japanese physicist who shared the 1973 Nobel Prize for his discovery of tunnelling in semiconductor diodes.

Eskola, Pentti Eelis 1883–1964
Finnish geologist important in the field of petrology.

Euclid lived *c.* 300 BC
Greek mathematician whose *Elements* provided the foundation of geometrical thought and study.

Eudoxus *c.* 408–*c.* 353 BC
also known as Eudoxus of Cnidus, Greek mathematician and astronomer who is said to have studied under Plato.

Euler, Leonhard 1707–1783
Swiss mathematician whose power of mental calculation was prodigious; he made fundamental contributions to analysis, mechanics, the calculus of variations, and other areas of mathematics and mathematical

physics and continued to work productively even after going blind in the 1760s.

Ewing, (William) Maurice 1906–1974
US geologist, a major innovator in modern geology, whose studies of the ocean floor provided crucial data for the development of plate tectonics.

Fabre, (Jean) Henri Casimir 1823–1915
French entomologist whose studies of insects, particularly their anatomy and behaviour, have become classics.

Fabricius ab Aquapendente, Hieronymus 1537–1619
Italian Girolamo Fabrizio, anatomist and embryologist who gave the first accurate description of the semilunar valves in the veins and whose pioneering studies of embryonic development helped to establish embryology as an independent discipline.

Fabry, Charles 1867–1945
French physicist who specialized in optics, devising methods for the accurate measurement of interference effects.

Fahrenheit, Gabriel Daniel 1686–1736
Polish-born Dutch physicist who invented the first accurate thermometers and devised the Fahrenheit scale of temperature.

Fajans, Kasimir 1887–1975
Polish-born US chemist, best known for his work on radioactivity and isotopes and for formulating rules that help to explain valence and chemical bonding.

Faraday, Michael 1791–1867
English physicist and chemist who made many discoveries in electricity and magnetism; the father of field theory.

Feller, William 1906–1970
Croatian-born US mathematician largely responsible for making the theory of probability accessible to students of subjects outside the field of mathematics through his two-volume textbook on the subject.

Fermat, Pierre de 1601–1665
French lawyer and magistrate for whom mathematics was an absorbing hobby. His famous 'last theorem' kept mathematicians occupied for almost 400 years.

Fermi, Enrico 1901–1954
Italian-born US physicist best known for bringing about the first controlled chain reaction (in a nuclear reactor) and for his part in the development of the atomic bomb.

Fessenden, Reginald Aubrey 1866–1932
Canadian physicist whose invention of radio-wave modulation paved the way for modern radio communication.

Feynman, Richard P(hillips) 1918–1988
US physicist who shared the 1965 Nobel prize for his role in the development of the theory of quantum electrodynamics.

Fibonacci, Leonardo c. 1180–c. 1250
also known as Leonardo of Pisa, Italian mathematician whose writings were influential in introducing and popularizing the Indo-Arabic numeral system, and whose work in algebra, geometry, and theoretical mathematics was far in advance of the contemporary European standards.

Field, George Brooks 1929–
US theoretical astrophysicist whose main research has been into the nature and composition of intergalactic matter and the properties of residual radiation in space.

Fischer, Emil Hermann 1852–1919
German organic chemist who analysed and synthesized many biologically important compounds.

Fischer, Ernst Otto 1918–
German inorganic chemist who shared the 1973 Nobel Prize for Chemistry with English inorganic chemist Geoffrey Wilkinson for pioneering work (carried out independently) on the organometallic compounds of the transition metals.

Fischer, Hans 1881–1945
German organic chemist who is best known for his determinations of the molecular structures of three important biological pigments: haemoglobin, chlorophyll, and bilirubin.

Fischer, Hermann Otto Laurenz 1888–1960
German organic chemist whose chief contribution to the science concerned the synthetic and structural chemistry of carbohydrates, glycerides, and inositols.

Fisher, Ronald Aylmer 1890–1962
English mathematical biologist whose work in the field of statistics resulted in the formulation of a methodology in which the analysis of results obtained using small samples produced interpretations that were objective and valid overall.

FitzGerald, George Francis 1851–1901
Irish theoretical physicist who worked on the electromagnetic theory of light and radio waves.

Fizeau, Armand Hippolyte Louis 1819–1896
French physicist who was the first to measure the velocity of light on the Earth's surface.

Flamsteed, John 1646–1719
English astronomer and writer who became the first Astronomer Royal based at Greenwich. He began systematic observations of the positions of stars, Moon, and planets.

Fleming, Alexander 1881–1955
Scottish bacteriologist who discovered penicillin, a substance produced by the mould *Penicillium notatum* and found to be effective in killing various pathogenic bacteria without harming the cells of the human body.

Fleming, Williamina Paton Stevens 1857–1911
Scottish-born US co-author (with Edward Pickering) of the first general catalogue classifying stellar spectra.

Flory, Paul John 1910–1985
US polymer chemist who was awarded the 1974 Nobel Prize for Chemistry for his investigations of synthetic and natural macromolecules.

Forbes, Edward 1815–1854
British naturalist who made significant contributions to oceanography.

Forsyth, Andrew Russell 1858–1942
Scottish mathematician who brought modern continental developments to the attention of British mathematicians, particularly through his *Theory of Functions* (1893), and helped modernize the teaching of mathematics at Cambridge University.

Foucault, (Jean Bernard) Léon 1819–1868
French physicist who invented the gyroscope, demonstrated the rotation of the Earth, and obtained the first accurate value for the velocity of light.

Fourier, (Jean Baptiste) Joseph 1768–1830
French mathematical physicist who devised the technique of 'Fourier series' to describe the transfer of heat in purely mathematical terms.

Fowler, William Alfred 1911–1995
US physicist and astronomer who, with Subrahmanyan Chandrasekhar, won the 1983 Nobel Prize for Physics for his work on the nuclear reactions that play a role in the formation of chemical elements in the universe.

Fraenkel, Abraham Adolf 1891–1965
German-born Israeli mathematician who is chiefly remembered for his research and perception in set theory, and for his many textbooks.

Franck, James 1882–1964
German-born US physicist who provided the experimental evidence for the quantum theory of Max Planck and the quantum model of the atom developed by Niels Bohr.

Franklin, Benjamin 1706–1790
US scientist and statesman who made an important contribution to physics by arriving at an understanding of the nature of electric charge as a presence or absence of electricity, introducing the terms 'positive' and 'negative' to describe charges.

Franklin, Rosalind Elsie 1920–1958
English chemist and X-ray crystallographer whose experimental measurements contributed crucially to recognition of the helical shape of DNA.

Fraunhofer, Joseph von 1787–1826
German physicist and optician who was the first person to investigate the dark lines in the spectra of the Sun and other stars, which are named Fraunhofer lines in his honour.

Fredholm, Erik Ivar 1866–1927
Swedish mathematician and mathematical physicist who founded the modern theory of integral equations, and in his work provided the foundations upon which much of the extremely important research later carried out by David Hilbert was based.

Fresnel, Augustin Jean 1788–1827
French physicist who established the transverse-wave theory of light in 1821.

Freundlich, Herbert Max Finlay 1880–1941
German physical chemist, best known for his extensive work on the nature of colloids, particularly sols and gels.

Friedel, Charles 1832–1899
French organic chemist and mineralogist, best remembered for his part in the discovery of the Friedel–Crafts reaction.

Friedman, Aleksandr Aleksandrovich 1888–1925
Russian mathematician who suggested that

Einstein's general theory of relativity implied an expanding universe.

Frisch, Otto Robert 1904–1979
Austrian-born British physicist who coined the term 'fission' to describe the splitting of a uranium nucleus when hit by a neutron.

Frobenius, (Ferdinand) Georg 1849–1917
German mathematician who is now chiefly remembered for his formulation of the concept of the abstract group – a theory that proposed what is now generally considered to be the first abstract structure of 'new' mathematics.

Fuchs, Immanuel Lazarus 1833–1902
German mathematician whose work on Bernhard Riemann's method for the solution of differential equations led to a study of the theory of functions that was later crucial to Henri Poincaré in his own important investigation of function theory.

Gabor, Dennis 1900–1979
Hungarian-born British physicist and electrical engineer, famous for his invention of holography – three-dimensional photography using lasers – for which he received the 1971 Nobel Prize for Physics.

Galileo 1564–1643
also known as Galileo Galilei, Italian physicist and astronomer who made the first important astronomical discoveries with the telescope, and who founded the modern scientific method of formulating mathematical laws to explain the results of observation and experiment.

Galle, Johann Gottfried 1812–1910
German astronomer who was the first to observe Neptune and recognize it as a new planet.

Galois, Evariste 1811–1832
French mathematician who, building on the work of Joseph Lagrange, Karl Gauss, Niels Abel, and Augustin Cauchy, greatly extended the understanding of the conditions in which an algebraic equation is solvable and, by his method of doing so, laid the foundations of modern group theory.

Galton, Francis 1822–1911
English scientist, inventor, and explorer who made contributions to several disciplines, including anthropology, meteorology, geography, and statistics, but who is best known as the initiator of the study of eugenics (a term he coined).

Galvani, Luigi 1737–1798
Italian anatomist whose discovery of 'animal electricity' stimulated the work of Alessandro Volta and others to discover and develop current electricity.

Gamow, George 1904–1968
born Georgi Antonovich Gamow, Russian-born US physicist who provided the first evidence for the Big Bang theory of the origin of the universe.

Gardner, Julia Anna 1882–1960
US geologist and palaeontologist internationally known for her work on stratigraphic palaeontology.

Gassendi, Pierre 1592–1655
French philosopher, physicist, and priest, best known for his opposition to Aristotelian philosophy and his revival of the Epicurean system.

Gauss, Carl Friedrich 1777–1855
German mathematician, physicist, and astronomer whose innovations in mathematics proved him to be the equal of Archimedes or Isaac Newton.

Gay-Lussac, Joseph Louis 1778–1850
French chemist who pioneered the quantitative study of gases and established the link between gaseous behaviour and chemical reactions.

Geiger, Hans Wilhelm 1882–1945
German physicist who invented the Geiger counter for detecting radioactivity.

Gell-Mann, Murray 1929–
US theoretical physicist who was awarded the 1969 Nobel Prize for Physics for his work on the classification and interactions of subatomic particles.

Germain, Sophie Marie 1776–1831
French mathematician who developed the modern theory of elasticity (the mathematical theory of the stress and strain that a material can sustain and still return to its original form) and made major contributions to numbers theory and acoustics.

Giacconi, Riccardo 1931–
Italian-born US physicist, the head of a team whose work has been fundamental in the development of X-ray astronomy.

Gibbs, (Josiah) Willard 1839–1903
US scientist who laid the foundations of modern chemical thermodynamics.

367

Gilbert, Walter 1932–
US molecular biologist who shared (with Paul Berg and Frederick Sanger) the 1980 Nobel Prize for Chemistry for his work in devising techniques for determining the sequence of bases in DNA.

Gilbert, William 1544–1603
English physician and physicist who performed fundamental pioneering research into magnetism and also helped to establish the modern scientific method.

Gill, David 1843–1914
Scottish astronomer whose precision and patience using old instruments brought him renown before he achieved even greater fame for his pioneering work in the use of photography to catalogue stars.

Gilman, Henry 1893–1986
US organic chemist best known for his work on organometallic compounds, particularly Grignard reagents.

Ginzburg, Vitalii Lazarevich 1916–
Russian astrophysicist whose use of quantum theory contributed to the development of nuclear physics.

Glaser, Donald Arthur 1926–
US physicist who invented the bubble chamber, an instrument much used in nuclear physics to study short-lived subatomic particles.

Glashow, Sheldon Lee 1932–
US particle physicist who proposed the existence of a fourth 'charmed' quark and later argued that quarks must be coloured.

Glauber, Johann Rudolf 1604–1670
German chemist whose pursuit of alchemical wisdom led to many practical advances in chemical techniques and to the discovery of new and useful compounds.

Gödel, Kurt 1906–1978
Austrian-born US philosopher and mathematician who, in his philosophical endeavour to establish the science of mathematics as totally consistent and totally complete, proved that it could never be.

Goeppert-Mayer, Maria 1906–1972
German-born US physicist who shared the 1963 Nobel Prize for Physics for discovering the shell model of nuclear structure.

Gold, Thomas 1920–
Austrian-born US astronomer and physicist who has carried out research in several fields but remains most famous for his share in formulating, with Fred Hoyle and Hermann Bondi, the steady-state theory regarding the creation of the universe.

Goldberg, Leo 1913–1987
US astrophysicist who carried out research, generally as one of a team, into the composition of stellar atmospheres and the dynamics of the loss of mass from cool stars.

Goldschmidt, Victor Moritz 1888–1947
Swiss-born Norwegian chemist who has been called the founder of modern geochemistry.

Goldstein, Eugen 1850–1930
German physicist who investigated electrical discharges through gases at low pressures.

Goodall, Jane 1934–
English zoologist best known for her studies of chimpanzees in the Gombe Stream Reserve in Tanzania.

Goodyear, Charles 1800–1860
US inventor who is generally credited with inventing the process for vulcanizing rubber.

Gosset, William Sealey 1876–1937
English industrial research scientist, famous for his work on statistics.

Gough, John 1757–1825
British scientist who discovered that stretching rubber produced heat and then crystallized if cooled while extended.

Gould, Stephen Jay 1941–
US palaeontologist and writer who teaches and researches in geology, evolutionary biology, and the history of science.

Graham, Thomas 1805–1869
Scottish physical chemist who pioneered the chemistry of colloids, but who is best known for his studies of the diffusion of gases, the principal law concerning which is named after him.

Grandi, Guido 1671–1742
Italian mathematician famous for his work on the definition of curves – particularly curves that are symmetrically pleasing to the eye.

Grassmann, Hermann Günther 1809–1877
German mathematician whose pioneering contributions to geometric analysis were largely ignored during his lifetime; he later turned to the study of linguistics.

Gray, Asa 1810–1888
US botanist who was the leading authority on botanical taxonomy in the USA in the

19th century and a pioneer of plant geography.

Green, George 1793–1841
English businessman and self-taught mathematician; he coined the term 'potential' and made fundamental contributions to the mathematical theory of electricity and magnetism.

Greenstein, Jesse Leonard 1909–
US astronomer who has made important discoveries by combining his observational skills with current theoretical ideas and techniques.

Grignard, (François Auguste) Victor 1871–1935
French organic chemist, best known for his work on organomagnesium compounds, or Grignard reagents.

Grimaldi, Francesco Maria 1618–1663
Italian physicist who discovered the diffraction of light and made various observations in physiology.

Guericke, Otto von 1602–1686
German physicist and politician who invented the air pump and carried out the classic experiment of the Magdeburg hemispheres, which demonstrated the pressure of the atmosphere.

Guettard, Jean-Etienne 1715–1786
French naturalist who pioneered geological mapping.

Gurdon, John Bertrand 1933–
English molecular biologist who is probably best known for his work on nuclear transplantation.

Haber, Fritz 1868–1934
German chemist who made contributions to physical chemistry and electrochemistry, but who is best remembered for the Haber process, a method of synthesizing ammonia by the direct catalytic combination of nitrogen and hydrogen.

Hadamard, Jacques Salomon 1865–1963
French mathematician who contributed to many fields, particularly number theory, where he proved a fundamental theorem on the distribution of prime numbers.

Haeckel, Ernst Heinrich Philipp Auguste 1834–1919
German zoologist well known for his genealogical trees of living organisms and for his early support of Darwin's ideas on evolution.

Hahn, Otto 1879–1968
German radiochemist who discovered nuclear fission.

Hale, George Ellery 1868–1938
US astronomer who spent much of his life arranging funding for the construction of large telescopes, including the 15-m/60-in and 2.5-m/100-in reflectors at Mount Wilson, and the 5-m/200-in Mount Palomar telescope.

Hales, Stephen 1677–1761
English clergyman who invented the pneumatic trough to facilitate the study of gases, particularly 'fixed air' (carbon dioxide).

Hall, Asaph 1829–1907
US astronomer who discovered the two Martian satellites, Deimos and Phobos.

Hall, James 1761–1832
Scottish geologist, one of the founders of experimental geology, who provided evidence in support of James Hutton's theories on the formation of the Earth's crust.

Hall, Philip 1904–1982
English mathematician who specialized in the study of group theory.

Halley, Edmond 1656–1742
English mathematician, physicist, and astronomer who not only identified the comet later to be known by his name, but also compiled a star catalogue, supervised the publication of Isaac Newton's *Principia*, and began a line of research that – after his death – resulted in a reasonably accurate calculation of the astronomical unit.

Hamilton, William Rowan 1805–1865
Irish mathematician, widely regarded in his time as a 'new Newton', who created a new system of algebra based on quaternions.

Hammick, Dalziel Llewellyn 1887–1966
English chemist whose major contributions were in the fields of theoretical and synthetic organic chemistry.

Hankel, Hermann 1839–1873
German mathematician and mathematical historian who made significant contributions to the study of complex and hypercomplex numbers and the theory of functions.

Harden, Arthur 1865–1940
English biochemist who investigated the mechanism of sugar fermentation and the role of enzymes in this process.

Hardy, Alister Clavering 1896–1985
English marine biologist who designed the Hardy plankton continuous recorder.

Hardy, Godfrey Harold 1877–1947
English mathematician whose work in analysis and number theory made him the leading English pure mathematician of his day.

Harrison, John 1693–1776
English horologist and instrumentmaker who made the first chronometers that were accurate enough to allow the precise determination of longitude at sea, and so permit reliable (and safe) navigation over long distances.

Harvey, Ethel Browne 1885–1965
US embryologist and cell biologist, particularly renowned for her discoveries of the mechanisms of cell division, using sea-urchin eggs as her experimental model.

Hausdorff, Felix 1868–1942
German mathematician and philosopher who is chiefly remembered for his development of the branch of mathematics known as topology, in which he formulated the theory of point sets.

Haüy, René-Just 1743–1822
French mineralogist, the founder of modern crystallography.

Hawking, Stephen (William) 1942–
English theoretical physicist and mathematician whose main field of research has been the nature of space-time and those anomalies in space-time known as 'black holes'.

Haworth, (Walter) Norman 1883–1950
English organic chemist whose researches concentrated on carbohydrates, particularly sugars.

Hayashi, Chushiro 1920–
Japanese physicist whose research in 1950 exposed a fallacy in the 'hot Big Bang' theory proposed two years earlier by Ralph Alpher, Hans Bethe, and George Gamow.

Heaviside, Oliver 1850–1925
English physicist who recast Maxwell's equations of the electromagnetic field into their now standard vector form and made pioneering contributions to the theory of signal propagation.

Heine, Heinrich Eduard 1812–1881
German mathematician who wrote extensively on the theory of spherical functions.

Heisenberg, Werner Karl 1901–1976
German physicist who founded quantum mechanics and the uncertainty principle.

Helmholtz, Hermann Ludwig Ferdinand von 1821–1894
German physicist and physiologist who made a major contribution to physics with the first precise formulation of the principle of conservation of energy and contributed to the theory of electromagnetism.

Helmont, Jan Baptist van 1579–1644
Flemish chemist and physician.

Henry, Joseph 1797–1878
US physicist who carried out early experiments in electromagnetic induction.

Henry, William 1774–1836
English chemist and physician, known for his study of the solubility of gases in liquids.

Heraklides of Pontus 388–315 BC
Greek philosopher and astronomer who is remembered particularly for his teaching that the Earth turns on its axis, from west to east, once every 24 hours.

Herbrand, Jacques 1908–1931
French mathematical prodigy who, in a life cut tragically short, still originated some innovatory concepts in the fields of mathematical logic and modern algebra.

Hermite, Charles 1822–1901
French mathematician who was a principal contributor to the development of the theory of algebraic forms, the arithmetical theory of quadratic forms, and the theories of elliptic and Abelian functions.

Hero of Alexandria lived c. AD 60
Greek mathematician and engineer, the greatest experimentalist of antiquity.

Herschel, Caroline Lucretia 1750–1848
German-born English astronomer who discovered 14 nebulae and 8 comets and worked on her brother William Herschel's catalogue of star clusters and nebulae.

Herschel, (Frederick) William 1738–1822
German-born English astronomer who discovered the planet Uranus, built large reflectors, and observed and speculated on the nature of stars and nebulae.

Herschel, John Frederick William
1792–1871
English astronomer, the first to carry out a systematic survey of the stars in the southern hemisphere and to attempt to measure, rather than estimate, the brightness of stars.

Hertzsprung, Ejnar 1873–1967
Danish astronomer and physicist who, having proposed the concept of the absolute magnitude of a star, went on to describe for the first time the relationship between the absolute magnitude and the temperature of a star, formulating his results in the form of a graphic diagram that has since become a standard reference.

Herzberg, Gerhard 1904–
German-born Canadian physicist who is best known for his work in determining – using spectrocopy – the electronic structure and geometry of molecules, especially free radicals (atoms or groups of atoms that possess a free, unbonded electron).

Hess, Germain Henri 1802–1850
German Ivanovich Gess, Swiss-born Russian chemist, best known for his pioneering work in thermochemistry and the law of constant heat summation named after him.

Hess, Harry Hammond 1906–1969
US geologist who played a key part in the plate tectonics revolution of the 1960s.

Hess, Victor Francis 1883–1964
Austrian-born US physicist who discovered cosmic rays, for which he was jointly awarded the 1936 Nobel Prize for Physics with Carl Anderson.

Hevelius, Johannes 1611–1687
German Johann Hewel or Howelcke, German astronomer, most famous for his careful charting of the surface of the Moon.

Hevesy, Georg Karl von 1885–1966
Hungarian-born Swedish chemist whose main achievements were the introduction of isotopic tracers (to follow chemical reactions) and the discovery of the element hafnium.

Hewish, Antony 1924–
English astronomer and physicist whose research into radio scintillation resulted in the discovery of pulsars.

Hey, James Stanley 1909–2000
English physicist whose work in radar led to pioneering research in radio astronomy.

Heyrovský, Jaroslav 1890–1967
Czech chemist who was awarded the 1959 Nobel Prize for Chemistry for his invention and development of polarography, an electrochemical technique of chemical analysis.

Hilbert, David 1862–1943
German mathematician, philosopher, and physicist whose work on the theory of invariants, algebraic number theory, and the foundations of mathematical knowledge proved very influential.

Hill, Robert 1899–1991
English biochemist who contributed greatly to modern knowledge of photosynthesis.

Hinshelwood, Cyril Norman 1897–1967
English physical chemist who made fundamental studies of the kinetics and mechanisms of chemical reactions.

Hipparchus *c.* 190–*c.* 120 BC
Greek astronomer and mathematician, and observer and compiler of a star catalogue.

Hippocrates *c.* 460–*c.* 377 BC
Greek physician, often known as the founder of medicine and, in ancient times, regarded as the greatest physician who had ever lived.

Hoagland, Mahlon Bush 1921–
US biochemist who was the first to isolate transfer RNA (tRNA), which plays an essential part in intracellular protein synthesis.

Hodgkin, Dorothy Mary Crowfoot
1910–1994
English chemist who used X-ray crystallographic analysis to determine the structures of numerous complex organic molecules, including penicillin and vitamin B_{12} (cyanocobalamin).

Hofmann, August Wilhelm von 1818–1892
German chemist who was one of the great organic chemists of the 19th century and had enormous influence on the development of the subject in both the UK and Germany.

Hofstadter, Robert 1915–1990
US physicist who shared the 1961 Nobel Prize for discovering that protons and neutrons contained smaller particles (now known to be quarks).

Hogben, Lancelot Thomas 1895–1975
English zoologist and geneticist who, somewhat surprisingly, wrote a very successful book entitled *Mathematics for the Millions*.

Hollerith, Herman 1860–1929
US mathematician and mechanical engineer who invented electrical tabulating machines.

Holmes, Arthur 1890–1965
English geologist who helped develop interest in the theory of continental drift.

Hooke, Robert 1635–1703
English physicist who was also active in many other branches of science. He studied elasticity and formulated Hooke's law.

Hooker, Joseph Dalton 1817–1911
English botanist who made many important contributions to botanical taxonomy but who is probably best known for introducing into the UK a range of previously unknown species of rhododendron and for his improvements to the Royal Botanical Gardens at Kew.

Hopkins, Frederick Gowland 1861–1947
English biochemist who was jointly awarded (with Christian Eijkman) the 1929 Nobel Prize for Physiology or Medicine for his work showing the necessity of certain dietary components – now known as vitamins – for the maintenance of health.

Hoyle, Fred(erick) 1915–2001
English cosmologist and astrophysicist, distinguished for his work on the evolution of stars, the development of the steady-state theory of the universe, and a new theory on gravitation.

Hubble, Edwin (Powell) 1889–1953
US astronomer who studied extragalactic nebulae and demonstrated them to be galaxies like our own.

Hubel, David Hunter 1926–
US neurophysiologist who worked with Torsten Wiesel on the physiology of vision and the way in which the higher centres of the brain process visual information.

Hückel, Erich Armand Arthur Joseph 1896–1980
German physical chemist who, with Peter Debye, developed the modern theory that accounts for the electrochemical behaviour of strong electrolytes in solution.

Huggins, William 1824–1910
English astronomer and pioneer of astrophysics. He revolutionized astronomy by using spectroscopy to determine the chemical make-up of stars and by using photography in stellar spectroscopy.

Humason, Milton Lasell 1891–1972
US astronomer famous for his investigations, at Mount Wilson Observatory, into distant galaxies.

Humboldt, (Friedrich Wilhelm Heinrich) Alexander 1769–1859
Baron von Humboldt, German geophysicist, botanist, geologist, and writer – a founder of ecology.

Hutton, James 1726–1797
Scottish natural philosopher who introduced ideas of uniformitarianism and Plutonism to geology.

Huxley, Hugh Esmor 1924–
English physiologist whose contribution to science has been concerned with the study of muscle cells.

Huxley, T(homas) H(enry) 1825–1895
English biologist who helped to break down the great barrier of traditional resistance to scientific advance during the mid-19th century, and did a great deal to popularize science.

Huygens, Christiaan 1629–1695
Dutch physicist and astronomer who proposed a wave theory of light, developed the pendulum clock, discovered polarization, and observed Saturn's ring.

Hyatt, John Wesley 1837–1920
US inventor who became famous for his invention of celluloid, the first artificial plastic.

Hyman, Libbie Henrietta 1888–1969
US zoologist renowned for her studies on the taxonomy (classification) and anatomy of invertebrates.

Hypatia c. 370–415
Greek natural philosopher who is credited with being the first female astronomer and mathematician of note.

Ibn Bajja (Latin **Avempace**) c.1095–c.1138
Spanish-born Arab philosopher, physician, and poet who wrote a number of commentaries on the works of Aristotle.

Ibn Rushd (Latin **Averroës**) 1126–1198
Spanish-born Arab philosopher and physician whose commentaries on Aristotle and Plato were highly influential.

Ingenhousz, Jan 1730–1799
Dutch biologist and physiologist who discovered photosynthesis and plant respiration.

Ingold, Christopher (Kelk) 1893–1970
English organic chemist who made a fundamental contribution to the theoretical aspects of the subject with his explanation for the mechanisms of organic reactions in terms of the behaviour of electrons in the molecules concerned.

Ipatieff, Vladimir Nikolayevich 1867–1952
Russian-born US organic chemist who is best known for his development of catalysis in organic chemistry, particularly in reactions involving hydrocarbons.

Jabir ibn Hayyan (Latin Geber)
c. 721–c. 815
Arabian alchemist, reputed to be the author of a large body of Arabic writings mainly on alchemical knowledge but also on a number of other subjects.

Jacobi, Karl Gustav Jacob 1804–1851
German mathematician and mathematical physicist, much of whose work was on the theory of elliptical functions, mathematical analysis, number theory, geometry, and mechanics.

Jansky, Karl Guthe 1905–1950
US radio engineer whose discovery of radio waves of extraterrestrial origin led to the development of radio astronomy.

Janssen, (Pierre) Jules César 1824–1907
French astronomer, famous for his work in physical astronomy and spectroscopy.

Jeans, James Hopwood 1877–1946
English mathematician and astrophysicist who made important contributions to cosmogony – particularly his theory of continuous creation of matter – and became known to a wide public through his popular books and broadcasts on astronomy.

Jenner, Edward 1749–1823
English biologist who was the first to prove by scientific experiment that cowpox inoculation gives immunity against smallpox.

Jensen, (Johannes) Hans Daniel
1907–1973
German physicist who shared the 1963 Nobel Prize for Physics with Maria Goeppert-Mayer and Eugene Wigner for work on the detailed characteristics of atomic nuclei.

John of Holywood (*Sacrobosco*)
*c.*1195–1256
English canon and mathematician whose textbooks on astronomy and arithmetic became standard parts of the university curriculum until the 17th century.

Jones, Harold Spencer 1890–1960
English astronomer, the tenth Astronomer Royal who was responsible for organizing the international campaign to redetermine the astronomical unit, by observing the asteroid Eros during its near approach in 1931.

Jordan, (Marie Ennemond) Camille
1838–1922
French mathematician, the greatest exponent of algebra of his day, who concentrated on research in topology, analysis, and group theory.

Josephson, Brian David 1940–
Welsh physicist who discovered the Josephson tunnelling effect in superconductivity.

Joule, James Prescott 1818–1889
English physicist who verified the principle of conservation of energy by making the first accurate determination of the mechanical equivalent of heat.

Joy, James Harrison 1882–1973
US astronomer, most famous for his work on stellar distances, the radial motions of stars, and variable stars.

Just, Ernest Everett 1833–1941
US biologist who became internationally known for his pioneering research into fertilization, cell physiology, and experimental embryology.

Kamerlingh Onnes, Heike 1853–1926
Dutch physicist who is particularly remembered for the contributions he made to the study of the properties of matter at low temperatures.

Kant, Immanuel 1724–1804
German philosopher whose theoretical work in astronomy inspired cosmological theories; many of his conjectures have been confirmed by observational evidence.

Kapitza, Peter Leonidovich 1894–1984
also known as Pyotr Kapitsa, Russian physicist best known for his work on the superfluidity of liquid helium.

Kapteyn, Jacobus Cornelius 1851–1922
Dutch astronomer who analysed the structure of the universe by studying the distribution of stars using photographic techniques.

Karrer, Paul 1889–1971
Swiss organic chemist, famous for his work on vitamins and vegetable dyestuffs.

Keeler, James Edward 1857–1900
US astrophysicist noted for his work on the rings of Saturn and on the abundance and structure of nebulae.

Kekulé von Stradonitz, Friedrich August 1829–1896
German organic chemist who founded structural organic chemistry and is best known for his 'ring' formula for benzene.

Kelvin, William Thomson 1824–1907
1st Baron Kelvin, Irish physicist who made pioneering contributions to thermodynamics, electromagnetic theory, and telegraphy; the degree of temperature on the absolute scale is named after him.

Kendrew, John Cowdery 1917–1997
English biochemist who shared the 1962 Nobel Prize for Chemistry with Max Perutz for his determination of the structure of the muscle protein myoglobin.

Kenyon, Joseph 1885–1961
English organic chemist, best known for his studies of optical activity, particularly of secondary alcohols.

Kepler, Johannes 1571–1630
German astronomer who greatly improved Copernicus's heliocentric theory, in particular by showing that the planets move in elliptic orbits around the Sun.

Kerr, John 1824–1907
Scottish physicist who discovered the Kerr effect, which produces double refraction in certain media on the application of an electric field.

Khorana, Har Gobind 1922–
Indian-born US chemist who has worked extensively on the chemistry of nucleic acids, for which he shared the Nobel Prize for Physiology or Medicine in 1968.

Kimura, Motoo 1924–
Japanese biologist who, as a result of his work on population genetics and molecular evolution, has developed a theory of neutral evolution that opposes the conventional neo-Darwinistic theory of evolution by natural selection.

King, Clarence 1842–1901
US geologist who helped organize and served as first director of the US Geological Survey.

Kipping, Frederick Stanley 1863–1949
English chemist who pioneered the study of the organic compounds of silicon; he invented the term 'silicone', which is now applied to an entire class of silicon and oxygen polymers.

Kirchhoff, Gustav Robert 1824–1887
German physicist who worked on electrical theory and the thermodynamics of radiation.

Kirkwood, Daniel 1814–1895
US astronomer who is known for his work on asteroids, meteors, and the evolution of the Solar System.

Klaproth, Martin Heinrich 1743–1817
German chemist famous for his discovery of several new elements and for pioneering analytical chemistry.

Klein, (Christian) Felix 1849–1925
German mathematician and mathematical physicist whose unification of the various Euclidean and non-Euclidean geometries was crucial to the future development of that branch of mathematics.

Koch, Robert 1843–1910
German bacteriologist who, with Louis Pasteur, is generally considered to be one of the two founders of modern bacteriology.

Kolbe, (Adolf Wilhelm) Hermann 1818–1884
German organic chemist, generally credited as the founder of modern organic chemistry with his synthesis of acetic acid (ethanoic acid) – an organic compound – from inorganic starting materials.

Kornberg, Arthur 1918–
US biochemist who in 1957 made the first synthetic molecules of DNA.

Kornberg, Hans Leo 1928–
German-born British biochemist who has made important contributions to the understanding of metabolic pathways and their regulation, especially in micro-organisms.

Krebs, Hans Adolf 1900–1981
German-born British biochemist famous for his outstanding work in elucidating the cyclical pathway involved in the intracellular metabolism of foodstuffs – a pathway known as the tricarboxylic cycle, the citric acid cycle, or the Krebs cycle.

Kronecker, Leopold 1823–1891
German mathematician who held that all

mathematics should be founded on whole numbers; he did much to unite diverse branches of mathematics.

Kuhn, Richard 1900–1967
Austrian-born German organic chemist who worked mainly with carbohydrates and was awarded (but not allowed immediately to accept) the 1938 Nobel Prize for Chemistry for his research on the synthesis of vitamins.

Kuiper, Gerard Peter 1905–1973
Dutch-born US astronomer best known for his studies of lunar and planetary surface features and his theoretical work on the origin of the planets in the Solar System.

Kummer, Ernst Eduard 1810–1893
German mathematician, famous for his work in higher arithmetic and geometry, who introduced 'ideal numbers' in the attempt to prove Fermat's last theorem.

Kundt, August Adolph Eduard Eberhard 1839–1894
German physicist who is best known for Kundt's tube, a simple device for measuring the velocity of sound in gases.

Lacaille, Nicolas Louis de 1713–1762
French scientist, a positional astronomer who also contributed to advances in geodesy.

Lagrange, Joseph Louis 1736–1813
born Giuseppe Lodovico Lagrange, Italian-born French mathematician who revolutionized the study of mechanics.

Lalande, Joseph Jérome le Français de 1732–1807
French astronomer noted for his planetary tables, his account of the transit of Venus, and numerous writings on astronomy.

Lamarck, Jean Baptiste Pierre Antoine de Monet 1744–1829
Chevalier de Lamarck, French naturalist best known for his alternative to Charles Darwin's theory of evolution and the distinction he made between vertebrate and invertebrate animals.

Lamb, Horace 1849–1934
English applied mathematician, noted for his many books on hydrodynamics, elasticity, sound, and mechanics.

Landau, Lev Davidovich 1908–1968
Russian-Azerbaijani theoretical physicist who developed a theory of condensed matter and explained the properties of helium.

Landé, Alfred 1888–1976
German-born US physicist known for the Landé splitting factor in quantum theory.

Langevin, Paul 1872–1946
French physicist who worked on relativity theory, magnetism, and the use of ultrasonic waves for echolocation.

Langmuir, Irving 1881–1957
US physical chemist who is best remembered for his studies of adsorption at surfaces and for his investigations of thermionic emission.

Laplace, Pierre Simon 1749–1827
marquis de Laplace, French astronomer and mathematician who contributed to the fields of celestial mechanics, probability, applied mathematics, and molecular physics.

Lapworth, Arthur 1872–1941
Scottish organic chemist whose most important work was the enunciation of the electronic theory of organic reactions (independently of Robert Robinson).

Laue, Max Theodor Felix von 1879–1960
German physicist who established that X-rays are electromagnetic waves by producing X-ray diffraction in crystals.

Lavoisier, Antoine Laurent 1743–1794
French chemist, universally regarded as the founder of modern chemistry. He proved that combustion needs only part of air, which he called oxygen.

Lawrence, Ernest O(rlando) 1901–1958
US physicist who was responsible for the concept and development of the cyclotron.

Leavitt, Henrietta Swan 1868–1921
US astronomer, an expert in the photographic analysis of the magnitudes of variable stars.

Lebedev, Pyotr Nikolayevich 1866–1912
Russian physicist whose most important contribution to science was the detection and measurement of the pressure that light exerts on bodies, an effect that had been predicted by James Clerk Maxwell in his electro-magnetic theory of light.

Lebesgue, Henri Léon 1875–1941
French mathematician who is known chiefly for the development of a new theory of integration named after him.

Leblanc, Nicolas 1742–1806
French industrial chemist who devised the first commercial process for the manufacture of soda (sodium carbonate), which became the general method of making the chemical for a hundred years.

Le Châtelier, Henri Louis 1850–1936
French physical chemist, best known for the principle named after him which states that if any constraint is applied to a system in chemical equilibrium, the system tends to adjust itself to counteract or oppose the constraint.

Leclanché, Georges 1839–1882
French engineer who invented the Leclanché battery, or dry cell, in 1866.

Leeuwenhoek, Anton van 1632–1723
Dutch microscopist, famous for the numerous detailed observations he made using his single-lens microscopes.

Legendre, Adrien-Marie 1752–1833
French mathematician who was particularly interested in number theory, celestial mechanics, and elliptic functions.

Leibniz, Gottfried 1646–1716
German philosopher and mathematician who was one of the founders of the differential calculus and symbolic logic.

Lemaître, Georges Edouard
1894–1966
Belgian cosmologist who – perhaps because he was also a priest – was fascinated by the Creation, the beginning of the universe, for which he devised a version of what later became known as the 'Big Bang' theory.

Lenard, Philipp Eduard Anton
1862–1947
Hungarian-born German physicist who devised a way of producing beams of cathode rays (electrons) in air, enabling electrons to be studied.

Lenz, Heinrich Friedrich Emil 1804–1865
Russian physicist, known for Lenz's law, which is a fundamental law of electromagnetism.

Leverrier, Urbain Jean Joseph
1811–1877
French astronomer who, from perturbations in the orbit of Uranus, calculated the position of a new planet, Neptune, which Galle then discovered in 1846.

Levi-Civita, Tullio 1873–1941
Italian mathematician skilled in both pure and applied mathematics whose greatest achievement was his development, in collaboration with Gregorio Ricci-Curbastro, of the absolute differential calculus.

Lewis, Gilbert Newton 1875–1946
US theoretical chemist who made important contributions to thermodynamics and the electronic theory of valency.

Libby, Willard Frank 1908–1980
US chemist best known for developing the technique of radiocarbon dating, for which he was awarded the 1960 Nobel Prize for Chemistry.

Lie, (Marius) Sophus 1842–1899
Norwegian mathematician who made valuable contributions to the theory of algebraic invariants and who is remembered for the Lie theorem and the Lie groups.

Liebig, Justus 1803–1873
Baron von Liebig, German organic chemist, one of the greatest influences on 19th-century chemistry. He introduced the theory of compound radicals and discovered chloral and chloroform, and demonstrated the use of fertilizers.

Lindblad, Bertil 1895–1965
Swedish expert on stellar dynamics whose chief contribution to astronomy lay in his use of the work of Jacobus Kapteyn and Harlow Shapley to demonstrate the rotation of our Galaxy.

Lindemann, (Carl Louis) Ferdinand von
1852–1939
German mathematician who is famous for his proof that π (pi) is a transcendental number, which laid to rest the old question of 'squaring the circle'.

Lindemann, Frederick Alexander
1886–1957
Viscount Cherwell, English physicist who was involved with Hermann Nernst in advancing quantum theory.

Linnaeus, Carolus 1707–1778
Swedish Carl von Linné, Swedish botanist who became famous for introducing the binomial system of biological nomenclature (which is named after him and is universally used today), and for formulating basic principles for classification.

Linnett, John Wilfred 1913–1975
English chemist of wide-ranging interests,
from spectroscopy to reaction kinetics and
molecular structure.

Liouville, Joseph 1809–1882
French mathematician who wrote prolifically
on problems of analysis, but who is famous
chiefly as the founder and first editor of the
learned journal popularly known as the *Journal
de Liouville*.

Lipschitz, Rudolf Otto Sigismund
1832–1903
German mathematician of wide-ranging
interests who is remembered for the so-called
Lipschitz algebra and the Lipschitz condition.

Lipscomb, William Nunn 1919–
US chemist whose main interest is in the
relationships between the geometric and
electronic structures of molecules and their
chemical and physical behaviour.

Lissajous, Jules Antoine 1822–1880
French physicist who developed Lissajous
figures for demonstrating wave motion.

Lobachevsky, Nikolai Ivanovich 1792–1856
Russian mathematician, one of the founders
of non-Euclidean geometry, whose system is
sometimes called Lobachevskian geometry.

Lockyer, (Joseph) Norman 1836–1920
English scientist whose interests and studies
were wide-ranging, but who is remembered
mainly for his pioneering work in spectroscopy, through which he discovered the existence of helium, although it was not to be isolated
in the laboratory until nearly 30 years later.

Lodge, Oliver Joseph 1851–1940
English physicist who was among the pioneers
of radio.

Longuet-Higgins, Hugh Christopher
1923–
English theoretical chemist whose main
contributions have involved the application
of precise mathematical analyses, particularly
statistical mechanics, to chemical problems.

Lonsdale, Kathleen 1903–1971
born Kathleen Yardley, Irish-born British
X-ray crystallographer who rose from the
humblest of backgrounds to become one of
the best-known workers in her field, being
among the first to determine the structures
of organic molecules.

Lorentz, Hendrik Antoon 1853–1928

Dutch physicist who pioneered in the
development of the electron theory and in the
study of the electrodynamics of moving
bodies; awarded (jointly with his pupil Pieter
Zeeman) the 1902 Nobel Prize for Physics.

Lorenz, Konrad Zacharias 1903–1989
Austrian zoologist who is generally considered
to be the founder of modern ethology.

Lorenz, Ludwig Valentin 1829–1891
Danish mathematician and physicist who
made important contributions to our
knowledge of heat, electricity, and optics.

Lovell, (Alfred Charles) Bernard 1913–
English radio astronomer and author who was
instrumental in the setting up of the Jodrell
Bank Experimental Station in 1951.

Lovelock, James Ephraim 1919–
English scientist, specializing in the atmospheric sciences, including the study of
chlorofluorocarbons (CFCs), and who
introduced the concept of the Earth as a
single organism – the Gaia hypothesis.

Lowell, Percival 1855–1916
US astronomer and mathematician, the
founder of an important observatory in
the USA, whose main field of research was
the planets of the Solar System.

Lummer, Otto Richard 1860–1925
German physicist who specialized in optics
and is particularly remembered for his work
on thermal radiation.

Lwoff, André Michael 1902–1994
French microbiologist who was awarded the
1965 Nobel Prize for Physiology or Medicine
for his research into the genetic control of
enzyme activity.

Lyell, Charles 1797–1875
Scottish geologist champion of
uniformitarianism – the theory that
geological causes now operating were the
same, in kind and degree, as those in the past.

Lyman, Theodore 1874–1954
US physicist famous for his spectroscopic work
in the ultraviolet region.

Lynden-Bell, Donald 1935–
English astrophysicist particularly interested in
the structure and dynamics of galaxies.

Lyot, Bernard Ferdinand 1897–1952
French astronomer and an exceptionally
talented designer and constructor of optical
instruments.

377

Lysenko, Trofim Denisovich 1898–1976
Ukrainian botanist who dominated biology in the Soviet Union from about the mid-1930s to 1965.

Lyttleton, Raymond Arthur 1911–1995
English astronomer and theoretical physicist whose main interest was stellar evolution and composition, although he extended this in order to investigate the nature of the Solar System.

MacArthur, Robert Helmer 1930–1972
Canadian-born US ecologist who did much to change ecology from a descriptive discipline to a quantitative, predictive science.

McBain, James William 1882–1953
Canadian physical chemist whose main researches were concerned with colloidal solutions, particularly soap solutions.

McClintock, Barbara 1902–1992
US geneticist who discovered jumping genes (genes that can change their position on a chromosome from generation to generation), thereby offering an explanation of how originally identical cells take on specialized functions as skin, muscle, bone, and nerve, and also how evolution can give rise to the multiplicity of species.

Mach, Ernst 1838–1916
Austrian physicist and philosopher who studied motion faster than the speed of sound and published searching critiques of problems of scientific knowledge, particularly the use of unexamined hypotheses.

Maclaurin, Colin 1698–1746
Scottish mathematician who first presented the correct theory for distinguishing between the maximum and minimum values of a function and who played a leading part in establishing the hegemony of the Newtonian calculus in 18th-century Britain.

Maiman, Theodore Harold 1927–
US physicist who is best known for constructing, in 1960, the first working laser.

Malthus, Thomas Robert 1766–1834
English economist who made the first serious study of human population trends, although his views of the future of the human race enraged many thinkers of his day.

Malus, Etienne Louis 1775–1812
French physicist who discovered the polarization of light by reflection.

Mandelbrot, Benoit B 1924–
Polish-born French mathematician who coined the term 'fractal' to describe geometrical figures in which an identical motif repeats itself on an ever-diminishing scale.

Marconi, Guglielmo 1874–1937
Italian electrical engineer who saw the possibility of using radio waves – long-wavelength electromagnetic radiation – for the transmission of information.

Markov, Andrei Andreyevich 1856–1922
Russian mathematician, famous for his work on the probability calculus and for the Markov chains.

Martin, Archer John Porter 1910–
English biochemist who shared the 1952 Nobel Prize for Chemistry with his co-worker Richard Synge for their development of paper chromatography.

Maskelyne, Nevil 1732–1811
English astronomer, who was founder of the *Nautical Almanac* and became Astronomer Royal at age 32.

Maury, Antonia Caetana de Paiva Pereira 1866–1952
US expert in stellar spectroscopy who specialized in the detection of binary stars.

Maury, Matthew Fontaine 1806–1873
US naval officer, hydrographer, and pioneer of oceanography, whose system of recording oceanographic data is still in use.

Maxwell, James Clerk 1831–1879
Scottish physicist who theorized that light consists of electromagnetic waves and established the kinetic theory of gases.

Mayer, Christian 1719–1783
Austrian astronomer, mathematician, and physicist who was the first to investigate and catalogue double stars.

Mayer, Johann Tobias 1723–1762
German cartographer, astronomer, and physicist who did much to improve standards of observation and navigation, particularly through his improvements in lunar theory, which could be used to find longitude.

Mayer, Julius Robert von 1814–1878
German physicist who was among the first to formulate the principle of conservation of energy and determine the mechanical equivalent of heat.

Medawar, Peter Brian 1915–1987
British zoologist best known for his contributions to immunology, for which he shared the 1960 Nobel Prize for Physiology or Medicine with Frank Macfarlane Burnet.

Meitner, Lise 1878–1968
Austrian-born Swedish physicist who was one of the first scientists to study radioactive decay and the radiations emitted during this process.

Mellanby, Kenneth 1908–1993
English entomologist and ecologist best known for his work on the environmental effects of pollution.

Mendel, Gregor Johann 1822–1884
Austrian geneticist and monk who discovered the basic laws of heredity, thereby laying the foundation of modern genetics – although the importance of his work was not recognized until after his death.

Mendeleyev, Dmitri Ivanovich 1834–1907
Russian chemist whose name will always be linked with his outstanding achievement, the development of the periodic table.

Menzel, Donald Howard 1901–1976
US physicist and astronomer whose work on the spectrum of the solar chromosphere revolutionized much of solar astronomy.

Mercator, Gerardus 1512–1594
born Gerhard Kremer, Flemish mathematician, geographer, instrumentmaker, and cartographer.

Messier, Charles 1730–1817
French astronomer whose work on the discovery of comets led to a compilation of the locations of nebulae and star clusters – the Messier catalogue – that is still of some relevance 200 years later.

Metchnikoff, Elie 1845–1916
Russian Ilya Ilich Mechnikov, Russian-born French zoologist who discovered phagocytes, amoebalike blood cells that engulf foreign bodies.

Meyer, (Julius) Lothar 1830–1895
German chemist who, independently of Dmitri Mendeleyev, produced a periodic law describing the properties of the chemical elements.

Meyer, Viktor 1848–1897
German organic chemist best known for the method of determining vapour densities (and hence molecular masses) named after him.

Michelson, Albert Abraham 1852–1931
German-born US physicist who made precision measurements of the speed of light and, with Edward Morley, tried unsuccessfully to detect the motion of the Earth through the 'ether'.

Midgley, Thomas 1889–1944
US industrial chemist and engineer who discovered that tetraethyl lead is an efficient antiknock additive to petrol (preventing pre-ignition in car engines) and introduced Freons (a group of chlorofluorocarbons) as the working gases in domestic refrigerators.

Miller, Stanley Lloyd 1930–
US chemist who carried out a key experiment that demonstrated how amino acids, the building blocks of life, might have arisen in the primeval oceans of the primitive Earth.

Millikan, Robert Andrews 1868–1953
US physicist who made the first determination of the charge of the electron and of Planck's constant.

Mills, William Hobson 1873–1959
English organic chemist famous for his work on stereochemistry and on the synthesis of cyanine dyes.

Milne, Edward Arthur 1896–1950
English astrophysicist, mathematician, and theoreticist, most famous for his formulation of a theory of relativity parallel to Albert Einstein's general theory, which he called kinematic relativity.

Milstein, César 1927–
Argentine-born British molecular biologist who has performed important research on the genetics, biosynthesis, and chemistry of immunoglobulins (antibody proteins), developing a technique for preparing chemically pure monoclonal antibodies.

Minkowski, Hermann 1864–1909
Lithuanian-born German mathematician whose introduction of the concept of space-time was essential to the genesis of the general theory of relativity.

Minkowski, Rudolph Leo 1895–1976
German-born US astrophysicist, responsible for the compilation of the incomparably valuable set of photographs found in every astronomical library, the National Geographic Society Palomar Observatory Sky Survey.

Mises, Richard von 1883–1953
Austrian-born US mathematician and

379

aerodynamicist who made valuable contributions to statistics and the theory of probability.

Mitchell, Maria 1818–1889
US astronomer and the first woman to become professor of astronomy in the USA and the first woman appointed to the American Academy of Arts and Sciences.

Mitchell, Peter Dennis 1920–1992
English biochemist who performed important research into the processes involved in the transfer of biological energy.

Mitscherlich, Eilhard 1794–1863
German chemist who discovered isomorphism (the phenomenon in which substances of analogous chemical composition crystallize in the same crystal form).

Möbius, August Ferdinand 1790–1868
German mathematician and theoretical astronomer who worked on the 'barycentric calculus' and devised the one-sided surface now named after him.

Mohs, Friedrich 1773–1839
German mineralogist who devised the Mohs scale, by which minerals are characterized by relative hardness.

Moivre, Abraham de 1667–1754
French mathematician who, despite being persecuted for his religious faith and subsequently leading a somewhat unstable life, pioneered the development of analytical trigonometry – for which he formulated his theorem regarding complex numbers – devised a means of research into the theory of probability, and was a friend of some of the greatest scientists of his age.

Molina, Mario 1943–
Mexican chemist who shared the Nobel Prize for Chemistry in 1995 with Paul Crutzen and F Sherwood Rowland for their work in atmospheric chemistry, particularly concerning the formation and decomposition of ozone.

Mond, Ludwig 1839–1909
German-born British chemist who established many industrial chemical processes in the UK.

Monge, Gaspard 1746–1818
French mathematician and chemist who was famous for his work in descriptive and analytical geometry and is generally regarded as the founder of descriptive geometry.

Monod, Jacques Lucien 1910–1976
French biochemist who is best known for his research into the way in which genes regulate intracellular activities.

Morgan, Thomas Hunt 1866–1945
US geneticist and embryologist famous for his pioneering work on the genetics of the fruit fly *Drosophila melanogaster* – now extensively used in genetic research – and for establishing the chromosome theory of heredity.

Morley, Edward Williams 1838–1923
US physicist and chemist who is best known for his collaboration with Albert Michelson in the classic Michelson–Morley experiment that failed to show evidence of Earth's motion through the 'ether'.

Moseley, Henry Gwyn Jeffreys 1887–1915
English physicist who first established the atomic numbers of the elements by studying their X-ray spectra.

Mössbauer, Rudolf (Ludwig) 1929–
German physicist who was awarded (jointly with Robert Hofstadter) the 1961 Nobel Prize for Physics for his discovery of the Mössbauer effect.

Mott, Nevill Francis 1905–1996
English physicist whose work on semi-conductors won him the 1977 Nobel Prize for Physics.

Muir, John 1838–1914
Scottish-born US conservationist who headed the campaign that led to the establishment of Yosemite National Park.

Muir, Thomas 1844–1934
Scottish mathematician, famous for his monumental and pioneering work in unravelling the history of determinants.

Muller, Hermann Joseph 1890–1967
US geneticist famous for his discovery that genetic mutations can be artificially induced by means of X-rays, for which he was awarded the 1946 Nobel Prize for Physiology or Medicine.

Müller, Johannes (also known as Regiomontanus) 1436–1476
German astronomer who compiled astronomical tables, translated Ptolemy's *Almagest* from Greek into Latin, and assisted in the reform of the Julian Calendar.

Müller, Paul Hermann 1899–1965
Swiss chemist, known for his development

of DDT as an insecticide, for which he was awarded the 1948 Nobel Prize for Physiology or Medicine.

Murchison, Roderick 1792–1871
Scottish geologist who named and described the Silurian, Devonian, and Permian systems of Paleozoic rocks.

Nagell, Trygve 1895–1988
Norwegian mathematician whose most important work was in the fields of abstract algebra and number theory.

Napier, John 1550–1617
8th Laird of Merchiston, Scottish mathematician who invented logarithmic tables.

Natta, Giulio 1903–1979
Italian chemist who shared the 1963 Nobel Prize for Chemistry with Karl Ziegler for his work on the production of polymers.

Needham, Joseph 1900–1995
English biochemist, science historian and orientalist whose most important scientific contribution was in the field of biochemical embryology.

Nernst, (Walther) Hermann 1864–1941
German physical chemist who made basic contributions to electrochemistry and is probably best known as the discoverer of the third law of thermodynamics.

Neugebauer, Gerry 1932–
born Gerald Neugebauer, US astronomer whose work has been crucial in establishing infrared astronomy.

Newcomb, Simon 1835–1909
Canadian-born US mathematician and astronomer who compiled charts and tables of astronomical data with phenomenal accuracy.

Newlands, John Alexander Reina 1837–1898
English chemist who preceded Dmitri Mendeleyev in formulating the concept of periodicity in the properties of chemical elements, although his ideas were not accepted at the time.

Newton, Isaac 1642–1727
English physicist and mathematician who is regarded as one of the greatest scientists ever to have lived.

Nicol, William 1768–1851
Scottish physicist and geologist who is best

known for inventing the first device for obtaining plane-polarized light – the Nicol prism.

Nobel, Alfred Bernhard 1833–1896
Swedish industrial chemist and philanthropist who invented dynamite and endowed the Nobel Foundation, which after 1901 awarded the annual Nobel prizes.

Noether, Emmy 1882–1935
born Amalie Noether, German mathematician who became one of the leading figures in modern abstract algebra.

Noether, Max 1844–1921
German mathematician who contributed to the development of 19th-century algebraic geometry and the theory of algebraic functions.

Norrish, Ronald George Wreyford 1897–1978
English physical chemist who studied fast chemical reactions, particularly those initiated by light.

Nüsslein-Volhard, Christiane 1942–
German geneticist who shared the Nobel Prize for Physiology or Medicine in 1995 with US geneticists Edward Lewis and Eric Wieschaus for her work on genes controlling the early embryonic development of *Drosophila melanogaster*, the fruit fly.

Nyholm, Ronald Sydney 1917–1971
Australian inorganic chemist famous for his work on the coordination compounds (complexes) of the transition metals.

Ochoa, Severo 1905–1993
Spanish-born US biochemist who reproduced in the laboratory the way in which cells synthesize nucleic acids by their use of enzymes.

Odum, Eugene Pleasants 1913–
US ecologist and educator who became one of the founders of a rigorously scientific approach to ecology.

Oersted, Hans Christian 1777–1851
Danish physicist who discovered that an electric current deflects a magnetized needle.

Ohm, Georg Simon 1789–1854
German physicist who is remembered for Ohm's law, which relates the current flowing through a conductor to the potential difference and the resistance.

381

Olbers, Heinrich Wilhelm Matthäus
1758–1840
German doctor, mathematician, and astronomer who discovered two asteroids, formulated a new method for calculating the orbits of comets, and discussed why, if there are stars in all directions, the night sky is nonetheless dark.

Onsager, Lars 1903–1976
Norwegian-born US theoretical chemist who was awarded the 1968 Nobel Prize for Chemistry for his work on reversible processes.

Oort, Jan Hendrik 1900–1992
Dutch astrophysicist whose main area of research was the composition of galaxies.

Oparin, Alexandr Ivanovich 1894–1980
Russian biochemist who made important contributions to evolutionary biochemistry, developing one of the first of the modern theories about the origin of life on Earth.

Öpik, Ernst Julius 1893–1985
Estonian astronomer whose work on the nature of meteors and comets was instrumental in the development of heat-deflective surfaces for spacecraft on their re-entry into the Earth's atmosphere.

Oppenheimer, J(ulius) Robert 1904–1967
US physicist who contributed significantly to the growth of quantum mechanics and played a critical role in the rapid development of the first atomic bombs.

Oppolzer, Theodor Egon Ritter von
1841–1886
Austrian mathematician and astronomer whose interest in asteroids and comets and eclipses led to his compiling meticulous lists of such bodies and events for the use of other astronomers.

Ore, Oystein 1899–1968
Norwegian mathematician whose studies, researches, and publications concentrated on the fields of abstract algebra, number theory, and the theory of graphs.

Osborn, Henry Fairfield 1857–1935
US palaeontologist who did much to promote the acceptance of evolutionism in the USA.

Ostwald, (Friedrich) Wilhelm 1853–1932
Latvian-born German physical chemist famous for his contributions to solution chemistry and to colour science.

Owen, Richard 1804–1892
English anatomist, palaeontologist, and leading naturalist who coined the term 'dinosaur' meaning terrible lizard.

Paneth, Friedrich Adolf 1887–1958
Austrian chemist known for his contribution to the development of radiotracer techniques and to organic chemistry.

Papin, Denis 1647–c. 1712
French physicist and technologist who invented a vessel that was the forerunner of the pressure cooker or autoclave.

Paracelsus, Philippus Aureolus
1493–1541
adopted name of Theophrastus Bombastus von Hohenheim, Swiss physician and chemist whose works did much to overthrow the accepted scientific authorities of his day (such as Galen) and to establish the importance of chemistry in medicine.

Parsons, William 1800–1867
3rd Earl of Rosse, Irish politician, engineer, and astronomer who built a very large reflecting telescope which he used to study nebulae.

Pascal, Blaise 1623–1662
French mathematician, physicist, and religious recluse who devised a mechanical calculator, studied probability and geometry, and did important experiments on air pressure and the vacuum.

Pasteur, Louis 1822–1895
French chemist and microbiologist who became world famous for originating the process of pasteurization and for establishing the validity of the germ theory of disease, although he also made many other scientific contributions.

Pauli, Wolfgang 1900–1958
Austrian-born Swiss physicist who made a substantial contribution to quantum theory with the Pauli exclusion principle.

Pauling, Linus Carl 1901–1994
US theoretical chemist and biologist whose achievements ranked among the most important of any in 20th-century science. His work on the nature of chemical bonds is fundamental to modern theories of molecular structure and secured him the Nobel Prize for Chemistry in 1954. He was awarded the Peace prize in 1962 for his campaigns against nuclear weapons.

Payne-Gaposchkin, Cecilia Helena
1900–1979
English-born US astronomer and author whose interest in stellar evolution and galactic structure led to important research in the study of variable, binary, and eclipsing stars.

Peano, Giuseppe 1858–1932
Italian mathematician who applied the rigorous and axiomatic methods used in mathematics to his study of logic.

Pearson, Karl 1857–1936
English mathematician and biometrician who is chiefly remembered for his crucial role in the development of statistics as applied to a wide variety of scientific and social topics.

Peierls, Rudolf Ernst 1907–1995
German-born British physicist who made contributions to quantum theory and to nuclear physics.

Pelletier, Pierre-Joseph 1788–1842
French chemist whose extractions of a range of biologically active compounds from plants founded the chemistry of the alkaloids.

Pennington, Mary Engle 1872–1952
US chemist known chiefly for her pioneering work in food refrigeration.

Pennycuick, Colin James 1933–
English biologist who is best known for his extremely detailed studies of flight.

Penrose, Roger 1931–
English mathematician who, through his theoretical work, has made important contributions to the understanding of astrophysical phenomena.

Penzias, Arno Allan 1933–
German-born US radio engineer who shared the 1978 Nobel Prize for Physics with Robert Wilson for their discovery of cosmic microwave background radiation.

Peregrinus, Petrus born c. 1220
adopted name of Peregrinus de Maricourt, French scientist and scholar about whom little is known except through his seminal work – based largely on experiment – on magnetism.

Perkin, William Henry 1838–1907
English chemist who achieved international fame for his accidental discovery of mauve, the first aniline dye and the first commercially significant synthetic dyestuff.

Perrin, Jean Baptiste 1870–1942
French physicist who made the first demonstration of the existence of atoms.

Perutz, Max Ferdinand 1914–
Austrian-born British molecular biologist who shared the 1962 Nobel Prize for Chemistry for his solution of the structure of the haemoglobin molecule; his co-worker John Kendrew, who had determined the structure of myoglobin, was the other winner of the prize.

Petit, Alexis(-Thérèse) 1791–1820
French scientist who worked mainly in physics but whose collaboration with Pierre Dulong resulted in a discovery that was to play an important part in chemistry in the determination of atomic weights (relative atomic masses).

Pfeffer, Wilhelm Friedrich Philipp
1845–1920
German physiological botanist who is best known for his contributions to the study of osmotic pressures, which is important in both biology and chemistry.

Piazzi, Giuseppe 1749–1826
Italian monk, originally trained in theology, philosophy, and mathematics, who established an observatory at Palermo where he compiled a star catalogue and discovered the first asteroid Ceres.

Picard, (Charles) Emile 1856–1941
French mathematician whose work in analysis – and particularly in analytical geometry – brought him deserved fame.

Pickering, Edward Charles 1846–1919
US astronomer, one of the most famous and hard-working of his time, who was a pioneer in three practical areas of astronomical research: visual photometry, stellar spectroscopy, and stellar photography.

Pippard, (Alfred) Brian 1920–
English physicist who has carried out important work in superconductivity.

Planck, Max Karl Ernst Ludwig
1858–1947
German physicist who first theorized that energy consists of fundamental indivisible units, which he called quanta.

Plaskett, John Stanley 1865–1941
Canadian astronomer and engineer whose work in instrument design and telescope

383

construction led to his becoming director of the Dominion Astrophysical Observatory in Victoria, British Columbia.

Plato *c.* 420–340 BC
Greek mathematician and philosopher who founded an influential school of learning which emphasized striving to find mathematical and intellectual harmony rather than pursuing empirical information.

Pliny the Elder *c.* AD 23–79
born Gaius Plinius Secundus, Roman scientific encyclopedist and historian who surveyed all known sciences of his day.

Plücker, Julius 1801–1868
German mathematician and physicist who made fundamental contributions to the field of analytical geometry and was a pioneer in the investigations of cathode rays.

Poincaré, (Jules) Henri 1854–1912
French mathematician and prolific mathematical writer who developed the theory of differential equations and was a pioneer in relativity theory.

Poisson, Siméon-Denis 1781–1840
French mathematician and physicist who formulated the Poisson distribution in probability.

Polanyi, Michael 1891–1976
Hungarian-born British physical chemist, particularly noted for his contributions to reaction kinetics.

Pólya, George 1887–1985
Hungarian mathematician best known for his work on function theory, probability, and applied mathematics.

Poncelet, Jean-Victor 1788–1867
French military engineer who, to pass the time during two years as a prisoner of war, revised all the mathematics he could remember and went on to make fresh discoveries, particularly in projective geometry.

Pond, John 1767–1836
English astronomer whose meticulous observations at his private observatory led to his discovering errors in data published by the Royal Observatory in Greenwich.

Pons, Jean-Louis 1761–1831
French astronomer who, in a career that began at a comparatively late age, nevertheless discovered 37 comets and became director of the Florence Observatory.

Porter, George 1920–
Baron Porter of Luddenham, English physical chemist who developed the technique of flash photolysis for the direct study of extremely fast chemical reactions, for which achievement he shared the 1967 Nobel Prize for Chemistry with Ronald Norrish and Manfred Eigen.

Powell, Cecil Frank 1903–1969
English physicist who developed photographic techniques for studying subatomic particles and who discovered the pi-meson (pion).

Powell, John Wesley 1834–1902
US geologist who pioneered exploration of the Colorado River and the Grand Canyon.

Poynting, John Henry 1852–1914
English physicist, mathematician, and inventor who formulated an equation by which the rate of flow of electromagnetic energy (now called the Poynting vector) is determined by the product of the electric and magnetic field intensities, and also devised a new way to measure Isaac Newton's gravitational constant.

Prandtl, Ludwig 1875–1953
German physicist who put fluid mechanics on a sound theoretical basis.

Prelog, Vladimir 1906–1998
Bosnian-born Swiss organic chemist famous for his studies of alkaloids and antibiotics, and for his work on stereochemistry.

Prévost, Pierre 1751–1839
Swiss physicist who first showed that all bodies radiate heat, no matter how hot or cold they are.

Priestley, Joseph 1733–1804
English chemist and theologian who discovered what later came to be called oxygen in 1774. He also discovered nitric oxide and isolated ammonia.

Prigogine, Ilya 1917–
Viscount Prigogine, Russian-born Belgian theoretical chemist who was awarded the 1977 Nobel Prize for Chemistry for widening the scope of thermodynamics from the purely physical sciences to ecological and sociological studies.

Pringsheim, Ernst 1859–1917
German physicist whose experimental work on the nature of thermal radiation led directly to the quantum theory.

Proust, (Joseph) Louis 1754–1826
French chemist who discovered the law of constant composition, sometimes called Proust's law, which states that every true chemical compound has exactly the same composition no matter how it is prepared.

Prout, William 1785–1850
English chemist who pioneered physiological chemistry, but who is best known for formulating Prout's hypothesis, which states that the atomic weights (relative atomic masses) of all elements are exact multiples of the atomic weight of hydrogen.

Ptolemy, (Claudius Ptolemaeus) lived AD 2nd century
Egyptian astronomer, astrologer, geographer, and philosopher whose *Almagest* encapsulated ancient geocentric, astronomical theory.

Pye, John David 1932–
English zoologist who has performed important research in the field of ultrasonic bio-acoustics, particularly in bats.

Pyman, Frank Lee 1882–1944
English organic chemist, famous for his contributions to pharmaceutics and chemotherapy.

Pythagoras lived *c.* 530 BC
Greek mathematician and philosopher, part of whose mystic beliefs entailed an intense study of whole numbers, the effect of which he sought to find in the workings of nature.

Quetelet, (Lambert) Adolphe Jacques 1796–1874
Belgian statistician, astronomer, and social scientist who applied statistical reasoning to social phenomena and was influential to the course of European social science.

Raman, Chandrasekhara Venkata 1888–1970
Indian physicist who discovered that light is scattered by the molecules in a gas, liquid, or solid so as to cause a change in its wavelength.

Ramanujan, Srinavasa Ayengar 1887–1920
Indian mathematician who, virtually unaided and untaught, made original contributions to function theory and number theory.

Ramsay, William 1852–1919
Scottish chemist famous for his discovery of the inert gases, for which achievement he was awarded the 1904 Nobel Prize for Chemistry.

Rankine, William John Macquorn 1820–1872
Scottish engineer and physicist who was one of the founders of the science of thermodynamics, especially in reference to the theory of steam engines.

Ray, John 1627–1705
English naturalist whose plant and animal classifications were the first significant attempts to produce a systematic taxonomy based on a variety of structural characteristics, including internal anatomy.

Rayet, George Antoine Pons 1839–1906
French astronomer who, in collaboration with Charles Wolf, detected a new class of peculiar white or yellowish stars whose spectra contain broad hydrogen and helium emission lines.

Rayleigh, John William Strutt 1842–1919
3rd Baron Rayleigh, English physicist who greatly advanced the theory of sound and, with William Ramsay, discovered the element argon.

Reber, Grote 1911–
US radio astronomer – indeed, at one time he was probably the world's only radio astronomer – who may truly be said to have pioneered the new aspect of astronomical science from its inception.

Redman, Roderick Oliver 1905–1975
English astronomer who was chiefly interested in stellar spectroscopy and solar physics.

Regnault, Henri Victor 1810–1878
French physical chemist who is best known for his work on the physical properties of gases.

Ricci-Curbastro, Gregorio 1853–1925
Italian mathematician who is chiefly remembered for his systematization of absolute differential calculus (also now called the Ricci calculus), which later enabled Albert Einstein to write his gravitational equations, to express the principle of the conservation of energy, and thereby fully to derive the theory of relativity.

Richards, Theodore William 1868–1928
US chemist who gained worldwide fame for his extremely accurate determinations of atomic weights (relative atomic masses).

Richter, Burton 1931–
US experimental particle physicist who designed the Stanford Positron–Electron

Accelerating Ring, which he used to produce a new subatomic molecule, the ψ meson.

Riemann, (Georg Friedrich) Bernhard
1826–1866
German mathematician whose work in geometry – both in combining the results of others and in his own crucial and innovative research – developed that branch of mathematics to a large degree.

Ritter, Johann Wilhelm 1776–1810
German physicist who carried out early work on electrolytic cells and who discovered ultraviolet radiation.

Robinson, Robert 1886–1975
English organic chemist who, during a long and distinguished career, made many contributions to the science.

Romer, Alfred Sherwood 1894–1973
US palaeontologist and comparative anatomist who is best known for his influential studies of vertebrate evolution.

Römer, Ole (or Olaus) Christensen
1644–1710
Danish astronomer and civil servant who, through his observations of the eclipses of the moons of Jupiter, first derived a value for the speed of light.

Röntgen, Wilhelm Konrad 1845–1923
German physicist who discovered X-rays.

Roux, Wilhelm 1850–1924
German anatomist and zoologist who is famous for his work on developmental mechanics in embryology.

Rowland, F Sherwood 1927–
US chemist who shared the Nobel Prize for Chemistry in 1995 with Mario Molina and Paul Crutzen for their work in atmospheric chemistry, particularly concerning the formation and decomposition of ozone.

Rowland, Henry Augustus 1848–1901
US physicist who is best known for the development of the concave diffraction grating, which heralded a new era in the analysis of spectra.

Ruffini, Paolo 1765–1822
Italian mathematician, philosopher, and doctor who made valuable contributions in all three disciplines.

Rumford, Benjamin Thompson 1753–1814
Count von Rumford, American-born British physicist who first demonstrated conclusively that heat is not a fluid but a form of motion.

Russell, Bertrand Arthur William
1872–1970
3rd Earl Russell, British philosopher and mathematician who, during a long and active life, made many contributions to mathematics and wrote about morals and politics, but is best remembered as one of the founders of modern logic.

Russell, Frederick Stratten 1897–1984
English marine biologist best known for his studies of the life histories and distribution of plankton.

Russell, Henry Norris 1877–1957
US astronomer who was chiefly interested in the nature of binary stars, but who is best remembered for his publication in 1913 of a diagram charting the absolute magnitude of stars plotted against their spectral type.

Rutherford, Ernest 1871–1937
1st Baron Rutherford of Nelson, New Zealand-born British physicist who first explained that radioactivity is produced by the disintegration of atoms and discovered that alpha particles consist of helium nuclei.

Rutherfurd, Lewis Morris 1816–1892
US spectroscopist and celestial photographer.

Rydberg, Johannes Robert 1854–1919
Swedish physicist who discovered a mathematical expression that gives the frequencies of spectral lines for elements.

Ryle, Martin 1918–1984
English astronomer who developed the technique of sky-mapping using 'aperture synthesis', combining smaller dish aerials to give the characteristics of one large one.

Sabatier, Paul 1854–1941
French organic chemist who investigated the actions of catalysts in gaseous reactions.

Sabine, Edward 1788–1883
Irish geophysicist who made important studies of terrestrial magnetism.

Sagan, Carl Edward 1934–1996
US astronomer and popularizer of astronomy whose main research was on planetary atmospheres, including that of the primordial Earth.

Sakharov, Andrei Dmitrievich 1921–1989
Russian physicist who led the development of Soviet thermonuclear weapons; later campaigned for peace and political reform.

Salam, Abdus 1926–1996
Pakistani theoretical physicist who shared the 1979 Nobel Prize for Physics for jointly developing a unified theory of weak and electromagnetic interactions.

Sanger, Frederick 1918–
English biochemist who worked out the sequence of amino acids in various protein molecules.

Scheele, Karl Wilhelm 1742–1786
Swedish chemist who isolated many elements and compounds for the first time, including what later came to be known as oxygen and chlorine.

Scheiner, Christoph 1573–1650
German astronomer who carried out one of the earliest and most meticulous studies of sunspots and who made significant improvements to the helioscope and the telescope.

Schiaparelli, Giovanni Virginio 1835–1910
Italian astronomer who carried out significant research into the nature of comets and the inner planets of the Solar System.

Schleiden, Matthias Jakob 1804–1881
German botanist who, with Theodor Schwann, is best known for the establishment of the cell theory.

Schmidt, Bernhard Voldemar 1879–1935
Estonian lens- and mirror-maker who devised a special sort of lens to work in conjunction with a spherical mirror in a reflecting telescope.

Schoenheimer, Rudolf 1898–1941
German-born US biochemist who first used isotopes as tracers to study biochemical processes.

Schonbein, Christian 1799–1868
German chemist who discovered ozone and also nitrocellulose or gun cotton.

Schrieffer, John Robert 1931–
US physicist who, with John Bardeen and Leon Cooper, was awarded the Nobel Prize for Physics in 1972 for developing the first satisfactory theory of superconductivity.

Schrödinger, Erwin 1887–1961
Austrian physicist who founded wave mechanics with the formulation of the Schrödinger wave equation to describe the behaviour of electrons in atoms.

Schwabe, Samuel Heinrich 1789–1875
German chemist and astronomer who was the first person to measure the periodicity of the sunspot cycle.

Schwann, Theodor 1810–1882
German physiologist who, with Matthias Schleiden, is credited with formulating the cell theory, one of the most fundamental of all concepts in biology.

Schwarzschild, Martin 1912–1997
German-born US astronomer whose most important work was in the field of stellar structure and evolution.

Seaborg, Glenn Theodore 1912–1999
US physical chemist who was best known for his researches on the synthetic transuranic elements.

Secchi, Pietro Angelo 1818–1878
Italian astronomer and physicist famous for his work on solar phenomena, stellar spectroscopy, and spectral classification.

Sedgwick, Adam 1785–1873
English geologist who contributed greatly to the understanding of the stratigraphy of the British Isles.

Segrè, Emilio Gino 1905–1989
Italian physicist who named and described the Cambrian and Devonian systems of Paleozoic rocks.

Seki Kowa c.1642–1708
or Seki Takakazu, Japanese mathematician who revised and extended Chinese mathematical techniques to solve problems in algebra and to find an approximate value of π (pi).

Semenov, Nikolai Nikolaevich 1896–1986
Russian physical chemist who studied chemical chain reactions, particularly branched-chain reactions that can accelerate with explosive velocity.

Seyfert, Carl Keenan 1911–1960
US astronomer and astrophysicist whose interests in photometry, the spectra of stars and galaxies, and the structure of the Milky Way resulted in the identification and study of the type of galaxy that now bears his name.

Shannon, Claude Elwood 1916–
US mathematical engineer whose work on technical and engineering problems within the communications industry led him to fundamental considerations on the nature of information and its meaningful transmission.

387

Shapley, Harlow 1885–1972
US astronomer who made what Otto von Struve called 'the most significant single contribution toward our understanding of the physical characteristics of the very close double stars'.

Shaw, (William) Napier 1854–1945
English meteorologist who, in the late 1800s and early 1900s, did much to establish the then young science of meteorology.

Shockley, William Bradford 1910–1989
US physicist who was awarded the Nobel Prize for Physics in 1956, for his work on semiconductors and his discovery of the transistor. He shared the award with John Bardeen and Walter Brattain.

Sidgwick, Nevil Vincent 1873–1952
English theoretical chemist best known for his contributions to the theory of valency and chemical bonding.

Simon, Franz Eugen 1893–1956
German-born British physicist who developed methods of achieving extremely low temperatures and who also established the validity of the third law of thermodynamics.

Simpson, George Clark 1878–1965
English meteorologist who studied atmospheric electricity and the effect of radiation on the polar ice, and standardized the Beaufort scale of wind speed.

Simpson, George Gaylord 1902–1984
US palaeontologist who studied the evolution of mammals and applied population genetics to the subject and to the migrations of animals between continents.

Simpson, Thomas 1710–1761
English mathematician who wrote influential textbooks on many mathematical topics, including Newton's fluxions and the laws of chance.

Skolem, Thoralf Albert 1887–1963
Norwegian mathematician who did important work on Diophantine equations and who helped to provide the axiomatic foundations for set theory in logic.

Slipher, Vesto Melvin 1875–1969
US astronomer whose important work in spectroscopy increased our knowledge of the universe and paved the way for some of the most important results obtained in more recent astrophysics.

Smalley, Richard E(rrett) 1943–
US chemist who, with his colleagues US chemist Robert Curl and English chemist Harold Kroto, discovered buckminster-fullerene C_{60} in 1985.

Smith, Francis Graham 1923–
English astronomer, one of the leaders in radio astronomy since its earliest post-war days.

Smith, William 1769–1839
called 'Strata Smith', English geologist who was one of the first to use characteristic fossils to identify and order rock formations.

Snel, Willebrord van Roijen 1580–1626
Dutch physicist who discovered the law of refraction.

Soddy, Frederick 1877–1956
English chemist who was responsible for major advances in the early developments of radiochemistry, being mainly concerned with radioactive decay and the study of isotopes.

Solvay, Ernest 1838–1922
Belgian industrial chemist who invented the ammonia–soda process, also known as the Solvay process, for making the alkali sodium carbonate (soda).

Somerville, Mary Greig 1780–1872
born Mary Fairfax, Scottish astronomer and mathematician who popularized astronomy and wrote several textbooks.

Sommerfeld, Arnold Johannes Wilhelm 1868–1951
German physicist who made an important contribution to the development of the quantum theory of atomic structure.

Sommerville, Duncan MacLaren Young 1879–1934
Scottish mathematician who made significant contributions to the study of non-Euclidean geometry.

Sorby, Henry Clifton 1826–1908
English geologist who made huge advances in the field of petrology.

South, James 1785–1867
English astronomer noted for the observatory that he founded and his observations of double stars.

Spallanzani, Lazzaro 1729–1799
Italian physiologist who is famous for disproving the theory of spontaneous generation.

Spemann, Hans 1869–1941
German embryologist who discovered the phenomenon now called embryonic induction – the influence exerted by various regions of an embryo that controls the subsequent development of cells into specific organs and tissues.

Spitzer, Lyman 1914–1997
US astrophysicist who made important contributions to cosmogony (the study of the origin and evolution of stars and planetary systems).

Stahl, Georg Ernst 1660–1734
German chemist and physician who founded the phlogiston theory of combustion.

Stark, Johannes 1874–1957
German physicist who is known for his discovery of the phenomenon (now called the Stark effect) of the division of spectral lines in an electric field, and for his discovery of the Doppler effect in canal rays (high-velocity rays of positively charged ions).

Stas, Jean Servais 1813–1891
Belgian analytical chemist who is remembered for making the first accurate determinations of atomic weights (relative atomic masses).

Staudinger, Hermann 1881–1965
German organic chemist who pioneered polymer chemistry.

Stebbins, Joel 1878–1966
US astronomer, the first to develop the technique of electric photometry in the study of stars.

Steele, Edward John 1948–
Australian immunologist whose research into the inheritance of immunity has lent a certain amount of support to the Lamarckian theory of the inheritance of acquired characteristics, thus challenging modern theories of heredity and evolution.

Stefan, Josef 1835–1893
Austrian physicist who first determined the relationship between the amount of energy radiated by a body and its temperature.

Steiner, Jakob 1796–1863
Swiss mathematician, a leading geometrician of the 19th century and the founder of modern synthetic, or projective, geometry.

Steno, Nicolaus 1638–1686
Latinized form of Niels Stensen, Danish naturalist widely regarded as one of the founders of stratigraphy.

Stern, Otto 1888–1969
German-born US physicist who showed that beams of atoms and molecules have wave properties.

Stevens, Nettie Maria 1861–1912
US biologist whose researches concentrated on the role of chromosomes and their relationship to heredity.

Stieltjes, Thomas Jan 1856–1894
Dutch-born French mathematician who contributed greatly to the theory of series and is often called the founder of analytical theory.

Stock, Alfred 1876–1946
German inorganic chemist best known for his preparations of the hydrides of boron (called boranes) and for his campaign for better safety measures in the use of mercury in chemistry and industry.

Stokes, George Gabriel 1819–1903
Irish physicist who did important experimental and mathematical work in optics, thermodynamics, and geodesy.

Strömgren, Bengt Georg Daniel 1908–1987
Swedish astronomer best known for his hypothesis about the so-called 'Strömgren spheres' – zones of ionized hydrogen gas surrounding hot stars embedded in interstellar gas clouds.

Struve, F(riedrich) G(eorg) W(ilhelm) von 1793–1864
German-born Russian astronomer who was an expert on double stars and one of the first astronomers to measure stellar parallax.

Struve, (Gustav Wilhelm) Ludwig (Ottovich) von 1858–1920
Russian astronomer, the younger brother of Hermann von Struve and son of Otto Wilhelm von Struve, who was an expert on the occultation of stars and stellar motion.

Struve, (Karl) Hermann (Ottovich) von 1854–1920
Russian astronomer, third in the line of famous astronomers, who was an expert on Saturn.

Struve, Otto von 1897–1963
Russian-born US astronomer, the last of four generations of a family of eminent astronomers. He contributed to many areas of stellar astronomy but was best known for his work on interstellar matter and stellar and nebular spectroscopy.

389

Struve, Otto Wilhelm von 1819–1905
Russian astronomer, an active collaborator with his father, F G W von Struve, in many astronomical and geodetic investigations.

Sturgeon, William 1783–1850
English physicist and inventor who made the first electromagnets.

Sturtevant, Alfred Henry 1891–1970
US geneticist who was the first to map the positions of genes on a chromosome.

Suess, Eduard 1831–1914
Austrian geologist who advocated catastrophism and introduced the nappe theory of mountain building.

Sutherland, Gordon Brims Black McIvor
1907–1980
Scottish physicist who is best known for his work in infrared spectroscopy, particularly the use of this technique for studying molecular structure.

Sutton-Pringle, John William 1912–1982
British zoologist best known for his studies of insect flight.

Svedberg, Theodor 1884–1971
Swedish physical chemist who invented the ultracentrifuge to facilitate his work on colloids.

Swammerdam, Jan 1637–1680
Dutch naturalist who investigated many aspects of biology but who is probably best known for his outstanding microscope observations, his detailed and accurate anatomical descriptions, and his studies of insects.

Swings, Pol(idore) F F 1906–1983
Belgian astrophysicist with a particular interest in cometary spectroscopy.

Sylow, Ludwig Mejdell 1832–1918
Norwegian mathematician who is remembered for his fundamental theorem on groups and for the special type of subgroups that are named after him.

Sylvester, James Joseph 1814–1897
English mathematician, one of the pre-eminent algebraists of the 19th century and the discoverer, with Arthur Cayley, of the theory of algebraic invariants.

Synge, Richard Laurence Millington
1914–1994
English biochemist who carried out research into methods of isolating and analysing proteins and related substances.

Szent-Györgyi, Albert von Nagyrapolt
1893–1986
Hungarian-born US biochemist who studied the physiology of muscle contraction and carried out research into cancer.

Szilard, Leo 1898–1964
Hungarian-born US physicist and one of the 20th century's most original minds. He was one of the first to realize that nuclear fission could lead to a chain reaction releasing enormous amounts of energy.

Tabor, David 1913–
English physicist who has worked mainly in tribology, the study of the effects between solid surfaces.

Tansley, Arthur George 1871–1955
English botanist who was a pioneer in the science of plant ecology.

Teller, Edward 1908–
Hungarian-born US physicist widely known as the 'father of the hydrogen bomb'. He worked on the Manhattan Project developing the first atomic bomb 1942–46, and on the hydrogen bomb 1946–52.

Tesla, Nikola 1856–1943
Croatian-born US physicist and electrical engineer who was one of the great pioneers of the use of alternating-current electricity.

Thales c. 624–c. 547 BC
also known as Thales of Miletus, Greek philosopher who was among the first early Greek philosophers to reject mythopoetic forms of thought for a basically scientific approach to the world.

Theophrastus c. 372–c. 287 BC
Greek thinker who wrote on a wide range of subjects – science, philosophy, law, literature, music, and poetry, for example – but who is best known as the founder of botany.

Thom, René Frédéric 1923–
French mathematician who is a leading specialist in the fields of differentiable manifolds and topology and is famous for his model popularly known as 'catastrophe theory'.

Thompson, D'Arcy Wentworth
1860–1948
Scottish biologist and classical scholar who is best known for his book *On Growth and Form*, first published in 1917.

Thomson, George Paget 1892–1975
English physicist who in 1927 demonstrated electron diffraction, a confirmation of the wave nature of the electron.

Thomson, James 1822–1892
Irish physicist and engineer who discovered that the melting point of ice decreases with pressure.

Thomson, J(oseph) J(ohn) 1856–1940
English physicist who is famous for discovering the electron and for his research into the conduction of electricity through gases, for which he was awarded the 1906 Nobel Prize for Physics.

Tinbergen, Niko(laas) 1907–1988
Dutch-born British zoologist who studied the courtship behaviour of sticklebacks and the social behaviour of gulls.

Todd, Alexander Robertus 1907–1997
Baron Todd, Scottish chemist who made outstanding contributions to the study of natural substances.

Tolansky, Samuel 1907–1973
English physicist who made important contributions to the fields of spectroscopy and interferometry.

Tombaugh, Clyde William 1906–1997
US astronomer whose painstaking work led to his discovery of Pluto.

Tomonaga, Sin-Itiro 1906–1979
Japanese theoretical physicist who was one of the first to develop a consistent theory of relativistic quantum electrodynamics.

Torricelli, Evangelista 1608–1647
Italian physicist and mathematician who is best known for his invention of the barometer.

Townes, Charles Hard 1915–
US physicist who is best known for his investigations into the theory, and subsequent invention in 1953, of the maser.

Townsend, John Sealy Edward 1868–1957
Irish mathematical physicist who was responsible for the development of the study of the kinetics of electrons and ions in gases.

Travers, Morris William 1872–1961
English chemist famous for his association with William Ramsay on the discovery of the inert, or noble, gases.

Trumpler, Robert Julius 1886–1956
Swiss-born US astronomer who is known for his studies and classification of star clusters found in our Galaxy.

Tsiolkovsky (or Tsiolkovskii), Konstantin Eduardovich 1857–1935
Russian theoretician and one of the pioneers of space rocketry, hailed by his country as the 'father of Soviet cosmonautics'.

Tswett, Mikhail Semyonovich 1872–1919
Italian-born Russian scientist who made an extensive study of plant pigments and developed the technique of chromatography to separate them.

Turing, Alan Mathison 1912–1954
English mathematician who worked in numerical analysis and played a major part in the early development of British computers.

Turner, Charles Henry 1867–1923
US biologist and teacher who was internationally known for his research into insect behaviour patterns.

Twort, Frederick William 1877–1950
English bacteriologist, the original discoverer of bacteriophages (often called phages), the relatively large viruses that attack and destroy bacteria.

Tyndall, John 1820–1893
Irish physicist who studied diamagnetism, the response of vapours to light and heat, and the propagation of sound; he was a great popularizer of science.

Ulugh Beg 1394–1449
born Muhammad Taragay (Turkish 'great prince'), Mongol mathematician and astronomer, ruler of Samarkand from 1409 and of the Mongol Empire from 1447.

Urey, Harold Clayton 1893–1981
US chemist who in 1931 discovered heavy water and deuterium, the isotope of hydrogen of mass 2.

Van Allen, James Alfred 1914–
US physicist who was closely involved with the early development of the US space programme and discovered the magnetosphere, the zone of high levels of radiation around the Earth caused by the presence of trapped charged particles.

van de Graaff, Robert Jemison 1901–1967
US physicist who designed and built the electrostatic high-voltage generator named after him.

Vandermonde, Alexandre-Théophile 1735–1796
French musician and musical theorist who

wrote original and influential papers on
algebraic equations and determinants.

van der Waals, Johannes Diderik
1837–1923
Dutch physicist who was awarded the Nobel
Prize for Physics in 1910 for his research on
the gaseous and liquid states of matter.

van't Hoff, Jacobus Henricus 1852–1911
Dutch theoretical chemist who made major
contributions to stereochemistry, reaction
kinetics, thermodynamics, and the theory of
solutions.

van Vleck, John Hasbrouck 1899–1980
US physicist who made important
contributions to our knowledge of magnetism,
as a result of which he is widely considered to
be one of the founders of modern magnetic
theory.

Vauquelin, Louis Nicolas 1763–1829
French chemist who worked mainly in the
inorganic field analysing minerals and is best
known for his discoveries of chromium and
beryllium.

Vening Meinesz, Felix Andries
1887–1966
Dutch geophysicist who originated the method
of making very precise gravity measurements
in the stable environment of a submarine.

Venn, John 1834–1923
English logician whose diagram, known as the
Venn diagram, is much used in the teaching
of elementary mathematics.

Vesalius, Andreas 1514–1564
Belgian physician who was a founder of
modern anatomy. He produced the first real
anatomy textbook, with his meticulous
illustrations drawn from postmortem
dissections.

Viète, François 1540–1603
French mathematician, the first to use letters
of the alphabet extensively to represent
numerical quantities and the foremost
algebraist of the 16th century.

Vogel, Hermann Carl 1841–1907
German astronomer who became the first
director of the Potsdam Astrophysical
Observatory.

Volhard, Jacob 1834–1910
German chemist who is best remembered for
various significant methods of organic
synthesis.

**Volta, Alessandro Giuseppe Antonio
Anastasio** 1745–1827
Count Volta, Italian physicist who discovered
how to produce electric current and built the
first electric battery.

Volterra, Vito 1860–1940
Italian mathematician whose chief work was
in the fields of function theory and differential
equations.

von Braun, Wernher Magnus Maximilian
1912–1977
German-born US rocket engineer who was
instrumental in the design and development of
German rocket weapons during World War II
and who, after the war, was a prime mover in
The early days of space rocketry in the USA.

von Neumann, John (or **Johann**)
1903–1957
Hungarian-born US physicist and
mathematician who originated games theory
and developed the fundamental concepts
involved in programming computers.

**Vorontsov-Vel'iaminov, Boris
Aleksandrovich** 1904–
Russian astronomer and astrophysicist who
demonstrated the occurrence of the
absorption of stellar light by interstellar
dust.

Vries, Hugo (Marie) de 1848–1935
Dutch botanist and geneticist who is best
known for his rediscovery (simultaneously
with Karl Correns and Erich Tschermak von
Seysenegg) of Gregor Mendel's laws of
heredity and for his studies of mutation.

Wallace, Alfred Russel 1823–1913
Welsh naturalist who is best known for
proposing a theory of evolution by natural
selection independently of Charles Darwin.

Wallis, John 1616–1703
English mathematician who made important
contributions to the development of algebra
and analytical geometry and who was one of
the founders of the Royal Society.

Walton, Ernest Thomas Sinton
1903–1995
Irish physicist best known for his work with
John Cockcroft on the development of the
first particle accelerator, which produced the
first artificial transmutation in 1932.

Warburg, Otto Heinrich 1883–1970
German biochemist who made several
important discoveries about metabolic

processes, particularly intracellular respiration and photosynthesis, and pioneered the use of physicochemical methods for investigating the biochemistry of cells.

Warming, Johannes Eugenius Bülow 1841–1924
Danish botanist whose pioneering studies of the relationships between plants and their natural environments established plant ecology as a new discipline within botany.

Waterston, John James 1811–1883
Scottish physicist who first formulated the essential features of the kinetic theory of gases.

Watson, James Dewey 1928–
US geneticist who, with Crick and Wilkins discovered the molecular structure of DNA, for which they shared the Nobel Prize for Physiology or Medicine in 1962.

Watson-Watt, Robert Alexander 1892–1973
Scottish physicist and engineer who was largely responsible for the early development of radar.

Weber, Heinrich 1842–1913
German mathematician whose chief work was in the fields of algebra and number theory.

Weber, Wilhelm Eduard 1804–1891
German physicist who made important advances in the measurement of electricity and magnetism by devising sensitive instruments and defining electric and magnetic units.

Wedderburn, Joseph Henry Maclagan 1882–1948
Scottish mathematician who opened new lines of thought in the subject of mathematical fields and who had a deep influence on the development of modern algebra.

Wegener, Alfred Lothar 1880–1930
German metereologist and geologist who introduced and developed the theory of continental drift.

Weierstrass, Karl Theodor Wilhelm 1815–1897
German mathematician who is remembered especially for deepening and broadening the understanding of functions.

Weil, André 1906–1998
French mathematician whose main fields of activity were number theory, group theory, and algebraic geometry.

Weinberg, Steven 1933–
US theoretical physicist best known for developing the unified electroweak theory.

Weismann, August Friedrich Leopold 1834–1914
German zoologist who is best known for his 'germ plasm' theory of heredity and for his opposition to Jean Baptiste Lamarck's doctrine of the inheritance of acquired characteristics.

Weizsäcker, Carl Friedrich von 1912–
German theoretical physicist who has made fundamental contributions to astronomy by investigating the way in which energy is generated in the cores of stars.

Welsbach, Carl Auer, Baron von Welsbach 1858–1929
Austrian chemist and engineer who discovered two rare-earth elements and invented the incandescent gas mantle.

Werner, Abraham Gottlob 1749–1817
German minerologist and geologist who developed the first influential paradigms of Earth structure and history.

Werner, Alfred 1866–1919
French-born Swiss chemist who founded the modern theory of coordination bonding in molecules (formerly inorganic coordination compounds were known by the generic term 'complexes').

Weyl, Hermann 1885–1955
German mathematician and mathematical physicist whose range of research and interests was remarkably wide, and found expression also in published works on philosophy, logic, and the history of mathematics.

Wheatstone, Charles 1802–1875
English physicist who was among the inventors of the electric telegraph; he also popularized the use of the Wheatstone bridge for measuring electrical resistance.

Whipple, George Hoyt 1878–1976
US pathologist, physician, and physiologist whose work on the formation of haemoglobin led to a cure for pernicious anaemia.

White, Gilbert 1720–1793
English naturalist and cleric who is remembered chiefly for his book *The Natural History and Antiquities of Selborne* (1789), a classic work in which White vividly records his acute

observations of the flora and fauna in the area of Selborne (now in Hampshire).

Whitehead, Alfred North 1861–1947
English mathematician and philosopher whose research in mathematics involved a highly original attempt – incorporating the principles of logic – to create an extension of ordinary algebra to universal algebra, a meticulous re-examination of the relativity theory of Albert Einstein, and, in work carried out together with Bertrand Russell, the production of *Principia Mathematica*.

Whitehead, John Henry Constantine
1904–1960
English mathematician who achieved eminence in the more abstract areas of diffential geometry, and of algebraic and geometrical topology.

Wieland, Heinrich Otto 1877–1957
German organic chemist, particularly noted for his work in determining the structures of steroids and related compounds.

Wiener, Norbert 1894–1964
US mathematician who worked on harmonic analysis and communication theory; founded the field of cybernetics.

Wigglesworth, Vincent Brian 1899–1994
English entomologist whose research covered many areas of insect physiology but who is best known for his investigations into the role of hormones in growth and metamorphosis.

Wilkes, Maurice Vincent 1913–
English mathematician who led the team at Cambridge University that built the EDSAC (electronic delay storage automatic calculator), one of the earliest of the British electronic computers.

Wilkins, Maurice Hugh Frederick 1916–
New Zealand-born British biophysicist who contributed to the discovery of the structure of DNA.

Wilkinson, Geoffrey 1921–1996
English inorganic chemist who shared the Nobel Prize for Chemistry in 1973 for his pioneering work on the organometallic compounds of the transition metals with Ernst Fischer.

Wilks, Samuel Stanley 1906–1964
US statistician whose work in data analysis enabled him to formulate methods of deriving valid information from small samples.

Williamson, Alexander William 1824–1904
British organic chemist who made significant discoveries concerning alcohols and ethers, catalysis, and reversible reactions.

Willstätter, Richard 1872–1942
German organic chemist best known for his investigations of alkaloids and plant pigments, such as chlorophyll, for which he was awarded the 1915 Nobel Prize for Chemistry.

Wilson, Charles Thomson Rees
1869–1959
Scottish physicist who invented the Wilson cloud chamber, the first instrument to detect the tracks of atomic particles.

Wilson, Edward Osborne 1929–
US biologist and a leading authority on ants who has also pioneered sociobiology.

Wilson, John Tuzo 1908–1993
Canadian geologist and geophysicist who established and brought about a general understanding of the concept of plate tectonics.

Wilson, Robert Woodrow 1936–
US radio astronomer who, with Arno Penzias, detected the cosmic microwave background radiation, which is thought to represent a residue of the primordial 'Big Bang' with which the universe is believed to have begun.

Withering, William 1741–1799
English physician, botanist, and mineralogist, best known for his work on the drug digitalis (from the foxglove plant), which he initially used as a diuretic to treat dropsy (oedema).

Wittig, Georg 1897–1987
German chemist, best known for his method of synthesizing olefins (alkenes) from carbonyl compounds, a reaction often termed the Wittig synthesis.

Wöhler, Friedrich 1800–1882
German chemist who is generally credited with having carried out the first laboratory synthesis of an organic compound, although his main interest was inorganic chemistry.

Wolf, Maximilian Franz Joseph Cornelius
1863–1932
German astronomer particularly noted for his application of photographic methods to observational astronomy.

Wollaston, William Hyde 1766–1828
English chemist and physicist who developed the technique of powder metallurgy and discovered rhodium and palladium, two elements similar to platinum.

Woodger, Joseph Henry 1894–1981
English biologist known principally for his theoretical work on the underlying philosophical basis of scientific methodology in biology, especially for his attempt to provide biology with a strict and logical foundation on which observations, theories, and methods could be based.

Woodward, Robert Burns 1917–1979
US organic chemist famous for his syntheses of complex biochemicals, including vitamin B_{12}.

Wright, Thomas 1711–1786
English astronomer and teacher who published speculative theories about the structure of the universe.

Wurtz, Charles Adolphe 1817–1884
French organic chemist, best known for his synthetic reactions and for discovering ethylamine and ethylene glycol (1,2-ethanediol).

Young, Charles Augustus 1834–1908
US astronomer who made some of the first spectroscopic investigations of the Sun.

Young, J(ohn) Z(achary) 1907–1997
English zoologist whose discovery of and subsequent work on the giant nerve fibres in squids contributed greatly to knowledge of nerve structure and function.

Young, Thomas 1773–1829
English physicist and physician who discovered the principle of interference of light, showing it to be caused by light waves.

Yukawa, Hideki 1907–1981
Japanese physicist famous for his important theoretical work on elementary particles and nuclear forces, particularly for predicting the existence of the pi-meson (or pion) and the short-range strong nuclear force associated with this particle.

Zassenhaus, Hans Julius 1912–1991
German mathematician whose main area of study lay in the fields of group theory and number theory.

Zeeman, (Erik) Christopher 1925–
English mathematician noted for his work in topology and for his research into models of social behaviour that accord with the relatively recent formulation of catastrophe theory.

Zeeman, Pieter 1865–1943
Dutch physicist who discovered the Zeeman effect, which is the splitting of spectral lines in an intense magnetic field.

Zel'dovich, Yakov Borisovich 1914–1987
Soviet astrophysicist who was originally a specialist in nuclear physics, but who became interested in particle physics and cosmology during the 1950s.

Zermelo, Ernst Friedrich Ferdinand 1871–1953
German mathematician who made many important contributions to the development of set theory, particularly in developing the axiomatic set theory that now bears his name.

Ziegler, Karl 1898–1973
German organic chemist famous for his studies of polymers, for which he shared the 1963 Nobel Prize for Chemistry with Giulio Natta.

Zsigmondy, Richard Adolf 1865–1929
Austrian-born German colloid chemist who invented the ultramicroscope.

Zwicky, Fritz 1898–1974
Swiss astronomer and astrophysicist who was distinguished for his discoveries of supernovae, dwarf galaxies, and clusters of galaxies, and also for his theory on the formation of neutron stars.

Zworykin, Vladimir Kosma 1889–1982
Russian-born US electronics engineer and inventor whose major inventions – the iconoscope television camera tube and the electron microscope – have had ramifications far outside the immediate field of electronics.

Further Reading

ASTRONOMY

Chapman, Allan
Astronomical Instruments and Their Users: Tycho Brahe to William Lassell, Ashgate Publishing Company, 1996
A collection of articles considering the impact of technology on astronomy demonstrated by key figures in the history of astronomy.

Christianson, Gale E
Edwin Hubble: Mariner of the Nebulae, Farrar, Straus & Giroux, 1995
A detailed biography of the man who discovered that the universe contained more than one galaxy and that it is constantly expanding.

Cook, Alan H
Edmond Halley: Charting the Heavens and the Seas, Clarendon Press, 1998
Although Halley is best remembered for the comet that bears his name, this extensive biography charts the rest of his important astronomical work as well as his personal life.

Gillispie, Charles Coulston; Fox, Robert; and Grattan-Guinness, I
Pierre-Simon Laplace, 1749–1827: A Life in Exact Science, Princeton University Press, 1997
Extensive biography of this influential French mathematician and astronomer, and his work on celestial mechanics.

Gingerich, Owen
The Eye of Heaven: Ptolemy, Copernicus, Kepler, Masters of Modern Physics series, Springer-Verlag, 1992
25 essays ordered chronologically by their subject matter, discussing the development of astronomy as a creative science.

Greene, Brian
The Elegant Universe: Superstrings, Hidden Dimensions, and the Quest for the Ultimate Theory, W W Norton & Company, 1999
An excellent introduction to the structure of the universe, from the largest to the smallest scales.

Machamer, Peter K (ed)
The Cambridge Companion to Galileo, Cambridge Companions to Philosophy series, Cambridge University Press, 1998
An in-depth collection of essays interpreting the influence of Galileo, and dealing particularly with his relationship with the Church.

Osterbrock, Donald E
Pauper and Prince: Ritchey, Hale, and Big American Telescopes, University of Arizona Press, 1993
Behind-the-scenes account of the people and processes involved in making some of the largest telescopes in the USA during the 20th century.

Rosen, Edward
Copernicus and the Scientific Revolution, Anvil series, Krieger Publishing Company, 1984
Investigation into the revolutionary heliocentric theory, and its implications for future scientific endeavour.

Strom, Robert G
Mercury: The Elusive Planet, Smithsonian Library of the Solar System, Smithsonian Institution Press, 1987
The story of the innermost planet and its exploration by *Mariner 10*.

Taton, Rene and Wilson, Curtis
Planetary Astronomy from the Renaissance to the Rise of Astrophysics. Part A: Tycho Brahe to Newton, Cambridge University Press, 1989
First part of a two-part reference set on the historical development of astronomy. It includes a useful technical glossary and extensive bibliography.

Thoren, Victor E
The Lord of Uraniborg: A Biography of Tycho Brahe, Cambridge University Press, 1990
Scholarly and detailed biography of the great astronomer, utilizing original source documents.

Wali, Kameshwar C (ed)
S Chandrasekhar: The Man Behind the Legend, Imperial College Press, 1997
A collection of essays by friends and colleagues of the Nobel laureate, focusing more on his personal life than his scientific achievements.

BIOLOGY

Bowler, Peter J
Charles Darwin: The Man and His Influence, Cambridge Science Biographies series, Cambridge University Press, 1996 v. 1
A synopsis of Charles Darwin's contributions to science and culture, including in-depth analyses of key moments in his scientific career, such as the publication of *On the Origin of Species*.

Brooks, John Langdon
Just Before the Origin: Alfred Russel Wallace's Theory of Evolution, Columbia University Press, 1999
Presentation of the theories of Alfred Russel Wallace on evolution, and comparison of his work with Darwin's.

Henig, Robin Marantz
The Monk in the Garden: The Lost and Found Genius of Gregor Mendel, Mariner Books, 2000
An accessible and engaging biography of Gregor Mendel, successful even though very little of Mendel's writings survive.

Holmes, Frederic L
Hans Krebs, v 1: The Formation of a Scientific Life, 1900–1933, Monographs in the History and Philosophy of Biology, Oxford University Press, 1991
Focusing on his early life and his discovery of the urea cycle, this biography charts the rise of this renowned biochemist.

Jones, Steve
Darwin's Ghost: The Origin of Species Updated, Random House, 2000
Whilst virtually everybody has heard of *On the Origin of Species*, very few people have actually read the book. Here, Steve Jones has attempted to make Darwin's arguments more accessible to a wider audience by rewriting the famous book with current knowledge but still maintaining the original chapter headings and ideas.

Judson, Horace
The Eighth Day of Creation: the makers of the revolution in biology, Cold Spring Harbor Laboratory, 1996
Judson's history of molecular biology, and its search to answer the question 'what is life?', features interviews with leading scientists, and biographical details give the reader a sense of 'being there' at the moment of discovery.

King-Hele, Desmond
Erasmus Darwin, 1731–1802: Master of Interdisciplinary Science, Lichfield Science and Engineering Society, 1991
Published to commemorate the first Erasmus Darwin Memorial Lecture given by Dr Richard Dawkins on 21 November 1990. Reproduced from *Interdisciplinary Science Review*, v. 10, pp 170–191 (1985).

Kolata, Gina
CLONE: The Road to Dolly and the Path Ahead, Penguin Books, 1998
Describes the events leading to the birth of Dolly the sheep, the world's first clone. Kolata also discusses the implications and ethics of what this advance in genetic technology means to human beings.

Levine, Louis (ed)
Genetics of Natural Populations: The Continuing Importance of Theodosius Dobzhansky, Columbia University Press, 1995
A collection of scientific papers, personal reflections, and intellectual legacies assessing the influence of Theodosius Dobzhansky on evolutionary genetics.

Lewontin, Richard C
The Triple Helix: Gene, Organism, and Environment, Harvard University Press, 2000
Lewontin explains the themes, controversies, and debates in biology over the quarter century since his classic *The Genetic Basis of Evolutionary Change* appeared. He emphasizes the reciprocal relationship of the three factors in the course of evolution, and warns against reducing evolution to a sequence of events predetermined by genetic programming.

Margulis, Lynn, and Schwartz, Karlene
Five Kingdoms: An Illustrated Guide to the Phyla of Life on Earth, W H Freeman, 1999
An excellent work on classification with an updated molecular analysis for the third edition.

Maynard Smith, John and Szathmáry, Eörs
The Origins of Life: From the Birth of Life to the Origins of Language, Oxford University Press, 1999
The authors present an original picture of evolution. Central to the book is a discussion of the way in which information is passed through generations, starting with the appearance of the first replicating molecules, through the evolution of cooperating animal societies, and finishing with the development of language in humans.

Tudge, Colin

The Variety of Life: A Survey and a Celebration of All the Creatures that Have Ever Lived, Oxford University Press, 2000

An explanation of how biologists have arrived at their present understanding of life's diversity. Developments in the study of fossils, ecology, molecular biology, and the technique of classification known as cladistics, have brought about a ground shift in our perception of how living things are classified.

CHEMISTRY

Donovan, Arthur

Antoine Lavoisier: Science, Administration, and Revolution, Cambridge Science Biographies series, Cambridge University Press, 1996

First issued by the Cambridge University Press in 1993. Comprehensive and accessible account of the life and science of one of France's greatest chemists leading up to the French Revolution.

Fant, Kene; Ruuth, Marianne (transl)

Alfred Nobel: A Biography, Arcade Publishing, 1996

Biography of the Swedish chemist and benefactor who instituted the Nobel prizes, drawing on original writings and correspondence.

Ferry, Georgina

Dorothy Hodgkin: A Life, Granta Books, 1998

Portrait of one of the few women to receive a Nobel prize (for her work on the structures of penicillin and vitamin B_{12}).

Mauskopf, Seymour H (ed)

Chemical Sciences in the Modern World, University of Philadelphia Press, 1993

Collection of 18 essays examining the last 150 years of chemistry, covering its social, industrial, and technical impact.

Melhado, Evan Marc

Jacob Berzelius: The Emergence of His Chemical System, Lychros-Bibliotek, Almqvist & Wiksell International, 1980 v. 34

Biography of the man and his work which revolutionized the modern chemical system.

Royal Society of Chemistry

The Age of the Molecule, Royal Society of Chemistry, 1999

Highly illustrated and readable look at how molecular research has changed our lives and the developments predicted to occur in the future.

Seaborg, Glenn T

A Scientist Speaks Out: A Personal Perspective on Science, Society and Change, World Scientific Publishing Co., 1996

Autobiography of the famed Nobel laureate, including many original speeches and articles.

Strathern, Paul

Mendeleyev's Dream, Hamish Hamilton, 2001

The story of the periodic table and how Mendeleyev succeeded in discovering an underlying order to the elements even to the extent of leaving gaps of elements yet to be discovered.

EARTH SCIENCES

Bowler, Peter J

The Environmental Sciences, Fontana Press, 1992

Comprehensive survey of the history of the earth sciences, including oceanography, meteorology and ecology.

Carruthers, M W and Clinton, S

Pioneers of Geology: Discovering Earth's Secrets, Franklin Watts, Inc., 2001

Biographies of six of the most important geologists from the 18th through the 20th centuries: James Hutton, Charles Lyell, G K Gilbert, Alfred Wegener, Harry Hess, and Gene Shoemaker.

Dean, Dennis R

James Hutton and the History of Geology, Cornell University Press, 1992

Useful commentary on the 18th-century geologist's life and writings, which mark him out as the founder of modern geology.

Faul, Henry and Carol

It Began with a Stone: A History of Geology from the Stone Age to the Age of Plate Tectonics, John Wiley and Sons, 1983

A thorough, interesting, and very readable account of nearly the entire history of geology.

Hallam, Anthony

A Revolution in the Earth Sciences: From Continental Drift to Plate Tectonics, Oxford University Press, 1973

An engaging illustrated history of the development of the theory of plate tectonics, from early observations on the matches of continents, through Alfred Wegener's theory of continental drift, to seafloor spreading and plate tectonics.

399

Lewis, Cherry
The Dating Game: One Man's Search for the Age of the Earth, Cambridge University Press, 2000
A fascinating biography of English geologist Arthur Holmes, and an account of his and his colleagues efforts during the first half of the 20th century to date rocks precisely, and thus determine the age of the Earth.

Mathez, Edmond
Earth: Inside and Out, New Press, 2001
An illustrated companion to the workings of the Earth, with scientific essays, profiles of historically important figures, and modern case studies.

Oldroyd, David R
Thinking about the Earth: A History of Ideas in Geology, Athlone Press, 1996
Best single-volume study of the history of the earth sciences from antiquity to the present.

Rudwick, M J S
Georges Cuvier: Fossil Bones and Geological Catastrophes: New Translations and Interpretations of the Primary Texts, University of Chicago Press, 1997
Re-evaluation of the work of the 18th-century French palaeontologist. It is argued that he anticipated modern research into 'punctuated equilibrium'.

Sigurdsson, H
Melting the Earth: the History of Ideas on Volcanic Eruption, Oxford University Press, 1999
A professional volcanologist's fascinating history of the science of volcanology and the evolution of our current understanding of volcanism, from the Stone Age to the present.

ECOLOGY

Cloudsley-Thompson, John
Teach Yourself Ecology, Hodder and Stoughton, 1997
Concise introduction to the basics of ecology.

Houghton, J
Global Warming: The Complete Briefing, Cambridge University Press, 1997
The causes, effects, and politics of global warming, written by an expert in the field.

Isenberg, Andrew C
The Destruction of the Bison: An Environmental History, 1759–1920, Cambridge University Press, 2000
Interdisciplinary investigation into the factors that lead to the destruction of more than 30 million bison during the 18th and 19th centuries.

Ponting, Clive
A Green History of the World: The Environment and the Collapse of Great Civilizations, Penguin Books, 1991
Ecological explanation for the fall of ancient, and not so ancient, societies. Charts the impacts of the development of agricultural communities to 20th-century industrialization.

Wall, Derek
Green History: A Reader in Environmental Literature, Philosophy and Politics, Routledge, 1994
Charts the origins of the modern environmental movement, from early environmental thinkers in science and the humanities.

Worster, Donald
Nature's Economy: A History of Ecological Ideas, Cambridge University Press, 1994
Charts the way in which nature and the science of ecology have been approached throughout history.

MATHEMATICS

Artmann, B
Euclid: The Creation of Mathematics, Springer-Verlag, 1999
Summary of, and commentary on, the 13 books of Euclid's *Elements*. It also includes a useful appendix covering items of general mathematical interest, such as squaring the circle, or platonic polyhedra.

Boyer, Carl B
A History of Mathematics, John Wiley and Sons, 1989
An extremely well-written account of the history of mathematical thought, offering many insights into numbers.

Cohen, Jack, and Stewart, Ian
The Collapse of Chaos, Penguin Books, 1994
What is the relation between simplicity and complexity in science? This book explores the new mathematics of complexity theory, intermingled with biology, physics, and evolution.

Devlin, Keith
Mathematics: The New Golden Age, Columbia University Press, 1999
Updated edition of Devlin's 1988 book,

charting the development of mathematics since the 1960s for the non-specialist reader. Topics include Fermat's last theorem, fractals, chaos theory, and the four-colour problem.

Greenberg, Martin Jay
Euclidean and Non-Euclidean Geometries, W H Freeman & Co., 1993
Shows that there are more geometries than we usually imagine, and parallel lines need not behave the way we usually think.

Kitcher, Philip
The Nature of Mathematical Knowledge, Oxford University Press, 1984
Philosophical analysis of the meaning and significance of mathematics, which portrays mathematics as evolving over time, like natural science.

Kline, Morris
Mathematical Thought From Ancient to Modern Times, Oxford University Press, 1990
Comprehensive, reliable, and accessible account in three volumes, from ancient Mesopotamia to the mid-20th century.

Maor, Eli
e, The Story of a Number, Princeton University Press, 1994
A popular account of the history of the universal constant, and its importance since its discovery.

Peterson, Ivars
The Mathematical Tourist, Freeman, 1988
An extremely readable survey of a variety of areas of frontier research in the mathematical sciences.

Rassias, G M
The Mathematical Heritage of C F Gauss, World Scientific, 1991
A collection of essays in fields to which C F Gauss has contributed. Aimed mainly at graduate students and researchers in mathematics.

Singh, Simon
Fermat's Last Theorem: The Story of a Riddle That Confounded the World's Greatest Minds for 358 Years, Fourth Estate, 1998
An intriguing history of one of the oldest mathematical problems, and how it came to be solved, finally, by Andrew Wiles in 1994.

Stewart, Ian
Concepts of Modern Mathematics, Dover Publications, 1995
An introductory text to the nature of modern mathematics, including groups, topology, Boolean algebra, and more.

Stewart, Ian
The Problems of Mathematics, Oxford University Press, 1987
A sweeping survey of today's mathematical frontiers, originally aimed at undergraduates.

Wang, Hao
Reflections on Kurt Gödel, MIT Press, 1990
Excellent introduction to the famous mathematician and philosopher and his work on mathematical logic.

Yoder, Joella G
Unrolling Time: Christiaan Huygens and the Mathematization of Nature, Cambridge University Press, 1989
An analysis of Christiaan Huygens's work in applied mathematics.

PHYSICS

Baird, Davis; Hughes, R I; and Nordmann, Alfred (eds)
Heinrich Hertz: Classical Physicist, Modern Philosopher, Boston Studies in the Philosophy of Science, Kluwer Academic Publishers, 1997 v. 198
A collection of essays paying homage to the man who is credited with giving the first description of electromagnetic waves. However, the book focuses not on this work, but on Hertz's 1894 publication *Principles of Mechanics*.

Bernstein, Jeremy
Albert Einstein and the Frontiers of Physics, Oxford Portraits in Science, Oxford University Press, 1996
A fascinating biography of one of the greatest thinkers of all time, covering theories and experiments as much as the man.

Bitbol, Michel
Schrödinger's Philosophy of Quantum Mechanics, Boston Studies in the Philosophy of Science series, Kluwer Academic Publishers, 1996 v. 188
An account of the successive interpretations of quantum physics by Erwin Schrödinger, one of the founding fathers of quantum theory. The book traces subtle changes in his ideas, culminating with his final views of the 1950s.

Brennan, Richard P
Heisenberg Probably Slept Here: The Lives, Times and Ideas of the Great Physicists of the 20th Century, Wiley-Liss, 1997

401

A chronicle of the lives of seven of the 20th century's most distinguished physicists including Albert Einstein, Max Planck, and Richard Feynman.

Burchfield, Joe D
Lord Kelvin and the Age of the Earth, University of Chicago Press, 1990
Engaging historical examination of the impact of Kelvin's application of the laws of thermodynamics to the age of the Earth during the 19th century.

Cahan, David (ed)
Hermann von Helmholtz and the Foundations of Nineteenth-Century Science, California Studies in the History of Science, University of California Press, 1993 v. 12
Biography of the German scientist who is remembered for co-discovering the first law of thermodynamics, charting his contributions to physical theory and the philosophy of science.

Christianson, Gale E
Isaac Newton and the Scientific Revolution, Oxford Portraits in Science series, Oxford University Press, 1998
Exploration of the life and scientific career of the English mathematician and physicist, revealing his genius and personal frailties.

Dahl, Per F
Flash of the Cathode Rays: A History of J J Thomson's Electron, Institute of Physics Publishing, 1997
The story of the discovery of the first subatomic particle, incorporating modern research and archival material.

Filkin, David
Stephen Hawking's Universe: The Cosmos Explained, Basic Books, 1998
Introductory text on modern cosmological thinking. Written for non-scientists, and well-illustrated, it discusses the origin and composition of the universe from quasars to quarks, strings to black holes.

Gleick, James
Genius: The Life and Science of Richard Feynman, Little Brown, 1992
Scientific biography of the brilliant physicist.

More is made of his personal life than his extraordinary work, but in a way that leads one to appreciate his work all the more.

Hakfoort, Casper
Optics in the Age of Euler: Conceptions of the Nature of Light, 1700–1795, Cambridge University Press, 1995
Useful and detailed history of theories of the nature of light in the 18th century, and the impact that Euler's wave theory of light had on the scientific concensus of the time.

Hunter, Michael Cyril William (ed)
Robert Boyle Reconsidered, Cambridge University Press, 1994
A series of essays scrutinizing the life and impact of Robert Boyle and his works.

Meyenn, Karl von
'Physics in the making of Pauli's Zurich' in: Sarlemijn, A and Sparnaay, M J (eds), *Physics in the Making: Essays on Developments in 20th Century Physics*, North-Holland, 1989 pp 93–130
Scholarly introduction to the background of physicist Wolfgang Pauli. Chapter within a general book on 20th century physics.

Mould, R F
A Century of X-rays and Radioactivity, Institute of Physics, 1993

Compellingly written history of the use of X-rays since their discovery in 1895, including therapy, industrial use, and protection.

Pais, Abraham (ed)
Paul Dirac: The Man and His Work, Cambridge University Press, 1998

A collection of lectures given in honour of Paul Dirac, recollecting his life and the influence of his work on quantum physics.

Rife, Patricia
Lise Meitner and the Dawn of the Nuclear Age, Birkhauser, 1997
Account of the life and work of Austrian scientist Lise Meitner, putting her discoveries in their historical context, and showing their impact on modern physics.

GENERAL SCIENCE

Academy of Achievement
http://www.achievement.org/frames.html
A 'gallery of achievement' profiling famous achievers in science. There is an interview, biography, and profile for each scientist, focusing on the inspiring ways in which the various scientists have displayed qualities necessary to achievement, such as perseverance, vision, and courage.

Discovery Channel
http://www.discoverychannel.com
Content-rich site containing information on a number of subjects under the headings of 'Animal planet', 'Home and leisure', 'Health', 'Civilization', 'Travel and adventure', and 'Sci-Trek'. There are additional features such as Webcams, and a 'Fun and games' section where you can play the 'EcoHome game' or 'Senet', an ancient Egyptian strategy game. Linked page contains advertising.

Mad Scientist Network
http://www.madsci.org/
Indispensable resource for those in need of answers to scientific questions. The Mad Scientist Network is a group of scientists prepared to field queries on a vast range of subjects. There is an online archive of previous questions and answers, and helpful guides to accessing other general interest science Web sites.

Nobel Prizes – Official Site
http://www.nobel.se/
The 'e-museum' of the Nobel foundation. The well-organized site gives the history of the Nobel prizes and a searchable catalogue of pieces on every winner. It also explains the different prizes (Physics, Chemistry, Physiology or Medicine, Economics, Literature, and Peace) and offers a virtual tour of the Nobel Foundation.

Scientific Misconduct
http://www.chem.vt.edu/ethics/vinny/
www_ethx.html
Anatomy of a serious problem confronted by the scientific community, that of scientific misconduct. The site offers celebrated cases, reports, retaliations, articles on ethics in science, thoughts, and other material on plagiarism.

Strange Science
http://www.turnpike.net/~mscott/index.htm
Subtitled 'The Rocky Road to Modern Palaeontology and Biology', this site examines some of the medieval discoveries that led to the growth of interest in modern-day science. The site is illustrated with images that clearly show how people's perception of the world differed, and how people made up for gaps in their knowledge with a little imagination!

The Talk.Origins Archive
http://www.talkorigins.org/
Collection of articles from the Usenet newsgroup 'Talk.origins', debating the biological and physical origins of the universe from a mainstream scientific perspective.

Walk Through Time: The Evolution of Time Measurement
http://physics.nist.gov/GenInt/Time/time.html
Designed by the National Institute of Standards and Technology (NIST) Physics Laboratory to provide an historic understanding of the evolution of time measurement. There are sections on 'Ancient calendars', 'Early clocks', 'World time scales', the 'Atomic age', and 'The revolution in timekeeping'. The site also explains Greenwich Mean Time and world time zones. Visitors can synchronize their clocks to NIST time.

Why Files
http://whyfiles.org/index.html
Well-designed and topical science site tackling a couple of recent issues in each bimonthly update. As well as having a lively presentation, it also includes a lot of useful background information on a wide range of topics. This site also retains many of its previous entries, which are fully searchable.

ASTRONOMY

Apollo 11
http://www.hq.nasa.gov/alsj/a11/a11.html
This NASA page relives the excitement of the *Apollo 11* mission, with recollections from the

participating astronauts, images, audio clips, access to key White House documents, and a bibliography.

Armagh Observatory Home Page

http://star.arm.ac.uk/home.html

Historical astronomical observatory in Armagh, Northern Ireland. Founded in 1790, the observatory was at the forefront of astronomy for over a hundred years, and still remains a highly active educational institution today.

Ask An Astronaut

http://www.nss.org/askastro/

Multimedia archive of the 1972 *Apollo 17* mission to the Moon, with video and audio files, plus mission details and astronaut biographies. Sponsored by the National Space Society, this site includes numerous images, a list of questions and answers, and links to related sites.

Ask the Space Scientist

http://image.gsfc.nasa.gov/poetry/ask/askmag.html

Well-organized NASA site with answers to any question about astronomy and space. The site endeavours to send a reply to any question about space and astronomy within three days. There are useful links to question archives, general files, and other space-related Web sites.

Asteroid and Comet Impact Hazards

http://impact.arc.nasa.gov/index.html

Overview of, and the latest news on, asteroid and comet impact hazards from NASA's Ames Space Science Division, with the last Spaceguard Survey Report and a list of future Near Earth Objects (NEOs).

Astro Weight

http://members.tripod.com/scifitimes/astrowt/astrowt.htm

Enter your weight and click on 'Calculate', and this site gives your weight on all nine planets of the Solar System plus the largest five moons of Jupiter. It also includes a link to a page where you can calculate your age in all nine planetary years.

Auroras: Paintings in the Sky

http://www.exploratorium.edu/learning_studio/auroras/

Take a tour of the auroras. This page provides an in-depth explanation of the aurora phenomenon, includes a number of images taken from Earth and space, and sheds light on where they are best viewed. You can view a video clip of the aurora borealis and listen to audio clips of David Stern from NASA.

Banneker, Benjamin

http://www.progress.org/banneker/bb.html

Profile of the self-taught astronomer and surveyor from the US institute that bears his name. In addition to biographical details, a puzzle of the kind that interested Banneker is used to illustrate the creative mind of this pioneer African American.

Comet Hale-Bopp Home Page

http://www.jpl.nasa.gov/comet/

Nearly 5,000 images of the comet, which was visible to the naked eye when it came within 190 million km/118 million mi of the Earth in 1997, information on its discovery and position, and details of current research findings.

Earth and Moon Viewer

http://www.fourmilab.ch/earthview/vplanet.html

View a map of the Earth showing the day and night regions at this moment, or view the Earth from the Sun, the Moon, or any number of other locations. Alternatively, take a look at the Moon from the Earth or the Sun, or from above various formations on the lunar surface.

Gagarin, Yuri

http://www.allstar.fiu.edu/aerojava/gagarin.htm

Biography of the peasant's son who became 'the Columbus of the cosmos'. It traces Gagarin's education and training as a pilot prior to describing the courage with which he piloted *Vostok 1*. There is also a picture of the intrepid cosmonaut.

Galileo Project Information

http://nssdc.gsfc.nasa.gov/planetary/galileo.html

Site dedicated to the Galileo Project and the opportunities it has offered scientists and astronomers to enhance our understanding of the universe. The site includes information about the mission's objectives, scientific results, images, and links to other relevant sites on the Web.

Gong Show: Ringing Truths About the Sun

http://www.sciam.com/explorations/072896explorations.html

Part of a larger site maintained by *Scientific American*, this page reports on the fledgling scientific field of helioseismology, the study of the Sun's interior. Find out what scientists are learning about the processes taking place deep within the Sun. The text includes hypertext

links to further information, and there is also a
list of related links at the bottom of the page.

IMSS – History of Science Multimedia Catalogue Galileo Galilei

http://galileo.imss.firenze.it/museo/4/index.html

Room IV of the Institute and Museum of
the History of Science, which is dedicated to
Galileo. This Web site in Florence, Italy,
includes a virtual visit (Quick Time format)
to this room in the museum along with a
multimedia catalogue that includes a descriptive
text of the artefacts on display, accompanied
by still, video, or animated images.
There is an extensive list of objects to choose
from, including, rather bizarrely, the middle
finger of Galileo's right hand.

J-Track Satellite Tracker

http://liftoff.msfc.nasa.gov/RealTime/JTrack/
Spacecraft.html

Real-time tracking system that displays on a
world map the current position and orbit
information for the space shuttle, Hubble
Space Telescope, and the UARS and COBE
satellites. Pressing the shift button on your
keyboard while clicking on a craft will take you
to a page with information about that craft.

Kennedy Space Centre

http://www.ksc.nasa.gov/

NASA's well-presented guide to the history
and current operations of the USA's gateway
to the universe. There is an enormous
quantity of textual and multimedia
information of interest to the general reader
and to those who are technically minded.

Lunar Eclipse Computer

http://aa.usno.navy.mil/AA/data/docs/
LunarEclipse.html

Part of a larger site on astronomical data
maintained by the US Naval Observatory, this
site provides data on recent and upcoming
lunar eclipses for any location around the
world. The data available includes the local
time of each eclipse 'event', the altitude and
azimuth of the Moon at each of the events,
plus the time of moonrise immediately
preceding, and the time of moonset
immediately following, the eclipse. Sections of
'Frequently Asked Questions' and research
information are also included.

Magellan Mission to Venus

http://www.jpl.nasa.gov/magellan/

Details of the NASA Magellan project that
sent a probe to Venus. It includes a full
mission overview, technical details about the
planet, many images, and an animated view
of Venus.

NASA Home Page

http://www.nasa.gov/

Latest news from NASA, plus the most recent
images from the Hubble Space Telescope.
This site also contains answers to questions
about NASA resources and the space
programme, and a gallery of video, audio
clips, and still images.

NASA Shuttle Web

http://spaceflight.nasa.gov/shuttle/index.html

Official NASA site for all shuttle missions.
There is extensive technical and non-technical
information, both textual and graphic.
Questions can be sent to shuttle crew
members during missions. There is an
extensive list of frequently asked questions.
There are helpful links to related sites and
even a plain-English explanation of NASA's
bewildering jargon and acronyms.

Nine Planets

http://www.nineplanets.org/

Multimedia tour of the Solar System, with
descriptions of each of the planets and major
moons. In addition, there are appendices on
such topics as astronomical names and how
they are assigned, the origin of the Solar
System, and hypothetical planets. Linked page
contains advertising.

Picture Gallery

http://southport.jpl.nasa.gov/pic.html

Small but spectacular collection of radar
images taken from onboard space shuttle
Endeavour in 1994 and offered by NASA.
The collection also includes 3D images,
videos, and animations. The high technology
enthusiast will also find ample details on the
imaging radar system used for the project.

Ptolemy, The Man

http://seds.lpl.arizona.edu/billa/psc/theman.html

Profile of the astronomer, mathematician, and
geographer. The few known biographical facts
are related, together with summaries of his
achievements. The text is supported by a
number of diagrams and a map. There are
links to a number of other Ptolemy,
astronomy, and Greek geography sites.

Scientific Revolution in Astronomy

http://history.idbsu.edu/westciv/science/

Part of a larger site on the history of Western
civilization maintained by Boise State

405

University, this page provides an introduction to the scientific revolution in astronomy during the 16th and 17th centuries. Information is organized into 17 brief articles on astronomy and early astronomers, including a look at the trial of Galileo, the problem of the planets, acceptance of Newton, the problem of the Copernican theory, and so on. There is also a list of references for further study.

SETI Institute
http://www.seti-inst.edu/
SETI (Search for Extra-Terrestrial Intelligence) Institute conducts research designed to answer the question 'Are we alone in the universe'? This site has information on current and past projects, including their connection with the films *Independence Day* and *Contact*. There is also the opportunity to devote some of your own computer's time to 'crunching' the enormous amounts of data the Institute collects and doing a little to help the search.

Solar System Live
http://www.fourmilab.ch/solar/solar.html
Take a look at the entire Solar System as it might be seen at different times and dates or from different viewpoints.

Space Telescope Electronic Information Service
http://www.stsci.edu/
Homepage of the Hubble Space Telescope, which includes an archive of past observations, a description of the instruments aboard, and a section for educators, students, and general reference. It also includes pictures, audio clips, and press releases.

Sputnik: The Beep Heard Round the World
http://www.sciam.com/explorations/100697sputnik/hall_1.html
Part of a larger site maintained by *Scientific American*, this page recalls the launching of Sputnik 1 and the dawn of the 'space race' between the USA and USSR. You will find information on the subsequent race to the Moon, the legacy of *Sputnik 1* (in the form of thousands of satellites that currently orbit the Earth), and more. The text includes hypertext links to further information, and there is also a list of related links.

United Space Alliance: Shuttle Tracking Monitor
http://www.unitedspacealliance.com/live/tracker.htm
Displays real-time animated space shuttle positioning based on a live telemetry feed from Mission Control in Houston. A display tells you the current mission and name of the space shuttle in orbit. There is a link to a diagram of the space shuttle; click on a particular element to learn about its various components. You can also take a virtual tour of the space shuttle facilities at Kennedy and Johnson centres.

Virtual Reality Moon Phase Pictures
http://tycho.usno.navy.mil/vphase.html
GIF image display of the current phase of the Moon, updated every four hours. You can also find the Moon phase for any date and time between 1800 and 2199 AD.

Welcome to the Mars Missions, Year 2000 and Beyond!
http://marsweb.jpl.nasa.gov/
Well-presented NASA site with comprehensive information on current and future missions to Mars. There are fascinating and well-written accounts of *Pathfinder* and *Global Surveyor*, and large numbers of images of the Red Planet.

Windows to the Universe
http://www.windows.ucar.edu/
Dramatic site containing lots of information about the universe. It contains sections on the Earth and the Solar System, plus myths about the universe. The articles within each section are available to read at three levels: beginner, intermediate, and advanced. There is also a 'Kid's space' section that contains games and an 'Ask a scientist' section.

BIOLOGY

Access Excellence
http://www.accessexcellence.org/
US-based site for biology teachers sponsored by a biotechnology company, this site has plenty to interest the casual browser as well – particularly its 'What's new?' section, with weekly science reports and interviews with scientists making the news, and 'About biotech', an in-depth look at the field of biotechnology.

Cells Alive

http://www.cellsalive.com/

Well-organized collection of images, video, and text describing living human and animal cells, and their interaction with bacteria and parasites. There is also a cell-cam, which displays a growing culture of cells. Further links are available to other cell sites, such as one dealing with the common cold.

Crick, Francis Harry Compton

http://kroeber.anthro.mankato.msus.edu/ information/biography/abcde/crick_francis.html

Profile of the life and achievements of the pioneer molecular biologist. It traces his upbringing and education and how he brought his knowledge of X-ray diffraction to his work with James Watson in unravelling the structure of DNA. There is a listing of Crick's major books and articles, and a bibliography.

Darwin, Charles

http://www.literature.org/Works/Charles-Darwin

Complete text of Darwin's seminal work *On the Origin of Species*.

Darwin, Charles

http://userwww.sfsu.edu/~rsauzier/Darwin.html

Informative biography of Charles Darwin, which takes into account his upbringing and education, as well as details of his work on evolution. There is also information on related figures, such as his cousin Francis Galton, and the naturalist Alfred Russel Wallace.

Dawkins, Richard

http://www.world-of-dawkins.com/default.asp

Biographical information about Richard Dawkins, plus quotes, interviews, papers, articles, and excerpts from his books. This is an unofficial site gathering together a whole host of regularly updated information from a wide variety of sources.

Dennis Kunkel's Microscopy

http://www.pbrc.hawaii.edu/~kunkel/

This site is a photomicrographer's dream – full of pictures taken with both light and electron microscopes. As well as several differing galleries of images, there is also information about microscopy and how the pictures were taken.

Dobzhansky, Theodosius

http://kroeber.anthro.mankato.msus.edu/ information/biography/abcde/dobzhansky_ theodosius.html

Profile of the pioneering geneticist. It traces his childhood interest in insects, the frustration of his ambitions in the Soviet Union, and his subsequent research in the USA. There are photos of Dobzhansky and a bibliography.

Dr Frankenstein, I presume?

http://www.salonmagazine.com/feb97/news/ news2970224.html

Interview with the man who made the first cloned mammal, Dolly. Embryologist Ian Wilmut speaks to Andrew Ross about his worries and his future projects, about the distinction between science fiction and human cloning, and about what could go wrong with researching this delicate area.

Hopkins, Sir Frederick Gowland

http://web.calstatela.edu/faculty/nthomas/ hopkins.htm

This page is devoted to the life and scientific work of Frederick Hopkins. Biographical information includes text and a timeline of important moments in Hopkins's life. You will also find a bibliography of books by and about Hopkins. A special 'science section' provides information on essential amino acids and vitamins. There is also a bibliography of texts about vitamins.

Human Genome Project Information

http://www.ornl.gov/TechResources/ Human_Genome/home.html

US-based site devoted to this mammoth project – with news, progress reports, a molecular genetics primer, and links to other relevant sites.

Live A Life Page

http://alife.fusebox.com/

Fascinating page that brings together interactive programs to simulate ecological and evolutionary processes. Included is an adaptation of Richard Dawkins's Biomorphs program, described in *The Blind Watchmaker*, which enables the user to select 'morphs' for certain qualities and watch as their offspring evolve.

MendelWeb

http://www.netspace.org/MendelWeb/

Hefty resource for anyone interested in Gregor Mendel, the origins of classical genetics, and the history and literature of science. View or download Mendel's original paper, browse through glossaries, biographical information, and exercises, or look up the essays, timeline, bibliography, and statistical tools.

Molecular Expressions: The Amino Acid Collection

http://micro.magnet.fsu.edu/aminoacids/index.html

Fascinating collection of images showing what all the known amino acids look like when photographed through a microscope. There is also a detailed article about the different amino acids.

Molecular Expressions: The DNA Collection

http://micro.magnet.fsu.edu/micro/gallery/dna/dna4.html

Spectacular gallery of DNA photographic representations in the laboratory as well as *in vivo*. This site also has links to several other sites offering photographs through a microscope of various substances, including computer chips and various pharmaceutical substances.

Molecular Expressions: The Vitamin Collection

http://micro.magnet.fsu.edu/vitamins/index.html

Fascinating collection of images showing what all the known vitamins look like when recrystallized and photographed through a microscope. There is also a brief article about vitamins.

Natural History of Genetics

http://gslc.genetics.utah.edu/

Through a combination of scientific experts and teachers, this site offers an accessible and well-designed introduction to genetics. It includes several guided projects with experiments and explanations aimed initially at young teenagers. However, this site also includes 'Intermediate' and 'Expert' sections allowing this page to be used by a wide variety of ages and levels of expertise. In addition to the experiments, the site also includes sections on such topics as 'Core genetics', 'Teacher workshops', and 'Fun stuff'.

Peeking and Poking at DNA

http://www.sciam.com/explorations/033197aps/033197gibbs.html

Part of a larger site maintained by *Scientific American*, this page reports on new microscopes that are allowing molecular biologists and other scientists to study ever smaller subjects. Find out how these instruments work and what exciting new things scientists are discovering as a result. The text includes hypertext links to further information, and there is also a list of related links at the bottom of the page, including one to a video clip of DNA.

Photosynthesis Directory

http://esg-www.mit.edu/esgbio/ps/psdir.html

Helpful introduction to photosynthesis, detailing its 'evolution and discovery' and including clear explanations of the light and dark reactions. Also includes information and a diagram on the structure and function of the chloroplast.

Rambling Road to Humanity

http://www.sciam.com/explorations/evolution/gibbs.html

Part of a larger site maintained by *Scientific American*, this page investigates research suggesting that human evolution has not experienced the kind of steady improvement of the species that has often been accepted. There are several pictures and a number of links, including links to an abstract of the researchers' results that appeared in *Nature* magazine.

CHEMISTRY

Analytical Chemistry Basics

http://www.chem.vt.edu/chem-ed/ac-basic.html

Detailed online course, designed for those at undergraduate level, that provides the user with an introduction to some of the fundamental concepts and methods of analytical chemistry. Some of the sections included are gravimetric analysis, titration, and spectroscopy.

Chemistry of Carbon

http://cwis.nyu.edu/pages/mathmol/modules/carbon/carbon1.html

Introduction to carbon, the element at the heart of life as we know it. This site is illustrated throughout and explains the three basic forms of carbon and the importance of the way scientists choose to represent these various structures.

Crystallography and Mineralogy

http://www.iumsc.indiana.edu/crystmin.html

Understand the shapes and symmetries of crystallography, with these interactive drawings of cubic, tetrahedral, octahedral, and dodecahedral solids (just drag your mouse over the figures to rotate them).

Discovery of Argon – A Case Study in Scientific Method

http://maple.lemoyne.edu/~giunta/acspaper.html

Account of the background to Ramsay and Raleigh's discovery of argon. This is part of a course designed to show the influence of scientific method on the progress of science. The reasons for the failure of earlier chemists to isolate the noble gas are also set out here.

Elementistory

http://smallfry.dmu.ac.uk/chem/periodic/elementi.html

Periodic table of elements showing historical rather than scientific information. The contents under the chemical links in the table are mainly brief in nature, providing primarily names and dates of discovery.

Libby, Willard Frank

http://kroeber.anthro.mankato.msus.edu/information/biography/klmno/libby_willard.html

Profile of the Nobel prizewinning US chemist. It traces his academic career and official appointments and the process which led to his discovery of the technique of radiocarbon dating.

Mendeleyev, Dmitri Ivanovich

http://www.chem.msu.su/eng/misc/mendeleev/welcome.html

Detailed biography of Dmitri (or Dmitry) Mendeleyev, the chemist who framed the periodic table of elements.

Molecule of the Month

http://www.bris.ac.uk/Depts/Chemistry/MOTM/motm.htm

Pages on interesting – and sometimes hypothetical – molecules, contributed by university chemistry departments throughout the world.

Royal Society of Chemistry

http://www.rsc.org/

Work of the UK society to promote understanding of chemistry and assist its advancement. There are full details of the society's research work, online and print publications, and comprehensive educational programme. All the resources of the largest UK chemistry library can be searched.

Virtual Experiments

http://neon.chem.ox.ac.uk/vrchemistry/labintro/newdefault.html

Series of interactive chemistry experiments for A-level and university-level students hosted by this site from Oxford University, England. As well as clear instructions and safety information, the site contains photos of key stages of each experiment. Please note that most of the experiments require specialist equipment and are not suitable for the home. However, the site does contain introductions

to various subject areas, such as superconductors and simple inorganic solids.

Web Elements

http://www.webelements.com/

Periodic table on the Web, with 12 different categories of information available for each element – from its physical and chemical characteristics to its electronic configuration.

EARTH SCIENCES

Bodleian Library Map Room – The Map Case

http://www.bodley.ox.ac.uk/nnj/mapcase.htm

Broad selection of images from the historical map collection of the Bodleian library in Oxford. Visitors can choose between rare maps of Oxfordshire, London, areas of Britain, New England, Canada, and more. The maps can be viewed by thumbnail and then selected in their full GIF or JPEG version.

Cracking the Ice Age

http://www.pbs.org/wgbh/nova/ice/

Companion to the US Public Broadcasting Service (PBS) television programme *Nova*, this page provides information about glaciation, the natural changes in climate over the past 60 million years, the greenhouse effect, global warming, and continental movement. There is also a list of related links.

Geologylink

http://www.geologylink.com/

Comprehensive information on geology featuring a daily update on current geologic events, virtual classroom tours, and virtual field trips to locations around the world. The site also provides an image gallery, maps, and a glossary giving clear and concise descriptions.

Georges Cuvier – Discourse on the Revolutionary Upheavals on the Surface of the Earth

http://www.mala.bc.ca/~johnstoi/cuvier.htm

Complete original text and English translation of Cuvier's celebrated *Discours sur les révolutions de la surface du globe*. This public domain version has translator's comments and explanatory footnotes to help illustrate Cuvier's argument.

409

Hydrology Primer

http://wwwdutslc.wr.usgs.gov/infores/hydrology.
primer.html

Information from the US Geological Survey
about all aspects of hydrology. The 'clickable'
chapters include facts about surface water and
ground water, the work of hydrologists, and
careers in hydrology. For answers to further
questions click on 'ask a hydrologist', which
provides links to other US national and
regional sources.

**International Global Atmospheric
Chemistry**

http://web.mit.edu/afs/athena.mit.edu/org/i/
igac/www/

Examination of the complex chemistry of the
atmosphere and how it is being affected by
human development. It includes a good
introduction with diagrams, as well as other
more academically-inclined information for
those wishing to make a serious study of this
topic.

Museum of Palaeontology

http://www.ucmp.berkeley.edu/exhibit/exhibits.
html

Large amount of detailed palaeontological
information in a carefully structured and
carefully cross-referenced site. You can explore
palaeontology through the areas of phylogeny,
geology, and evolution.

NOAA

http://www.noaa.gov/

The US government National Oceanic and
Atmospheric Administration (NOAA) site is
beautiful and user-friendly, with information
and the latest news on weather, climate,
oceanography, and remote sensing.

Plate Tectonics

http://www.seismo.unr.edu/ftp/pub/louie/class/
100/plate-tectonics.html

Well-illustrated site on this geological
phenomenon. As well as plentiful illustrations,
this site also has a good, clear manner of
explaining the way the plates of the Earth's
crust interact to produce seismic activity.

San Andreas Fault and Bay Area

http://sepwww.stanford.edu/oldsep/joe/
fault_images/BayAreaSanAndreasFault.html

Detailed tour of the San Andreas Fault and
the San Francisco Bay area, with information
on the origination of the fault. The site is
supported by a full range of area maps.

Story of the Richter Scale

http://www.dkonline.com/science/private/
earthquest/contents/hall2.html

Biography of the US seismologist Charles
Francis Richter (1900–1985), and the story of
his famous earthquake scale. Part of a much
larger site on the earth sciences, the profile
shows how the Richter scale, which was
developed in 1935, differs from its predecessors
in measuring the magnitude of an earthquake
at its epicentre, rather than the amount of
shaking it produces at ground level.

**This Dynamic Earth: The Story of Plate
Tectonics**

http://pubs.usgs.gov/publications/text/dynamic.html

Electronic version of a book published by the
US Geological Survey.

Tsunami!

http://www.geophys.washington.edu/tsunami/
intro.html

Description of many aspects of tsunamis.
Included are details on how a tsunami is
generated and how it propagates, how they
have affected humans, and how people in
coastal areas are warned about them. The
site also discusses if and how you may
protect yourself from a tsunami and
provides 'near real-time' tsunami information
bulletins.

ECOLOGY

Acid Rain

http://qlink.queensu.ca/%7E4lrm4/table.htm

Thorough and readable explanation of acid
rain and its consequences. The page begins
with a chemical definition and goes on to
describe the causes of acid rain and its effects
on the aquatic environment, trees, soils, and
human beings. Fascinating nuggets of
information include an account of an acid
rain attack in New England, USA, where the
rain was as acidic as vinegar.

Biosphere 2 Centre

http://www.bio2.edu/

All about the Biosphere 2 project in Oracle,
Arizona, including a virtual tour of the site.
There are full details of the research
programmes being conducted on the 'seven
wilderness ecosystems' contained within the
dome. The site also contains tailored
educational resources for schoolchildren.

Chernobyl: Ten Years On

http://www.nea.fr/html/rp/chernobyl/chernobyl.
html

This site examines the disaster at the Chernobyl nuclear reactor in April 1986 and what has happened since the accident. In addition to learning about the events leading up to the explosion, you can find out about the health impacts of the radiation leak, and learn how the nuclear reactors at Chernobyl were meant to work. This official document, produced by the OECD (Organization for Economic Cooperation and Development), contains the most recent news regarding the accident and its after-effects.

Fossey, Dian

http://kroeber.anthro.mankato.msus.edu/
information/biography/fghij/fossey_dian.html

Profile of the life and death of the US primatologist. It traces how the young occupational therapist developed an interest in gorillas, taught herself biology, was the first to live closely with gorillas, went on to establish academic credentials, and was able to highlight the plight of gorillas.

Introduction to the Ecosystem Concept

http://www.geog.ouc.bc.ca/physgeog/contents/
9j.html

Clear and well-organized hyperlinked explanation of ecosystems, supported by a diagram showing the components of an ecosystem and their inter-relatedness. Part of a much larger site 'Fundamentals of Physical Geography' set up by a Canadian University, the page covers many aspects of the area of study.

John Muir Exhibit

http://www.sierraclub.org/john_muir_exhibit/

Full details of the life and legacy of the pioneer US conservationist. Contents include a biography, quotations from his writings, a large selection of pictures, and full listing of resources available for geography teachers.

Life in Extreme Environments

http://www.reston.com/astro/extreme.html

Fully searchable database of information on the scientific research into plants and animals living in extreme conditions. The site includes sections on organisms that can survive extremes of darkness, cold, radiation, and heat.

Molecular Expressions: The Pesticide Collection

http://micro.magnet.fsu.edu/pesticides/index.html

Fascinating collection of images showing what pesticides look like when recrystallized and photographed through a microscope. There is also an informative article about pesticides.

Rachel Carson Homestead

http://www.rachelcarson.org/

Information on the life and legacy of the pioneering ecologist from the trust preserving Carson's childhood home. There is a biography of Carson, details of books by and about Carson, and full details of the work of the conservation organizations continuing her work.

Technosphere III

http://www.technosphere.org.uk/

Innovative Web site that allows you to create an artificial life form, and then monitor its progress in 'Technosphere', the environment in which these creatures live. You will receive e-mails from your creation, and can communicate with it as it grows, evolves, and finally dies in its 3D environment.

MATHEMATICS

Alan Turing Home Page

http://www.turing.org.uk/turing/

Authoritative illustrated biography of the computer pioneer, plus links to related sites. This site contains information on his origins and his code-breaking work during World War II, as well as several works written by Turing himself.

Arabic Numerals

http://www.islam.org/Mosque/ihame/Ref6.htm

Account of how numbers developed from their ancient Indian origins to the modern Arabic numerals that are generally used today. There is a brief description of the numbering systems used by the ancient Egyptians, Greeks, and Romans, and some of the problems these systems presented. The page also includes links to aspects of Islamic and Middle Eastern history.

Archimedes Home Page

http://www.mcs.drexel.edu/~crorres/Archimedes/
contents.html

Extensive material on the ancient Greek philosopher and mathematician Archimedes. The site includes sections on his life and

411

travels, extracts from ancient sources about his inventions, and some famous problems associated with him.

Calculators Online Centre

http://www-sci.lib.uci.edu/~martindale/
RefCalculators.html

A page of over 10,000 links to a range of pages providing conversion and calculation tools for virtually anything, divided into mathematics, statistics, chemistry, physics, engineering, and computing. The site is not superbly well constructed, and the lack of a search engine means that it can be a bit of trawl, despite the alphabetical and subject-specific categorization, but it is nevertheless a useful resource.

Dance of Chance

http://polymer.bu.edu/museum/

Examination of fractal patterns in nature, including those of bacterial growth, erosion, metal deposition, and termite trails. Are they dominated by chance or are other factors involved?

Fibonacci Numbers and the Golden Section

http://www.mcs.surrey.ac.uk/Personal/R.Knott/
Fibonacci/fib.html

Fascinating facts and puzzles involving the Fibonacci series and the golden section. This site also includes biographical information about Fibonacci and some practical applications of his mathematical theories.

Frege, Gottlob

http://mally.stanford.edu/frege.html

Photograph, chronology, and brief biography of the mathematician and philosopher. His advances in logic are discussed, and there is a bibliography and large guide to further study.

Frequently Asked Questions in Mathematics

http://www.cs.unb.ca/~alopez-o/math-faq/
mathtext/math-faq.html

Expert answers to frequently asked questions, from 'Why is there no Nobel prize in mathematics?' to 'What is the current status of Fermat's last theorem?'

Introduction to the Works of Euclid

http://www.obkb.com/dcljr/euclid.html

Thorough introduction to the principles of Euclidean geometry. There is an emphasis on the *Elements*, but other works associated with, or attributed to, Euclid are discussed. The little that is known of the life of the

mathematician is also presented. There are a number of sources of further information.

Pythagoras

http://www.utm.edu/research/iep/p/pythagor.htm

Profile of the legendary mathematician, scientist, and philosopher. An attempt is made to disentangle the known facts of his life from the mass of legends about him. His theories of the tripartite soul, transmigration, and the cosmological limit are summarized, alongside his interest in geometry.

What's New in Mathematics?

http://www.ams.org/new-in-math/

Section of the American Mathematical Society's Web site aimed at the general public, with features on latest developments and buzz words.

Whitehead, Alfred North

http://plato.stanford.edu/entries/whitehead/

Good summary of the life and legacy of the English mathematician and logician. His influence in mathematics, metaphysics, and the philosophy of science is examined. There is a full list of his publications and a good bibliography.

PHYSICS

About Rainbows

http://unidata.ucar.edu/staff/blynds/rnbw.html

An explanation of rainbows. The site gives a good explanation of the refraction of white light into the seven colours of the visible spectrum. Linked page contains advertising.

Abu Ali Hasan, Ibn Al-Haitham

http://www.mala.bc.ca/~mcneil/haithamt.htm

Good biography of the eminent Arab physicist. It traces the main events of his life and his contribution to optics, mathematics, physics, medicine, and the development of scientific method.

Beam Me Up: Photons and Teleportation

http://www.sciam.com/explorations/
122297teleport/

Part of a larger site maintained by *Scientific American*, this page reports on the amazing research conducted by physicists at the University of Innsbruck, who have turned science fiction into reality by teleporting the properties of one photon particle to another. Learn how quantum teleportation was accomplished and find out what the likelihood is that you will soon be teleporting yourself to those distant vacation spots around the world. The text includes hypertext links to further information, and there is also a list of related links at the bottom of the page.

Biographies of Physicists
http://hermes.astro.washington.edu/scied/physics/physbio.html
Valuable compilation of biographies of the most famous physicists of all time, including Aristotle, da Vinci, Kepler, Galileo, Newton, Franklin, Curie, Feynman, and Oppenheimer. Be careful to bookmark it properly because you can easily get lost in the many pages of this site.

Doppler Effect
http://www.ncsa.uiuc.edu/Cyberia/Bima/doppler.html
Explanation of the Doppler effect and links to other related sites.

Einstein's Legacy
http://www.ncsa.uiuc.edu/Cyberia/NumRel/EinsteinLegacy.html
Illustrated introduction to the man and his greatest legacy – relativity and the concept of space-time. There is a video- and audio-clip version of the page courtesy of a US scientist and details about how current research is linked to Einstein's revolutionary ideas.

European Laboratory for Particle Physics
http://www.cern.ch/
Information about CERN, the world-class physics laboratory in Geneva. As well as presenting committees, groups, and associations hosted by the Laboratory, this official site offers important scientific material and visual evidence on the current activities and projects. Visitors will also find postings about colloquia, schools, meetings, and other services offered there. A special section is devoted to the history of the World Wide Web and the pioneering contribution of CERN in its conception and expansion.

Explanation of Temperature Related Theories
http://www.unidata.ucar.edu/staff/blynds/tmp.html
Detailed explanatory site on the laws and theories of temperature. It explains what temperature actually is, what a thermometer is, and the development of both, complete with illustrations and links to pioneers in the field. There is a temperature conversion facility and explanations of associated topics such as kinetic theory and thermal radiation.

Guided Tour of Fermilab Exhibit
http://www.fnal.gov/pub/about/tour/index.html
Guided tour of the particle physics laboratory Fermilab in Illinois, USA. It includes an explanation of the principles of particle physics, a guide to particle accelerators, and an insight into the experiments currently being conducted at the lab.

Hawking, Stephen
http://www.norfacad.pvt.k12.va.us/project/hawking/hawking.htm
Stephen Hawking's own home page, with a brief biography, disability advice, and a selection of his lectures, including 'The beginning of time' and a series debating the nature of space and time.

History of Physics
http://www.aip.org/history/
Site whose stated mission is 'to preserve and make known the history of physics and allied fields'. Part of the American Institute of Physics, the site has a searchable catalogue of sources and an exhibit hall with information about and pictures of famous scientists and important events.

Hunting of the Quark
http://researchmag.asu.edu/stories/quark.html
Commentary on the research carried out at Arizona State University, USA, into subatomic particles. There is also a subsidiary page on the complexities of the subatomic world.

Leo Szilard Home Page
http://www.dannen.com/szilard.html
Home page for the Hungarian-born physicist who was one of the first to realize the significance of splitting the atom. This site includes an illustrated biography, audio clips of people who worked with him, and documentation relating to the USA's decision to develop and then use the atom bomb.

Life and Theories of Albert Einstein
http://www.pbs.org/wgbh/nova/einstein/index.html
Heavily illustrated site on the life and theories of Einstein. There is an illustrated biographical chart, including a summary of his major achievements and their importance to science. The theory of relativity gets an understandably more in-depth coverage, along with photos and illustrations. The pages on his theories on light and time include illustrated explanations and an interactive test. There is also a 'Time-traveller' game demonstrating these theories.

413

LightForest
http://web.mit.edu/museum/lightforest/
holograms.html
This site is centred around an artist's exhibition of superimposed holographic images of leaves. The pages explaining how the holograms were created are both well designed and informative.

Look Inside the Atom
http://www.aip.org/history/electron/jjhome.htm
Part of the American Institute of Physics site, this page examines J J Thomson's 1897 experiments that led to the discovery of a fundamental building block of matter, the electron. The site includes images and quotes and a section on the legacy of his discovery. Also included is a section on suggested readings and related links.

Nuclear Energy: Frequently Asked Questions
http://www-formal.stanford.edu/jmc/progress/
nuclear-faq.html
Answers to the most commonly asked questions about nuclear energy, particularly with a view to sustaining human progress. It contains many links to related pages and is a personal opinion that openly asks for comment from visitors.

Nuclear Physics
http://www.scri.fsu.edu/~jac/Nuclear/
'Hyper-textbook' of nuclear physics, with an introduction that includes a graphical description of the size and shape of nuclei and their other properties. The site also includes information about the work of nuclear physicists, and the uses and applications of nuclear physics from medicine, through energy, to smoke detectors.

Origin of the Celsius Temperature Scale
http://www.santesson.com/engtemp.html
Brief but comprehensive description of how Swedish astronomer Anders Celsius devised his temperature scale. The Web site is available in three languages, English, Swedish and Dutch, and the Swedish cover of Celsius's paper on his temperature scale is also shown here.

Physics 2000
http://www.colorado.edu/physics/2000/
Stylish site on physics covering a range of topics under the four main headings of 'What is it?', 'Einstein's legacy', 'The atomic lab', and 'Science trek'. Each one is well illustrated and makes good use of interactive Java applets, bringing complex ideas to life.

Radioactivity in Nature
http://www.physics.isu.edu/radinf/natural.htm
Detailed explanation of the different types of radiation found naturally on Earth and in its atmosphere, as well as those produced by humans. It includes tables of the breakdown of nuclides commonly found in soil, the oceans, the air, and even the human body.

Schrödinger's Cation
http://www.sciam.com/explorations/
061796explorations.html
Part of a larger site maintained by *Scientific American*, this page features an explanation of the quantum mechanics paradox known as 'Schrödinger's Cat', an experiment devised by Erwin Schrödinger to illustrate the difference between the quantum and macroscopic worlds. You will also find information about another experiment that showed an atom actually existing in two states at one time. The text includes hypertext links to further information about quantum mechanics, and there is also a list of related links at the bottom of the page.

Usenet Relativity FAQ
http://math.ucr.edu/home/baez/physics/
relativity.html
Concise answers to some of the most common questions about relativity. The speed of light and its relation to mass, dark matter, black holes, time travel, and the Big Bang are some of the things covered by this illuminating series of articles based both on Usenet discussions and good reference sources. The site also directs the visitors to appropriate discussion groups where they can pose more questions, and also solicits more articles on themes not yet covered by the 'Frequently Asked Questions'.

Appendices

Nobel Prizes

The Nobel prizes were first awarded in 1901 under the will of Alfred B Nobel (1833–1896), a Swedish chemist, who invented dynamite. The interest on the Nobel endowment fund is divided annually among the persons who have made the greatest contributions in the fields of physics, chemistry, medicine (or physiology), literature, and world peace. The first four are awarded by academic committees based in Sweden. The prizes have a large cash award and are given at an awards ceremony held on 10 December each year, the anniversary of Nobel's death.

Nobel Prize for Chemistry

Year	Winner(s)[1]	Awarded for
1901	Jacobus van't Hoff (Netherlands)	laws of chemical dynamics and osmotic pressure
1902	Emil Fischer (Germany)	sugar and purine syntheses
1903	Svante Arrhenius (Sweden)	theory of electrolytic dissociation
1904	William Ramsay (UK)	discovery of rare gases in air and their locations in the periodic table
1905	Adolf von Baeyer (Germany)	work in organic dyes and hydroaromatic compounds
1906	Henri Moissan (France)	isolation of fluorine and adoption of electric furnace
1907	Eduard Buchner (Germany)	biochemical research and discovery of cell-free fermentation
1908	Ernest Rutherford (UK)	work in atomic disintegration, and the chemistry of radioactive substances
1909	Wilhelm Ostwald (Germany)	work in catalysis, and principles of equilibria and rates of reaction
1910	Otto Wallach (Germany)	work in alicyclic compounds
1911	Marie Curie (France)	discovery of radium and polonium, and the isolation and study of radium
1912	Victor Grignard (France)	discovery of Grignard reagents
	Paul Sabatier (France)	finding method of catalytic hydrogenation of organic compounds
1913	Alfred Werner (Switzerland)	work in bonding of atoms within inorganic molecules
1914	Theodore Richards (USA)	accurate determination of the atomic masses of many elements
1915	Richard Willstätter (Germany)	research into plant pigments, especially chlorophyll
1916	no award	
1917	no award	
1918	Fritz Haber (Germany)	synthesis of ammonia from its elements
1919	no award	
1920	Walther Nernst (Germany)	work in thermochemistry

Year	Winner(s)[1]	Awarded for
1921	Frederick Soddy (UK)	work in radioactive substances, especially isotopes
1922	Francis Aston (UK)	work in mass spectrometry of isotopes of radioactive elements, and enunciation of the whole-number rule
1923	Fritz Pregl (Austria)	method of microanalysis of organic substances
1924	no award	
1925	Richard Zsigmondy (Austria)	elucidation of heterogeneity of colloids
1926	Theodor Svedberg (Sweden)	investigation of dispersed systems
1927	Heinrich Wieland (Germany)	research on constitution of bile acids and related substances
1928	Adolf Windaus (Germany)	research on constitution of sterols and related vitamins
1929	Arthur Harden (UK) and Hans von Euler-Chelpin (Sweden)	work on fermentation of sugar, and fermentative enzymes
1930	Hans Fischer (Germany)	analysis of haem (the iron-bearing group in haemoglobin) and chlorophyll, and the synthesis of haemin (a compound of haem)
1931	Carl Bosch (Germany) and Friedrich Bergius (Germany)	invention and development of chemical high-pressure methods
1932	Irving Langmuir (USA)	discoveries and investigations in surface chemistry
1933	no award	
1934	Harold Urey (USA)	discovery of deuterium (heavy hydrogen)
1935	Irène and Frédéric Joliot-Curie (France)	synthesis of new radioactive elements
1936	Peter Debye (Netherlands)	work in molecular structures by investigation of dipole moments and the diffraction of X-rays and electrons in gases
1937	Norman Haworth (UK)	work in carbohydrates and ascorbic acid (vitamin C)
	Paul Karrer (Switzerland)	work in carotenoids, flavins, retinol (vitamin A), and riboflavin (vitamin B_2)
1938	Richard Kuhn (Germany) (declined)	carotenoids and vitamins research
1939	Adolf Butenandt (Germany) (declined)	work in sex hormones
	Leopold Ružička (Switzerland)	polymethylenes and higher terpenes
1940	no award	
1941	no award	
1942	no award	
1943	Georg von Hevesy (Hungary)	use of isotopes as tracers in chemical processes
1944	Otto Hahn (Germany)	discovery of nuclear fission
1945	Artturi Virtanen (Finland)	work in agriculture and nutrition, especially fodder preservation
1946	James Sumner (USA)	discovery of crystallization of enzymes
	John Northrop (USA) and Wendell Stanley (USA)	preparation of pure enzymes and virus proteins

Year	Winner(s)[1]	Awarded for
1947	Robert Robinson (UK)	investigation of biologically important plant products, especially alkaloids
1948	Arne Tiselius (Sweden)	researches in electrophoresis and adsorption analysis, and discoveries concerning serum proteins
1949	William Giauque (USA)	work in chemical thermodynamics, especially at very low temperatures
1950	Otto Diels (West Germany) and Kurt Alder (West Germany)	discovery and development of diene synthesis
1951	Edwin McMillan (USA) and Glenn Seaborg (USA)	discovery and work in chemistry of transuranic elements
1952	Archer Martin (UK) and Richard Synge (UK)	development of partition chromatography
1953	Hermann Staudinger (West Germany)	discoveries in macromolecular chemistry
1954	Linus Pauling (USA)	study of nature of chemical bonds, especially in complex substances
1955	Vincent Du Vigneaud (USA)	investigations into biochemically important sulphur compounds, and the first synthesis of a polypeptide hormone
1956	Cyril Hinshelwood (UK) and Nikolai Semenov (USSR)	work in mechanism of chemical reactions
1957	Alexander Todd (UK)	work in nucleotides and nucleotide coenzymes
1958	Frederick Sanger (UK)	determination of the structure of proteins, especially insulin
1959	Jaroslav Heyrovský (Czechoslovakia)	discovery and development of polarographic methods of chemical analysis
1960	Willard Libby (USA)	development of radiocarbon dating in archaeology, geology, and geography
1961	Melvin Calvin (USA)	study of assimilation of carbon dioxide by plants
1962	Max Perutz (UK) and John Kendrew (UK)	determination of structures of globular proteins
1963	Karl Ziegler (West Germany) and Giulio Natta (Italy)	chemistry and technology of producing high polymers
1964	Dorothy Crowfoot Hodgkin (UK)	crystallographic determination of the structures of biochemical compounds, notably penicillin and cyanocobalamin (vitamin B_{12})
1965	Robert Woodward (USA)	synthesis of organic substances, particularly chlorophyll
1966	Robert Mulliken (USA)	molecular orbital theory of chemical bonds and structures
1967	Manfred Eigen (West Germany), Ronald Norrish (UK), and George Porter (UK)	investigation of rapid chemical reactions by means of very short pulses of light energy
1968	Lars Onsager (USA)	discovery of reciprocal relations, fundamental for the thermodynamics of irreversible processes
1969	Derek Barton (UK) and Odd Hassel (Norway)	concept and applications of conformation in chemistry
1970	Luis Leloir (Argentina)	discovery of sugar nucleotides and their role in carbohydrate biosynthesis

Year	Winner(s)[1]	Awarded for
1971	Gerhard Herzberg (Canada)	research on electronic structure and geometry of molecules, particularly free radicals
1972	Christian Anfinsen (USA), Stanford Moore (USA), and William Stein (USA)	work in amino-acid structure and biological activity of the enzyme ribonuclease
1973	Ernst Fischer (West Germany) and Geoffrey Wilkinson (UK)	work in chemistry of organometallic sandwich compounds
1974	Paul Flory (USA)	studies of physical chemistry of macromolecules
1975	John Cornforth (UK)	work in stereochemistry of enzyme-catalysed reactions
	Vladimir Prelog (Switzerland)	work in stereochemistry of organic molecules and their reactions
1976	William Lipscomb (USA)	study of structure and chemical bonding of boranes (compounds of boron and hydrogen)
1977	Ilya Prigogine (Belgium)	work in thermodynamics of irreversible and dissipative processes
1978	Peter Mitchell (UK)	formulation of a theory of biological energy transfer and chemiosmotic theory
1979	Herbert Brown (USA) and Georg Wittig (West Germany)	use of boron and phosphorus compounds, respectively, in organic syntheses
1980	Paul Berg (USA)	biochemistry of nucleic acids, especially recombinant DNA
	Walter Gilbert (USA) and Frederick Sanger (UK)	base sequences in nucleic acids
1981	Kenichi Fukui (Japan) and Roald Hoffmann (USA)	theories concerning chemical reactions
1982	Aaron Klug (UK)	determination of crystallographic electron microscopy: structure of biologically important nucleic-acid–protein complexes
1983	Henry Taube (USA)	study of electron-transfer reactions in inorganic chemical reactions
1984	Bruce Merrifield (USA)	development of chemical syntheses on a solid matrix
1985	Herbert Hauptman (USA) and Jerome Karle (USA)	development of methods of determining crystal structures
1986	Dudley Herschbach (USA), Yuan Lee (USA), and John Polanyi (Canada)	development of dynamics of chemical elementary processes
1987	Donald Cram (USA), Jean-Marie Lehn (France), and Charles Pedersen (USA)	development of molecules with highly selective structure-specific interactions
1988	Johann Deisenhofer (West Germany), Robert Huber (West Germany), and Hartmut Michel (West Germany)	discovery of three-dimensional structure of the reaction centre of photosynthesis
1989	Sidney Altman (USA) and Thomas Cech (USA)	discovery of catalytic function of RNA
1990	Elias James Corey (USA)	new methods of synthesizing chemical compounds
1991	Richard Ernst (Switzerland)	improvements in the technology of nuclear magnetic resonance (NMR) imaging

Year	Winner(s)[1]	Awarded for
1992	Rudolph Marcus (USA)	theoretical discoveries relating to reduction and oxidation reactions
1993	Kary Mullis (USA)	invention of the polymerase chain reaction technique for amplifying DNA
	Michael Smith (Canada)	invention of techniques for splicing foreign genetic segments into an organism's DNA in order to modify the proteins produced
1994	George Olah (USA)	development of technique for examining hydrocarbon molecules
1995	F Sherwood Rowland (USA), Mario Molina (USA), and Paul Crutzen (Netherlands)	explaining the chemical process of the ozone layer
1996	Robert Curl Jr (USA), Harold Kroto (UK), and Richard Smalley (USA)	discovery of fullerenes
1997	John Walker (UK), Paul Boyer (USA), and Jens Skou (Denmark)	study of the enzymes involved in the production of adenosine triphospate (ATP), which acts as a store of energy in bodies called mitochondria inside cells
1998	Walter Kohn (USA), John Pople (USA)	research into quantum chemistry
1999	Ahmed Zewail (USA)	studies of the transition states of chemical reactions using femtosecond spectroscopy
2000	Alan J Heeger (USA), Alan G MacDiarmid (New Zealand), and Hideki Shirakawa (Japan)	their roles in the development of electrically conductive polymers
2001	William S Knowles (USA), Ryoji Noyori (Japan), and K Barry Sharpless (USA)	research into chirally catalysed oxidation

[1] Nationality given is the citizenship of recipient at the time award was made.

Nobel Prize for Physics

Year	Winner(s)[1]	Awarded for
1901	Wilhelm Röntgen (Germany)	discovery of X-rays
1902	Hendrik Lorentz (Netherlands) and Pieter Zeeman (Netherlands)	influence of magnetism on radiation phenomena
1903	Henri Becquerel (France)	discovery of spontaneous radioactivity
	Pierre Curie (France) and Marie Curie (France)	research on radiation phenomena
1904	John Strutt (Lord Rayleigh, UK)	densities of gases and discovery of argon
1905	Philipp von Lenard (Germany)	work on cathode rays
1906	Joseph J Thomson (UK)	theoretical and experimental work on the conduction of electricity by gases
1907	Albert Michelson (USA)	measurement of the speed of light through the design and application of precise optical instruments such as the interferometer
1908	Gabriel Lippmann (France)	photographic reproduction of colours by interference
1909	Guglielmo Marconi (Italy) and Karl Ferdinand Braun (Germany)	development of wireless telegraphy
1910	Johannes van der Waals (Netherlands)	equation describing the physical behaviour of gases and liquids
1911	Wilhelm Wien (Germany)	laws governing radiation of heat
1912	Nils Dalén (Sweden)	invention of light-controlled valves, which allow lighthouses and buoys to operate automatically
1913	Heike Kamerlingh Onnes (Netherlands)	studies of properties of matter at low temperatures
1914	Max von Laue (Germany)	discovery of diffraction of X-rays by crystals
1915	William Bragg (UK) and Lawrence Bragg (UK)	X-ray analysis of crystal structures
1916	no award	
1917	Charles Barkla (UK)	discovery of characteristic X-ray emission of the elements
1918	Max Planck (Germany)	formulation of quantum theory
1919	Johannes Stark (Germany)	discovery of Doppler effect in rays of positive ions, and splitting of spectral lines in electric fields
1920	Charles Guillaume (Switzerland)	discovery of anomalies in nickel–steel alloys
1921	Albert Einstein (Switzerland)	theoretical physics, especially law of photoelectric effect
1922	Niels Bohr (Denmark)	discovery of the structure of atoms and radiation emanating from them
1923	Robert Millikan (USA)	discovery of the electric charge of an electron, and study of the photoelectric effect
1924	Karl Siegbahn (Sweden)	X-ray spectroscopy
1925	James Franck (Germany) and Gustav Hertz (Germany)	discovery of laws governing the impact of an electron upon an atom
1926	Jean Perrin (France)	confirmation of the discontinuous structure of matter

Year	Winner(s)[1]	Awarded for
1927	Arthur Compton (USA)	transfer of energy from electromagnetic radiation to a particle
	Charles Wilson (UK)	invention of the Wilson cloud chamber, by which the movement of electrically charged particles may be tracked
1928	Owen Richardson (UK)	work on thermionic phenomena and associated law
1929	Louis de Broglie (France)	discovery of the wavelike nature of electrons
1930	Chandrasekhara Raman (India)	discovery of the scattering of single-wavelength light when it is passed through a transparent substance
1931	no award	
1932	Werner Heisenberg (Germany)	creation of quantum mechanics
1933	Erwin Schrödinger (Austria) and Paul Dirac (UK)	development of quantum mechanics
1934	no award	
1935	James Chadwick (UK)	discovery of the neutron
1936	Victor Hess (Austria)	discovery of cosmic radiation
	Carl Anderson (USA)	discovery of the positron
1937	Clinton Davisson (USA) and George Thomson (UK)	diffraction of electrons by crystals
1938	Enrico Fermi (Italy)	use of neutron irradiation to produce new elements, and discovery of nuclear reactions induced by slow neutrons
1939	Ernest Lawrence (USA)	invention and development of the cyclotron, and production of artificial radioactive elements
1940	no award	
1941	no award	
1942	no award	
1943	Otto Stern (USA)	molecular-ray method of investigating elementary particles, and discovery of magnetic moment of proton
1944	Isidor Rabi (USA)	resonance method of recording the magnetic properties of atomic nuclei
1945	Wolfgang Pauli (Austria)	discovery of the exclusion principle
1946	Percy Bridgman (USA)	development of high-pressure physics
1947	Edward Appleton (UK)	physics of the upper atmosphere
1948	Patrick Blackett (UK)	application of the Wilson cloud chamber to nuclear physics and cosmic radiation
1949	Hideki Yukawa (Japan)	theoretical work predicting existence of mesons
1950	Cecil Powell (UK)	use of photographic emulsion to study nuclear processes, and discovery of pions (pi mesons)
1951	John Cockcroft (UK) and Ernest Walton (Ireland)	transmutation of atomic nuclei by means of accelerated subatomic particles
1952	Felix Bloch (USA) and Edward Purcell (USA)	precise nuclear magnetic measurements
1953	Frits Zernike (Netherlands)	invention of phase-contrast microscope
1954	Max Born (UK)	statistical interpretation of wave function in quantum mechanics

Year	Winner(s)[1]	Awarded for
1954	Walther Bothe (West Germany)	coincidence method of detecting the emission of electrons
1955	Willis Lamb (USA)	structure of hydrogen spectrum
	Polykarp Kusch (USA)	determination of magnetic moment of the electron
1956	William Shockley (USA), John Bardeen (USA), and Walter Houser Brattain (USA)	study of semiconductors, and discovery of the transistor effect
1957	Tsung-Dao Lee (China) and Chen Ning Yang (China)	investigations of weak interactions between elementary particles
1958	Pavel Cherenkov (USSR), Ilya Frank (USSR), and Igor Tamm (USSR)	discovery and interpretation of Cherenkov radiation
1959	Emilio Segrè (USA) and Owen Chamberlain (USA)	discovery of the antiproton
1960	Donald Glaser (USA)	invention of the bubble chamber
1961	Robert Hofstadter (USA)	scattering of electrons in atomic nuclei, and structure of protons and neutrons
	Rudolf Mössbauer (West Germany)	resonance absorption of gamma radiation
1962	Lev Landau (USSR)	theories of condensed matter, especially liquid helium
1963	Eugene Wigner (USA)	discovery and application of symmetry principles in atomic physics
	Maria Goeppert-Mayer (USA) and Hans Jensen (Germany)	discovery of the shell-like structure of atomic nuclei
1964	Charles Townes (USA), Nikolay Basov (USSR), and Aleksandr Prokhorov (USSR)	work on quantum electronics leading to construction of oscillators and amplifiers based on maser–laser principle
1965	Sin-Itiro Tomonaga (Japan), Julian Schwinger (USA), and Richard Feynman (USA)	basic principles of quantum electrodynamics
1966	Alfred Kastler (France)	development of optical pumping, whereby atoms are raised to higher energy levels by illumination
1967	Hans Bethe (USA)	theory of nuclear reactions, and discoveries concerning production of energy in stars
1968	Luis Alvarez (USA)	elementary-particle physics, and discovery of resonance states, using hydrogen bubble chamber and data analysis
1969	Murray Gell-Mann (USA)	classification of elementary particles, and study of their interactions
1970	Hannes Alfvén (Sweden)	work in magnetohydrodynamics and its applications in plasma physics
	Louis Néel (France)	work in antiferromagnetism and ferromagnetism in solid-state physics
1971	Dennis Gabor (UK)	invention and development of holography
1972	John Bardeen (USA), Leon Cooper (USA), and John Robert Schrieffer (USA)	theory of superconductivity

Year	Winner(s)[1]	Awarded for
1973	Leo Esaki (Japan) and Ivar Giaever (USA)	tunnelling phenomena in semiconductors and superconductors
	Brian Josephson (UK)	theoretical predictions of the properties of a supercurrent through a tunnel barrier
1974	Martin Ryle (UK) and Antony Hewish (UK)	development of radioastronomy, particularly the aperture-synthesis technique, and the discovery of pulsars
1975	Aage Bohr (Denmark), Ben Mottelson (Denmark), and James Rainwater (USA)	discovery of connection between collective motion and particle motion in atomic nuclei, and development of theory of nuclear structure
1976	Burton Richter (USA) and Samuel Ting (USA)	discovery of the psi meson
1977	Philip Anderson (USA), Nevill Mott (UK), and John Van Vleck (USA)	contributions to understanding electronic structure of magnetic and disordered systems
1978	Peter Kapitza (USSR)	invention and application of low-temperature physics
	Arno Penzias (USA) and Robert Wilson (USA)	discovery of cosmic background radiation
1979	Sheldon Glashow (USA), Abdus Salam (Pakistan), and Steven Weinberg (USA)	unified theory of weak and electromagnetic fundamental forces, and prediction of the existence of the weak neutral current
1980	James W Cronin (USA) and Val Fitch (USA)	violations of fundamental symmetry principles in the decay of neutral kaon mesons
1981	Nicolaas Bloembergen (USA) and Arthur Schawlow (USA)	development of laser spectroscopy
	Kai Siegbahn (Sweden)	high-resolution electron spectroscopy
1982	Kenneth Wilson (USA)	theory for critical phenomena in connection with phase transitions
1983	Subrahmanyan Chandrasekhar (USA)	theoretical studies of physical processes in connection with structure and evolution of stars
	William Fowler (USA)	nuclear reactions involved in the formation of chemical elements in the universe
1984	Carlo Rubbia (Italy) and Simon van der Meer (Netherlands)	contributions to the discovery of the W and Z particles (weakons)
1985	Klaus von Klitzing (West Germany)	discovery of the quantized Hall effect
1986	Ernst Ruska (West Germany)	electron optics, and design of the first electron microscope
	Gerd Binnig (West Germany) and Heinrich Rohrer (Switzerland)	design of scanning tunnelling microscope
1987	Georg Bednorz (West Germany) and Alex Müller (Switzerland)	superconductivity in ceramic materials
1988	Leon M Lederman (USA), Melvin Schwartz (USA), and Jack Steinberger (USA)	neutrino-beam method, and demonstration of the doublet structure of leptons through discovery of muon neutrino
1989	Norman Ramsey (USA)	measurement techniques leading to discovery of caesium atomic clock
	Hans Dehmelt (USA) and Wolfgang Paul (Germany)	ion-trap method for isolating single atoms

Year	Winner(s)[1]	Awarded for
1990	Jerome Friedman (USA), Henry Kendall (USA), and Richard Taylor (Canada)	experiments demonstrating that protons and neutrons are made up of quarks
1991	Pierre-Gilles de Gennes (France)	work on disordered systems including polymers and liquid crystals; development of mathematical methods for studying the behaviour of molecules in a liquid on the verge of solidifying
1992	Georges Charpak (France)	invention and development of detectors used in high-energy physics
1993	Joseph Taylor (USA) and Russell Hulse (USA)	discovery of first binary pulsar (confirming the existence of gravitational waves)
1994	Clifford Shull (USA) and Bertram Brockhouse (Canada)	development of technique known as 'neutron scattering' which led to advances in semiconductor technology
1995	Frederick Reines (USA)	discovery of the neutrino
	Martin Perl (USA)	discovery of the tau lepton
1996	David Lee (USA), Douglas Osheroff (USA), and Robert Richardson (USA)	discovery of superfluidity in helium-3
1997	Claude Cohen-Tannoudji (France), William Phillips (USA), and Steven Chu (USA)	discovery of a way to slow down individual atoms using lasers for study in a near-vacuum
1998	Robert B Laughlin (USA), Horst L Störmer (USA), and Daniel C Tsui (USA)	discovery of a new form of quantum fluid with fractionally charged excitations
1999	Gerardus 't Hooft (Netherlands) and Martinus Veltman (Netherlands)	elucidating the quantum structure of electroweak interactions in physics
2000	Zhores I Alferov (Russia) and Herbert Kroemer (Germany)	development of semiconductor heterostructures, which lead to faster transistors and more efficient laser diodes
	Jack St Clair Kilby (USA)	co-invention of the integrated circuit
2001	Eric A Cornell (USA), Wolfgang Ketterle (Germany), and Carl E Wieman (USA)	achievement of Bose-Einstein condensation in dilute gases of alkali atoms, and for early fundamental studies of the properties of the condensates

[1] Nationality given is the citizenship of recipient at the time award was made.

Nobel Prize for Physiology or Medicine

Year	Winner(s)[1]	Awarded for
1901	Emil von Behring (Germany)	discovery that the body produces antitoxins, and development of serum therapy for diseases such as diphtheria
1902	Ronald Ross (UK)	work on the role of the *Anopheles* mosquito in transmitting malaria
1903	Niels Finsen (Denmark)	discovery of the use of ultraviolet light to treat skin diseases
1904	Ivan Pavlov (Russia)	discovery of the physiology of digestion
1905	Robert Koch (Germany)	investigations and discoveries in relation to tuberculosis
1906	Camillo Golgi (Italy) and Santiago Ramón y Cajal (Spain)	discovery of the fine structure of the nervous system
1907	Charles Laveran (France)	discovery that certain protozoa can cause disease
1908	Ilya Mechnikov (Russia) and Paul Ehrlich (Germany)	work on immunity
1909	Emil Kocher (Switzerland)	work on the physiology, pathology, and surgery of the thyroid gland
1910	Albrecht Kossel (Germany)	study of cell proteins and nucleic acids
1911	Allvar Gullstrand (Sweden)	work on the refraction of light through the different components of the eye
1912	Alexis Carrel (France)	work on the techniques for connecting severed blood vessels and transplanting organs
1913	Charles Richet (France)	work on allergic responses
1914	Robert Bárány (Austria-Hungary)	work on the physiology and pathology of the equilibrium organs of the inner ear
1915	no award	
1916	no award	
1917	no award	
1918	no award	
1919	Jules Bordet (Belgium)	work on immunity
1920	August Krogh (Denmark)	discovery of the mechanism regulating the dilation and constriction of blood capillaries
1921	no award	
1922	Archibald Hill (UK)	work in the production of heat in contracting muscle
	Otto Meyerhof (Germany)	work in the relationship between oxygen consumption and metabolism of lactic acid in muscle
1923	Frederick Banting (Canada) and John Macleod (UK)	discovery and isolation of the hormone insulin
1924	Willem Einthoven (Netherlands)	invention of the electrocardiograph
1925	no award	
1926	Johannes Fibiger (Denmark)	discovery of a parasite *Spiroptera carcinoma* that causes cancer

Year	Winner(s)[1]	Awarded for
1927	Julius Wagner-Jauregg (Austria)	use of induced malarial fever to treat paralysis caused by mental deterioration
1928	Charles Nicolle (France)	work on the role of the body louse in transmitting typhus
1929	Christiaan Eijkman (Netherlands)	discovery of a cure for beriberi, a vitamin-deficiency disease
	Frederick Hopkins (UK)	discovery of trace substances, now known as vitamins, that stimulate growth
1930	Karl Landsteiner (USA)	discovery of human blood groups
1931	Otto Warburg (Germany)	discovery of respiratory enzymes that enable cells to process oxygen
1932	Charles Sherrington (UK) and Edgar Adrian (UK)	discovery of function of neurons (nerve cells)
1933	Thomas Morgan (USA)	work on the role of chromosomes in heredity
1934	George Whipple (USA), George Minot (USA), and William Murphy (USA)	work on treatment of pernicious anaemia by increasing the amount of liver in the diet
1935	Hans Spemann (Germany)	organizer effect in embryonic development
1936	Henry Dale (UK) and Otto Loewi (Germany)	chemical transmission of nerve impulses
1937	Albert Szent-Györgyi (Hungary)	investigation of biological oxidation processes and of the action of ascorbic acid (vitamin C)
1938	Corneille Heymans (Belgium)	mechanisms regulating respiration
1939	Gerhard Domagk (Germany)	discovery of the first antibacterial sulphonamide drug
1940	no award	
1941	no award	
1942	no award	
1943	Henrik Dam (Denmark)	discovery of vitamin K
	Edward Doisy (USA)	chemical nature of vitamin K
1944	Joseph Erlanger (USA) and Herbert Gasser (USA)	transmission of impulses by nerve fibres
1945	Alexander Fleming (UK)	discovery of the bactericidal effect of penicillin
	Ernst Chain (UK) and Howard Florey (Australia)	isolation of penicillin and its development as an antibiotic drug
1946	Hermann Muller (USA)	discovery that X-ray irradiation can cause mutation
1947	Carl Cori (USA) and Gerty Cori (USA)	production and breakdown of glycogen (animal starch)
	Bernardo Houssay (Argentina)	function of the pituitary gland in sugar metabolism
1948	Paul Müller (Switzerland)	discovery of the first synthetic contact insecticide DDT
1949	Walter Hess (Switzerland)	mapping areas of the midbrain that control the activities of certain body organs
	Antonio Egas Moniz (Portugal)	therapeutic value of prefrontal leucotomy in certain psychoses

Year	Winner(s)[1]	Awarded for
1950	Edward Kendall (USA), Tadeus Reichstein (Switzerland), and Philip Hench (USA)	structure and biological effects of hormones of the adrenal cortex
1951	Max Theiler (South Africa)	discovery of a vaccine against yellow fever
1952	Selman Waksman (USA)	discovery of streptomycin, the first antibiotic effective against tuberculosis
1953	Hans Krebs (UK)	discovery of the Krebs cycle
	Fritz Lipmann (USA)	discovery of coenzyme A, a nonprotein compound that acts in conjunction with enzymes to catalyse metabolic reactions leading up to the Krebs cycle
1954	John Enders (USA), Thomas Weller (USA), and Frederick Robbins (USA)	cultivation of the polio virus in the laboratory
1955	Hugo Theorell (Sweden)	work on the nature and action of oxidation enzymes
1956	André Cournand (USA), Werner Forssmann (West Germany), and Dickinson Richards (USA)	work on the technique for passing a catheter into the heart for diagnostic purposes
1957	Daniel Bovet (Italy)	discovery of synthetic drugs used as muscle relaxants in anaesthesia
1958	George Beadle (USA) and Edward Tatum (USA)	discovery that genes regulate precise chemical effects
	Joshua Lederberg (USA)	work on genetic recombination and the organization of bacterial genetic material
1959	Severo Ochoa (USA) and Arthur Kornberg (USA)	discovery of enzymes that catalyse the formation of RNA (ribonucleic acid) and DNA (deoxyribonucleic acid)
1960	Macfarlane Burnet (Australia) and Peter Medawar (UK)	acquired immunological tolerance of transplanted tissues
1961	Georg von Békésy (USA)	investigations into the mechanism of hearing within the cochlea of the inner ear
1962	Francis Crick (UK), James Watson (USA), and Maurice Wilkins (UK)	discovery of the double-helical structure of DNA and of the significance of this structure in the replication and transfer of genetic information
1963	John Eccles (Australia), Alan Hodgkin (UK), and Andrew Huxley (UK)	ionic mechanisms involved in the communication or inhibition of impulses across neuron (nerve cell) membranes
1964	Konrad Bloch (USA) and Feodor Lynen (West Germany)	work on the cholesterol and fatty-acid metabolism
1965	François Jacob (France), André Lwoff (France), and Jacques Monod (France)	genetic control of enzyme and virus synthesis
1966	Peyton Rous (USA)	discovery of tumour-inducing viruses
	Charles Huggins (USA)	hormonal treatment of prostatic cancer
1967	Ragnar Granit (Sweden), Haldan Hartline (USA), and George Wald (USA)	physiology and chemistry of vision

Year	Winner(s)[1]	Awarded for
1968	Robert Holley (USA), Har Gobind Khorana (USA), and Marshall Nirenberg (USA)	interpretation of genetic code and its function in protein synthesis
1969	Max Delbrück (USA), Alfred Hershey (USA), and Salvador Luria (USA)	replication mechanism and genetic structure of viruses
1970	Bernard Katz (UK), Ulf von Euler (Sweden), and Julius Axelrod (USA)	work on the storage, release, and inactivation of neurotransmitters
1971	Earl Sutherland (USA)	discovery of cyclic AMP, a chemical messenger that plays a role in the action of many hormones
1972	Gerald Edelman (USA) and Rodney Porter (UK)	work on the chemical structure of antibodies
1973	Karl von Frisch (Austria), Konrad Lorenz (Austria), and Nikolaas Tinbergen (UK)	work in animal behaviour patterns
1974	Albert Claude (USA), Christian de Duve (Belgium), and George Palade (USA)	work in structural and functional organization of the cell
1975	David Baltimore (USA), Renato Dulbecco (USA), and Howard Temin (USA)	work on interactions between tumour-inducing viruses and the genetic material of the cell
1976	Baruch Blumberg (USA) and Carleton Gajdusek (USA)	new mechanisms for the origin and transmission of infectious diseases
1977	Roger Guillemin (USA) and Andrew Schally (USA)	discovery of hormones produced by the hypothalamus region of the brain
	Rosalyn Yalow (USA)	radioimmunoassay techniques by which minute quantities of hormone may be detected
1978	Werner Arber (Switzerland), Daniel Nathans (USA), and Hamilton Smith (USA)	discovery of restriction enzymes and their application to molecular genetics
1979	Allan Cormack (USA) and Godfrey Hounsfield (UK)	development of the computer axial tomography (CAT) scan
1980	Baruj Benacerraf (USA), Jean Dausset (France), and George Snell (USA)	work on genetically determined structures on the cell surface that regulate immunological reactions
1981	Roger Sperry (USA)	functional specialization of the brain's cerebral hemispheres
	David Hubel (USA) and Torsten Wiesel (Sweden)	work on visual perception
1982	Sune Bergström (Sweden), Bengt Samuelsson (Sweden), and John Vane (UK)	discovery of prostaglandins and related biologically active substances
1983	Barbara McClintock (USA)	discovery of mobile genetic elements
1984	Niels Jerne (Denmark-UK), Georges Köhler (West Germany), and César Milstein (Argentina)	work on immunity and discovery of a technique for producing highly specific, monoclonal antibodies
1985	Michael Brown (USA) and Joseph L Goldstein (USA)	work on the regulation of cholesterol metabolism

Year	Winner(s)[1]	Awarded for
1986	Stanley Cohen (USA) and Rita Levi-Montalcini (USA-Italy)	discovery of factors that promote the growth of nerve and epidermal cells
1987	Susumu Tonegawa (Japan)	work on the process by which genes alter to produce a range of different antibodies
1988	James Black (UK), Gertrude Elion (USA), and George Hitchings (USA)	work on the principles governing the design of new drug treatments
1989	Michael Bishop (USA) and Harold Varmus (USA)	discovery of oncogenes, genes carried by viruses that can trigger cancerous growth in normal cells
1990	Joseph Murray (USA) and Donnall Thomas (USA)	pioneering work in organ and cell transplants
1991	Erwin Neher (Germany) and Bert Sakmann (Germany)	discovery of how gatelike structures (ion channels) regulate the flow of ions into and out of cells
1992	Edmond Fischer (USA) and Edwin Krebs (USA)	isolating and describing the action of the enzyme responsible for reversible protein phosphorylation, a major biological control mechanism
1993	Phillip Sharp (USA) and Richard Roberts (UK)	discovery of split genes (genes interrupted by nonsense segments of DNA)
1994	Alfred Gilman (USA) and Martin Rodbell (USA)	discovery of a family of proteins (G-proteins) that translate messages – in the form of hormones or other chemical signals – into action inside cells
1995	Edward Lewis (USA), Eric Wieschaus (USA), and Christiane Nüsslein-Volhard (Germany)	discovery of genes which control the early stages of the body's development
1996	Peter Doherty (Australia) and Rolf Zinkernagel (Switzerland)	discovery of how the immune system recognizes virus-infected cells
1997	Stanley Prusiner (USA)	discoveries, including the 'prion' theory, that could lead to new treatments of dementia-related diseases, including Alzheimer's and Parkinson's diseases
1998	Robert Furchgott (USA), Ferid Murad (USA), and Louis Ignarro (USA)	discovery that nitric oxide (NO) acts as a key chemical messenger between cells
1999	Günter Blobel (USA)	discovery that proteins have intrinsic signals that govern their transport and localization in the cell
2000	Arvid Carlsson (Sweden), Paul Greegard (USA), and Eric Kandel (USA)	elucidation of how signals are transmitted between nerve cells
2001	Leland H Hartwell (USA), R Timothy Hunt (UK), and Paul M Nurse (UK)	discovery of key regulators of the cell cycle

[1] Nationality given is the citizenship of recipient at the time award was made.

Fields Medal

This international prize for achievement in the field of mathematics is awarded every four years by the International Mathematical Union.

Year	Winner(s)
1936	Lars Ahlfors (Finland); Jesse Douglas (USA)
1950	Atle Selberg (USA); Laurent Schwartz (France)
1954	Kunihiko Kodaira (USA); Jean-Pierre Serre (France)
1958	Klaus Roth (UK); René Thom (France)
1962	Lars Hörmander (Sweden); John Milnor (USA)
1966	Michael Atiyah (UK); Paul J Cohen (USA); Alexander Grothendieck (France); Stephen Smale (USA)
1970	Alan Baker (UK); Heisuke Hironaka (USA); Sergei Novikov (USSR); John G Thompson (USA)
1974	Enrico Bombieri (Italy); David Mumford (USA)
1978	Pierre Deligne (Belgium); Charles Fefferman (USA); G A Margulis (USSR); Daniel Quillen (USA)
1982	Alain Connes (France); William Thurston (USA); S T Yau (USA)
1986	Simon Donaldson (UK); Gerd Faltings (West Germany); Michael Freedman (USA)
1990	Vladimir Drinfeld (USSR); Vaughan F R Jones (USA); Shigefumi Mori (Japan); Edward Witten (USA)
1994	L J Bourgain (USA/France); P-L Lions (France); J-C Yoccoz (France); E I Zelmanov (USA)
1998	Richard E Borcherds (UK); W Timothy Gowers (UK); Maxim Kontsevich (Russia); Curtis T McMullen (USA)

Discovery of the Elements
(– = not applicable)

Date	Element (symbol)	Discoverer
Ancient knowledge		
	antimony (Sb)	–
	arsenic (As)	–
	bismuth (Bi)	–
	carbon (C)	–
	copper (Cu)	–
	gold (Au)	–
	iron (Fe)	–
	lead (Pb)	–
	mercury (Hg)	–
	silver (Ag)	–
	sulphur (S)	–
	tin (Sn)	–
	zinc (Zn)	
1669	phosphorus (P)	Hennig Brand
1737	cobalt (Co)	Georg Brandt
1748	platinum (Pt)	Antonio de Ulloa (identified as an element by William Brownrigg in 1750)
1751	nickel (Ni)	Axel Cronstedt
1755	magnesium (Mg)	Joseph Black (oxide isolated by Humphry Davy in 1808; pure form isolated by Antoine-Alexandre-Brutus Bussy in 1828)
1766	hydrogen (H)	Henry Cavendish
1771	fluorine (F)	Karl Scheele (isolated by Henri Moissan in 1886)
1772	nitrogen (N)	Daniel Rutherford
1774	chlorine (Cl)	Karl Scheele (identified as an element by Humphry Davy in 1810)
	manganese (Mn)	Karl Scheele (isolated by Johan Gahn in the same year)
	oxygen (O)	Joseph Priestley and Karl Scheele, independently
1778	molybdenum (Mo)	named by Karl Scheele (isolated by Peter Jacob Hjelm in 1782)
1782	tellurium (Te)	Franz Müller von Reichenstein
1783	tungsten (W)	isolated by Juan José de Elhuyar and Fausto de Elhuyar
1789	uranium (U)	Martin Klaproth (isolated by Eugène Péligot in 1841)
	zirconium (Zr)	Martin Klaproth
1791	titanium (Ti)	William Gregor
1794	yttrium (Y)	Johan Gadolin (isolated by Jöns Berzelius in 1824)
1797	chromium (Cr)	Louis-Nicolas Vauquelin
1798	beryllium (Be)	Louis-Nicolas Vauquelin (isolated by Friedrich Wöhler and Antoine-Alexandre-Brutus Bussy in 1828)
1801	vanadium (V)	Andrés del Rio (disputed)
	niobium (Nb)	Charles Hatchett
1802	tantalum (Ta)	Anders Ekeberg

Date	Element (symbol)	Discoverer
1803	cerium (Ce)	Jöns Berzelius and Wilhelm Hisinger, and independently by Martin Klaproth
	palladium (Pd)	William Wollaston
	rhodium (Rh)	William Wollaston
1804	iridium (Ir)	Smithson Tennant
	osmium (Os)	Smithson Tennant
1807	potassium (K)	Humphry Davy
	sodium (Na)	Humphry Davy
1808	barium (Ba)	Humphry Davy
	boron (B)	Humphry Davy, and independently by Joseph Gay-Lussac and Louis-Jacques Thénard
	calcium (Ca)	Humphry Davy
	strontium (Sr)	Humphry Davy
1811	iodine (I)	Bernard Courtois
1817	cadmium (Cd)	Friedrich Stromeyer
	lithium (Li)	Johan Arfwedson
	selenium (Se)	Jöns Berzelius
1823	silicon (Si)	Jöns Berzelius
1825	aluminium (Al)	Hans Oersted (also attributed to Friedrich Wöhler in 1827)
1826	bromine (Br)	Antoine-Jérôme Balard
1827	ruthenium (Ru)	Gottfried Wilhelm Osann (isolated by Karl Klaus in 1844)
1828	thorium (Th)	Jöns Berzelius
1839	lanthanum (La)	Carl Mosander
1843	erbium (Er)	Carl Mosander
	terbium (Tb)	Carl Mosander
1860	caesium (Cs)	Robert Bunsen and Gustav Kirchhoff
1861	rubidium (Rb)	Robert Bunsen and Gustav Kirchhoff
	thallium (Tl)	William Crookes (isolated by William Crookes and Claude August Lamy independently of each other in 1862)
1863	indium (In)	Ferdinand Reich and Hieronymus Richter
1868	helium (He)	Pierre Janssen and Norman Lockyer
1875	gallium (Ga)	Paul-Emile Lecoq de Boisbaudran
1878	holmium (Ho)	Jacques Soret and Marc Delafontaine
	ytterbium (Yb)	Jean Charles de Marignac
1879	scandium (Sc)	Lars Nilson
	thulium (Tm)	Per Cleve
1880	gadolinium (Gd)	Jean-Charles de Marignac
	samarium (Sm)	Paul-Emile Lecoq de Boisbaudran
1885	neodymium (Nd)	Carl von Welsbach
	praseodymium (Pr)	Carl von Welsbach
1886	dysprosium (Dy)	Paul-Emile Lecoq de Boisbaudran
	germanium (Ge)	Clemens Winkler
1894	argon (Ar)	John Strutt (Lord Rayleigh) and William Ramsay

Date	Element (symbol)	Discoverer
1898	krypton (Kr)	William Ramsay and Morris Travers
	neon (Ne)	William Ramsay and Morris Travers
	polonium (Po)	Marie and Pierre Curie
	radium (Ra)	Marie Curie
	xenon (Xe)	William Ramsay and Morris Travers
1899	actinium (Ac)	André Debierne
1900	radon (Rn)	Friedrich Dorn
1901	europium (Eu)	Eugène Demarçay
1907	lutetium (Lu)	Georges Urbain and Carl von Welsbach, independently of each other
1917	protactinium (Pa)	Otto Hahn and Lise Meitner
1923	hafnium (Hf)	Dirk Coster and Georg von Hevesy
1925	rhenium (Re)	Walter Noddack, Ida Tacke, and Otto Berg
1937	technetium (Tc)	Carlo Perrier and Emilio Segrè
1939	francium (Fr)	Marguérite Perey
1940	astatine (At)	Dale R Corson, K R MacKenzie, and Emilio Segrè
	neptunium (Np)	Edwin McMillan and Philip Abelson
	plutonium (Pu)	Glenn Seaborg, Edwin McMillan, Joseph Kennedy, and Arthur Wahl
1944	americium (Am)	Glenn Seaborg, Ralph James, Leon Morgan, and Albert Ghiorso
	curium (Cm)	Glenn Seaborg, Ralph James, and Albert Ghiorso
1945	promethium (Pm)	J A Marinsky, Lawrence Glendenin, and Charles Coryell
1949	berkelium (Bk)	Glenn Seaborg, Stanley Thompson, and Albert Ghiorso
1950	californium (Cf)	Glenn Seaborg, Stanley Thompson, Kenneth Street Jr, and Albert Ghiorso
1952	einsteinium (Es)	Albert Ghiorso and co-workers
	fermium (Fm)	Albert Ghiorso and co-workers
1955	mendelevium (Md)	Albert Ghiorso, Bernard G Harvey, Gregory Choppin, Stanley Thompson, and Glenn Seaborg
1958	nobelium (No)	Albert Ghiorso, Torbjørn Sikkeland, J R Walton, and Glenn Seaborg
1961	lawrencium (Lr)	Albert Ghiorso, Torbjørn Sikkeland, Almon Larsh, and Robert Latimer
1967	dubnium (Db)	claimed by Soviet scientist Georgii Flerov and co-workers (disputed by US workers)
1969	rutherfordium (Rf)	claimed by US scientist Albert Ghiorso and co-workers (disputed by Soviet workers)
1970	dubnium (Db)	claimed by Albert Ghiorso and co-workers (disputed by Soviet workers)
1974	seaborgium (Sg)	claimed by Georgii Flerov and co-workers, and independently by Albert Ghiorso and co-workers
1976	bohrium (Bh)	Georgii Flerov and Yuri Oganessian
1982	meitnerium (Mt)	Peter Armbruster and co-workers
1984	hassium (Hs)	Peter Armbruster and co-workers
1994	ununnilium (Uun)	team at GSI heavy-ion cyclotron, Darmstadt, Germany

Date	Element (symbol)	Discoverer
1995	unununium (Uuu)	team at GSI heavy-ion cyclotron, Darmstadt, Germany
1996	ununbium (Uub)	team at GSI heavy-ion cyclotron, Darmstadt, Germany
1999	Ununquadium (Uuq)	team at Dubna (Joint Institute for Nuclear Research), Russia
	Ununhexium (Uuh)	team at Lawrence Berkeley National Laboratory, University of California, and Oregon State University, Corvallis
	Ununoctium (Uuo)	team at Lawrence Berkeley National Laboratory, University of California, and Oregon State University, Corvallis

Glossary

Abelian functions functions of the form $\int f(x, y)\,dx$, where y is an algebraic function of x, and $f(x, y)$ is an algebraic function of x and y.

aberration of starlight apparent displacement of a star from its true position, due to the combined effects of the speed of light and the speed of the Earth in orbit around the Sun (about 30 km per second/18.5 mi per second).

aberration, optical any of a number of defects that impair the image in an optical instrument. Aberration occurs because of minute variations in lenses and mirrors, and because different parts of the light spectrum are reflected or refracted by varying amounts.

abiotic factor non-living variable within the ecosystem, affecting the life of organisms. Examples include temperature, light, and water. Abiotic factors can be harmful to the environment, as when sulphur dioxide emissions from power stations produce acid rain.

ablation in astronomy, progressive burning away of the outer layers (for example, of a meteor) by friction with the atmosphere.

abscissa x-coordinate of a point – that is, the horizontal distance of that point from the vertical or y-axis. For example, a point with the coordinates (4, 3) has an abscissa of 4. The y-coordinate of a point is known as the ordinate.

absolute magnitude measure of the intrinsic brightness of a celestial body in contrast to its apparent brightness or magnitude as seen from Earth.

absolute value or *modulus*, value, or magnitude, of a number irrespective of its sign. The absolute value of a number n is written $|n|$ (or sometimes as mod n), and is defined as the positive square root of n^2. For example, the numbers -5 and 5 have the same absolute value:

$$|5| = |-5| = 5$$

absolute weight weight of a body considered apart from all modifying influences such as the atmosphere. To determine its absolute weight, the body must, therefore, be weighed in a vacuum or allowance must be made for buoyancy.

absolute zero lowest temperature theoretically possible according to kinetic theory, zero kelvin (0 K), equivalent to $-273.15°C/-459.67°F$, at which molecules are in their lowest energy state. Although the third law of thermodynamics indicates the impossibility of reaching absolute zero in practice, temperatures of less than a billionth of a degree above absolute zero have been produced. Near absolute zero, the physical properties of some materials change substantially; for example, some metals lose their electrical resistance and become superconducting.

absorption taking up of matter or energy of a substance by another, such as a liquid by a solid (ink by blotting paper) or a gas by a liquid (ammonia by water). Absorption is the phenomenon by which a substance retains the energy of radiation of particular wavelengths; for example, a piece of blue glass absorbs all visible light except the wavelengths in the blue part of the spectrum. It also refers to the partial loss of energy resulting from light and other electromagnetic waves passing through a medium. In nuclear physics, absorption is the capture by elements, such as boron, of neutrons produced by fission in a reactor.

absorption spectroscopy or *absorptiometry*, technique for determining the identity or amount present of a chemical substance by measuring the amount of electromagnetic radiation the substance absorbs at specific wavelengths. See spectroscopy.

abundance of helium see helium abundance.

abundant number natural number that is less than the sum of its divisors (factors).

acceleration rate of change of the velocity of a moving body. It is usually measured in metres per second per second ($m\,s^{-2}$) or feet per second per second ($ft\,s^{-2}$). Because velocity is a vector quantity (possessing both magnitude and direction) a body travelling at constant speed may be said to be accelerating if its direction of motion changes. According to Newton's second law of motion, a body will accelerate only if it is acted upon by an unbalanced, or resultant, force. Acceleration due to gravity is the acceleration of a body falling freely under the influence of the Earth's gravitational field; it varies slightly at different latitudes and altitudes. The value adopted internationally for gravitational acceleration is $9.806\,m\,s^{-2}/32.174\,ft\,s^{-2}$.

accelerator device to bring charged particles (such as protons and electrons) up to high speeds and energies, at which they can be of use in industry, medicine, and pure physics. At low energies, accelerated particles can be used to produce the image on a television screen and generate X-rays (by means of a cathode-ray tube), destroy tumour cells, or kill bacteria. When high-energy particles collide with other particles, the fragments formed reveal the nature of the fundamental forces.

accelerator nerve nerve that conducts impulses to the heart. On stimulation of the cardiac sympathetic nerves, the rate and strength of the heartbeat increases.

accretion process by which an object gathers up surrounding material by gravitational attraction, so simultaneously increasing in mass and releasing gravitational energy. Accretion on to compact objects such as white dwarfs, neutron stars, and black holes can release large amounts of gravitational energy, and is believed to be the power source for active galaxies. Accreted material falling towards a star may form a swirling disc of material known as an accretion disc that can be a source of X-rays.

accretion disc flattened ring of gas and dust orbiting an object in space, such as a star or black hole. The orbiting material is accreted (gathered in) from a neighbouring object such as another star. Giant accretion discs are thought to exist at the centres of some galaxies and quasars.

accumulator storage battery – that is, a group of rechargeable secondary cells. A familiar example is the lead–acid car battery.

acetylcholine *ACh*, chemical that serves as a neurotransmitter, communicating nerve impulses between the cells of the nervous system. It is largely associated with the transmission of impulses across the synapse (junction) between nerve and muscle cells, causing the muscles to contract.

acetyl coenzyme A or *acetyl CoA*, compound active in processes of metabolism. It is a heat-stable coenzyme with an acetyl group ($-COCH_3$) attached by sulphur linkage. This linkage is a high-energy bond and the acetyl group can easily be donated to other compounds. Acetyl groups donated in this way play an important part in glucose breakdown as well as in fatty acid and steroid synthesis. It is involved in the Krebs cycle, the cyclical pathway involved in the intracellular metabolism of foodstuffs.

acetylene common name for ethyne.

achondrite type of meteorite. They make up about 15% of all meteorites and lack the **chondrules** (silicate spheres) found in chondrites.**achromatic lens** lens (or combination of lenses) that brings different wavelengths within a ray of light to a single focus, thus overcoming chromatic aberration.

acid compound that releases hydrogen ions (H^+ or protons) in the presence of an ionizing solvent (usually water). Acids react with bases to form salts, and they act as solvents. Strong acids are corrosive; dilute acids have a sour or sharp taste, although in some organic acids this may be partially masked by other flavour characteristics. The strength of an acid is measured by its hydrogen-ion concentration, indicated by the pH value. All acids have a pH below 7.0.

acid amide any organic compound that may be regarded as being derived from ammonia by the substitution of acid or acyl groups for atoms of hydrogen. They are described as primary, secondary, tertiary, and so on, according to the number of atoms of hydrogen displaced. Thus the general formula for a primary amide is $RCONH_2$. The main acid amides are ethanamide and methanamide.

acid rain acidic precipitation thought to be caused principally by the release into the atmosphere of sulphur dioxide (SO_2) and oxides of nitrogen, which dissolve in pure rainwater making it acidic. Sulphur dioxide is formed by the burning of fossil fuels, such as coal, that contain high quantities of sulphur; nitrogen oxides are contributed from various industrial activities and from car exhaust fumes.

acquired character feature of the body that develops during the lifetime of an individual, usually as a result of repeated use or disuse, such as the enlarged muscles of a weightlifter.

actin protein that occurs in muscles. See myosin.

actinium (Greek *aktis* 'ray') white, radioactive, metallic element, the first of the actinide series, symbol Ac, atomic number 89 relative atomic mass 227; it is a weak emitter of high energy alpha particles.

action potential change in the potential difference (voltage) across the membrane of a nerve cell when an impulse passes along it. A change in potential (from about -60 to $+45$ millivolts) accompanies the passage of sodium and potassium ions across the membrane.

activation analysis technique used to reveal the presence and amount of minute impurities in a substance or element. A sample of a material that may contain traces of a certain element is irradiated with neutrons, as in a reactor. Gamma rays emitted by the material's radioisotopes have unique energies and relative intensities, similar to the spectral lines from a luminous gas. Measurements and interpretation of the gamma-ray spectrum, using data from standard samples for comparison, provide information on the amounts of impurities present.

activation energy energy required in order to start a chemical reaction. Some elements and compounds will react together merely by bringing them into contact (spontaneous reaction). For others it is necessary to supply energy in order to start the reaction, even if there is ultimately a net output of energy. This initial energy is the activation energy.

adaptation (Latin *adaptare* 'to fit to') any change in the structure or function of an organism that allows it to survive and reproduce more effectively in its environment. In evolution, adaptation is thought to occur as a result of random variation in the genetic make-up of organisms coupled with natural selection. Species become extinct when they are no longer adapted to their environment.

adaptive radiation formation of several species, with adaptations to different ways of life, from a single ancestral type. Adaptive radiation is likely to occur whenever members of a species migrate to a new habitat with unoccupied ecological niches. It is thought that the lack of competition in such niches allows sections of the migrant population to develop new adaptations, and eventually to become new species.

The colonization of newly formed volcanic islands has led to the development of many unique species. The 13 species of Darwin's finch on the Galapagos Islands, for example, are probably descended from a single species from the South American mainland. The parent stock evolved into different species that now occupy a range of diverse niches.

addition reaction chemical reaction in which the atoms of an element or compound react with a double bond or triple bond in an organic compound by opening up one of the bonds and becoming attached to it, for example

$$CH_2{=}CH_2 + HCl \rightarrow CH_3CH_2Cl$$

An example is the addition of hydrogen atoms to unsaturated compounds in vegetable oils to produce margarine.

adenosine triphosphate compound present in cells. See ATP.

adiabatic change process that takes place without any heat entering or leaving the system.

adipose tissue type of connective tissue of vertebrates that serves as an energy reserve, and also pads some organs. It is commonly called fat tissue, and consists of large, spherical cells filled with fat. In mammals, major layers are in the inner layer of skin and around the kidneys and heart.

ADP abbreviation for ***adenosine diphosphate***, the chemical product formed in cells when ATP breaks down to release energy.

adsorption taking up of a gas or liquid at the surface of another substance, most commonly a solid (for example, activated charcoal adsorbs gases). It involves molecular attraction at the surface, and should be distinguished from absorption (in which a uniform solution results from a gas or liquid being incorporated into the bulk structure of a liquid or solid).

aeon or ***eon***, in astronomical terms, 1,000 million years.

aerial oxidation reaction in which air is used to oxidize another substance, as in the contact process for the manufacture of sulphuric acid:

$$2SO_2 + O_2 \rightleftharpoons 2SO_3$$

and in the souring of wine.

aerobic term used to describe those organisms that require oxygen (usually dissolved in water) for the efficient release of energy contained in food molecules, such as glucose. They include almost all organisms (plants as well as animals) with the exception of certain bacteria, yeasts, and internal parasites.

aerodynamics branch of fluid physics that studies the forces exerted by air or other gases in motion. Examples include the airflow around bodies moving at speed through the atmosphere (such as land vehicles, bullets, rockets, and aircraft), the behaviour of gas in engines and furnaces, air conditioning of buildings, the deposition of snow, the operation of air-cushion vehicles (hovercraft), wind loads on buildings and bridges, bird and insect flight, musical wind instruments, and meteorology.

aerosol particles of liquid or solid suspended in a gas. Fog is a common natural example. Aerosol cans contain a substance such as

437

scent or cleaner packed under pressure with a device for releasing it as a fine spray. Most aerosols used chlorofluorocarbons (CFCs) as propellants until these were found to cause destruction of the ozone layer in the stratosphere.

aestivation state of inactivity and reduced metabolic activity, similar to hibernation, that occurs during the dry season in species such as lungfish and snails. In botany, the term is used to describe the way in which flower petals and sepals are folded in the buds. It is an important feature in plant classification.

aether alternative spelling of ether, the hypothetical medium once believed to permeate all of space.

affinity force of attraction (see bond) between atoms that helps to keep them in combination in a molecule. The term is also applied to attraction between molecules, such as those of biochemical significance (for example, between enzymes and substrate molecules). This is the basis for affinity chromatography, by which biologically important compounds are separated.

afforestation planting of trees in areas that have not previously held forests. (**Reafforestation** is the planting of trees in deforested areas.) Trees may be planted (1) to provide timber and wood pulp; (2) to provide firewood where this is an energy source; (3) to bind soil together and prevent soil erosion; and (4) to act as windbreaks.

agglutination clumping together of antigens, such as red blood cells or bacteria, to form larger, visible clumps, under the influence of antibodies. As each antigen clumps only in response to its particular antibody, agglutination provides a way of determining blood groups and the identity of unknown bacteria.

aggression behaviour used to intimidate or injure another organism (of the same or of a different species), usually for the purposes of gaining territory, a mate, or food. Aggression often involves an escalating series of threats aimed at intimidating an opponent without having to engage in potentially dangerous physical contact. Aggressive signals include roaring by red deer, snarling by dogs, the fluffing-up of feathers by birds, and the raising of fins by some species of fish.

agonist muscle that contracts and causes a movement. Contraction of an agonist is complemented by relaxation of its antagonist. For example, the biceps (in the front of the upper arm) bends the elbow whilst the triceps (lying behind the biceps) straightens the arm.

agrochemical artificially produced chemical used in modern, intensive agricultural systems. Agrochemicals include nitrate and phosphate fertilizers, pesticides, some animal-feed additives, and pharmaceuticals.

air mixture of gases making up the Earth's atmosphere.

airglow faint and variable light in the Earth's atmosphere produced by chemical reactions (the recombination of ionized particles) in the ionosphere.

air mass large body of air with particular characteristics of temperature and humidity. An air mass forms when air rests over an area long enough to pick up the conditions of that area. When an air mass moves to another area it affects the weather of that area, but its own characteristics become modified in the process. For example, an air mass formed over the Sahara will be hot and dry, becoming cooler as it moves northwards. Air masses that meet form **fronts**.

air pump device used to pump air from one vessel to another, or to evacuate a vessel altogether to produce a vacuum.

albedo fraction of the incoming light reflected by a body such as a planet. A body with a high albedo, near 1, is very bright, while a body with a low albedo, near 0, is dark. The Moon has an average albedo of 0.12, Venus 0.76, Earth 0.37.

alcohol any member of a group of organic chemical compounds characterized by the presence of one or more aliphatic OH (hydroxyl) groups in the molecule, and which form esters with acids. The main uses of alcohols are as solvents for gums, resins, lacquers, and varnishes; in the making of dyes; for essential oils in perfumery; and for medical substances in pharmacy. The alcohol produced naturally in the fermentation process and consumed as part of alcoholic beverages is called ethanol.

aldehyde any of a group of organic chemical compounds prepared by oxidation of primary alcohols, so that the OH (hydroxyl) group loses its hydrogen to give an oxygen joined by a double bond to a carbon atom (the aldehyde group, with the formula CHO).

algebra branch of mathematics in which the general properties of numbers are studied by using symbols, usually letters, to represent variables and unknown quantities. For example, the algebraic statement

$$(x + y)^2 = x^2 + 2xy + y^2$$

is true for all values of x and y. Substitution of $x = 7$ and $y = 3$, for instance:

$$(7 + 3)^2 = 7^2 + 2(7 \times 3) + 3^2 = 100$$

An algebraic expression that has one or more variables (denoted by letters) is a polynomial equation. Algebra is used in many areas of mathematics – for example, arithmetic progression, or number sequences, and Boolean algebra (the latter is used in working out the logic for computers).

algebraic curve geometrical curve that can be precisely described by an (algebraic) equation.

algebraic fraction fraction in which letters are used to represent numbers – for example, a/b, xy/z^2, and $1/(x + y)$. Like numerical fractions, algebraic fractions may be simplified or factorized. Two equivalent algebraic fractions can be cross-multiplied; for example, if

$$\frac{a}{b} = \frac{c}{d}$$

then $ad = bc$

algebraic numbers numbers that satisfy a polynomial equation with rational coefficients: for example, $\sqrt{2}$ solves $x^2 - 2 = 0$. Real numbers that are not algebraic are called transcendental numbers. Although there is an infinity of algebraic numbers, there is in fact a 'larger' infinity of transcendental numbers.

algebraic topology study of surfaces and similar but more general objects in higher dimensions, using algebraic techniques. It is based upon homology.

alginate salt of alginic acid, $(C_6H_8O_6)_n$, obtained from brown seaweeds and used in textiles, paper, food products, and pharmaceuticals.

algorithm procedure or series of steps that can be used to solve a problem.

In computer science, it describes the logical sequence of operations to be performed by a program. A flow chart is a visual representation of an algorithm.

aliphatic compound any organic chemical compound in which the carbon atoms are joined in straight chains, as in hexane (C_6H_{14}), or in branched chains, as in 2-methylpentane $(CH_3CH(CH_3) CH_2CH_2CH_3)$.

alkali base that is soluble in water. Alkalis neutralize acids and are soapy to the touch. The strength of an alkali is measured by its hydrogen-ion concentration, indicated by the pH value. They may be divided into strong and weak alkalis: a strong alkali (for example, potassium hydroxide, KOH) ionizes completely when disssolved in water, whereas a weak alkali (for example, ammonium hydroxide, NH_4OH) exists in a partially

ionized state in solution. All alkalis have a pH above 7.0.

The hydroxides of metals are alkalis. Those of sodium and potassium are chemically powerful; both were historically derived from the ashes of plants.

alkali metal any of a group of six metallic elements with similar chemical properties: lithium, sodium, potassium, rubidium, caesium, and francium. They form a linked group (group 1) in the periodic table of the elements. They are univalent (have a valency of one) and of very low density (lithium, sodium, and potassium float on water); in general they are reactive, soft, low-melting-point metals. Because of their reactivity they are only found as compounds in nature.

alkaline-earth metal any of a group of six metallic elements with similar chemical properties: beryllium, magnesium, calcium, strontium, barium, and radium. They form a linked group in the periodic table of the elements. They are strongly basic, bivalent (have a valency of two), and occur in nature only in compounds.

alkaloid any of a number of physiologically active and frequently poisonous substances contained in some plants. They are usually organic bases and contain nitrogen. They form salts with acids and, when soluble, give alkaline solutions.

alkane member of the group of hydrocarbons having the general formula C_nH_{2n+2}, commonly known as **paraffins**. As they contain only single covalent bonds, alkanes are said to be saturated. Lighter alkanes, such as methane, ethane, propane, and butane, are colourless gases; heavier ones are liquids or solids. In nature they are found in natural gas and petroleum.

alkene member of the group of hydrocarbons having the general formula C_nH_{2n}, formerly known as **olefins**. Alkenes are unsaturated compounds, characterized by one or more double bonds between adjacent carbon atoms. Lighter alkenes, such as ethene and propene, are gases, obtained from the cracking of oil fractions. Alkenes react by addition, and many useful compounds, such as polyethene and bromoethane, are made from them.

alkyl any radical of the formula C_nH_{2n+1}; the chief members are methyl, ethyl, propyl, butyl, and amyl. These radicals are not stable in the free state but are found combined in a large number of types of organic compounds such as alcohols, esters, aldehydes, ketones, and halides.

alkyne member of the group of hydrocarbons with the general formula C_nH_{2n-2}, formerly known as **acetylenes**. They are unsaturated compounds, characterized by one or more triple bonds between adjacent carbon atoms. Lighter alkynes, such as ethyne, are gases; heavier ones are liquids or solids.

allantois bladder in the embryo of reptiles, birds, and mammals, that grows outside the embryo into the wall of the yolk sac of reptiles and birds and under the chorion of mammals. In mammals, blood vessels in the allantois carry blood to the placenta; in reptiles and birds the blood vessels permit respiration. As they develop, the vessels become the umbilical vein and arteries.

allele one of two or more alternative forms of a gene at a given position (locus) on a chromosome, caused by a difference in the sequence of DNA. Blue and brown eyes in humans are determined by different alleles of the gene for eye colour.

allene any of a class of dienes with adjacent double bonds. The simplest example is $CH_2{=}C{=}CH_2$, allene itself. Because of the stereochemistry of the double bonds, the terminal hydrogen atoms lie in planes mutually at right angles. Allenes behave mainly as typical unsaturated compounds, but are less stable than dienes with nonadjacent double bonds.

allosteric effect regulatory effect on an enzyme which takes place at a site distinct from that enzyme's catalytic site. For example, in a chain of enzymes the end product may act on an enzyme in the chain to regulate its own production.

allotropy property whereby an element can exist in two or more forms (allotropes), each possessing different physical properties but the same state of matter (gas, liquid, or solid). The allotropes of carbon are diamond, fullerene, and graphite. Sulphur has several allotropes (flowers of sulphur, plastic, rhombic, and monoclinic). These solids have different crystal structures, as do the white and grey forms of tin and the black, red, and white forms of phosphorus.

alloy metal blended with some other metallic or nonmetallic substance to give it special qualities, such as resistance to corrosion, greater hardness, or tensile strength. Useful alloys include bronze, brass, cupronickel, duralumin, German silver, gunmetal, pewter, solder, steel, and stainless steel.

allyl unsaturated organic radical corresponding to the formula $CH_2{=}CHCH_2{-}$.

allyl–metal complex coordination compound of a metal and the organic allyl ($CH_2{=}CHCH_2{-}$) group. Allyl-metal complexes are very reactive to oxygen, inflaming in air. The allyl groups are easily replaced by other ligands such as the carbonyl group, CO.

alpha chain particular secondary structure of the polypeptide chain (of a protein) brought about by hydrogen bonding between adjacent peptide units. Alpha chains occur, for example, in the haemoglobin molecule.

alpha decay spontaneous alteration of the nucleus of a radioactive atom, which transmutes the atom from one atomic number to another through the emission of a helium nucleus (known as an alpha particle). As a result, the atomic number decreases by two and the atomic weight decreases by four. See also radioactivity.

alpha particle positively charged, high-energy particle emitted from the nucleus of a radioactive atom. It is one of the products of the spontaneous disintegration of radioactive elements (see radioactivity) such as radium and thorium, and is identical to the nucleus of a helium atom – that is, it consists of two protons and two neutrons. The process of emission, **alpha decay**, transforms one element into another, decreasing the atomic (or proton) number by two and the atomic mass (or nucleon number) by four.

altazimuth astronomical instrument designed for observing the altitude and azimuth of a celestial object. It is essentially a large precision theodolite; that is, an instrument for the measurement of horizontal and vertical angles.

alternating current *AC*, electric current that flows for an interval of time in one direction and then in the opposite direction; that is, a current that flows in alternately reversed directions through or around a circuit. Electric energy is usually generated as alternating current in a power station, and alternating currents may be used for both power and lighting.

altitude measurement of height, usually given in metres above sea level.

alum any double sulphate of a monovalent metal or radical (such as sodium, potassium, or ammonium) and a trivalent metal (such as aluminium, chromium, or iron). The commonest alum is the double sulphate of potassium and aluminium, $K_2Al_2(SO_4)_4 \cdot 24H_2O$, a white crystalline powder that is readily soluble in water. It is used in curing animal skins. Other alums are used in papermaking and to fix dye in the textile industry.

alumina or ***corundum***, Al_2O_3, oxide of aluminium, widely distributed in clays, slates, and shales. It is formed by the decomposition of the feldspars in granite and used as an abrasive. Typically it is a white powder, soluble in most strong acids or caustic alkalis but not in water. Impure alumina is called 'emery'. Rubies, sapphires, and topaz are corundum gemstones.

aluminium lightweight, silver-white, ductile and malleable, metallic element, symbol Al, atomic number 13, relative atomic mass 26.9815, melting point 658°C/1,216°F. It is the third most abundant element (and the most abundant metal) in the Earth's crust, of which it makes up about 8.1% by mass. It is non-magnetic, an excellent conductor of electricity, and oxidizes easily, the layer of oxide on its surface making it highly resistant to tarnish.

aluminium hydroxide or ***alumina cream***, $Al(OH)_3$, gelatinous precipitate formed when a small amount of alkali solution is added to a solution of an aluminium salt:

$$Al_{(aq)} + 3OH_{(aq)} \rightarrow Al(OH)_{3(s)}$$

It is an amphoteric compound as it readily reacts with both acids and alkalis.

amalgam any alloy of mercury with other metals. Most metals will form amalgams, except iron and platinum. Amalgam is used in dentistry for filling teeth, and usually contains copper, silver, and zinc as the main alloying ingredients. This amalgam is pliable when first mixed and then sets hard, but the mercury leaches out and may cause a type of heavy-metal poisoning.

amatol explosive consisting of ammonium nitrate and TNT (trinitrotoluene) in almost any proportions.

amine any of a class of organic chemical compounds in which one or more of the hydrogen atoms of ammonia (NH_3) have been replaced by other groups of atoms.

amino acid water-soluble organic molecule, mainly composed of carbon, oxygen, hydrogen, and nitrogen, containing both a basic amino group (NH_2) and an acidic carboxyl (COOH) group. They are small molecules able to pass through membranes. When two or more amino acids are joined together, they are known as peptides; proteins are made up of peptide chains folded or twisted in characteristic shapes.

ammeter instrument that measures electric current (flow of charge per unit time), usually in amperes, through a conductor. It should not be confused with a voltmeter, which measures potential difference between two points in a circuit. The ammeter is placed in series (see series circuit) with the component through which current is to be measured, and is constructed with a low internal resistance in order to prevent the reduction of that current as it flows through the instrument itself. A common type is the moving-coil meter, which measures direct current (DC), but can, in the presence of a rectifier, measure alternating current (AC) also. Hot-wire, moving-iron, and dynamometer ammeters can be used for both DC and AC.

ammonia, NH_3, colourless pungent-smelling gas, lighter than air and very soluble in water. It is made on an industrial scale by the Haber (or Haber–Bosch) process, and used mainly to produce nitrogenous fertilizers, nitric acid, and some explosives.

ammonium carbonate, $(NH_4)_2CO_3$, white, crystalline solid that readily sublimes at room temperature into its constituent gases: ammonia, carbon dioxide, and water. It was formerly used in smelling salts.

ammonium chloride or ***sal ammoniac***, NH_4Cl, volatile salt that forms white crystals around volcanic craters. It is prepared synthetically for use in 'dry-cell' batteries, fertilizers, and dyes.

ammonium nitrate, NH_4NO_3, colourless, crystalline solid, prepared by neutralization of nitric acid with ammonia; the salt is crystallized from the solution. It sublimes on heating.

amnion innermost of three membranes that enclose the embryo within the egg (reptiles and birds) or within the uterus (mammals). It contains the amniotic fluid that helps to cushion the embryo.

ampere SI unit (symbol A) of electrical current. Electrical current is measured in a similar way to water current, in terms of an amount per unit time; one ampere (amp) represents a flow of one coulomb per second, which is about 6.28×10^{18} electrons per second.

Ampère's rule rule developed by French physicist André-Marie Ampère connecting the direction of an electric current and its associated magnetic currents. It states that around a wire carrying a current towards the observer, the magnetic field curls in the anticlockwise direction. This assumes the conventional direction of current flow (from the positive to the negative terminal).

amphoteric term used to describe the ability of some chemical compounds to behave either as an acid or as a base depending on their environment. For example, the metals aluminium and zinc, and their oxides and

441

hydroxides, act as bases in acidic solutions and as acids in alkaline solutions.

amplifier electronic device that magnifies the strength of a signal, such as a radio signal. The ratio of output signal strength to input signal strength is called the gain of the amplifier. As well as achieving high gain, an amplifier should be free from distortion and able to operate over a range of frequencies. Practical amplifiers are usually complex circuits, although simple amplifiers can be built from single transistors or valves.

amplitude modulation *AM*, method by which radio waves are altered for the transmission of broadcasting signals. AM waves are constant in frequency, but the amplitude of the transmitting wave varies in accordance with the signal being broadcast.

amylase one of a group of enzymes that break down starches into their component molecules (sugars) for use in the body. It occurs widely in both plants and animals. In humans, it is found in saliva and in pancreatic juices.

anaerobic not requiring oxygen for the release of energy from food molecules such as glucose. Anaerobic organisms include many bacteria, yeasts, and internal parasites. Anaerobic respiration in humans is less efficient than aerobic respiration at releasing energy, but releases energy faster. See respiration.

anaesthetic drug that produces loss of sensation or consciousness; the resulting state is **anaesthesia**, in which the patient is insensitive to stimuli. Anaesthesia may also happen as a result of nerve disorder.

analgesic agent for relieving pain. Opiates alter the perception or appreciation of pain and are effective in controlling 'deep' visceral (internal) pain. Non-opiates, such as aspirin, paracetamol, and NSAIDs (nonsteroidal anti-inflammatory drugs), relieve musculo-skeletal pain and reduce inflammation in soft tissues.

analogous term describing a structure that has a similar function to a structure in another organism, but not a similar evolutionary path. For example, the wings of bees and of birds have the same purpose – to give powered flight – but have different origins. Compare homologous.

analysis in chemistry, determination of the composition of substances. See analytical chemistry.

analysis in mathematics, branch concerned with limiting processes on axiomatic number systems; calculus of variations and infinitesimal calculus are now called analysis.

analytical chemistry branch of chemistry that deals with the determination of the chemical composition of substances. **Qualitative analysis** determines the identities of the substances in a given sample; **quantitative analysis** determines how much of a particular substance is present.

AND rule rule used for finding the combined probability of two or more independent events both occurring. If two events E_1 and E_2 are independent (have no effect on each other) and the probabilities of their taking place are p_1 and p_2, respectively, then the combined probability p that both E_1 and E_2 will happen is given by:

$$p = p_1 \times p_2$$

anemometer device for measuring wind speed and liquid flow. The most basic form, the **cup-type anemometer**, consists of cups at the ends of arms, which rotate when the wind blows. The speed of rotation indicates the wind speed.

anemophily type of pollination in which the pollen is carried on the wind. Anemophilous flowers are usually unscented, have either very reduced petals and sepals or lack them altogether, and do not produce nectar. In some species they are borne in catkins. Male and female reproductive structures are commonly found in separate flowers. The male flowers have numerous exposed stamens, often on long filaments; the female flowers have long, often branched, feathery stigmas.

angiosperm flowering plant in which the seeds are enclosed within an ovary, which ripens into a fruit. Angiosperms are divided into monocotyledons (single seed leaf in the embryo) and dicotyledons (two seed leaves in the embryo). They include the majority of flowers, herbs, grasses, and trees except conifers.

angle of incidence angle between a ray of light striking a mirror (incident ray) and the normal to that mirror. It is equal to the angle of reflection.

angle of parallelism in non-Euclidean geometry involving the application of classical geometrical principles to nonflat surfaces, the angle at which a line perpendicular to one of two parallel lines meets the other – which may also be 90°, or may be less.

angle of reflection angle between a ray of light reflected from a mirror and the normal to that mirror. It is equal to the angle of incidence.

angle of refraction angle between a refracted ray of light and the normal to the surface at which refraction occurred. When a ray passes from air into a denser medium such as glass, it is bent towards the normal so that the angle of refraction is less than the angle of incidence.

angular momentum type of momentum.

anhydride chemical compound obtained by the removal of water from another compound, usually a dehydrated acid. For example, sulphur (VI) oxide (sulphur trioxide, SO_3) is the anhydride of sulphuric acid (H_2SO_4).

anhydrite naturally occurring anhydrous calcium sulphate ($CaSO_4$). It is used commercially for the manufacture of plaster of Paris and builders' plaster.

anhydrous of a chemical compound, containing no water. If the water of crystallization is removed from blue crystals of copper (II) sulphate, a white powder (anhydrous copper sulphate) results. Liquids from which all traces of water have been removed are also described as being anhydrous.

aniline (Portuguese *anil* 'indigo'), $C_6H_5NH_2$ or phenylamine, one of the simplest aromatic chemicals (a substance related to benzene, with its carbon atoms joined in a ring). When pure, it is a colourless oily liquid; it has a characteristic odour, and turns brown on contact with air. It occurs in coal tar, and is used in the rubber industry and to make drugs and dyes. It is highly poisonous.

anion ion carrying a negative charge. During electrolysis, anions in the electrolyte move towards the anode (positive electrode).

anisotropic or *aelotropic*, describing a substance that has different physical properties in different directions. Some crystals, for example, have different refractive indices in different directions.

annelid any segmented worm of the phylum Annelida. Annelids include earthworms, leeches, and marine worms such as lugworms.

annular eclipse solar eclipse in which the Moon does not completely obscure the Sun and a thin ring of sunlight remains visible. Annular eclipses occur when the Moon is at its furthest point from the Earth.

anode in chemistry, the positive electrode of an electrolytic cell, towards which negative particles (anions), usually in solution, are attracted. See electrolysis.

anode in electronics, the positive electrode of a thermionic valve, cathode ray tube, or similar device, towards which electrons are drawn after being emitted from the cathode.

antagonist muscle that relaxes in response to the contraction of its agonist muscle. The biceps, in the front of the upper arm, bends the elbow while the triceps, lying behind the biceps, straightens the arm.

antheridium organ producing the male gametes, antherozoids, in algae, bryophytes (mosses and liverworts), and pteridophytes (ferns, club mosses, and horsetails). It may be either single-celled, as in most algae, or multicellular, as in bryophytes and pteridophytes.

antherozoid motile (or independently moving) male gamete produced by algae, bryophytes (mosses and liverworts), pteridophytes (ferns, club mosses, and horsetails), and some gymnosperms (notably the cycads). Antherozoids are formed in an antheridium and, after being released, swim by means of one or more flagella, to the female gametes. Higher plants have nonmotile male gametes contained within pollen grains.

anthracene white, glistening, crystalline, tricyclic, aromatic hydrocarbon with a faint blue fluorescence when pure. Its melting point is about 216°C/421°F and its boiling point 351°C/664°F. It occurs in the high-boiling-point fractions of coal tar, where it was discovered in 1832 by the French chemists Auguste Laurent and Jean Dumas.

anthropic principle idea that 'the universe is the way it is because if it were different we would not be here to observe it'. The principle arises from the observation that if the laws of science were even slightly different, it would have been impossible for intelligent life to evolve. For example, if the strengths of the fundamental forces were only slightly different, stars would have been unable to burn hydrogen and produce the chemical elements that make up our bodies.

anthropometry science dealing with the measurement of the human body, particularly stature, body weight, cranial capacity, and length of limbs, in samples of living populations, as well as the remains of buried and fossilized humans.

antibiotic drug that kills or inhibits the growth of bacteria and fungi. It is derived from living organisms such as fungi or bacteria, which distinguishes it from synthetic antimicrobials.

antibody protein molecule produced in the blood by lymphocytes in response to the presence of foreign or invading substances (antigens); such substances include the proteins carried on the surface of infecting micro-organisms. Antibody production is only one aspect of immunity in vertebrates.

anticline rock layers or beds folded to form a convex arch (seldom preserved intact) in which older rocks make up the core. Where relative ages of the rock layers, or

443

stratigraphic ages, are not known, convex upward folded rocks are referred to as antiforms.

antiferromagnetic material material with a very low magnetic susceptibility that increases with temperature up to a certain temperature, called the Néel temperature. Above the Néel temperature, the material is only weakly attracted to a strong magnet.

antigen any substance that causes the production of antibodies by the body's immune system. Common antigens include the proteins carried on the surface of bacteria, viruses, and pollen grains. The proteins of incompatible blood groups or tissues also act as antigens, which has to be taken into account in medical procedures such as blood transfusions and organ transplants.

antilogarithm or *antilog*, the inverse of logarithm, or the number whose logarithm to a given base is a given number. If $y = \log_a x$, then $x = \text{antilog}_a y$.

antimatter a form of matter in which most of the attributes (such as electrical charge, magnetic moment, and spin) of elementary particles are reversed. Such particles (antiparticles) can be created in particle accelerators, such as those at CERN in Geneva, Switzerland, and at Fermilab in the USA. In 1996 physicists at CERN created the first atoms of antimatter: nine atoms of antihydrogen survived for 40 nanoseconds.

antiparticle in nuclear physics, a particle corresponding in mass and properties to a given elementary particle but with the opposite electrical charge, magnetic properties, or coupling to other fundamental forces. For example, an electron carries a negative charge whereas its antiparticle, the positron, carries a positive one. When a particle and its antiparticle collide, they destroy each other, in the process called 'annihilation', their total energy being converted to lighter particles and/or photons. A substance consisting entirely of antiparticles is known as antimatter.

antiseptic any substance that kills or inhibits the growth of micro-organisms. The use of antiseptics was pioneered by Joseph Lister. He used carbolic acid (phenol), which is a weak antiseptic; antiseptics such as TCP are derived from this.

aperture size of the opening admitting light into an optical instrument such as a telescope or camera.

aperture synthesis technique used in radio astronomy in which several small radio dishes are linked together over an area that can be many kilometres in diameter. This is done to simulate the performance of one very large radio telescope.

aphelion point at which an object, travelling in an elliptical orbit around the Sun, is at its furthest from the Sun. The Earth is at its aphelion on 5 July.

Apollonius' problem problem set by Apollonius of Perga, to describe a circle touching three other given circles.

aposematic coloration technical name for warning coloration markings that make a dangerous, poisonous, or foul-tasting animal particularly conspicuous and recognizable to a predator. Examples include the yellow and black stripes of bees and wasps, and the bright red or yellow colours of many poisonous frogs and snakes. See also mimicry.

apparent observed, used, for example, in **apparent magnitude** and **apparent star place** – the observed position which will only yield the mean position given in star catalogues when it has been corrected for the effects of aberration, precession, and nutation.

apparent depth depth that a transparent material such as water or glass appears to have when viewed from above. This is less than its real depth because of the refraction that takes place when light passes into a less dense medium. The ratio of the real depth to the apparent depth of a transparent material is equal to its refractive index.

Appleton layer or *F layer*, band containing ionized gases in the Earth's upper atmosphere, at a height of 150–1,000 km/ 94–625 mi, above the E layer (formerly the Kennelly–Heaviside layer). It acts as a dependable reflector of radio signals as it is not affected by atmospheric conditions, although its ionic composition varies with the sunspot cycle.

approximation rough estimate of a given value. For example, for pi (which has a value of 3.1415926 correct to seven decimal places), 3 is an approximation to the nearest whole number.

aqueous humour watery fluid found in the chamber between the cornea and lens of the vertebrate eye. Similar to blood serum in composition, it is constantly renewed.

arc section of a curved line or circle. A circle has three types of arc: a **semicircle**, which is exactly half of the circle; **minor arcs**, which are less than the semicircle; and **major arcs**, which are greater than the semicircle.

Archaean or *Archaeozoic*, widely used term for the earliest era of geological time; the first part of the Precambrian **Eon**, spanning the interval from the formation of Earth to about 2,500 million years ago.

Archimedes' principle principle that the weight of the liquid displaced by a floating body is equal to the weight of the body. The principle is often stated in the form: 'an object totally or partially submerged in a fluid displaces a volume of fluid that weighs the same as the apparent loss in weight of the object (which, in turn, equals the upwards force, or upthrust, experienced by that object).' It was discovered by the Greek mathematician Archimedes.

arc minute, arc second units for measuring small angles, used in geometry, surveying, map-making, and astronomy. An arc minute (symbol ′) is one-sixtieth of a degree, and an arc second (symbol ″) is one-sixtieth of an arc minute. Small distances in the sky, as between two close stars or the apparent width of a planet's disc, are expressed in minutes and seconds of arc.

Argand diagram in mathematics, a method for representing complex numbers by Cartesian coordinates (x, y). Along the x-axis (horizontal axis) are plotted the real numbers, and along the y-axis (vertical axis) the nonreal, or imaginary, numbers.

armillary sphere earliest known astronomical device, in use from the 3rd century BC. It showed the Earth at the centre of the universe, surrounded by a number of movable metal rings representing the Sun, Moon, and planets. The armillary sphere was originally used to observe the heavens and later for teaching navigators about the arrangements and movements of the heavenly bodies.

aromatic compound organic chemical compound in which some of the bonding electrons are delocalized (shared among several atoms within the molecule and not localized in the vicinity of the atoms involved in bonding). The commonest aromatic compounds have ring structures, the atoms msking up the ring being either all carbon or containing one or more different atoms (usually nitrogen, sulphur, or oxygen). Typical examples are benzene (C_6H_6) and pyridine (C_5H_5N).

aryl any organic radical derived from an aromatic compound; for example, the phenyl radical ($-C_6H_5$) is derived from benzene (C_6H_6).

associative operation an operation in which the outcome is independent of the grouping of the numbers or symbols concerned. For example, multiplication is associative, as $4 \times (3 \times 2) = (4 \times 3) \times 2 = 24$; however, division is not, as $12 \div (4 \div 2) = 6$, but $(12 \div 4) \div 2 = 1.5$. Compare commutative operation and distributive operation.

asteroid any of many thousands of small bodies, composed of rock and iron, that orbit the Sun. Most lie in a belt between the orbits of Mars and Jupiter, and are thought to be fragments left over from the formation of the Solar System. About 100,000 may exist, but their total mass is only a few hundredths the mass of the Moon.

astrolabe ancient navigational instrument, forerunner of the sextant. Astrolabes usually consisted of a flat disc with a sighting rod that could be pivoted to point at the Sun or bright stars.

From the altitude of the Sun or star above the horizon, the local time could be estimated.

astronomical colour index difference in a star's brightness when measured on two selected wavelengths, in order to determine the star's temperature. Cooler stars emit more light at longer wavelengths (and so appear redder than hot stars). Modern methods involve photoelectric filtering and the UBV photometry system.

astronomical unit unit (symbol AU) equal to the mean distance of the Earth from the Sun: 149.6 million km/92.96 million mi. It is used to describe planetary distances. Light travels this distance in approximately 8.3 minutes.

astrophysics study of the physical nature of stars, galaxies, and the universe. It began with the development of spectroscopy in the 19th century, which allowed astronomers to analyse the composition of stars from their light. Astrophysicists view the universe as a vast natural laboratory in which they can study matter under conditions of temperature, pressure, and density that are unattainable on Earth.

asymptote straight line that a curve approaches progressively more closely but never reaches. The x and y axes are asymptotes to the graph of $xy =$ constant (a rectangular hyperbola).

atmosphere mixture of gases surrounding a planet. Planetary atmospheres are prevented from escaping by the pull of gravity. Atmospheric pressure, the density of gases in the atmosphere, decreases with altitude. In its lowest layer, the Earth's atmosphere consists of nitrogen (78%) and oxygen (21%), both in molecular form (two atoms bonded together), and 1% argon. Small quantities of other gases are important to the

445

chemistry and physics of the Earth's atmosphere, including water, carbon dioxide, and ozone.

atmosphere or **standard atmosphere**, unit (symbol atm) of pressure equal to 760 torr, 1013.25 millibars, or 1.01325×10^5 pascals, or newtons per square metre. The actual pressure exerted by the atmosphere fluctuates around this value, which is assumed to be standard at sea level and 0°C/32°F, and is used when dealing with very high pressures.

atmospheric pressure pressure at any point on the Earth's surface that is due to the weight of the column of air above it; it therefore decreases as altitude increases, simply because there is less air above. At sea level the average pressure is 101 kilopascals (1,013 millibars, 760 mmHg, or 14.7 lb per sq in, or 1 atmosphere). Changes in atmospheric pressure, measured with a barometer, are used in weather forecasting. Areas of relatively high pressure are called anticyclones; areas of low pressure are called depressions.

atom (Greek *atomos* 'undivided') smallest unit of matter that can take part in a chemical reaction, and which cannot be broken down chemically into anything simpler. An atom is made up of protons and neutrons in a central nucleus surrounded by electrons. The atoms of the various elements differ in atomic number, relative atomic mass, and chemical behaviour.

atom, electronic structure of arrangement of electrons around the nucleus of an atom, in distinct energy levels, also called orbitals or shells (see orbital, atomic). These shells can be regarded as a series of concentric spheres, each of which can contain a certain maximum number of electrons; the noble gases have an arrangement in which every shell contains this number (see noble gas structure). The energy levels are usually numbered beginning with the shell nearest to the nucleus. The outermost shell is known as the valency shell as it contains the valence electrons.

atomic force microscope *AFM*, microscope developed in the late 1980s that produces a magnified image using a diamond probe, with a tip so fine that it may consist of a single atom, dragged over the surface of a specimen to 'feel' the contours of the surface. In effect, the tip acts like the stylus of a record player, reading the surface. The tiny up-and-down movements of the probe are converted to an image of the surface by computer and displayed on a screen. The AFM is useful for examination of biological specimens since, unlike the scanning tunnelling microscope, the specimen does not have to be electrically conducting.

atomic heat product of relative atomic mass and specific heat capacity. It is approximately constant for many solid elements and equal to 6 calories per gram atom per degree (25.2 joules per mole per Kelvin).

atomicity number of atoms of an element that combine together to form a molecule. A molecule of oxygen (O_2) has atomicity 2; sulphur (S_8) has atomicity 8.

atomic mass see relative atomic mass.

atomic mass unit or **dalton**, unit (symbol u) of mass that is used to measure the relative mass of atoms and molecules. It is equal to one-twelfth of the mass of a carbon-12 atom, which is approximately the mass of a proton or 1.66×10^{-27} kg. The relative atomic mass of an atom has no units; thus oxygen-16 has an atomic mass of 16 daltons but a relative atomic mass of 16.

atomic number or **proton number**, number (symbol Z) of protons in the nucleus of an atom. It is equal to the positive charge on the nucleus. In a neutral atom, it is also equal to the number of electrons surrounding the nucleus. The chemical elements are arranged in the periodic table of the elements according to their atomic number.

atomic orbital region in space occupied by an electron associated with the nucleus of an atom. Atomic orbitals have various shapes, depending on the energy level of the electron and the degree of hybridization. Atomic orbitals overlap to form molecular orbitals, or chemical bonds between atoms.

atomic radiation energy given out by disintegrating atoms during radioactive decay, whether natural or synthesized. The energy may be in the form of fast-moving particles, known as alpha particles and beta particles, or in the form of high-energy electromagnetic waves known as gamma radiation. Overlong exposure to atomic radiation can lead to radiation sickness.

atomic radius effective radius of an atom. Atomic radii vary periodically with atomic number, being largest for the alkali metals and smallest for the rare gases.

atomic size or **atomic radius**, size of an atom expressed as the radius in angstroms or other units of length.

The sodium atom has an atomic radius of 1.57 angstroms (1.57×10^{-8} cm). For metals, the size of the atom is always greater than the size of its ion. For non-metals the reverse is true.

atomic structure internal structure of an atom.

the nucleus The core of the atom is the **nucleus**, a dense body only one ten-thousandth the diameter of the atom itself. The simplest nucleus, that of hydrogen, comprises a single, stable, positively charged particle, the **proton**. Nuclei of other elements contain more protons and additional particles, called **neutrons**, of about the same mass as the proton but with no electrical charge. Each element has its own characteristic nucleus with a unique number of protons, the atomic number. The number of neutrons may vary. Where atoms of a single element have different numbers of neutrons, they are called isotopes. Although some isotopes tend to be unstable and exhibit radioactivity, they all have identical chemical properties.

electrons The nucleus is surrounded by a number of moving **electrons**, each of which has a negative charge equal to the positive charge on a proton, but which weighs only $1/1,836$ times as much. In a neutral atom, the nucleus is surrounded by the same number of electrons as it contains protons. According to quantum theory, the position of an electron is uncertain; it may be found at any point. However, it is more likely to be found in some places than others. The region of space in which an electron is most likely to be found is called an orbital (see orbital, atomic). The chemical properties of an element are determined by the ease with which its atoms can gain or lose electrons.

atomic volume volume of one gram-atom of an element.

atomic weight another name for relative atomic mass.

ATP abbreviation for *adenosine triphosphate*, nucleotide molecule found in all cells. It can yield large amounts of energy, and is used to drive the thousands of biological processes needed to sustain life, growth, movement, and reproduction. Green plants use light energy to manufacture ATP as part of the process of photosynthesis. In animals, ATP is formed by the breakdown of glucose molecules, usually obtained from the carbohydrate component of a diet, in a series of reactions termed respiration. It is the driving force behind muscle contraction and the synthesis of complex molecules needed by individual cells.

attenuation reduction in the virulence of a pathogenic micro-organism by culturing it in unfavourable conditions, by drying or heating it, or by subjecting it to chemical treatment. Attenuated viruses, for example, are used in vaccines.

aurora spectacular array of light in the night sky, caused by charged particles from the Sun entering the Earth's upper atmosphere. The **aurora borealis** is seen in the north of the northern hemisphere, the **aurora australis** in the south of the southern.

autoclave reactor vessel constructed to allow chemical reactions to take place at high temperature and pressure. Such vessels are used, for example, in the manufacture of chemicals and materials for the construction industry.

autogyro V/STOL aircraft able to take off and land without the use of a runway. It uses horizontal rotor to achieve take-off and sustain height, and forward propeller to provide forward motion through the air.

automorphic function function that in relation to a group of transformations has a value on the transformed point identical with the value on the original point.

autophagy process in which a cell synthesizes substances and then metabolizes and absorbs them for its own sustenance.

Avogadro's hypothesis law stating that equal volumes of all gases, when at the same temperature and pressure, have the same number of molecules. It was first propounded by Amedeo Avogadro.

Avogadro's number or *Avogadro's constant*, number of carbon atoms in 12 g of the carbon-12 isotope (6.022045×10^{23}). The relative atomic mass of any element, expressed in grams, contains this number of atoms. It is named after Amedeo Avogadro.

axiom statement that is assumed to be true and upon which theorems are proved by using logical deduction; for example, two straight lines cannot enclose a space. The Greek mathematician Euclid used a series of axioms that he considered could not be demonstrated in terms of simpler concepts to prove his geometrical theorems.

axis plural *axes*, one of the reference lines by which a point on a graph may be located. The horizontal axis is usually referred to as the x-axis, and the vertical axis as the y-axis. The term is also used to refer to the imaginary line about which an object may be said to be symmetrical (**axis of symmetry**) – for example, the diagonal of a square – or the line about which an object may revolve (**axis of rotation**).

axon long threadlike extension of a nerve cell that conducts electrochemical impulses away from the cell body towards other nerve

447

cells, or towards an effector organ such as a muscle. Axons terminate in synapses, junctions with other nerve cells, muscles, or glands.

azimuth angular distance of an object eastwards along the horizon, measured from due north, between the astronomical meridian (the vertical circle passing through the centre of the sky and the north and south points on the horizon) and the vertical circle containing the celestial body whose position is to be measured.

background radiation radiation that is always present in the environment. By far the greater proportion (87%) of it is emitted from natural sources. Alpha and beta particles, and gamma radiation are radiated by the traces of radioactive minerals that occur naturally in the environment and even in the human body, and by radioactive gases, such as radon, which are found in soil and may seep upwards into buildings. Radiation from space (cosmic radiation) also contributes to the background level.

bacteria singular *bacterium*, microscopic single-celled organisms lacking a nucleus. Bacteria are widespread, being present in soil, air, and water, and as parasites on and in other living things. Some parasitic bacteria cause disease by producing toxins, but others are harmless and can even benefit their hosts. Bacteria usually reproduce by binary fission (dividing into two equal parts), and, on average, this occurs every 20 minutes. Only 4,000 species of bacteria are known (in 1998), although bacteriologists believe that around 3 million species may actually exist. Certain types of bacteria are vital in many food and industrial processes, while others play an essential role in the nitrogen cycle, which maintains soil fertility.

bacteriology the study of bacteria.

bacteriophage virus that attacks bacteria, commonly called a phage. Such viruses are of use in genetic engineering.

bacteriotropin or *opsonin*, substance in blood serum that helps to make bacteria more vulnerable to leucocytes (white blood cells), which can then engulf and destroy them.

Baily's beads bright spots of sunlight seen around the edge of the Moon for a few seconds immediately before and after a total eclipse of the Sun, caused by sunlight shining between mountains at the Moon's edge. Sometimes one bead is much brighter than the others, producing the so-called **diamond ring** effect. The effect was described in 1836 by the English astronomer Francis Baily, a wealthy stockbroker who retired in 1825 to devote himself to astronomy.

balance of nature idea that there is an inherent equilibrium in most ecosystems, with plants and animals interacting so as to produce a stable, continuing system of life on Earth. The activities of human beings can, and frequently do, disrupt the balance of nature.

Balmer series visible spectrum of hydrogen, consisting of a series of distinct spectral lines with wavelengths in the visible region.

bandwidth range of frequencies over which the capability of a receiver or other electric device does not differ from its peak by a given amount.

barium (Greek *barytes* 'heavy') soft, silver-white, metallic element, symbol Ba, atomic number 56, relative atomic mass 137.33. It is one of the alkaline-earth metals, found in nature as barium carbonate and barium sulphate. As the sulphate it is used in medicine: taken as a suspension (a 'barium meal'), its movement along the gut is followed using X-rays. The barium sulphate, which is opaque to X-rays, shows the shape of the gut, revealing any abnormalities of the alimentary canal. Barium is also used in alloys, pigments, and safety matches and, with strontium, forms the emissive surface in cathode-ray tubes. It was first discovered in barytes or heavy spar.

barometer instrument that measures atmospheric pressure as an indication of weather. Most often used are the **mercury barometer** and the **aneroid barometer**.

barycentric calculus coordinate geometry calculations using a coordinate system devised by August Möbius in which numerical coefficients are assigned to points on a plane, giving the position of a general point by reference to four or more non-coplanar points. Described in this way, the general point thus represents a centre of gravity for a distribution of mass at the four (or more) points proportional to the assigned numbers.

base substance that accepts protons. Bases can contain negative ions such as the hydroxide ion (OH^-), which is the strongest base, or be molecules such as ammonia (NH_3). Ammonia is a weak base, as only some of its molecules accept protons.

$$OH^- + H^+_{(aq)} \rightarrow H_2O_{(l)}$$
$$NH_3 + H_2O \rightleftharpoons NH_4^+ + OH^-$$

Bases that dissolve in water are called alkalis.

base pair linkage of two base (purine or pyrimidine) molecules that join the complementary strands of DNA. Adenine forms a base pair with thymine (or uracil in RNA) and cytosine pairs with guanine in a double-stranded nucleic acid molecule.

battery any energy-storage device allowing release of electricity on demand. It is made up of one or more electrical cells. Primary-cell batteries are disposable; secondary-cell batteries, or accumulators, are rechargeable. Primary-cell batteries are an extremely uneconomical form of energy, since they produce only 2% of the power used in their manufacture. It is dangerous to try to recharge a primary-cell battery.

Bernoulli numbers or *B numbers*, sequence of rational numbers that may be represented by the symbolic form $B_n = (B+1)^n$, corresponding to the sequence:

$$B_0 = 1 \qquad B_1 = -\tfrac{1}{2}$$
$$B_3 = 0 \qquad B_4 = -\tfrac{1}{30}$$
$$B_6 = \tfrac{1}{42} \qquad B_7 = 0$$

and so on. *B* numbers of odd order (except B_1) are zero; *B* numbers of even order alternate to sign. They are named after Swiss mathematician Jean Bernoulli and were first discussed in his work on probability *Ars Conjectandi*.

Bernoulli's theorem sum of the pressure, potential energy, and kinetic energy of a fluid flowing along a tube is constant, provided the flow is steady, incompressible and non-viscous.

Bessemer process first cheap method of making steel, invented by Henry Bessemer in England in 1856. It has since been superseded by more efficient steel-making processes, such as the basic–oxygen process. In the Bessemer process compressed air is blown into the bottom of a converter, a furnace shaped like a cement mixer, containing molten pig iron. The excess carbon in the iron burns out, other impurities form a slag, and the furnace is emptied by tilting.

beta decay disintegration of the nucleus of an atom to produce a beta particle, or high-speed electron, and an electron antineutrino. During beta decay, a neutron in the nucleus changes into a proton, thereby increasing the atomic number by one while the mass number stays the same. The mass lost in the change is converted into kinetic (movement) energy of the beta particle. Beta decay is caused by the weak nuclear force, one of the fundamental forces of nature operating inside the nucleus.

beta particle electron ejected with great velocity from a radioactive atom that is undergoing spontaneous disintegration. Beta particles do not exist in the nucleus but are created on disintegration, beta decay, when a neutron converts to a proton by emitting an electron.

Betti numbers numbers characterizing the connectivity of a variety. They are named after Italian mathematician Enrico Betti.

Bhabha scattering scattering process involving electrons and positrons, first determined by Indian physicist Homi Bhabha in 1935.

Big Bang hypothetical 'explosive' event that marked the origin of the universe as we know it. At the time of the Big Bang, the entire universe was squeezed into a hot, superdense state. The Big Bang explosion threw this compact material outwards, producing the expanding universe (see red shift). The cause of the Big Bang is unknown; observations of the current rate of expansion of the universe suggest that it took place about 10–20 billion years ago. The Big Bang theory began modern cosmology.

bile brownish alkaline fluid produced by the liver. Bile is stored in the gall bladder and is intermittently released into the duodenum (small intestine) to aid digestion. Bile consists of bile salts, bile pigments, cholesterol, and lecithin. **Bile salts** assist in the breakdown and absorption of fats; **bile pigments** are the breakdown products of old red blood cells that are passed into the gut to be eliminated with the faeces.

binary star pair of stars moving in orbit around their common centre of mass. Observations show that most stars are binary, or even multiple – for example, the nearest star system to the Sun, Rigil Kent (Alpha Centauri).

biochemistry science concerned with the chemistry of living organisms: the structure and reactions of proteins (such as enzymes), nucleic acids, carbohydrates, and lipids.

biodegradable capable of being broken down by living organisms, principally bacteria and fungi. In biodegradable substances, such as food and sewage, the natural processes of decay lead to compaction and liquefaction, and to the release of nutrients that are then recycled by the ecosystem.

biofeedback modification or control of a biological system by its results or effects. For example, a change in the position or trophic level of one species affects all levels above it.

bioluminescence production of light by living organisms. It is a feature of many deep-sea

449

fishes, crustaceans, and other marine animals. On land, bioluminescence is seen in some nocturnal insects such as glow-worms and fireflies, and in certain bacteria and fungi. Light is usually produced by the oxidation of luciferin, a reaction catalyzed by the enzyme luciferase. This reaction is unique, being the only known biological oxidation that does not produce heat. Animal luminescence is involved in communication, camouflage, or the luring of prey, but its function in some organisms is unclear.

biomass total mass of living organisms present in a given area. It may be specified for a particular species (such as earthworm biomass) or for a general category (such as herbivore biomass). Estimates also exist for the entire global plant biomass. Measurements of biomass can be used to study interactions between organisms, the stability of those interactions, and variations in population numbers. Where dry biomass is measured, the material is dried to remove all water before weighing.

biosynthesis synthesis of organic chemicals from simple inorganic ones by living cells – for example, the conversion of carbon dioxide and water to glucose by plants during photosynthesis.

Other biosynthetic reactions produce cell constituents including proteins and fats.

biotic factor organic variable affecting an ecosystem – for example, the changing population of elephants and its effect on the African savannah.

biquaternions type of quaternions devised by English mathematician William Clifford to use specifically in association with linear algebra to represent motions in three-dimensional non-Euclidean space.

black body hypothetical object that completely absorbs all electromagnetic radiation striking it. It is also a perfect emitter of thermal radiation.

black-body radiation radiation of a thermal nature emitted by a theoretically perfect emitter of radiation at a certain temperature.

black hole object in space whose gravity is so great that nothing can escape from it, not even light. Thought to form when massive stars shrink at the end of their lives, a black hole sucks in more matter, including other stars, from the space around it. Matter that falls into a black hole is squeezed to infinite density at the centre of the hole. Black holes can be detected because gas falling towards them becomes so hot that it emits X-rays.

blast furnace smelting furnace used to extract metals from their ores, chiefly pig iron from iron ore. The temperature is raised by the injection of an air blast.

blastocyst in mammals, the hollow ball of cells which is an early stage in the development of the embryo, roughly equivalent to the blastula of other animal groups.

blastoderm sheet of cells that grows on the surface of a fertilized ovum. In mammals it forms a disc of cells that eventually develops into the embryo between the amniotic cavity and the yolk sac. The endoderm, mesoderm, and ectoderm also develop from the blastoderm.

blastula early stage in the development of a fertilized egg, when the egg changes from a solid mass of cells (the morula) to a hollow ball of cells (the blastula), containing a fluid-filled cavity (the blastocoel). See also embryology.

blood group any of the types into which blood is classified according to the presence or otherwise of certain antigens on the surface of its red cells. Red blood cells of one individual may carry molecules on their surface that act as antigens in another individual whose red blood cells lack these molecules. The two main antigens are designated A and B. These give rise to four blood groups: having A only (A), having B only (B), having both (AB), and having neither (O). Each of these groups may or may not contain the rhesus factor. Correct typing of blood groups is vital in transfusion, since incompatible types of donor and recipient blood will result in coagulation, with possible death of the recipient.

blue dwarf, blue giant high-temperature stars (as opposed to red stars). Blue giants are generally on or near the main sequence of the Hertzsprung–Russell diagram; blue dwarfs represent the very dense, but very small, near-final form of what was once a red giant.

B number abbreviation for Bernoulli number.

Bode's law or *Titius–Bode law*, numerical sequence that gives the approximate distances, in astronomical units (distance between Earth and Sun = one astronomical unit), of the planets from the Sun by adding 4 to each term of the series 0, 3, 6, 12, 24, . . . and then dividing by 10. Bode's law predicted the existence of a planet between Mars and Jupiter, which led to the discovery of the asteroids.

Bohr model model of the atom conceived by Danish physicist Neils Bohr in 1913. It assumes that the following rules govern the behaviour of electrons: (1) electrons revolve in orbits of specific radius around the nucleus without emitting radiation; (2) within each orbit, each electron has a fixed amount of energy; electrons in orbits farther away from the nucleus have greater energies; (3) an electron may 'jump' from one orbit of high energy to another of lower energy causing the energy difference to be emitted as a photon of electromagnetic radiation such as light; (4) an electron may absorb a photon of radiation and jump from a lower-energy orbit to a higher-energy one. The Bohr model has been superseded by wave mechanics (see quantum theory).

boiling point for any given liquid, the temperature at which the application of heat raises the temperature of the liquid no further, but converts it into vapour.

Bok's globule small, circular dark spot in a nebula, with a mass comparable to that of the Sun. Bok's globules are possibly gas clouds in the process of condensing into stars. They were first discovered by Dutch astrophysicist Bart Bok.

bond result of the forces of attraction that hold together atoms of an element or elements to form a molecule. The principal types of bonding are ionic, covalent, metallic, and intermolecular (such as hydrogen bonding).

Boolean algebra set of algebraic rules, named after English mathematician George Boole, in which TRUE and FALSE are equated to 1 and 0, respectively. Boolean algebra includes a series of operators (AND, OR, NOT, NAND (NOT AND), NOR, and XOR (exclusive OR)), which can be used to manipulate TRUE and FALSE values. It is the basis of computer logic because the truth values can be directly associated with bits.

Bose–Einstein statistics treatment of particles which have integral spin in statistical mechanics, enabling properties of systems made up of such particles to be calculated.

boson elementary particle whose spin can only take values that are whole numbers or zero. Bosons may be classified as gauge bosons (carriers of the four fundamental forces) or mesons. All elementary particles are either bosons or fermions.

boundary value natural phenomena in a given region may be described by functions that satisfy certain differential equations in the interior of the region and take specific values on the boundary of the region. The latter are referred to as boundary values.

Bourdon gauge instrument for measuring pressure, patented by French watchmaker Eugène Bourdon in 1849. The gauge contains a C-shaped tube, closed at one end. When the pressure inside the tube increases, the tube uncurls slightly causing a small movement at its closed end. A system of levers and gears magnifies this movement and turns a pointer, which indicates the pressure on a circular scale. Bourdon gauges are often fitted to cylinders of compressed gas used in industry and hospitals.

Boyle's law law stating that the volume of a given mass of gas at a constant temperature is inversely proportional to its pressure. For example, if the pressure on a gas doubles, its volume will be reduced by a half, and vice versa. The law was discovered in 1662 by Irish physicist and chemist Robert Boyle.

Bragg's law law that states that the maximum intensity of X-rays diffracted (see X-ray diffraction) through a crystal occurs when the sine of the complement of the angle of incidence of the X-rays onto the crystal satisfies the relation:

$$n\lambda = 2d \sin \theta$$

where λ is the wavelength of the radiation, d is the lattice spacing, and n is an integer. The equation was determined by Australian-born British physicist Lawrence Bragg.

braid theory part of the study of nodes in three-dimensional space, first devised by Austrian mathematician Emil Artin.

Bremsstrahlung electromagnetic radiation produced by the rapid deceleration of charged particles such as electrons, as occurs in the collison between electrons and nuclei.

Brewster's law law that states that the refractive index of a medium is given by the tangent of the angle at which maximum polarization occurs. It was determined by Scottish physicist David Brewster.

Brownian movement continuous random motion of particles in a fluid medium (gas or liquid) as they are subjected to impact from the molecules of the medium. The phenomenon was explained by German physicist Albert Einstein in 1905 but was observed as long ago as 1827 by the Scottish botanist Robert Brown. Brown was looking at pollen grains in water under a microscope when he noticed the pollen grains were in constant, haphazard motion. The motion of these particles was due to the impact of moving water molecules. It provides evidence for the kinetic theory of matter.

bubble chamber device for observing the nature and movement of atomic particles, and their interaction with radiation. It is a vessel filled with a superheated liquid through which ionizing particles move and

451

collide. The paths of these particles are shown by strings of bubbles, which can be photographed and studied. By using a pressurized liquid medium instead of a gas, it overcomes drawbacks inherent in the earlier cloud chamber. It was invented by US physicist Donald Glaser in 1952. See particle detector.

Bunsen burner gas burner used in laboratories, consisting of a vertical metal tube through which a fine jet of fuel gas is directed. Air is drawn in through airholes near the base of the tube and the mixture is ignited and burns at the tube's upper opening.

Burali–Forte's paradox paradox stating that to every collection of ordinal numbers (numbers in the series first, second, third and so on, relating to order) there corresponds an ordinal number greater than any element of the collection. In particular it would follow that the collection of all ordinal numbers is itself an ordinal number. This contradiction demonstrated the need for a rigorous exposition of set theory in which not all collections may be accepted as valid subjects of discourse. There is thus no such thing as a 'set of all sets', nor a 'set of all ordinals' (indicating that the foundations of mathematics cannot be expressed in purely logical terms). One therefore distinguishes between sets, which may be manipulated, and classes, which may not (except in the simplest of circumstances).

calculus (Latin 'pebble') branch of mathematics which uses the concept of a derivative (see differentiation) to analyse the way in which the values of a function vary. Calculus is probably the most widely used part of mathematics. Many real-life problems are analysed by expressing one quantity as a function of another – position of a moving object as a function of time, temperature of an object as a function of distance from a heat source, force on an object as a function of distance from the source of the force, and so on – and calculus is concerned with such functions.

calculus of variations method of calculation for solving problems in which one of the unknowns cannot be expressed as a number or a finite set of numbers, but is representable as a curve, a function, or a system of functions. (A classic problem in the subject is to show that a circle, among all curves of fixed length, encloses the maximum area.)

calibre internal diameter of a bore or pipe.

caloric theory theory that heat consists of a fluid called 'caloric' that flows from hotter to colder bodies. It was abandoned by the mid-19th century.

calorimeter instrument used in physics to measure various thermal properties, such as heat capacity or the heat produced by fuel. A simple calorimeter consists of a heavy copper vessel that is polished (to reduce heat losses by radiation) and covered with insulating material (to reduce losses by convection and conduction).

calorimetry measurement of the heat-related constants of material, for example, thermal capacity and latent heat of vaporization.

Cambrian period period of geological time roughly 570–510 million years ago, the first period of the Palaeozoic era. All invertebrate animal life appeared, and marine algae were widespread. The **Cambrian Explosion** 530–520 million years ago saw the major radiaton in the fossil record of modern animal phyla; the earliest fossils with hard shells, such as trilobites, date from this period.

camera obscura ('dark room') darkened box with a tiny hole for projecting the inverted image of the scene outside onto a screen inside.

canal rays streams of positively charged ions produced from an anode in a discharge tube, in which gas is subjected to an electric discharge.

capacitance, electrical property of a capacitor that determines how much charge can be stored in it for a given potential difference between its terminals. It is equal to the ratio of the electrical charge stored to the potential difference. The SI unit of capacitance is the farad, but most capacitors have much smaller capacitances, and the microfarad (a millionth of a farad) is the commonly used practical unit.

carbohydrate chemical compound composed of carbon, hydrogen, and oxygen, with the basic formula $C_m(H_2O)_n$, and related compounds with the same basic structure but modified functional groups. As sugar and starch, carbohydrates are an important part of a balanced human diet, providing energy for life processes including growth and movement. Excess carbohydrate intake can be converted into fat and stored in the body.

Carboniferous period period of geological time roughly 362.5 to 290 million years ago, the fifth period of the Palaeozoic era. In the USA it is divided into two periods: the Mississippian (lower) and the Pennsylvanian (upper).

Typical of the lower-Carboniferous rocks are shallow-water limestones, while upper-

Carboniferous rocks have delta deposits with coal (hence the name). Amphibians and land plants were abundant, and reptiles evolved during this period.

carbon–nitrogen cycle use of carbon and nitrogen as intermediates in the nuclear fusion process of the Sun. Cooler stars undergo the proton–proton cycle.

carboxylic acid organic acid containing the carboxyl group (–COOH) attached to another group (R), which can be hydrogen (giving methanoic acid, HCOOH) or a larger molecule (up to 24 carbon atoms). When R is a straight-chain alkyl group (such as CH_3 or CH_3CH_2), the acid is known as a fatty acid.

carburation any process involving chemical combination with carbon, especially the mixing or charging of a gas, such as air, with volatile compounds of carbon (petrol, kerosene, or fuel oil) in order to increase potential heat energy during combustion. Carburation applies to combustion in the cylinders of reciprocating petrol engines of the types used in aircraft, road vehicles, or marine vessels. The device by which the liquid fuel is atomized and mixed with air is called a **carburettor**.

carcinogenesis means by which the changes responsible for the development of cancer are brought about.

Carnot cycle series of changes in the physical condition of a gas in a reversible heat engine, necessarily in the following order: (1) isothermal expansion (without change of temperature), (2) adiabatic expansion (without change of heat content), (3) isothermal compression, and (4) adiabatic compression.

Carnot's theorem theorem in thermo-dynamics stating that the efficiency of any (reversible) heat engine depends only on the temperature range through which the machine operates. It was determined by French physicist Sadi Carnot.

carrier wave wave of electromagnetic radiation of constant frequency and amplitude used in radio communication. Modulation of the wave allows information to be carried by it.

carrying capacity maximum number of animals of a given species that a particular habitat can support. When the carrying capacity is exceeded, there is insufficient food (or other resources) for the members of the population. The population may then be reduced by emigration, reproductive failure, or death through starvation.

Cassegrain telescope or ***Cassegrain reflector***, type of reflecting telescope in which light collected by a concave primary mirror is reflected onto a convex secondary mirror, which in turn directs it back through a hole in the primary mirror to a focus behind it. As a result, the telescope tube can be kept short, allowing equipment for analysing and recording starlight to be mounted behind the main mirror. All modern large astronomical telescopes are of the Cassegrain type.

catalyst substance that alters the speed of, or makes possible, a chemical or biochemical reaction but remains unchanged at the end of the reaction. Enzymes are natural biochemical catalysts. In practice most catalysts are used to speed up reactions.

catastrophism theory that regards the variations in fossils from different geological strata as having resulted from a series of natural catastrophes that gave rise to new species.

cathode in chemistry, the negative electrode of an electrolytic cell, towards which positive particles (cations), usually in solution, are attracted. See electrolysis.

cathode in electronics, the part of an electronic device in which electrons are generated. In a thermionic valve, electrons are produced by the heating effect of an applied current; in a photocell, they are produced by the interaction of light and a semi-conducting material. The cathode is kept at a negative potential relative to the device's other electrodes (anodes) in order to ensure that the liberated electrons stream away from the cathode and towards the anodes.

cathode ray stream of fast-moving electrons that travel from a cathode (negative elec-trode) towards an anode (positive electrode) in a vacuum tube. They carry a negative charge and can be deflected by electric and magnetic fields. Cathode rays focused into fine beams of fast electrons are used in cathode-ray tubes, the electrons' kinetic energy being converted into light energy as they collide with the tube's fluorescent screen.

cation ion carrying a positive charge. During electrolysis, cations in the electrolyte move to the cathode (negative electrode).

caustic curve curve formed by the points of intersection of rays of light reflected or refracted from a curved surface.

cavitation formation of cavities containing a partial vacuum in fluids at high velocities, produced by propellers or other machine

453

parts in hydraulic engines, in accordance with Bernoulli's principle. When these cavities collapse, pitting, vibration, and noise can occur in the metal parts in contact with the fluids.

celestial mechanics branch of astronomy that deals with the calculation of the orbits of celestial bodies, their gravitational attractions (such as those that produce the Earth's tides), and also the orbits of artificial satellites and space probes. It is based on the laws of motion and gravity laid down in the 17th century by English physicist and mathematician Isaac Newton.

celestial sphere imaginary sphere surrounding the Earth, on which the celestial bodies seem to lie. The positions of bodies such as stars, planets, and galaxies are specified by their coordinates on the celestial sphere. The equivalents of latitude and longitude on the celestial sphere are called declination and right ascension (which is measured in hours from 0 to 24). The **celestial poles** lie directly above the Earth's poles, and the **celestial equator** lies over the Earth's Equator. The celestial sphere appears to rotate once around the Earth each day, actually a result of the rotation of the Earth on its axis.

cell basic structural unit of life. It is the smallest unit capable of independent existence which can reproduce itself exactly. All living organisms – with the exception of viruses – are composed of one or more cells. Single-cell organisms such as bacteria, protozoa, and other micro-organisms are termed **unicellular**, while plants and animals which contain many cells are termed **multicellular** organisms. Highly complex organisms such as human beings consist of billions of cells, all of which are adapted to carry out specific functions – for instance, groups of these specialized cells are organized into tissues and organs. Although these cells may differ widely in size, appearance, and function, their essential features are similar.

Cells divide by mitosis, or by meiosis when gametes are being formed.

cell, electrical or *voltaic cell* or *galvanic cell*, device in which chemical energy is converted into electrical energy; the popular name is 'battery', but this strictly refers to a collection of cells in one unit. The reactive chemicals of a **primary cell** cannot be replenished, whereas **secondary cells** – such as storage batteries – are rechargeable: their chemical reactions can be reversed and the original condition restored by applying an electric current. It is dangerous to attempt to recharge a primary cell.

cell, electrolytic device to which electrical energy is applied in order to bring about a chemical reaction. See electrolysis.

cell theory theory that regards all living things as being composed of cells and that their replication and growth result from cell division. The theory was proposed by German botanist Matthias Schleiden and German physiologist Theodor Schwann in 1838–39.

celluloid transparent or translucent, highly flammable, plastic material (a thermoplastic) made from cellulose nitrate and camphor. It was once used for toilet articles, novelties, and photographic film, but has now been replaced by the nonflammable substance cellulose acetate.

cellulose complex carbohydrate composed of long chains of glucose units, joined by chemical bonds called glycosidic links. It is the principal constituent of the cell wall of higher plants, and a vital ingredient in the diet of many herbivores. Molecules of cellulose are organized into long, unbranched microfibrils that give support to the cell wall. No mammal produces the enzyme cellulase, necessary for digesting cellulose; mammals such as rabbits and cows are only able to digest grass because the bacteria present in their gut can manufacture it.

Cenozoic Era or *Caenozoic*, era of geological time that began 65 million years ago and continues to the present day. It is divided into the Tertiary and Quaternary periods. The Cenozoic marks the emergence of mammals as a dominant group, and the rearrangment of continental masses towards their present positions.

centre of gravity point in an object about which its weight is evenly balanced. In a uniform gravitational field, this is the same as the centre of mass.

centre of mass point in or near an object at which the whole mass of the object may be considered to be concentrated. A symmetrical homogeneous object such as a sphere or cube has its centre of mass at its geometrical centre; a hollow object (such as a cup) may have its centre of mass in space inside the hollow.

centrifugal force apparent force arising for an observer moving with a rotating system. For an object of mass m moving with a velocity v in a circle of radius r, the centrifugal force F equals mv^2/r (outward).

centrifuge apparatus that rotates containers at high speeds, creating centrifugal forces. One use is for separating mixtures of substances of different densities.

centripetal force force that acts radially inward on an object moving in a curved path. For example, with a weight whirled in a circle at the end of a length of string, the centripetal force is the tension in the string. For an object of mass m moving with a velocity v in a circle of radius r, the centripetal force F equals mv^2/r (inward). The reaction to this force is the centrifugal force.

centrosome cell body that contains the centrioles. During cell division the centrosomes organize the microtubules to form the spindle that divides the chromosomes into daughter cells. Centrosomes were first described in 1887, independently by German biologist Theodor Boveri and Belgian biologist Edouard van Beneden.

Cepheid variable yellow supergiant star that varies regularly in brightness every few days or weeks as a result of pulsations. The time that a Cepheid variable takes to pulsate is directly related to its average brightness; the longer the pulsation period, the brighter the star.

chain reaction in chemistry, a succession of reactions, usually involving free radicals, where the products of one stage are the reactants of the next. A chain reaction is characterized by the continual generation of reactive substances.

chain reaction in nuclear physics, a fission reaction that is maintained because neutrons released by the splitting of some atomic nuclei themselves go on to split others, releasing even more neutrons. Such a reaction can be controlled (as in a nuclear reactor) by using moderators to absorb excess neutrons. Uncontrolled, a chain reaction produces a nuclear explosion (as in an atom bomb).

Chandrasekhar limit or *Chandrasekhar mass*, maximum possible mass of a white dwarf star. The limit depends slightly on the composition of the star but is equivalent to 1.4 times the mass of the Sun. A white dwarf heavier than the Chandrasekhar limit would collapse under its own weight to form a neutron star or a black hole. The limit is named after the Indian–US astrophysicist Subrahmanyan Chandrasekhar who developed the theory of white dwarfs in the 1930s.

charge see electric charge.

charge conservation feature of quantum mechanics, in which reactions between elementary particles occur in such a way that there is no change in the total charge of the system after the event has occurred.

Charles's law law stating that the volume of a given mass of gas at constant pressure is directly proportional to its absolute temperature (temperature in kelvins). It was discovered by French physicist Jacques Charles in 1787, and independently by French chemist Joseph Gay-Lussac in 1802.

chemical oxygen demand *COD*, measure of water and effluent quality, expressed as the amount of oxygen (in parts per million) required to oxidize the reducing substances present.

chemotherapy any medical treatment with chemicals. It usually refers to treatment of cancer with cytotoxic and other drugs. The term was coined by the German bacteriologist Paul Ehrlich for the use of synthetic chemicals against infectious diseases.

Cherenkov detector apparatus through which it is possible to observe the existence and velocity of high-speed particles, important in experimental nuclear physics and in the study of cosmic radiation. It was originally built to investigate the Cherenkov radiation effect, in which charged particles travel through a medium at a speed greater than that of light in that medium.

chi-squared function or *χ^2 function*, function that in probability theory provides a test for deviation from a null hypothesis. It is usually represented as being made up of:

$$\frac{(\text{observed frequency of result} - \text{expected frequency of result})^2}{\text{expected frequency of result}}$$

in which the top line indicates the (squared) deviation from the expected.

chlorophyll group of pigments including chlorophyll a and chlorophyll b, the green pigments in plants; it is responsible for the absorption of light energy during photosynthesis. The pigment absorbs the red and blue-violet parts of sunlight but reflects the green, thus giving plants their characteristic colour. Other chlorophylls include chlorophyll c (in brown algae) and chlorophyll d (found in red algae).

chloroplast structure (organelle) within a plant cell containing the green pigment chlorophyll. Chloroplasts occur in most cells of the green plant that are exposed to light, often in large numbers. Typically, they are flattened and disclike, with a double

455

membrane enclosing the stroma, a gel-like matrix. Within the stroma are stacks of fluid-containing cavities, or vesicles, where photosynthesis occurs.

chondrite type of meteorite characterized by **chondrules**, small spheres, about 1 mm/0.04 in in diameter, made up of the silicate minerals olivine and orthopyroxene.

chordate animal belonging to the phylum Chordata, which includes vertebrates, sea squirts, amphioxi, and others. All these animals, at some stage of their lives, have a supporting rod of tissue (notochord or backbone) running down their bodies.

chorion outermost of the three membranes enclosing the embryo of reptiles, birds, and mammals; the amnion is the innermost membrane.

chromatic aberration optical effect commonly found in simple lens instruments in which coloured fringes are seen around an image as a result of the wavelength dependence of the refractive index for glass.

chromatin nucleoprotein found in chromosomes and thought to be the molecular substance of heredity. It is readily stained by basic dyes and is therefore easily identified and studied under the microscope.

chromatography (Greek *chromos* 'colour') technique for separating or analysing a mixture of gases, liquids, or dissolved substances. This is brought about by means of two immiscible substances, one of which (**the mobile phase**) transports the sample mixture through the other (**the stationary phase**). The mobile phase may be a gas or a liquid; the stationary phase may be a liquid or a solid, and may be in a column, on paper, or in a thin layer on a glass or plastic support. The components of the mixture are absorbed or impeded by the stationary phase to different extents and therefore become separated. The technique is used for both qualitative and quantitive analyses in biology and chemistry.

chromosome structure in a cell nucleus that carries the genes. Each chromosome consists of one very long strand of DNA, coiled and folded to produce a compact body. The point on a chromosome where a particular gene occurs is known as its locus. Most higher organisms have two copies of each chromosome, together known as a **homologous pair** (they are diploid) but some have only one (they are haploid). There are 46 chromosomes in a normal human cell. See also mitosis and meiosis.

chromosome map description of the position of genes along a chromosome.

chromosphere (Greek 'colour' and 'sphere') layer of mostly hydrogen gas about 10,000 km/6,000 mi deep above the visible surface of the Sun (the photosphere). It appears pinkish red during eclipses of the Sun.

circuit arrangement of electrical components through which a current can flow. There are two basic circuits, series and parallel. In a series circuit, the components are connected end to end so that the current flows through all components one after the other. In a parallel circuit, components are connected side by side so that part of the current passes through each component. A circuit diagram shows in graphical form how components are connected together, using standard symbols for the components.

cis- prefix used in stereochemistry to distinguish an isomer that has two substituents or groupings on the same side of the main axis or plane of the molecule. The isomer with the two on opposite sides is denoted by the prefix **trans-**.

class used in biological classification, it is a subdivision of phylum and forms a group of related orders. For example, all mammals belong to the class Mammalia and all birds to the class Aves. Among plants, all class names end in 'idae' (such as Asteridae) and among fungi in 'mycetes'; there are no equivalent conventions among animals. Related classes are grouped together in a phylum.

class in mathematics, another name for a set.

class field theory theory involving the mathematical structure known as a field, dealing specifically with those that extend a given field in a special kind of way.

climax community assemblage of plants and animals that is relatively stable in its environment. It is brought about by ecological succession, and represents the point at which succession ceases to occur.

clone exact replica. In genetics, any one of a group of genetically identical cells or organisms. An identical twin is a clone; so, too, are bacteria living in the same colony. 'Clone' also describes genetically engineered replicas of DNA sequences. The term has also been adopted by computer technology to describe a (nonexistent) device that mimics an actual one to enable certain software programs to run correctly.

closed curve curve of which the end point coincides with the initial point; for example, a circle or an ellipse.

cluster group of stars or of galaxies, usually with some recognizably systematic configuration. It appears that both types

456

of cluster are a structural feature of the universe, which form over the passage of time.

coacervate collection of particles in an emulsion that can be reversed into droplets of liquid before they flocculate.

coccyx lowermost component of the spine. It consists of four vestigial vertebrae fused to form a single triangular bone.

coefficient of expansion quantity that describes the amount of expansion undergone by a material for a degree rise in temperature. It is expressed as the increase in length per unit length, per degree Celsius. For example, the metal with the lowest coefficient of expansion is Invar, at 2.3×10^{-6} m of expansion for every metre in length, per degree Celsius rise in temperature.

coenzyme small, organic nonprotein compound that attaches to an enzyme and is necessary for its correct functioning. Tightly bound coenzymes are known as prosthetic groups; more loosely bound ones are called cofactors. The coenzyme itself is not usually changed during a reaction. If it is, it is usually converted rapidly back to its original form. Well-known coenzymes include NAD, ATP, and coenzyme A.

coherence property of two or more waves of a beam of light or other electromagnetic radiation having the same frequency and the same phase, or a constant phase difference.

cohomology or *cohomology theory*, algebraic study, using group theory, of geometric objects with specific reference to the operation of finding a boundary. Cohomology theory represents a modification of homology in which it is possible both to add and to multiply classes.

coke solid, porous material produced from the carbonization of coal, all the volatile material having been driven off. It is used in the production of steel.

collinear lying on the same straight line.

colloid substance composed of extremely small particles of one material (the dispersed phase) evenly and stably distributed in another material (the continuous phase). The size of the dispersed particles (1–1,000 nanometres across) is less than that of particles in suspension but greater than that of molecules in true solution. Colloids involving gases include **aerosols** (dispersions of liquid or solid particles in a gas, as in fog or smoke) and **foams** (dispersions of gases in liquids).

Those involving liquids include **emulsions** (in which both the dispersed and the continuous phases are liquids) and **sols** (solid particles dispersed in a liquid). Sols in which both phases contribute to a molecular three-dimensional network have a jellylike form and are known as **gels**; gelatin, starch 'solution', and silica gel are common examples.

colon main part of the large intestine, between the caecum and rectum. Water and mineral salts are absorbed from undigested food in the colon, and the residue passes as faeces towards the rectum.

colonization spread of species into a new habitat, such as a freshly cleared field, a new motorway verge, or a recently flooded valley. The first species to move in are called **pioneers**, and may establish conditions that allow other animals and plants to move in (for example, by improving the condition of the soil or by providing shade). Over time a range of species arrives and the habitat matures; early colonizers will probably be replaced, so that the variety of animal and plant life present changes. This is known as **succession**.

colour index measure of the colour of a star made by comparing its brightness through different coloured filters. It is defined as the difference between the magnitude of the star measured through two standard photometric filters. Colour index is directly related to the surface temperature of a star and its spectral classification.

coma hazy cloud of gas and dust that surrounds the nucleus of a comet.

combustion burning, defined in chemical terms as the rapid combination of a substance with oxygen, accompanied by the evolution of heat and usually light. A slow-burning candle flame and the explosion of a mixture of petrol vapour and air are extreme examples of combustion. Combustion is an exothermic reaction as heat energy is given out.

comet small, icy body orbiting the Sun, usually on a highly elliptical path. A comet consists of a central nucleus a few kilometres across, and has been likened to a dirty snowball because it consists mostly of ice mixed with dust. As a comet approaches the Sun its nucleus heats up, releasing gas and dust which form a tenuous coma, up to 100,000 km/60,000 mi wide, around the nucleus. Gas and dust stream away from the coma to form one or more tails, which may extend for millions of kilometres.

community assemblage of plants, animals, and other organisms living within a circum-scribed area. Communities are usually named by reference to a dominant feature

457

such as characteristic plant species (for example, a beech-wood community), or a prominent physical feature (for example, a freshwater-pond community).

commutative operation operation that is independent of the order of the numbers or symbols concerned. For example, addition is commutative: the result of adding $4 + 2$ is the same as that of adding $2 + 4$; subtraction is not as $4 - 2 = 2$, but $2 - 4 = -2$. Compare associative operation and distributive operation.

compact spaces or ***bicompact spaces***, special kinds of topological space exhibiting the property that, internally, every family of open sets whose union is the whole space necessarily contains a finite subfamily whose union is already the whole space. An alternative definition, first formulated with regard to a special class of such spaces, requires every sequence of points to have a converging subsequence. Russian mathematicians initially used the word **bicompact** (not now used in the West) to distinguish between the specialized and the general definitions.

companion star either one of a binary star system (although usually the less massive), sometimes only detectable by spectroscopy.

competitive exclusion principle principle of natural selection whereby similar species are forced to specialize ever more minutely so as not to overlap with each other in a particular niche; if they do not specialize adequately, they die and become extinct.

complement set of the elements within the universal set that are not contained in the designated set. For example, if the universal set is the set of all positive whole numbers and the designated set S is the set of all even numbers, then the complement of S (denoted S') is the set of all odd numbers.

complex any of a class of substances with a characteristic structure in which a central metal atom (often a transition element) is surrounded by – and bonded to – several nonmetallic atoms or groups of atoms (ligands). Complexes are also called coordination compounds.

complex number number written in the form $a + ib$, where a and b are real numbers and i is the square root of -1 (that is, $i^2 = -1$); ib used to be known as the 'imaginary' part of the complex number. Some equations in algebra, such as those of the form

$$x^2 + 5 = 0$$

cannot be solved without recourse to complex numbers, because the real numbers do not include square roots of negative numbers.

complex number astrophysics basis of twistor theory.

compound chemical substance made up of two or more elements bonded together, so that they cannot be separated by physical means. Compounds are held together by ionic or covalent bonds.

Compton effect increase in wavelength (loss of energy) of a photon by its collision with a free electron (**Compton scattering**). The Compton effect was first demonstrated with X-rays and provided early evidence that electromagnetic waves consist of particles – photons – which carry both energy and momentum. It is named after US physicist Arthur Compton.

concave of a surface, curving inwards, or away from the eye. For example, a bowl appears concave when viewed from above. In geometry, a concave polygon is one that has an interior angle greater than 180°. Concave is the opposite of convex.

concentric circles two or more circles that share the same centre.

conchoid curve algebraic curve represented by an equation of the general form:

$$x^2 y^2 = (x - a)^2 (c^2 - x^2)$$

conditioned stimulus originally neutral stimulus applied in conditioning experiments that evokes a trained or conditioned response. In Pavlov's classical experiments with dogs, the neutral stimulus (the sound of a bell) originally evoked no salivation reflex; but after being presented for a time with an unconditioned stimulus (food), it became the conditioned stimulus which evoked the conditioned response.

conduction, electrical flow of charged particles through a material giving rise to electric current. Conduction in metals involves the flow of negatively charged free electrons. Conduction in gases and some liquids involves the flow of ions that carry positive charges in one direction and negative charges in the other. Conduction in a semiconductor such as silicon involves the flow of electrons and positive holes.

conduction, heat flow of heat energy through a material without the movement of any part of the material itself (compare conduction, electrical). Heat energy is present in all materials in the form of the kinetic energy of their vibrating molecules, and may be conducted from one molecule to the next in the form of this mechanical vibration. In the case of metals, which are particularly good conductors of heat, the free electrons within the material carry heat around very quickly.

conductivity, thermal (unit $W\,m^{-1}\,K^{-1}$) measure of how well a material conducts heat. A good conductor, such as a metal, has a high conductivity; a poor conductor, called an insulator, has a low conductivity. See also U-value.

conductor any material that conducts heat or electricity (as opposed to an insulator, or nonconductor). A good conductor has a high electrical or heat conductivity, and is generally a substance rich in free electrons such as a metal. A poor conductor (such as the nonmetals, glass, and porcelain) has few free electrons. Carbon is exceptional in being nonmetallic and yet (in some of its forms) a relatively good conductor of heat and electricity. Substances such as silicon and germanium, with intermediate conductivities that are improved by heat, light, or impurities, are known as semiconductors.

congruence two sets are congruent if either can be transformed by translations and rotations into the other.

conics study initiated by Apollonius of Perga of how a cone can be 'cut' so as to produce circles, ellipses, parabolas, and hyperbolas; he stated 'a conic section is the locus of a point that moves so that the ratio of its distance *f* from a fixed point, to its distance *d* from a straight line, is constant'. Whether the constant *c* is greater than, equal to, or less than one determines the type of a curve the section represents.

conic section curve obtained when a conical surface is intersected by a plane. If the intersecting plane cuts both extensions of the cone, it yields a hyperbola; if it is parallel to the side of the cone, it produces a parabola. Other intersecting planes produce circles or ellipses.

conjugation alternation of double (or triple) and single carbon–carbon bonds in a molecule – for example, in penta-1,3-diene, $H_2C{=}CH{-}CH{=}CH{-}CH_3$. Conjugation imparts additional stability as the double bonds are less reactive than isolated double bonds.

conservation of energy principle that states that in a chemical reaction, the total amount of energy in the system remains unchanged.

conservation of mass principle that states that in a chemical reaction the sum of all the masses of the substances involved in the reaction (reactants) is equal to the sum of all of the masses of the substances produced by the reaction (products) – that is, no matter is gained or lost.

constant of precession see precession of the equinoxes.

container habitat small, self-contained ecosystem, such as a water pool accumulating in a hole in a tree. Some ecologists believe that much can be learned about larger ecosystems through studying the dynamics of container habitats, which can contain numerous leaf-litter feeders and their predators.

continental drift theory that, about 250–200 million years ago, the Earth consisted of a single large continent (Pangaea), which subsequently broke apart to form the continents known today. The theory was proposed in 1912 by German meteorologist Alfred Wegener, but such vast continental movements could not be satisfactorily explained or even accepted by geologists until the 1960s.

continued fraction development of any real number in the form of a sequence of integers from which approximations to the number may be calculated successively; for example:

$$k_1 + \cfrac{1}{k_1 + \cfrac{1}{k_2 + \cfrac{1}{k_3 + \cfrac{1}{k_4}}}}$$

and so on. The sequence can be finite or infinite. The development will be finite in the case of a rational number, and the calculation will then terminate on reaching the rational number. In the case of an irrational number, a termination will be reached only as the limit of the sequence of values calculated.

continuous function or *continuity*, representation of continuous motion, uniform variation of transformations. More precisely, a function *f* is said to be continuous at an argument value *x* if the function value for arguments close to *x* can be held down to a value as near to $f(x)$ as required by keeping the argument close enough to *x*.

convection heat energy transfer that involves the movement of a fluid (gas or liquid). Fluid in contact with the source of heat expands and tends to rise within the bulk of the fluid. Cooler fluid sinks to take its place, setting up a convection current. This is the principle of natural convection in many domestic hot-water systems and space heaters.

convection process in the Sun (and possibly other stars) perhaps caused by solar rotation, that produces the immensely powerful electrical and magnetic fields associated with sunspots.

convergence property of a series of numbers in which the difference between consecutive

459

terms gradually decreases. The sum of a converging series approaches a limit as the number of terms tends to infinity.

convergent evolution or *convergence*, the independent evolution of similar structures in species (or other taxonomic groups) that are not closely related, as a result of living in a similar way. Thus, birds and bees have wings, not because they are descended from a common winged ancestor, but because their respective ancestors independently evolved flight.

convex of a surface, curving outwards, or towards the eye. For example, the outer surface of a ball appears convex. In geometry, the term is used to describe any polygon possessing no interior angle greater than 180°. Convex is the opposite of concave.

coordination compound any of a diverse group of complex compounds characterized by a structure in which several ligands surround – and are covalently bonded to – a central metal atom. Such compounds may be electrically neutral, or positive or negative ions. Similarly the central metal atom may be neutral, anionic, or, rarely, cationic, but it is always one that is able to accept an electron pair(s) to form a coordinate bond(s). The total number of bonds between the central atom and the ligands is the coordination number, which, in general, ranges from two to twelve; four and six are the most common.

coordination number see coordination compound.

Copernican model of the universe model of the universe with the Sun at its centre (heliocentric) that replaced the Earth-centred (geocentric) Ptolemaic model, and was thus a considerable improvement. The model, however, still involved epicycles and the spheres.

Coriolis effect effect of the Earth's rotation on the atmosphere, oceans, and theoretically all objects moving over the Earth's surface. In the northern hemisphere it causes moving objects and currents to be deflected to the right; in the southern hemisphere it causes deflection to the left. The effect is named after its discoverer, French mathematician Gaspard de Coriolis.

corona faint halo of hot (about 2,000,000°C/ 3,600,000°F) and tenuous gas around the Sun, which boils from the surface.

coronagraph device for studying the solar corona at any time of the day. It was first invented by French astronomer Bernard Lyot.

corpuscle small biological body. The cellular components of blood are sometimes referred to as corpuscles. See white blood cell.

cortex outer part of a structure such as the brain, kidney, or adrenal gland. In botany the cortex includes nonspecialized cells lying just beneath the surface cells of the root and stem.

cosecant function of an angle in a right-angled triangle found by dividing the length of the hypotenuse (the longest side) by the length of the side opposite the angle. Thus the cosecant of an angle A, usually shortened to cosec A, is always greater than (or equal to) 1. It is the reciprocal of the sine of the angle: cosec $A = 1/\sin A$.

cosine function of an angle in a right-angled triangle found by dividing the length of the side adjacent to the angle by the length of the hypotenuse (the longest side). It is usually shortened to **cos**.

cosmic censorship theory that the hidden interior within all event horizons is the same and is always, necessarily, hidden.

cosmic radiation streams of high-energy particles and electromagnetic radiation from outer space, consisting of electrons, protons, alpha particles, light nuclei, and gamma rays, which collide with atomic nuclei in the Earth's atmosphere, and produce secondary nuclear particles (chiefly mesons, such as pions and muons) that shower the Earth.

cosmic year time the Sun takes to 'orbit' in galactic rotation: about 225 million years.

cosmological principle hypothesis that any observer anywhere in the universe has the same view that we have; that is, that the universe is not expanding from any centre but all galaxies are moving away from one another.

cosmology branch of astronomy that deals with the structure and evolution of the universe as an ordered whole. Cosmologists construct 'model universes' mathematically and compare their large-scale properties with those of the observed universe.

coulomb SI unit (symbol C) of electrical charge. One coulomb is the quantity of electricity conveyed by a current of one ampere in one second.

Coulomb field field of force surrounding an electric charge. Its intensity can be deduced from Coulomb's law.

Coulomb's law law that states that the force between two charged bodies varies directly as the product of the two charges, and inversely as the square of the distance between them. It was determined by French physicist Charles Coulomb.

covalent bond chemical bond produced when two atoms share one or more pairs of electrons (usually each atom contributes an electron). The bond is often represented by a single line drawn between the two atoms. Covalently bonded substances include hydrogen (H_2), water (H_2O), and most organic substances.

CP violation breaking of a fundamental quantum theory conservation rule by some unstable particles undergoing decay into other particles. C represents charge conjugation that relates particles to antiparticles and P stands for parity.

Cramer's paradox paradox that although two different cubic curves intersect at nine points, part of the definition of a single cubic curve is that it is itself determined by nine points. It was observed by Swiss mathematician Gabriel Cramer.

Cramer's rule method of solving a simultaneous system of linear equations by using determinants. It is named after Swiss mathematician Gabriel Cramer.

cresol or *hydroxytoluene*, an important constituent of explosives, plastics, and dyestuff intermediates.

Cretaceous period period of geological time approximately 143–65 million years ago. It is the last period of the Mesozoic era, during which angiosperm (seed-bearing) plants evolved, and dinosaurs reached a peak. The end of the Cretaceous period is marked by a mass extinction of many lifeforms, most notably the dinosaurs. The north European chalk, which forms the white cliffs of Dover, England, was deposited during the latter half of the Cretaceous, hence the name Cretaceous, which comes from the Latin *creta*, 'chalk'.

critical temperature temperature above which a particular gas cannot be converted into a liquid by pressure alone. It is also the temperature at which a magnetic material loses its magnetism (the Curie temperature or point).

critical velocity velocity above which the flow of a liquid ceases to be smooth following streamlines, and becomes turbulent.

Crookes's radiometer instrument consisting of an evacuated glass dome in which sits a freely rotating system of vanes, whose opposite sides are white and black respectively. The rotation of the vanes when put near a source of heat is a demonstration of the kinetic theory of gases. It was devised by English physicist William Crookes.

crossing over process that occurs during meiosis. While homologous chromosomes are lying alongside each other in pairs, each partner may twist around the other and exchange corresponding chromosomal segments. It is a form of genetic recombination, which increases variation and thus allows offspring to vary from their parents.

cross ratio ratio expressing a relationship between two other ratios determined by four points on a given line, namely the ratio:

CA/CB:DA/DB

For the purposes of projection, the 'point at infinity' may be any one of A, B, C, and D; if the given line is then projected onto another, the cross ratio of the projected points will remain the same – hence the importance of the cross ratio.

crust rocky outer layer of Earth, consisting of two distinct parts, the oceanic crust and the continental crust. The **oceanic** crust is on average about 10 km/6 mi thick and consists mostly of basaltic rock overlain by muddy sediments. By contrast, the **continental** crust is largely of granitic composition and is more complex in its structure. Because it is continually recycled back into the mantle by the process of subduction, the oceanic crust is in no place older than about 200 million years. However, parts of the continental crust are over 3.5 billion years old.

cryogenics science of very low temperatures (approaching absolute zero), including the production of very low temperatures and the exploitation of special properties associated with them, such as the disappearance of electrical resistance (superconductivity).

crystal lattice regular system of points in space (for example, the corners of a cube) about which atoms, molecules, or ions in solids vibrate.

crystalloid substance that, when dissolved in a solvent, can pass through a semipermeable membrane (as opposed to a colloid, which cannot).

cubic curve geometrical curve in three-dimensional space. It may be parametrized after a change of variables as $(x, y, z) = (at^3, bt^2, ct)$ and may thus be said to be determined by nine points. See also Cramer's paradox.

current, alternating see alternating current.

current, direct see direct current.

cybernetics study of how communication and control mechanisms in machines can be made to imitate those of human beings.

cyclotron circular type of particle accelerator.

cyst hollow cavity in an animal or plant, lined with epithelium and usually filled with fluid. Cysts may be normal, for example the urinary bladder, or pathological, for example an ovarian cyst.

cytochrome protein responsible for part of the process of respiration by which food molecules are broken down in aerobic organisms. Cytochromes are part of the electron transport chain, which uses energized electrons to reduce molecular oxygen (O_2) to oxygen ions (O^{2-}). These combine with hydrogen ions (H^+) to form water (H_2O), the end product of aerobic respiration. As electrons are passed from one cytochrome to another, energy is released and used to make ATP.

cytoplasm part of the cell outside the nucleus. Strictly speaking, this includes all the organelles (mitochondria, chloroplasts, and so on), but often cytoplasm refers to the jellylike matter in which the organelles are embedded (correctly termed the cytosol). The cytoplasm is the site of protein synthesis.

Daniell cell primary cell that uses a cathode of zinc, immersed in sulphuric acid contained in a porous pot that itself stands in a container of copper sulphate, in which the copper anode stands. An electromotive force of about 1.1 volts is produced by this cell.

dark reaction series of reactions in photosynthesis that do not require light. During the dark reaction, carbon dioxide is incorporated into three-carbon sugar phosphate molecules; this reaction is dependent on the light reaction which does require light.

deamination removal of the amino group ($-NH_2$) from an unwanted amino acid. This is the nitrogen-containing part, and it is converted into ammonia, uric acid, or urea (depending on the type of animal) to be excreted in the urine.

In vertebrates, deamination occurs in the liver.

de Broglie hypothesis cornerstone of quantum physics which relates the wave nature of systems to their particlelike characteristics. For a particle with velocity v and mass m, one can associate a de Broglie wave of wavelength λ given by the de Broglie equation:

$$\lambda = h/mv$$

where h is Planck's constant. It was determined by French physicist Louis de Broglie.

declination coordinate on the celestial sphere (imaginary sphere surrounding the Earth) that corresponds to latitude on the Earth's surface. Declination runs from 0° at the celestial equator to 90° at the north and south celestial poles.

decomposition process whereby a chemical compound is reduced to its component substances. In biology, it is the destruction of dead organisms either by chemical reduction or by the action of decomposers, such as bacteria and fungi.

Dedekind's cuts mathematical device by which irrational numbers can be referred to by means of sets of fractions (rational numbers). It was defined by German mathematician Julius Dedekind.

deflection of light gravitational effect that bends a ray of light. Such an effect was predicted within the general theory of relativity, although previously considered impossible.

degradation breaking down of compounds into simpler molecules; for example, the action of enzymes brings about the degradation of proteins to amino acids.

delta δ, Δ, as Δ, the term means 'difference'; Δx represents the difference between consecutive x values according to context. German mathematician Leopold Kronecker's delta is a symbol used in the evaluations of determinants (in matrix theory), to the effect that $\delta(i, j) = 1$ if $i = j$, otherwise it equals zero. (It thus measures whether i and j are different.) The d used by Gottfried Leibniz in his notation for differential calculus, as in dy/dx, was based on an intended association with the delta.

As a lower-case (small) letter, another delta (∂) commonly represents a partial derivative in partial differential calculus.

de Moivre's equation statement that for integers n $(\cos z + i \sin z)^n = \cos nz + i \sin z$ where $i = \sqrt{-1}$. The equation was determined by French mathematician Abraham de Moivre.

dendrite part of a nerve cell or neuron. The dendrites are slender filaments projecting from the cell body. They receive incoming messages from many other nerve cells and pass them on to the cell body.

If the combined effect of these messages is strong enough, the cell body will send an electrical impulse along the axon (the threadlike extension of a nerve cell). The tip of the axon passes its message to the dendrites of other nerve cells.

denominator bottom number of a fraction, so called because it names the family of the fraction. The top number, or numerator, specifies how many unit fractions are to be taken.

density measure of the compactness of a substance; it is equal to its mass per unit volume and is measured in kg per cubic metre/lb per cubic foot. Density is a scalar quantity. The average density D of a mass m occupying a volume V is given by the formula:

$$D = m/V$$

Relative density is the ratio of the density of a substance to that of water at 4°C/39.2°F.

derivative or ***differential coefficient***, limit of the gradient of a chord linking two points on a curve as the distance between the points tends to zero; for a function of a single variable, $y = f(x)$, it is denoted by $f'(x)$, $Df(x)$, or dy/dx, and is equal to the gradient of the curve.

derivative function derived from another by the application of differentiation or partial differentiation. A derivative of a derivative is called a derivative of the second order.

descriptive geometry branch of mathematics in which three-dimensional objects are represented as two-dimensional (plane) figures, using any of many types of projection.

determinant array of elements written as a square, and denoted by two vertical lines enclosing the array. For a 2×2 matrix, the determinant is given by the difference between the products of the diagonal terms. Determinants are used to solve sets of simultaneous equations by matrix methods.

determinates 'known' values, as opposed to indeterminates, 'unknown', values.

deuterium naturally occurring heavy isotope of hydrogen, mass number 2 (one proton and one neutron), discovered by US chemist Harold Urey in 1932. It is sometimes given the symbol D. In nature, about one in every 6,500 hydrogen atoms is deuterium. Combined with oxygen, it produces 'heavy water' (D_2O), used in the nuclear industry.

Devonian period period of geological time 408–360 million years ago, the fourth period of the Palaeozoic era. Many desert sandstones from North America and Europe date from this time. The first land plants flourished in the Devonian period, corals were abundant in the seas, amphibians evolved from air-breathing fish, and insects developed on land.

diagmagnetism form of magnetism induced in one substance by the magnetic field of another. Its basic cause lies in the shift of the orbital motion of the electrons of a substance resulting from the external magnetic field of the other substance. It occurs in all materials.

dialysis technique for removing waste products from the blood. There are two main methods, haemodialysis and peritoneal dialysis.

diaphragm thin membrane found in telephone receivers and microphones onto which sound waves impinge, being converted into electrical impulses by a device that takes its input from the diaphragm movements.

diastole relaxation of a hollow organ. In particular, the term is used to indicate the resting period between beats of the heart when blood is flowing into it.

dicotyledon major subdivision of the angiosperms, containing the great majority of flowering plants. Dicotyledons are characterized by the presence of two seed leaves, or cotyledons, in the embryo, which is usually surrounded by the endosperm. They generally have broad leaves with netlike veins.

dielectric insulator or nonconductor of electricity, such as rubber, glass, and paraffin wax. An electric field in a dielectric material gives rise to no net flow of electricity. However, the applied field causes electrons within the material to be displaced, creating an electric charge on the surface of the material. This reduces the field strength within the material by a factor known as the dielectric constant (or relative permittivity) of the material. Dielectrics are used in capacitors, to reduce dangerously strong electric fields, and have optical applications.

dielectric constant capacitance of a capacitor (condenser) with a certain dielectric divided by the capacitance of the same capacitor with a vacuum (or, in practice, air) as a dielectric.

difference equation equation that relates the value of a function at time t to its values at a specified number of past times, from among $t-1, t-2, t-3, \ldots$

differential calculus branch of calculus involving applications such as the determination of maximum and minimum points and rates of change.

differential equations equations involving derivatives (see differential calculus). In a linear differential equation, the unknown function and its derivatives never appear in a power other than one. Partial differential equations involve unknown functions of several variables, and partial derivatives do therefore appear.

differential gear gear that allows two shafts to rotate at different rates. Such a system is used in cars to allow the wheels on the outside edge of a corner to rotate relative to those on the inside edge.

differential geometry investigation of geometrical surfaces using differential calculus.

differential rotation of a stellar cluster or galaxy, the different rates of 'orbiting' of stars. Those nearer the centre move faster than those at the edge. It is also used to describe the axial rotation of equatorial latitudes faster than polar latitudes of a single body (such as the Sun or a gaseous planet).

differentiation in embryology, the process by which cells become increasingly different and specialized, giving rise to more complex structures that have particular functions in the adult organism. For instance, embryonic cells may develop into nerve, muscle, or bone cells.

differentiation in mathematics, a procedure for determining the derivative or gradient of the tangent to a curve $f(x)$ at any point x.

diffraction spreading out of waves when they pass through a small gap or around a small object, resulting in some change in the direction of the waves. In order for this effect to be observed the size of the object or gap must be comparable with or smaller than the wavelength of the waves. Diffraction occurs with all forms of progressive waves – electromagnetic, sound, and water waves – and explains such phenomena as why long-wave radio waves can bend round hills better than short-wave radio waves.

diffusion spontaneous and random movement of molecules or particles in a fluid (gas or liquid) from a region in which they are at a high concentration to a region of lower concentration, until a uniform concentration is achieved throughout. The difference in concentration between two such regions is called the **concentration gradient**. No mechanical mixing or stirring is involved. For instance, if a drop of ink is added to water, its molecules will diffuse until their colour becomes evenly distributed throughout. Diffusion occurs more rapidly across a higher concentration gradient and at higher temperature.

dimorphism property of a chemical substance that allows it to crystallize in two different forms.

diode combination of a cold anode and a heated cathode, or the semiconductor equivalent, which incorporates a p–n junction; see semiconductor diode. Either device allows the passage of direct current in one direction only, and so is commonly used in a rectifier to convert alternating current (AC) to direct current (DC).

Diophantine equations algebraic equations involving one or more unknowns (indeterminates) with integers (whole numbers) as coefficients, to which one or more solutions are sought, also in integers. The classic form is:

$$ax + by = c$$

Part of the significance of this is that even if not enough information is given to derive a single solution, enough is given to reduce the answer to a definite type. Diophantus, who lived in the third century AD, thus began the investigations into number theory that still continue.

dioxin any of a family of over 200 organic chemicals, all of which are heterocyclic hydrocarbons. The term is commonly applied, however, to only one member of the family, 2,3,7,8-tetrachlorodibenzo-p-dioxin (2,3,7,8-TCDD), a highly toxic chemical that occurs, for example, as an impurity in the defoliant Agent Orange, used in the Vietnam War, and sometimes in the weedkiller 2,4,5-T. It has been associated with a disfiguring skin complaint (chloracne), birth defects, miscarriages, and cancer.

diphosphate chemical compound containing two phosphate groups, as in adenosine diphosphate (ADP).

dipole uneven distribution of magnetic or electrical characteristics within a molecule or substance so that it behaves as though it possesses two equal but opposite poles or charges, a finite distance apart.

direct current *DC*, electric current that flows in one direction, and does not reverse its flow as alternating current does. The electricity produced by a battery is direct current.

discharge, electric release of a stored electric charge.

discharge tube device containing (usually two) electrodes and a vacuum or gas at low pressure; a (high) voltage applied to the electrodes causes an electric discharge to take place between them. A gas-filled tube (such as a neon tube) may emit visible light and other forms of radiation.

discriminants special functions of the coefficients of an equation, used to find roots of a polynomial equation.

disproportionation or dismutation splitting of a molecule into two or more simpler molecules.

dissociation process whereby a single compound splits into two or more smaller products, which may be capable of recombining to form the reactant.

distributive law law expressing the principal operative in the equation:

$$a(b + c) = ab + ac$$

divergent of a sequence or series, when the fact that there is no limit for it to approach; it is the opposite of convergent.

DNA abbreviation for ***deoxyribonucleic acid***, molecular basis of heredity. A complex giant molecule that contains, in chemically coded form, the information needed for a cell to make proteins. DNA is a ladderlike double-stranded nucleic acid which forms the basis of genetic inheritance in all organisms, except for a few viruses that have only RNA. DNA is organized into chromosomes and, in organisms other than bacteria, it is found only in the cell nucleus.

domain set of objects within a mathematical structure on which operations are to be performed. In a simpler context it is the set of arguments (inputs) for which a function is defined.

dominance masking of one allele (an alternative form of a gene) by another allele. For example, if a heterozygous person has one allele for blue eyes and one for brown eyes, his or her eye colour will be brown. The allele for blue eyes is described as recessive and the allele for brown eyes as dominant.

Doppler effect change in the observed frequency (or wavelength) of waves due to relative motion between the wave source and the observer. The Doppler effect is responsible for the perceived change in pitch of a siren as it approaches and then recedes, and for the red shift of light from distant galaxies. It is named after the Austrian physicist Christian Doppler.

double refraction or *birefringence*, in a crystal, when an unpolarized ray of light entering the crystal is split into two polarized rays, one of which does obey Snell's law of refraction, and the other does not. Calcite is such a crystal.

double star 'system' of two stars that appear – because of coincidental alignment when viewed from Earth – to be close together; it is, however, an optical effect only, and therefore not the same as a binary star system. Before the 20th century there were few means of distinguishing double and binary stars.

double theta functions elliptic functions in the form of theta functions of higher degree.

dry cell source of electromotive force which does not contain a liquid. Usually taken to imply the Leclanché type of metal and paste cell.

duality principle that a law or theorem remains valid if one particular element within that law is exchanged for another equally pertinent element. In projective geometry, a statement of two-dimensional proposition remains valid if the word 'point' is exchanged for 'line' (and vice versa); a three-dimensional proposition likewise if 'point' is exchanged for 'plane'.

ductile material material that can sustain large deformations beyond its elastic limit (see elasticity) without fracture. Metals are very ductile, and may be pulled out into wires, or hammered or rolled into thin sheets without breaking.

dynamics or *kinetics*, mathematical and physical study of the behaviour of bodies under the action of forces that produce changes of motion in them.

dynamometer instrument for measuring the power generated by a device.

dynamo theory theory of the origin of the magnetic fields of the Earth and other plants having magnetic fields in which the rotation of the planet as a whole sets up currents within the planet capable of producing a weak magnetic field.

e symbol for Euler's number.

earthquake abrupt motion that propagates through the Earth and along its surfaces. Earthquakes are caused by the sudden release in rocks of strain accumulated over time as a result of tectonics. The study of earthquakes is called seismology. Most earthquakes occur along faults (fractures or breaks) and Benioff zones. Plate tectonic movements generate the major proportion: as two plates move past each other they can become jammed. When sufficient strain has accumulated, the rock breaks, releasing a series of elastic waves (seismic waves) as the plates spring free. The force of earthquakes (magnitude) is measured on the Richter scale, and their effect (intensity) on the Mercalli scale. The point at which an earthquake originates is the **seismic focus** or **hypocentre**; the point on the Earth's surface directly above this is the **epicentre**.

eccentricity extent to which an elliptical orbit departs from a circular one. It is usually expressed as a decimal fraction, regarding a circle as having an eccentricity of 0.

echolocation or *biosonar*, method used by certain animals, notably bats, whales, and dolphins, to detect the positions of objects by using sound. The animal emits a stream of high-pitched sounds, generally at ultrasonic frequencies (beyond the range of human hearing), and listens for the returning echoes reflected off objects to determine their exact location.

eclipse passage of an astronomical body through the shadow of another. The term is usually used for solar and lunar eclipses, which may be either partial or total, but may also refer to other bodies, for example, to an eclipse of one of Jupiter's satellites by Jupiter

465

itself. An eclipse of a star by a body in the Solar System is also called an occultation.

eclipsing binary binary (double) star in which the two stars periodically pass in front of each other as seen from Earth.

ecliptic path, against the background of stars, that the Sun appears to follow each year as it is orbited by the Earth. It can be thought of as the plane of the Earth's orbit projected on to the celestial sphere (imaginary sphere around the Earth).

ectoderm outer layer of cells in an embryo and all the tissues that it gives rise to.

Edison effect electrical conduction between a negatively charged filament, and a positively charged electrode kept together, though separated, in a vacuum chamber.

efferent nerve nerve that conducts impulses away from the central nervous system (brain and spinal cord). Most efferent nerves are motor nerves and run to effector organs.

efficiency general term indicating the degree to which a process or device can convert energy from one form to another without loss. It is normally expressed as a fraction or percentage, where 100% indicates conversion with no loss. The efficiency of a machine, for example, is the ratio of the work done by the machine to the energy put into the machine; in practice it is always less than 100% because of frictional heat losses. Certain electrical machines with no moving parts, such as transformers, can approach 100% efficiency.

efflux combination of combustion products and air forming the propulsive medium of a jet or rocket engine.

eigenvalue for a matrix A, the number λ is said to be an eigenvalue of the matrix if there is a nonzero vector x such that $Ax = \lambda x$. Eigenvalues are used to derive a change of base to simplify the matrix to one that has entries only on its diagonal. More generally, the number λ is an eigenvalue of a linear transformation T if there is a nonzero vector x so that $T(x) = \lambda x$.

eightfold-way scheme for the classification of elementary particles into families, grouped according to common properties as expressed by various quantum numbers.

elasticity ability of a solid to recover its shape once deforming forces (stresses modifying its dimensions or shape) are removed. An elastic material obeys Hooke's law, which states that its deformation is proportional to the applied stress up to a certain point, called the **elastic limit**, beyond which additional stress will deform it permanently. Elastic materials include metals and rubber;

however, all materials have some degree of elasticity.

electrical conduction see conduction, electrical.

electrical energy form of energy carried by an electric current. It may be converted into other forms of energy such as heat, light, and motion. The electrical energy per second, or power, W watts converted in a circuit component through which a current I amperes passes and across which there is a potential difference of V volts is given by the formula $W = IV$.

electric cell device in which chemical energy is converted into electrical energy; see cell, electrical.

electric charge property of some bodies that causes them to exert forces on each other. Two bodies both with positive or both with negative charges repel each other, whereas bodies with opposite or 'unlike' charges attract each other. Electrons possess a negative charge, and protons an equal positive charge. The SI unit of electric charge is the coulomb (symbol C).

electricity all phenomena caused by electric charge, whether static or in motion. Electric charge is caused by an excess or deficit of electrons in the charged substance, and an electric current is the movement of charge through a material. Substances may be electrical conductors, such as metals, that allow the passage of electricity through them readily, or insulators, such as rubber, that are extremely poor conductors. Substances with relatively poor conductivities that increase with a rise in temperature or when light falls on the material, are known as semiconductors.

electrochemical series or *electromotive series*, list of chemical elements arranged in descending order of the ease with which they can lose electrons to form cations (positive ions). An element can be displaced (displacement reaction) from a compound by any element above it in the series.

electrochemistry branch of science that studies chemical reactions involving electricity. The use of electricity to produce chemical effects, electrolysis, is employed in many industrial processes, such as the manufacture of chlorine and the extraction of aluminium. The use of chemical reactions to produce electricity is the basis of electrical cells, such as the dry cell and the Leclanché cell.

electrode any terminal by which an electric current passes in or out of a conducting substance; for example, the anode or cathode in a battery or the carbons in an arc lamp. The terminals that emit and collect the flow of electrons in thermionic valves (electron tubes) are also called electrodes; for example, cathodes, plates, and grids.

electrode potential electric potential between an element and its ions in solution.

electrodynamics branch of physics dealing with electric charges, electric currents, and associated forces. Quantum electrodynamics (QED) studies the interaction between charged particles and their emission and absorption of electromagnetic radiation. This subject combines quantum theory and relativity theory, making accurate predictions about subatomic processes involving charged particles such as electrons and protons.

electrolysis production of chemical changes by passing an electric current through a solution or molten salt (the electrolyte), resulting in the migration of ions to the electrodes: positive ions (cations) to the negative electrode (cathode) and negative ions (anions) to the positive electrode (anode).

electrolyte solution or molten substance in which an electric current is made to flow by the movement and discharge of ions in accordance with Faraday's laws of electrolysis.

electrolytic cell device in which an externally applied voltage brings about a chemical reaction; see electrolysis.

electromagnet coil of wire wound around a soft iron core that acts as a magnet when an electric current flows through the wire. Electromagnets have many uses: in switches, electric bells, solenoids, and metal-lifting cranes.

electromagnetic force one of the four fundamental forces of nature, the other three being the gravitational force or gravity, the weak nuclear force, and the strong nuclear force. The particle that is the carrier for the electromagnetic force is the photon.

electromagnetic induction production of an electromotive force (emf) in a circuit by a change of magnetic flux through the circuit or by relative motion of the circuit and the magnetic flux. In a closed circuit an induced current will be produced. All dynamos and generators make use of this effect. When magnetic tape is driven past the playback head (a small coil) of a tape-recorder, the moving magnetic field induces an emf in the head, which is then amplified to reproduce the recorded sounds.

electromagnetic induction, Faraday's law of law stating that the induced electromotive force is equal to the rate of decrease of magnetic flux. The law was determined by English physicist Michael Faraday.

electromagnetic interaction interaction between two charged particles (for example, an electron and a proton) which appears as a force (attractive if the two charges are different in sign). In quantum theory, the electromagnetic interaction is carried between particles by photons.

electromagnetic radiation transfer of energy in the form of electromagnetic waves.

electromagnetic spectrum complete range, over all wavelengths and frequencies, of electromagnetic waves. These include radio and television waves, infrared radiation, visible light, ultraviolet light, X-rays, and gamma radiation.

electromagnetic waves oscillating electric and magnetic fields travelling together through space at a speed of nearly 300,000 km/186,000 mi per second. The (limitless) range of possible wavelengths and frequencies of electromagnetic waves, which can be thought of as making up the **electromagnetic spectrum**, includes radio waves, infrared radiation, visible light, ultraviolet radiation, X-rays, and gamma rays.

electromotive force *emf*, loosely, the voltage produced by an electric battery or generator in an electrical circuit or, more precisely, the energy supplied by a source of electric power in driving a unit charge around the circuit. The unit is the volt.

electron stable, negatively charged elementary particle; it is a constituent of all atoms, and a member of the class of particles known as leptons. The electrons in each atom surround the nucleus in groupings called shells; in a neutral atom the number of electrons is equal to the number of protons in the nucleus. This electron structure is responsible for the chemical properties of the atom (see atomic structure).

electron acceptor compound that can accept an electron and is reduced in doing so; it can therefore take part in oxidation–reduction reactions.

electronegativity the ease with which an atom can attract electrons to itself. Electronegative elements attract electrons, so forming negative ions.

electronics branch of science that deals with the emission of electrons from conductors and semiconductors, with the subsequent manipulation of these electrons, and with the construction of electronic devices. The first electronic device was the thermionic valve, or vacuum tube, in which electrons moved in

467

a vacuum, and led to such inventions as radio, television, radar, and the digital computer. Replacement of valves with the comparatively tiny and reliable transistor from 1948 revolutionized electronic development. Modern electronic devices are based on minute integrated circuits (silicon chips), wafer-thin crystal slices holding tens of thousands of electronic components.

electron microscope means of achieving very high magnification of objects. It uses high-energy electrons, whose characteristic wavelength is much shorter than that of visible light used in ordinary microscopes. Its resolving power is thus far higher.

electrophilic describing a reagent that readily accepts electrons during a chemical reaction. Such reagents, therefore, typically react at centres of high electron density.

electrophoresis movement of electrically charged solute particles in a colloid towards the oppositely charged electrode when a pair of electrodes is immersed in the colloidal solution and connected to an external source of direct-current electricity. It is used as an analytical technique similar to chromatography.

electrophorus means of inducing a charge in one body into another. Usually consists of a metal plate with an insulating handle, the plate being capable of carrying electrostatic charges.

electroplating deposition of metals upon metallic surfaces by electrolysis for decorative and/or protective purposes. It is used in the preparation of printers' blocks, 'master' audio discs, and in many other processes.

electropositivity measure of the ability of elements (mainly metals) to donate electrons to form positive ions. The greater the metallic character, the more electropositive the element.

electrostatics study of stationary electric charges and their fields (not currents).

element substance that cannot be split chemically into simpler substances. The atoms of a particular element all have the same number of protons in their nuclei (their atomic number). Elements are classified in the periodic table of the elements. Of the known elements, 92 are known to occur in nature (those with atomic numbers 1–92). Those elements with atomic numbers above 96 do not occur in nature and are synthesized only, produced in particle accelerators. Of the elements, 81 are stable; all the others, which include atomic numbers 43, 61, and from 84 up, are radioactive.

element in mathematics, a member of a set.

elementary particle subatomic particle that is not known to be made up of smaller particles, and so can be considered one of the fundamental units of matter. There are three groups of elementary particles: quarks, leptons, and gauge bosons.

ellipse curve joining all points (loci) around two fixed points (foci) such that the sum of the distances from those points is always constant. The diameter passing through the foci is the major axis, and the diameter bisecting this at right angles is the minor axis. An ellipse is one of a series of curves known as conic sections. A slice across a cone that is not made parallel to, and does not pass through, the base will produce an ellipse.

elliptic function integral of the general form $\int f(x, \sqrt{R})\,dx$, where f is any rational function of x and R is a quartic polynomial corresponding to:

$$a_0 x^4 + a_1 x^3 + a_2 x^2 + a_3 x + a_4$$

with no multiple roots. French mathematician Adrien Legendre proved that any elliptic integral can be reduced to the sum of an elementary function and of scalar multiples of three special functions. Abelian functions and theta functions are both extensions. Elliptic functions are used in the integration of the square root of a cubic or a quartic, and are thus important to many mathematical operations.

elliptic geometry system of non-Euclidean geometry developed as the initial form of Riemann geometry, and regarding all geometrical operations as carried out in 'curved' space; for example, as though on the surface of an ellipsoid or sphere. A 'straight line' is thus defined (then) as the shortest curve (geodesic) on the curved surface joining two points.

ellipticity of the shape of a planet or galaxy, the amount of distortion by which it departs from a perfect sphere. The overall ellipticity of the Earth is given as $1/299$. One class of galaxy is defined in terms of ellipticity, subdivided E0 to E7, according to degree.

elliptic modular functions functions defined in the upper half of an Argand plane that are automorphic relative to a group of modular transformations; that is, transformations T such as:

$$T(z) = az + b/cz + d$$

elution washing of an adsorbed substance from the adsorbing material; it is used, for example, in the separation processes of chromatography and electrophoresis.

empirical formula chemical formula of a substance in which only the relative proportions of each of its constituent elements are given. The empirical formula does not necessarily reflect a substance's molecular formula or its structure. The empirical formula of benzene, for example, is CH, whereas its molecular formula is C_6H_6.

emulsion stable dispersion of a liquid in another liquid – for example, oil and water in some cosmetic lotions.

enantiomorph or *antimers*, or *optical antipodes*, or *enantisomers*, compound that has two asymmetric structures, each a mirror image of the other. Enantio morphs, such as the optically active forms of lactic acid, have identical chemical and physical properties, except in reactions with other enantiomorphs or in inter-actions with polarized light. See optical isomerism.

endangered species plant or animal species whose numbers are so few that it is at risk of becoming extinct. Officially designated endangered species are listed by the World Conservation Union.

endocytosis ingestion of material by a cell, including phagocytosis and pinocytosis. **Phagocytosis** is the engulfment and ingestion by a white blood cell of bacteria or other foreign particles; **pinocytosis** involves the absorption and ingestion by a cell of surrounding fluid by the folding-in of the cell membrane to form a vesicle which (eventually) releases some of its contents into the cell's cytoplasm.

endoderm innermost of the three germ layers of an embryo.

endoplasmic reticulum *ER*, membranous system of tubes, channels, and flattened sacs that form compartments within eukaryotic cells. It stores and transports proteins within cells and also carries various enzymes needed for the synthesis of fats. The ribosomes, or the organelles that carry out protein synthesis, are sometimes attached to parts of the ER.

endplate mass of motor nerve endings that penetrate a muscle fibre.

end-point point during a titration when the two reagents involved are at exact equivalence; that is, when all of each of the reagents has reacted and there is no excess of either.

energy, electrical see electrical energy.

energy, kinetic see kinetic energy.

energy, law of conservation of fundamental principle of physics stating that energy can neither be created nor destroyed, but only changed from one form to another in a closed system.

energy levels, electronic series of specific, discrete energy states that electrons orbiting a nucleus can occupy. In certain processes an electron may absorb external energy and move to a higher energy level (in which case the electron is said to be excited) or it may release energy (usually in the form of light) and move to a lower energy level. Because the energy levels are discrete, these movements of electrons to different energy levels involve specific amounts (quanta) of energy. See quantum theory.

energy, magnetic see magnetic energy.

energy, potential see potential energy.

engineering application of science to the design, construction, and maintenance of works, machinery, roads, railways, bridges, harbour installations, engines, ships, aircraft and airports, spacecraft and space stations, and the generation, transmission, and use of electrical power. The main divisions of engineering are aerospace, chemical, civil, computer, electrical, electronic, gas, marine, materials, mechanical, mining, production, radio, and structural.

enthalpy alternative term for energy of reaction, the heat energy associated with a chemical change.

entropy parameter representing the state of disorder of a system at the atomic, ionic, or molecular level; the greater the disorder, the higher the entropy. Thus the fast-moving disordered molecules of water vapour have higher entropy than those of more ordered liquid water, which in turn have more entropy than the molecules in solid crystalline ice.

enzyme biological catalyst produced in cells, and capable of speeding up the chemical reactions necessary for life. They are large, complex proteins, and are highly specific, each chemical reaction requiring its own particular enzyme. The enzyme's specificity arises from its active site, an area with a shape corresponding to part of the molecule with which it reacts (the substrate). The enzyme and the substrate slot together forming an enzyme–substrate complex that allows the reaction to take place, after which the enzyme falls away unaltered.

enzyme induction stimulation of enzyme formation by the presence of its substrate or a derivative of the substrate.

epicycle circular orbit of a body round a point that is itself in a circular orbit round a parent body. Such a system was formulated to explain some planetary orbits in the Solar System before they were known to be elliptical.

epithelium in animals, tissue of closely packed cells that forms a surface or lines a cavity or tube. Epithelium may be protective (as in the skin) or secretory (as in the cells lining the wall of the gut).

epitrochoid or hypotrochoid locus of a point on a rolling circle moving round the circumference of another circle, that is not on the circumference of the rolling circle. The locus is of importance in the Wankel engine.

equilibrium unchanging condition in which an undisturbed system can remain indefinitely in a state of balance. In a **static equilibrium**, such as an object resting on the floor, there is no motion. In a **dynamic equilibrium**, in contrast, a steady state is maintained by constant, though opposing, changes. For example, in a sealed bottle half full of water, the constancy of the water level is a result of molecules evaporating from the surface and condensing on to it at the same rate.

equilibrium constant numerical value that expresses the position of a chemical equilibrium at a given temperature and pressure. It is given by the product of the concentrations of the reactants divided by the product of the concentrations of the products.

equinox time when the Sun is directly overhead at the Earth's Equator and consequently day and night are of equal length at all latitudes. This happens twice a year: 21 March is the spring, or vernal, equinox, and 23 September is the autumn equinox.

equivalent weight mass of a substance that exactly reacts with, or replaces, an arbitrarily fixed mass of another substance in a particular reaction. The combining proportions (by mass) of substances are in the ratio of their equivalent masses (or a multiple of that ratio) and a common standard has been adopted: for elements, the equivalent weight is the quantity that reacts with, or replaces, 1.00797 g/0.035279 oz of hydrogen or 7.9997 g/0.28215 oz of oxygen, or the mass of an element liberated during electrolysis by the passage of 1 faraday (96,487 coulombs per mole) of electricity. The equivalent weight of an element is given by its gram atomic mass divided by its valency. For oxidizing and reducing agents, the equivalent weight is the gram molecular mass divided by the number of electrons gained or lost by each molecule. Some substances have several equivalent weights, depending on the specific reaction in which they are involved.

ergodics study of the mathematical principles involved in the kinetic theory of gases.

Erlangen programme expression used by German mathematician Felix Klein to denote his unification and classification of geometries Euclidean and non-Euclidean as 'members' of one 'family', corresponding to the transformations found in each.

erosion wearing away of the Earth's surface, caused by the breakdown and transportation of particles of rock or soil (by contrast, weathering does not involve transportation). Agents of erosion include the sea, rivers, glaciers, and wind.

Water, consisting of sea waves and currents, rivers, and rain; ice, in the form of glaciers; and wind, hurling sand fragments against exposed rocks and moving dunes along, are the most potent forces of erosion.

People also contribute to erosion by bad farming practices and the cutting down of forests, which can lead to the formation of dust bowls.

error theory or ***theory of errors***, method of evaluating the effects and the significance of errors, for example when obtaining a mean value from a small sample.

escape velocity minimum velocity with which an object must be projected for it to escape from the gravitational pull of a planetary body. In the case of the Earth, the escape velocity is 11.2 kps/6.9 mps; the Moon, 2.4 kps/1.5 mps; Mars, 5 kps/3.1 mps; and Jupiter, 59.6 kps/37 mps.

ester organic compound formed by the reaction between an alcohol and an acid, with the elimination of water. Unlike salts, esters are covalent compounds.

etalon instrument that uses interference phenomena to make possible very high-resolution observations of spectral lines.

ether in chemistry, any of a series of organic chemical compounds having an oxygen atom linking the carbon atoms of two hydrocarbon radical groups (general formula $R–O–R'$); also the common name for ethoxyethane $C_2H_5OC_2H_5$ (also called diethyl ether). This is used as an anaesthetic and as an external cleansing agent before surgical operations. It is also used as a solvent, and in the extraction of oils, fats, waxes, resins, and alkaloids.

ether or ***aether***, in the history of science, a hypothetical medium permeating all of space. The concept originated with the Greeks, and has been revived on several occasions to explain the properties and propagation of light. It was supposed that light and other electromagnetic radiation – even in outer space – needed a medium, the ether, in which to travel. The idea was abandoned with the acceptance of relativity.

ethology comparative study of animal behaviour in its natural setting. Ethology is concerned with the causal mechanisms (both the stimuli that elicit behaviour and the physiological mechanisms controlling it), as well as the development of behaviour, its function, and its evolutionary history.

ethyne common name **acetylene**, CH≡CH colourless inflammable gas produced by mixing calcium carbide and water. It is the simplest member of the alkyne series of hydrocarbons. It is used in the manufacture of the synthetic rubber neoprene, and in oxyacetylene welding and cutting.

Euclid's fifth postulate states that parallel lines meet only at infinity.

eugenics (Greek *eugenes* 'well-born') study of ways in which the physical and mental characteristics of the human race may be improved. The eugenic principle was abused by the Nazi Party in Germany during the 1930s and early 1940s to justify the attempted extermination of entire social and ethnic groups and the establishment of selective breeding programmes. Modern eugenics is concerned mainly with the elimination of genetic disease.

Euler's number (e) the limit of the sequence:

$$a_n = 1 + 1/1! + 1/2! + 1/3! + \ldots + 1/n!$$

An irrational number introduced originally by Swiss mathematician Leonhard Euler, e may be represented to the sixth decimal place as 2.718282; it has useful theoretical properties in differential calculus and serves as a natural base for logarithms (known as 'natural logarithms').

eutrophication excessive enrichment of rivers, lakes, and shallow sea areas, primarily by nitrate fertilizers washed from the soil by rain, by phosphates from fertilizers, and from nutrients in municipal sewage, and by sewage itself. These encourage the growth of algae and bacteria which use up the oxygen in the water, thereby making it uninhabitable for fishes and other animal life.

event horizon 'edge' of a black hole; the interface between four-dimensional space and a singularity.

evolutionary toxicology study of the effects of pollution on evolution. A polluted habitat may cause organisms to select for certain traits, as in **industrial melanism** for example, where some insects, such as the peppered moth, evolve to be darker in polluted areas, and therefore better camouflaged against predation.

excitation injection of energy into an atom (which may be part of a radical or molecule) which raises it or one of its components to a higher energy level.

excited state, electronic condition of an electron that has absorbed external energy and, as a result, moved from its normal, ground state energy level to a higher energy level. The excitation energy is the difference in energy between the ground state and the excited state.

excluded middle, law of the law in logic, that a statement is either true or false, leaving no room for any further alternatives. There are nonclassical systems of logic that distinguish between true = proven true and false = proven false, and so allow intermediate values (such as 'possibly true'). Boolean-valued logic systems attach probability values to statements that may therefore also have intermediate values other than merely true or false.

exclusion principle principle of atomic structure originated by Austrian–US physicist Wolfgang Pauli. It states that no two electrons in a single atom may have the same set of quantum numbers.

Hence, it is impossible to pack together certain elementary particles, such as electrons, beyond a certain critical density, otherwise they would share the same location and quantum number. A white dwarf star, which consists of electrons and other elementary particles, is thus prevented from contracting further by the exclusion principle and never collapses.

exocytosis ejection from a cell of undigested remnants of material.

exothermic describing a process or reaction that involves the release of energy (usually in the form of heat). Combustion, for example, is an exothermic reaction.

exponent or **index**, superscript number that indicates the number of times a term is multiplied by itself; for example $x^2 = x \times x$, $4^3 = 4 \times 4 \times 4$.

exteroceptive describing receptors that receive stimuli from outside the body, such as those of the ear and the eye.

extinction complete disappearance of a species or higher taxon. Extinctions occur when a species becomes unfit for survival in its natural habitat usually to be replaced by another, better-suited species. An organism becomes ill-suited for survival because its environment is changed or because its relationship to other organisms is altered. For example, a predator's fitness for survival depends upon the availability of its prey.

471

factorial of a positive number, the product of all the whole numbers (integers) inclusive between 1 and the number itself. A factorial is indicated by the symbol '!'. Thus $6! = 1 \times 2 \times 3 \times 4 \times 5 \times 6 = 720$. Factorial zero, 0!, is defined as 1.

factorization reduction into constituent factors (which when multiplied together produce the original number or expression).

faculae bright areas on the face of the Sun, commonly in the vicinity of sunspots. Named by German astronomer Johannes Hevelius, they are thought to be caused by luminous hydrogen clouds close to the photosphere. They last on average about 15 Earth days.

family group of related genera (see genus). Family names are not printed in italic (unlike genus and species names), and by convention they all have the ending -idae (animals) or -aceae (plants and fungi). For example, the genera of hummingbirds are grouped in the hummingbird family, Trochilidae. Related families are grouped together in an order.

fat in the broadest sense, a mixture of lipids – chiefly triglycerides (lipids containing three fatty acid molecules linked to a molecule of glycerol). More specifically, the term refers to a lipid mixture that is solid at room temperature (20°C); lipid mixtures that are liquid at room temperature are called **oils**. The higher the proportion of saturated fatty acids in a mixture, the harder the fat.

fatty acid or *carboxylic acid*, organic compound consisting of a hydrocarbon chain of an even number of carbon atoms, with a carboxyl group (–COOH) at one end. The covalent bonds between the carbon atoms may be single or double; where a double bond occurs the carbon atoms concerned carry one instead of two hydrogen atoms. Chains with only single bonds have all the hydrogen they can carry, so they are said to be saturated with hydrogen. Chains with one or more double bonds are said to be unsaturated (see polyunsaturate). Fatty acids are produced in the small intestine when fat is digested.

Fermat's last theorem theorem that states that the equation:

$$x^n + y^n = z^n$$

is not solvable in the integers if n is greater than 2. Fermat's own proof has never been found and the theorem remained unproved until 1994, when English mathematician Andrew Wiles verified it.

fermentation breakdown of sugars by bacteria and yeasts using a method of respiration without oxygen (anaerobic). Fermentation processes have long been utilized in baking bread, making beer and wine, and producing cheese, yogurt, soy sauce, and many other foodstuffs.

Fermi–Dirac statistics mathematical treatment of particles with half-integer spin in statistical mechanics, enabling properties of systems made up by such particles to be calculated. They are named after Italian-born US physicist Enrico Fermi and English physicist Paul Dirac.

fermion subatomic particle whose spin can only take values that are half-odd-integers, such as 1/2 or 3/2. Fermions may be classified as leptons, such as the electron, and hadrons, such as the proton, neutron, mesons, and so on. All elementary particles are either fermions or bosons.

ferromagnetism form of magnetism that can be acquired in an external magnetic field and usually retained in its absence, so that ferromagnetic materials are used to make permanent magnets. A ferromagnetic material may therefore be said to have a high magnetic permeability and susceptibility (which depends upon temperature). Examples are iron, cobalt, nickel, and their alloys.

Fibonacci series sequence in which each term after the first two is the sum of the two terms immediately preceding it; it begins 1, 1, 2, 3, 5, 8, 13, 21, . . . and has a variety of important applications (for example, in search algorithms).

field region of space in which an object exerts a force on another separate object because of certain properties they both possess. For example, there is a force of attraction between any two objects that have mass when one is in the gravitational field of the other.

field strength see magnetic field strength.

field theory theory involving the mathematical structure known as a field, which displays the operations of addition and multiplication and their inverses (subtraction and division). An elementary example other than the real numbers is constituted by the rational numbers; there are many other types. A ring is much like a field but does not include the inverse operations. A group is a more restricted concept still.

fine structure splitting of individual spectral lines seen when viewed under high resolution. This can be produced by external magnetic fields acting on atoms, for example.

finite having a countable number of elements, the opposite of infinite.

first order in differential equations, involving only the first derivative.

fission splitting of a heavy atomic nucleus into two or more major fragments. It is accompanied by the emission of two or three neutrons and the release of large amounts of nuclear energy.

flare star dim red dwarf star that suddenly lights up with great – but brief – luminosity, corresponding to an equally powerful but short-lived burst of radio emission. The cause is thought to be a sudden and intense outburst of radiation on or above the star's surface.

fluid any substance, either liquid or gas, in which the molecules are relatively mobile and can 'flow'.

fluorescence short-lived luminescence (a glow not caused by high temperature). Phosphorescence lasts a little longer.

flux density see magnetic flux density.

fluxions another name for calculus. It is no longer used.

food chain sequence showing the feeding relationships between organisms in a particular ecosystem. Each organism depends on the next lowest member of the chain for its food. A pyramid of numbers can be used to show the reduction in food energy at each step up the food chain.

force any influence that tends to change the state of rest or the uniform motion in a straight line of a body. The action of an unbalanced or resultant force results in the acceleration of a body in the direction of action of the force, or it may, if the body is unable to move freely, result in its deformation (see Hooke's law). Force is a vector quantity, possessing both magnitude and direction; its SI unit is the newton.

force, centrifugal see centrifugal force.

force, centripetal see centripetal force.

force, electromagnetic see electromagnetic force.

force, electromotive see electromotive force.

fossil (Latin *fossilis* 'dug up') cast, impression, or the actual remains of an animal or plant preserved in rock. Fossils were created during periods of rock formation, caused by the gradual accumulation of sediment over millions of years at the bottom of the sea bed or an inland lake. Fossils may include footprints, an internal cast, or external impression. A few fossils are preserved intact, as with mammoths fossilized in Siberian ice, or insects trapped in tree resin that is today amber. The study of fossils is called palaeontology. Palaeontologists are able to deduce much of the geological history of a region from fossil remains.

Foucault pendulum pendulum consisting of a long wire to which is attached a heavy weight, which is then free to swing in any plane. Once it starts swinging in one particular plane, the plane slowly rotates, this being the result of the rotation of the Earth beneath the pendulum. This phenomenon was observed by French physicist Léon Foucault in 1851.

foundation part of the structure that ensures stability by providing mechanical fixing with solid earth.

foundations of mathematics or *foundations of arithmetic*, subject of attempts to derive the basic precepts of elementary mathematics from a standpoint of pure logic (thence to derive the more complex principles).

Fourier analysis see Fourier theorem.

Fourier series series in which the terms comprise multiples of the cosine and/or sine of multiple angles. Represented by the formula:

$$\frac{1}{2}a_0 + \Sigma(a_n \cos nx + b_n \sin nx)$$

it is used to analyse periodic functions (that is, functions whose graphs repeat themselves periodically).

Fourier theorem adaptation of the process developed by French mathematician Joseph Fourier as the Fourier series to the investigation of energy propagated in the form of waves (particularly heat, sound, and light). A further developed version of this method is known as harmonic analysis. Use of the theorem in investigating wave forms is known as Fourier analysis.

fraction (from Latin *fractus* 'broken') number that indicates one or more equal parts of a whole. Usually, the number of equal parts into which the unit is divided (denominator) is written below a horizontal or diagonal line, and the number of parts making up the fraction (numerator) is written above; for example, 2/3 has numerator 2 and denominator 3. Such fractions are called vulgar fractions or **simple fractions**. The denominator can never be zero.

frame of reference set of axes fixed in such a way as to define uniquely the position of an object in space.

Fraunhofer lines dark lines crossing the solar spectrum. They are caused by the absorption of light from hot regions of the Sun's surface by gases in the cooler, outer regions. They were first investigated by German physicist Joseph von Fraunhofer.

free energy sum of enthalpy and entropy; that is, the capacity of a system to perform work.

473

The change in free energy accompanying a chemical reaction is a measure of its completeness.

freezing point for any given liquid, the temperature at which any further removal of heat will convert the liquid into the solid state. The temperature remains at this point until all the liquid has solidified. It is invariable under similar conditions of pressure – for example, the freezing point of water under standard atmospheric pressure is 0°C/32°F.

frequency number of periodic oscillations, vibrations, or waves occurring per unit of time. The SI unit of frequency is the hertz (Hz), one hertz being equivalent to one cycle per second. Frequency is related to wavelength and velocity by the relationship

$$f = v/\lambda$$

where f is frequency, v is velocity, and λ is wavelength. Frequency is the reciprocal of the period T:

$$f = 1/T.$$

friction force that opposes the relative motion of two bodies in contact. The **coefficient of friction** is the ratio of the force required to achieve this relative motion to the force pressing the two bodies together.

front boundary between two air masses of different temperature or humidity. A **cold front** marks the line of advance of a cold air mass from below, as it displaces a warm air mass; a **warm front** marks the advance of a warm air mass as it rises up over a cold one. Frontal systems define the weather of the mid-latitudes, where warm tropical air is continually meeting cold air from the poles.

fuel cell cell converting chemical energy directly to electrical energy. It works on the same principle as a battery but is continually fed with fuel, usually hydrogen. Fuel cells are silent and reliable (no moving parts) but expensive to produce.

fulcrum point of support of a lever, about which the lever can pivot in lifting and lowering loads applied to it.

function a function f associates each element x of a set X with a single element y of a set Y. Functions are used in all branches of mathematics, physics, and science generally. For example, in the equation

$$y = 2x + 1,$$

y is a function of the symbol x. This can be written as $y = f(x)$.

functions of real variables, theory of involves functions of which the arguments are real numbers.

function theory or ***theory of functions***, use of functions, primarily in order to denote mathematical relationships, but also in application to other sciences. Functional analysis, for example, considers problems in which an unknown function is to be found; at this stage variables may represent functions as opposed to numerical values.

furnace device in which very high temperatures are produced to bring about chemical reactions.

fuse device that prevents the passage of electric current above a predetermined level in a circuit, by melting due to the consequent temperature rise, and thus breaking the circuit.

fusion fusing of the nuclei of light elements, such as hydrogen, into those of a heavier element, such as helium. The resultant loss in their combined mass is converted into energy. Stars and thermonuclear weapons are powered by nuclear fusion.

galactic centres phenomena that are now thought to comprise black holes – which would explain why the centre of our Galaxy appears strangely obscure, and emits only infrared radiation.

galactic rotation revolving of a galaxy round its central nucleus even as it continues its proper motion. Such rotation, however, is not uniform but differential. One revolution of the Sun within our own Galaxy takes about 225 million years, or 1 cosmic year.

galaxy congregation of millions or billions of stars, held together by gravity. **Spiral galaxies**, such as the Milky Way, are flattened in shape, with a central bulge of old stars surrounded by a disc of younger stars, arranged in spiral arms like a Catherine wheel.

 Barred spirals are spiral galaxies that have a straight bar of stars across their centre, from the ends of which the spiral arms emerge. The arms of spiral galaxies contain gas and dust from which new stars are still forming.

 Elliptical galaxies contain old stars and very little gas. They include the most massive galaxies known, containing a trillion stars. At least some elliptical galaxies are thought to be formed by mergers between spiral galaxies. There are also irregular galaxies. Most galaxies occur in clusters, containing anything from a few to thousands of members.

galvanometer instrument for detecting small electric currents by their magnetic effect.

gamete cell that functions in sexual reproduction by merging with another gamete to form a zygote. Examples of gametes include sperm and egg cells. In most organisms, the gametes are haploid (they contain half the number of chromosomes of the parent), owing to reduction division or meiosis.

game theory group of mathematical theories, developed in 1944 by German-born US mathematician Oscar Morgenstern and Hungarian-born US mathematician John Von Neumann, that seeks to abstract from invented game-playing scenarios and their outcome the essence of situations of conflict and/or cooperation in the real political, business, and social world.

gamma globulin one of a group of proteins or immunoglobulins in the blood that act as antibodies to specific infections. Gamma globulins extracted from the blood of a patient who has recovered from an infection may be used as vaccines to stimulate artificial immunity in others.

gamma radiation very high-frequency electromagnetic radiation, similar in nature to X-rays but of shorter wavelength, emitted by the nuclei of radioactive substances during decay or by the interactions of high-energy electrons with matter. Cosmic gamma rays have been identified as coming from pulsars, radio galaxies, and quasars, although they cannot penetrate the Earth's atmosphere.

gas a form of matter, such as air, in which the molecules move randomly in otherwise empty space, filling any size or shape of container into which the gas is put.

gas mantle illumination device made from impregnating a dome-shaped piece of rayon with compounds of thorium and cerium, which are decomposed by heat.

gastrula stage in embryonic development following the blastula stage in which gastrulation occurs.

gastrulation cell movements during embryonic development (after cleavage) in which cells move to the positions in which they eventually give rise to the organs of the growing embryo.

gauge distance between the inside edges of a railway for trains. Also, the diameter or wires, rods, and so on.

gauge boson any of the particles that carry the four fundamental forces of nature. Gauge bosons are elementary particles that cannot be subdivided, and include the photon, graviton, the gluons, and W^+, W^-, and Z particles.

gauss cgs unit (symbol Gs) of magnetic induction or magnetic flux density, replaced by the SI unit, the tesla, but still commonly used. It is equal to one line of magnetic flux per square centimetre. The Earth's magnetic field is about 0.5 Gs, and changes to it over time are measured in gammas (one gamma equals 10^{-5} gauss).

Gaussian distribution another name for normal distribution.

gegenschein faint oval patch of light, opposite the Sun, visible from Earth only at certain times of the year. Its nature and cause are still not known. It is sometimes known as 'counter-glow'.

Geiger counter any of a number of devices used for detecting nuclear radiation and/or measuring its intensity by counting the number of ionizing particles produced (see radioactivity). It detects the momentary current that passes between electrodes in a suitable gas when a nuclear particle or a radiation pulse causes the ionization of that gas.

gelatine soluble protein-based substance with the ability to form a jelly on cooling. It is used in photography and the making of glues.

gemmule in early genetic theory, minute particles thought to consist of miniature copies of all parts of the body carried in the blood to the gametes and from which their larger forms eventually developed. In modern usage, a gemmule is a bud formed on a sponge that may break free and develop into a new animal.

gene unit of inherited material, encoded by a strand of DNA and transcribed by RNA. In higher organisms, genes are located on the chromosomes. A gene consistently affects a particular character in an individual – for example, the gene for eye colour. Also termed a Mendelian gene, after Austrian biologist Gregor Mendel, it occurs at a particular point, or locus, on a particular chromosome and may have several variants, or alleles, each specifying a particular form of that character – for example, the alleles for blue or brown eyes. Some alleles show dominance. These mask the effect of other alleles, known as recessive.

gene pool total sum of alleles (variants of genes) possessed by all the members of a given population or species alive at a particular time.

generator machine that produces electrical energy from mechanical energy, as opposed to an electric motor, which does the opposite. A simple generator (known as a dynamo in the UK) consists of a wire-wound coil (armature) that is rotated between the poles of a permanent magnet. The movement of the wire in the magnetic field induces a current in the coil by electromagnetic induction, which can be fed by means of a commutator as a continuous direct current into an external circuit. Slip rings instead of a commutator produce an alternating current, when the generator is called an alternator.

genetic engineering all-inclusive term that describes the deliberate manipulation of genetic material by biochemical techniques. It is often achieved by the introduction of new DNA, usually by means of a virus or plasmid. This can be for pure research, gene therapy, or to breed functionally specific plants, animals, or bacteria. These organisms with a foreign gene added are said to be transgenic. At the beginning of 1995 more than 60 plant species had been genetically engineered, and nearly 3,000 transgenic crops had been field-tested.

genetics branch of biology concerned with the study of heredity and variation; it attempts to explain how characteristics of living organisms are passed on from one generation to the next. The science of genetics is based on the work of Austrian biologist Gregor Mendel whose experiments with the cross-breeding (hybridization) of peas showed that the inheritance of characteristics and traits takes place by means of discrete 'particles' (genes). These are present in the cells of all organisms, and are now recognized as being the basic units of heredity. All organisms possess genotypes (sets of variable genes) and phenotypes (characteristics produced by certain genes). Modern geneticists investigate the structure, function, and transmission of genes.

genome full complement of genes carried by a single (haploid) set of chromosomes. The term may be applied to the genetic information carried by an individual or to the range of genes found in a given species. The human genome is made up of approximately 30–40,000 genes.

genotype particular set of alleles (variants of genes) possessed by a given organism. The term is usually used in conjunction with phenotype, which is the product of the genotype and all environmental effects. See also nature–nurture controversy.

genus plural ***genera***, group of one or more species with many characteristics in common. Thus all doglike species (including dogs, wolves, and jackals) belong to the genus *Canis* (Latin 'dog').

Species of the same genus are thought to be descended from a common ancestor species. Related genera are grouped into families.

geocentric having the Earth at the centre.

geodesic shortest route between two points on any surface.

geodesy science of measuring and mapping Earth's surface for making maps and correlating geological, gravitational, and magnetic measurements. Geodetic surveys, formerly carried out by means of various measuring techniques on the surface, are now commonly made by using radio signals and laser beams from orbiting satellites (see global positioning system).

geological time time scale embracing the history of the Earth from its physical origin to the present day. Geological time is traditionally divided into eons (Archaean or Archaeozoic, Proterozoic, and Phanerozoic in ascending chronological order), which in turn are subdivided into eras, periods, epochs, ages, and finally chrons.

geometric curve curve that can be precisely expressed by an equation (unlike a mechanical curve); for example, a circle, parabola, or hyperbola.

germ plasm theory in early genetics, the theory that of the two tissue types in multicellular animals (somatoplasm in body cells and germ plasm in reproductive cells) only the integrity of germ plasm is necessary for the inheritance of characteristics.

glacial trough or ***U-shaped valley***, steep-sided, flat-bottomed valley formed by a glacier. The erosive action of the glacier and of the debris carried by it results in the formation not only of the trough itself but also of a number of associated features, such as truncated spurs (projections of rock that have been sheared off by the ice) and hanging valleys (smaller glacial valleys that enter the trough at a higher level than the trough floor). Features characteristic of glacial deposition, such as drumlins and eskers, are commonly found on the floor of the trough, together with linear lakes called ribbon lakes.

gland specialized organ of the body that manufactures and secretes enzymes, hormones, or other chemicals. In animals, glands vary in size from small (for example, tear glands) to large (for example, the pancreas), but in plants they are always small, and may consist of a single cell. Some glands discharge their products internally, endocrine glands, and others, externally, exocrine glands. Lymph nodes are sometimes wrongly called glands.

globular cluster spherical or near-spherical star cluster containing from approximately 10,000 to millions of stars. About 120 globular clusters are distributed in a spherical halo around our Galaxy. They consist of old stars, formed early in the Galaxy's history. Globular clusters are also found around other galaxies.

gluon gauge boson that carries the strong nuclear force, responsible for binding quarks together to form the strongly interacting subatomic particles known as hadrons. There are eight kinds of gluon.

glycolysis conversion of glucose to lactic acid. It takes place in the cytoplasm of cells as part of the process of cellular respiration.

Golgi apparatus or *Golgi body*, stack of flattened membranous sacs found in the cells of eukaryotes. Many molecules travel through the Golgi apparatus on their way to other organelles or to the endoplasmic reticulum. Some are modified or assembled inside the sacs. The Golgi apparatus is named after the Italian physician Camillo Golgi.

Gondwanaland or *Gondwana*, southern landmass formed 200 million years ago by the splitting of the single world continent Pangaea. (The northern landmass was Laurasia.) It later fragmented into the continents of South America, Africa, Australia, and Antarctica, which then drifted slowly to their present positions. The baobab tree found in both Africa and Australia is a relic of this ancient land mass.

graph pictorial representation of numerical data, such as statistical data, or a method of showing the mathematical relationship between two or more variables by drawing a diagram.

gravitational constant fundamental constant (symbol g) of physics that relates the force of gravity produced by a body to the masses involved and the separation. According to some physicists, its present value of 6.67×10^{-11} m³/(kg s²) is decreasing at a rate proportional to the age of the universe.

gravity force of attraction that arises between objects by virtue of their masses. On Earth, gravity is the force of attraction between any object in the Earth's gravitational field and the Earth itself. It is regarded as one of the four fundamental forces of nature, the other three being the electromagnetic force, the strong nuclear force, and the weak nuclear force. The gravitational force is the weakest of the four forces, but it acts over great distances. The particle that is postulated as the carrier of the gravitational force is the graviton.

gravity, centre of see centre of gravity.

Great Red Spot prominent oval feature, 14,000 km/8,500 mi wide and some 30,000 km/20,000 mi long, in the atmosphere of the planet Jupiter, south of the Equator. It was first observed in 1664. Space probes show it to be an anticlockwise vortex of cold clouds, coloured possibly by phosphorus.

greenhouse effect phenomenon of the Earth's atmosphere by which solar radiation, trapped by the Earth and re-emitted from the surface as infrared radiation, is prevented from escaping by various gases in the atmosphere. Greenhouse gases trap heat because they readily absorb infrared radiation. The result of the greenhouse effect is a rise in the Earth's temperature (global warming). The main greenhouse gases are carbon dioxide, methane, and chlorofluorocarbons (CFCs) as well as water vapour. Fossil-fuel consumption and forest fires are the principal causes of carbon dioxide build-up; methane is a by product of agriculture (rice, cattle, sheep).

grid electrode in the form of a mesh that, when placed between the anode and cathode of a thermionic valve, controls the flow of electrons.

ground state (electronic) the state of an electron in its lowest energy level. When all the electrons orbiting a nucleus are in their lowest energy levels, the atom as a whole has its minimum possible energy and is therefore in its most stable state.

group theory investigation and classification of the properties of the mathematical structures known as groups. A group possesses two operations – 'multiplication' and 'inverting of an element' – and a further designated element called 'unity'. An example is provided by the set of nonzero real numbers with ordinary multiplication and reciprocation (that is, being multiplied or becoming reciprocals).

gunpowder or *black powder*, oldest known explosive, a mixture of 75% potassium nitrate (saltpetre), 15% charcoal, and 10% sulphur. Sulphur ignites at a low temperature, charcoal burns readily, and the potassium nitrate provides oxygen for the explosion.

gymnosperm (Greek 'naked seed') any plant whose seeds are exposed, as opposed to the structurally more advanced angiosperms, where they are inside an ovary. The group includes conifers and related plants such as cycads and ginkgos, whose seeds develop in

477

cones. Fossil gymnosperms have been found in rocks about 350 million years old.

gyroscope mechanical instrument, used as a stabilizing device and consisting, in its simplest form, of a heavy wheel mounted on an axis fixed in a ring that can be rotated about another axis, which is also fixed in a ring capable of rotation about a third axis. Applications of the gyroscope principle include the gyrocompass, the gyropilot for automatic steering, and gyro-directed torpedoes.

Haber process or *Haber–Bosch process*, industrial process by which ammonia is manufactured by direct combination of its elements, nitrogen and hydrogen. The reaction is carried out at 400–500°C/752–932°F and at 200 atmospheres pressure. The two gases, in the proportions of 1:3 by volume, are passed over a catalyst of finely divided iron.

Around 10% of the reactants combine, and the unused gases are recycled. The ammonia is separated either by being dissolved in water or by being cooled to liquid form.

habitat localized environment in which an organism lives, and which provides for all (or almost all) of its needs. The diversity of habitats found within the Earth's ecosystem is enormous, and they are changing all the time. Many can be considered inorganic or physical, for example, the Arctic ice cap, a cave, or a cliff face. Others are more complex, for instance, a woodland, or a forest floor. Some habitats are so precise that they are called **microhabitats**, such as the area under a stone where a particular type of insect lives.

haematite principal ore of iron, consisting mainly of iron (III) oxide, Fe_2O_3. It occurs as **specular haematite** (dark, metallic lustre), **kidney ore** (reddish radiating fibres terminating in smooth, rounded surfaces), and a red earthy deposit.

haemoglobin protein used by all vertebrates and some invertebrates for oxygen transport because the two substances combine reversibly. In vertebrates it occurs in red blood cells (erythrocytes), giving them their colour.

haemoprotein protein containing an iron porphyrin group. The green plant pigment chlorophyll and the red blood pigment haemoglobin are both haemoproteins.

halide any compound produced by the combination of a halogen, such as chlorine or iodine, with a less electronegative element (see electronegativity). Halides may be formed by ionic bonds or by covalent bonds.

Hall effect production of a voltage across a conductor or semiconductor carrying a current at a right angle to a surrounding magnetic field. It was discovered in 1879 by the US physicist Edwin Hall. It is used in the **Hall probe** for measuring the strengths of magnetic fields and in magnetic switches.

Halley's comet comet that orbits the Sun roughly every 75 years, named after English mathemetician, physicist, and astronomer Edmond Halley, who calculated its orbit. It is the brightest and most conspicuous of the periodic comets. Recorded sightings go back over 2,000 years. It travels around the Sun in the opposite direction to the planets. Its orbit is inclined at almost 20° to the main plane of the Solar System and ranges between the orbits of Venus and Neptune. It will next reappear in 2061.

halo nebulous quality round a celestial body (particularly round a red giant); the galactic halo, however, describes the spherical collection of stars forming a surrounding 'shell' for our otherwise compact, discoid Galaxy.

halogen any of a group of five nonmetallic elements with similar chemical bonding properties: fluorine, chlorine, bromine, iodine, and astatine. They form a linked group in the periodic table of the elements, descending from fluorine, the most reactive, to astatine, the least reactive. They combine directly with most metals to form salts, such as common salt (NaCl). Each halogen has seven electrons in its valence shell, which accounts for the chemical similarities displayed by the group.

halogenoalkane any of a group of organic compounds (formerly called alkyl halides) formed by the halogenation of (addition of a halogen to) an alkane. In the presence of ultraviolet light, alkanes react with halides by substitution.

harmonic analysis see Fourier theorem.

head of water vertical distance between the top of a water stream and the point at which its energy is to be extracted.

heat form of energy possessed by a substance by virtue of the vibrational movement (kinetic energy) of its molecules or atoms. Heat energy is transferred by conduction, convection, and radiation. It always flows from a region of higher temperature (heat intensity) to one of lower temperature. Its effect on a substance may be simply to raise its temperature, or to cause it to expand, melt (if a solid), vaporize (if a liquid), or increase its pressure (if a confined gas).

heat capacity quantity of heat required to raise the temperature of an object by one degree centigrade or Kelvin. The **specific heat capacity** of a substance is the heat capacity per unit of mass, measured in joules per kilogram per kelvin ($J\,kg^{-1}\,K^{-1}$).

heat death possible fate of the universe in which it continues expanding indefinitely while all the stars burn out and no new ones are formed. See critical density.

heliocentric having the Sun at the centre.

heliograph device for recording the positions of sunspots.

heliometer instrument to measure the apparent diameter of the Sun at different seasons; also used to measure angular distances between stars.

helium (Greek *helios* 'Sun') colourless, odourless, gaseous, non-metallic element, symbol He, atomic number 2, relative atomic mass 4.0026. It is grouped with the inert gases, is nonreactive, and forms no compounds. It is the second most abundant element (after hydrogen) in the universe, and has the lowest boiling ($-268.9°C/-452°F$) and melting points ($-272.2°C/-458°F$) of all the elements. It is present in small quantities in the Earth's atmosphere from gases issuing from radioactive elements (from alpha decay) in the Earth's crust; after hydrogen it is the second lightest element.

helix three-dimensional curve resembling a spring, corkscrew, or screw thread. It is generated by a line that encircles a cylinder or cone at a constant angle.

hermaphrodite organism that has both male and female sex organs. Hermaphroditism is the norm in such species as earthworms and snails, and is common in flowering plants. Cross-fertilization is common among hermaphrodites, with the parents functioning as male and female simultaneously, or as one or the other sex at different stages in their development. Human hermaphrodites are extremely rare.

hertz SI unit (symbol Hz) of frequency (the number of repetitions of a regular occurrence in one second). Radio waves are often measured in megahertz (MHz), millions of hertz, and the clock rate of a computer is usually measured in megahertz. The unit is named after German physicist Heinrich Hertz.

Hertzsprung–Russell diagram graph on which the surface temperatures of stars are plotted against their luminosities. Most stars, including the Sun, fall into a narrow band called the main sequence. When a star grows old it moves from the main sequence to the upper right part of the graph, into the area of the giants and supergiants. At the end of its life, as the star shrinks to become a white dwarf, it moves again, to the bottom left area. It is named after the Danish astronomer Ejnar Hertzsprung and the US astronomer Henry Russell, who independently devised it in the years 1911–13.

heterodyne effect superimposition of two waves of different frequency in a radio receiver, one of which is being received, the other transmitted within the device itself, producing an intermediate frequency that can be demodulated.

heterozygous in a living organism, having two different alleles for a given trait. In homozygous organisms, by contrast, both chromosomes carry the same allele. In an outbreeding population an individual organism will generally be heterozygous for some genes but homozygous for others.

high-energy particles particles of electromagnetic radiation that contain high energies, measured in terms of electronvolts. The energy in gamma radiation is of the order of 8×10^{7} to 8×10^{5} electronvolts and in X-rays of 8×10^{3} to 8×10^{1} electronvolts.

hodometer device that enables the acceleration of an object moving with known velocity over a path to be determined.

holography method of producing three-dimensional (3-D) images, called holograms, by means of laser light. Holography uses a photographic technique (involving the splitting of a laser beam into two beams) to produce a picture, or hologram, that contains 3-D information about the object photographed. Some holograms show meaningless patterns in ordinary light and produce a 3-D image only when laser light is projected through them, but reflection holograms produce images when ordinary light is reflected from them (as found on credit cards).

hominids humans and their humanlike ape predecessors, which together constitute the family Hominidae.

homologous series any of a number of series of organic chemicals with similar chemical properties in which members differ by a constant relative molecular mass.

homology branch of topology involving the study of closed curves, closed surfaces and similar geometric arrangements in two- to *n*-dimensional space, and investigating the ways in which such spatial structures may be dissected. The formulation of the homological theory of dimensionality led to several basic laws of duality (relating to topological properties of an additional part of space).

homozygous in a living organism, having two identical alleles for a given trait. Individuals homozygous for a trait always breed true; that is, they produce offspring that resemble them in appearance when bred with a genetically similar individual; inbred varieties or species are homozygous for almost all traits. Recessive alleles are only expressed in the homozygous condition. Heterozygous organisms have two different alleles for a given trait.

Hooke's law law stating that the deformation of a body is proportional to the magnitude of the deforming force, provided that the body's elastic limit (see elasticity) is not exceeded. If the elastic limit is not reached, the body will return to its original size once the force is removed. The law was discovered by English physicist Robert Hooke in 1676.

hormone chemical secretion of the ductless endocrine glands and specialized nerve cells concerned with control of body functions. The major glands are the thyroid, parathyroid, pituitary, adrenal, pancreas, ovary, and testis. There are also hormone-secreting cells in the kidney, liver, gastrointestinal tract, thymus (in the neck), pineal (in the brain), and placenta. Hormones bring about changes in the functions of various organs according to the body's requirements. The hypothalamus, which adjoins the pituitary gland at the base of the brain, is a control centre for overall coordination of hormone secretion, the thyroid hormones determine the rate of general body chemistry, the adrenal hormones prepare the organism during stress for 'fight or flight', and the sexual hormones such as oestrogen and testosterone govern reproductive functions.

hot Big Bang later, but fundamental, concept within the Big Bang model, that the primordial explosion occurred in terms of almost unimaginable heat. The concept, formulated by George Gamow, led to considerable study of thermonuclear reactions and the search for background radiation.

Hubble expansion apparent recession of galaxies as seen from any point within the universe, the velocity of recession being proportional to the distance of the galaxy from the observer.

Hubble's law law that relates a galaxy's distance from us to its speed of recession as the universe expands, announced in 1929 by US astronomer Edwin Hubble. He found that galaxies are moving apart at speeds that increase in direct proportion to their distance apart. The rate of expansion is known as Hubble's constant.

hydration special type of solvation in which water molecules are attached – either by electrostatic forces or by coordinate (covalent) bonds – to ions or molecules of a solute. Some salts, called hydrates, retain associated water molecules in the solid state (this water is called the water of crystallization); in solid copper (II) sulphate, for example, the hydrated ion is

$$[Cu(H_2O)_4]^{2+} \text{ or } [Cu(H_2O)_6]^{2+}$$

hydraulic press device that uses an incompressible fluid, such as water or oil, to transmit a small downward force applied to a piston of small area to a larger area piston, which then produces a proportionately large upward force. Such a press is a demonstration of Pascal's law of pressures.

hydraulics field of study concerned with utilizing the properties of water and other liquids, in particular the way they flow and transmit pressure, and with the application of these properties in engineering. It applies the principles of hydrostatics and hydrodynamics. The oldest type of hydraulic machine is the **hydraulic press**, invented by Joseph Bramah in England in 1795. The hydraulic principle of pressurized liquid increasing a force is commonly used on vehicle braking systems, the forging press, and the hydraulic systems of aircraft and excavators.

hydrocarbon any of a class of chemical compounds containing only hydrogen and carbon (for example, the alkanes and alkenes). Hydrocarbons are obtained industrially principally from petroleum and coal tar.

hydrodynamics branch of physics dealing with fluids (liquids and gases) in motion.

hydroelectric power electricity generated by moving water. In a typical scheme, water stored in a reservoir, often created by damming a river, is piped into water turbines, coupled to electricity generators. In pumped storage plants, water flowing through the turbines is recycled. A tidal power station exploits the rise and fall of the tides. About one-fifth of the world's electricity comes from hydroelectric power.

hydrogen (Greek *hydro* + *gen* 'water generator') colourless, odourless, gaseous, nonmetallic element, symbol H, atomic number 1, relative atomic mass 1.00797. It is the lightest of all the elements and occurs on Earth, chiefly in combination with oxygen, as water.

Hydrogen is the most abundant element in the universe, where it accounts for 93% of the total number of atoms and 76% of the total mass. It is a component of most stars, including the Sun, whose heat and light are produced through the nuclear fusion process that converts hydrogen into helium. When subjected to a pressure 500,000 times greater than that of the Earth's atmosphere, hydrogen becomes a solid with metallic properties, as in one of the inner zones of Jupiter.

hydrogen bomb bomb that works on the principle of nuclear fusion. Large-scale explosion results from the thermonuclear release of energy when hydrogen nuclei are fused to form helium nuclei. The first hydrogen bomb was exploded at Enewetak Atoll in the Pacific Ocean by the USA in 1952.

hydrogen bond weak electrostatic bond that forms between covalently bonded hydrogen atoms and a strongly electronegative atom with a lone electron pair (for example, oxygen, nitrogen, and fluorine). Hydrogen bonds (denoted by a dashed line) are of great importance in biochemical processes, particularly the N–H······H bond, which enables proteins and nucleic acids to form the three-dimensional structures necessary for their biological activity.

hydrogen carrier compound that accepts hydrogen ions in biochemical reactions and is therefore important in oxidation-reduction reactions such as the intracellular use of oxygen.

hydrogen ion concentration number of grams of hydrogen ions per litre of solution; denoted by $[H^+]$. It is a measure of the acidity of a solution, in which context it is normally expressed in terms of pH values, given by $pH = \log_{10}(1/[H^+])$.

hydrography study and charting of Earth's surface waters in seas, lakes, and rivers.

hydrology study of the location and movement of inland water, both frozen and liquid, above and below ground. It is applied to major civil engineering projects such as irrigation schemes, dams, and hydroelectric power, and in planning water supply.

hydrolysis chemical reaction in which the action of water or its ions breaks down a substance into smaller molecules. Hydrolysis occurs in certain inorganic salts in solution, in nearly all nonmetallic chlorides, in esters, and in other organic substances. It is one of the mechanisms for the breakdown of food by the body, as in the conversion of starch to glucose.

hydrometer instrument used to measure the relative density of liquids (the density compared with that of water). A hydrometer consists of a thin glass tube ending in a sphere that leads into a smaller sphere, the latter being weighted so that the hydrometer floats upright, sinking deeper into less dense liquids than into denser liquids. Hydrometers are used in brewing and to test the strength of acid in car batteries.

hydroxyl group an atom of hydrogen and an atom of oxygen bonded together and covalently bonded to an organic molecule. Common compounds containing hydroxyl groups are alcohols and phenols.

hygrometer instrument for measuring the humidity, or water vapour content, of a gas (usually air). A wet and dry bulb hygrometer consists of two vertical thermometers, with one of the bulbs covered in absorbent cloth dipped into water. As the water evaporates, the bulb cools, producing a temperature difference between the two thermometers. The amount of evaporation, and hence cooling of the wet bulb, depends on the relative humidity of the air.

hyperbola curve formed by cutting a right circular cone with a plane so that the angle between the plane and the base is greater than the angle between the base and the side of the cone. All hyperbolae are bounded by two asymptotes (straight lines which the hyperbola moves closer and closer to but never reaches).

A hyperbola is a member of the family of curves known as conic sections.

hyperbolic geometry system of non-Euclidean geometry developed by German mathematician Bernhard Riemann, complementary to elliptic geometry and comprising the geometry of geodesics in the neighbourhood of a point on the (curved) surface at which a tangential plane intersects the surface in a hyperbolic curve.

hypercomplex numbers numbers that expand on complex numbers, for example, quaternions.

hyperfine structure very fine splitting of individual lines in a spectrum, which can be the result of the presence of different isotopes of an element in the source.

hyperon any of a group of highly unstable elementary particles that includes all the baryons with the exception of protons and neutrons. They are all composed of three quarks. The lambda, xi, sigma, and omega particles are hyperons.

ice age any period of extensive glaciation occurring in the Earth's history, but

481

particularly that in the Pleistocene epoch, immediately preceding historic times. On the North American continent, glaciers reached as far south as the Great Lakes, and an ice sheet spread over northern Europe, leaving its remains as far south as Switzerland. There were several glacial advances separated by interglacial stages, during which the ice melted and temperatures were higher than today.

ideal describes a point (one on every line) at infinity in such a way that the point has a coordinate position. In number theory, the term describes a collection of elements in a ring that has specific properties within a universal set, that is, that form a closed system under addition (among themselves) and under scaling by any element from the universal set. For example, the even numbers form an ideal within the universal set of integers in that when added (or multiplied) together, even numbers result.

ideal gas or ***perfect gas***, gas obeying the gas laws of Boyle, Charles, and Joule exactly. This would imply that the gas consists of perfectly elastic molecules, each of which has zero volume and no attractive or repulsive interaction with its neighbours.

igneous rock rock formed from the solidification of molten rock called magma. The acidic nature of this rock type means that areas with underlying igneous rock are particularly susceptible to the effects of acid rain. Igneous rocks that crystallize slowly from magma below the Earth's surface have large crystals. Examples include dolerite and granite.

imaginary number term often used to describe the non-real element of a complex number. For the complex number $(a + ib)$, ib is the imaginary number, where $i = \sqrt{(-1)}$, and b any real number.

immunity the protection that organisms have against foreign micro-organisms, such as bacteria and viruses, and against cancerous cells. The cells that provide this protection are called white blood cells, or leucocytes, and make up the immune system. They include neutrophils and macrophages, which can engulf invading organisms and other unwanted material, and natural killer cells that destroy cells infected by viruses and cancerous cells. Some of the most important immune cells are the B cells and T cells. Immune cells coordinate their activities by means of chemical messengers or lymphokines, including the antiviral messenger interferon. The lymph nodes play a major role in organizing the immune response.

immunization conferring immunity to infectious disease by artificial methods. The most widely used technique is vaccination. Immunization is an important public health measure. If most of the population has been immunized against a particular disease, it is impossible for an epidemic to take hold.

immunoglobulin human globulin protein that can be separated from blood and administered to confer immediate immunity on the recipient. It participates in the immune reaction as the antibody for a specific antigen (disease-causing agent).

imprinting process whereby a young animal learns to recognize both specific individuals (for example, its mother) and its own species.

impulse in mechanics, the product of a force and the time over which it acts. An impulse applied to a body causes its momentum to change and is equal to that change in momentum. It is measured in newton seconds (N s).

impulse in nerves, the electrical signal that is transmitted along a nerve fibre which has been sufficiently stimulated.

incandescence emission of light from a substance in consequence of its high temperature. The colour of the emitted light from liquids or solids depends on their temperature, and for solids generally the higher the temperature the whiter the light. Gases may become incandescent through ionizing radiation, as in the glowing vacuum discharge tube.

inclination angle between the ecliptic and the plane of the orbit of a planet, asteroid, or comet. In the case of satellites orbiting a planet, it is the angle between the plane of orbit of the satellite and the equator of the planet.

incoherent light light that is not of a single phase. Daylight is an example.

incompressible fluid fluid that resists changes in density – for example, oil.

indeterminate problems problems involving one or more unknown or variable quantities. See also Diophantine equations.

indicator compound that changes its structure and colour in response to its environment. The commonest chemical indicators detect changes in pH (for example, litmus and universal indicator), or in the oxidation state of a system (redox indicators).

indicator species plant or animal whose presence or absence in an area indicates certain environmental conditions, such as soil type, high levels of pollution, or, in rivers, low levels of dissolved oxygen. Many plants show a preference for either alkaline or acid soil conditions, while certain trees require aluminium, and are found only in soils where it is present.

induction alteration in the physical properties of a body that is brought about by the influence of a field. See electromagnetic induction and magnetic induction.

induction motor device that produces rotation by induction. An alternating current is fed to a winding of wires which thus induces electrical currents to flow in a second set of windings in a central rotor. Interaction between the two currents and the magnetic flux involved causes rotation.

inertia tendency of an object to remain in a state of rest or uniform motion until an external force is applied, as described by Isaac Newton's first law of motion (see Newton's laws of motion).

infinitesimal number that is not zero but is less than any finite number. Infinitesimal numbers clearly do not exist in the conventional system of real numbers, but modern developments in logic allow the use of an extended system of numbers that includes infinitesimal numbers. Calculations with these adhere to certain restrictions and require an understanding of limits.

infinitesimal calculus original name for 'calculus'; that is, differential and integral calculus, and so called because it was thought to rely on 'infinitely small' quantities. (It is now seen to be based upon a precise theory of limits.)

infinity mathematical entity that is larger than any fixed assignable quantity, symbol ∞. By convention, the result of dividing any number by zero is regarded as infinity.

infrared radiation invisible electromagnetic radiation of wavelength between about 0.75 micrometres and 1 millimetre – that is, between the limit of the red end of the visible spectrum and the shortest microwaves. All bodies above the absolute zero of temperature absorb and radiate infrared radiation. Infrared radiation is used in medical photography and treatment, and in industry, astronomy, and criminology.

insulator any poor conductor of heat, sound, or electricity. Most substances lacking free (mobile) electrons, such as non-metals, are electrical or thermal insulators. Usually, devices of glass or porcelain, called insulators, are used for insulating and supporting overhead wires.

integer any whole number. Integers may be positive or negative; 0 is an integer, and is often considered positive. Formally, integers are members of the set

$$Z = \{\ldots -3, -2, -1, 0, 1, 2, 3, \ldots\}$$

This is the integer set of the number line.

integral calculus branch of mathematics using the process of integration. It is concerned with finding volumes and areas and summing infinitesimally small quantities.

integral equations equations involving integrals of the unknown function.

integration method in calculus of determining the solutions of definite or indefinite integrals. An example of a definite integral can be thought of as finding the area under a curve (as represented by an algebraic expression or function) between particular values of the function's variable. In practice, integral calculus provides scientists with a powerful tool for doing calculations that involve a continually varying quantity (such as determining the position at any given instant of a space rocket that is accelerating away from Earth). Its basic principles were discovered in the late 1660s independently by the German philosopher Gottfried Leibniz and the English scientist Isaac Newton.

intensity interferometry use of two telescopes linked by computer to study the intensity of light received from a star. Analysis of the combined results has enabled measurement of the diameters of stars as apparently small as 2×10^{-4} seconds of arc.

interference phenomenon of two or more wave motions interacting and combining to produce a resultant wave of larger or smaller amplitude (depending on whether the combining waves are in or out of phase with each other).

intergalactic matter hypothetical material within a cluster of galaxies, whose gravitational effect is to maintain the equilibrium of the cluster. Theoretically comprising 10–30 times the mass of the galaxies themselves (in order to have the observed effect), it has yet to be detected in any form – although the most likely form is as hydrogen.

internal combustion transformation of the chemical energy of a fuel into mechanical energy in controlled combustion in an enclosed cylinder sealed at one end by a piston.

interstellar space space between the stars of a galaxy. It is generally not, however, a void vacuum, and is the subject of considerable spectral research.

intrinsic geometry study of a surface without reference to any point, condition, or space outside it. All measurements and operations carried out on the surface are therefore in terms of its own (intrinsic) form.

intuitional mathematics or *intuitionism*, alternative foundational basis for mathematics that adopts a stricter logic in its approach to proofs concerning the infinite. For example, it dismisses the law of the excluded middle, and so disregards 'proofs' derived by double negatives if the relevant positive statement has not actually been demonstrated to be true. Although this may be considered a sort of philosophical puritanism, the outlook leads to a more refined classification of proof material than merely 'true' or 'false'. The fact that such a critical attitude can itself be formalized in a mathematically sound system is an important achievement. Nevertheless, most practising mathematicians remain unconcerned by this logical analysis.

invagination formation of an inner pocket within a layer of cells by part of the layer pushing inwards to form a cavity that remains open to the original surface. See also endocytosis.

invariant as a general term, describes a property that is preserved through specified mathematical operations.

inverse function function that exactly reverses the transformation produced by a function f; it is usually written as f^{-1}. For example, $3x + 2$ and $(x-2)/3$ are mutually inverse functions. Multiplication and division are inverse operations (see reciprocals).

invertebrate animal without a backbone. The invertebrates form all of the major divisions of the animal kingdom called phyla, with the exception of vertebrates. Invertebrates include the sponges, coelenterates, flatworms, nematodes, annelids, arthropods, molluscs, and echinoderms. Primitive aquatic chordates such as sea squirts and lancelets, which only have notochords and do not possess a vertebral column of cartilage or bone, are sometimes called invertebrate chordates, but this is misleading, since the notochord is the precursor of the backbone in advanced chordates.

***in vitro* process** biological experiment or technique carried out in a laboratory, outside the body of a living organism (literally 'in glass', for example in a test tube). By contrast, an *in vivo* process takes place within the body of an organism.

***in vivo* process** biological experiment or technique carried out within a living organism; by contrast, an *in vitro* process takes place outside the organism, in an artificial environment such as a laboratory.

ion atom, or group of atoms, that is either positively charged (cation) or negatively charged (anion), as a result of the loss or gain of electrons during chemical reactions or exposure to certain forms of radiation. In solution or in the molten state, ionic compounds such as salts, acids, alkalis, and metal oxides conduct electricity. These compounds are known as electrolytes.

ionic bond or *electrovalent bond*, bond produced when atoms of one element donate electrons to atoms of another element, forming positively and negatively charged ions respectively. The attraction between the oppositely charged ions constitutes the bond. Sodium chloride (Na^+Cl^-) is a typical ionic compound.

ionic radius effective radius of an ion. In positively charged cations, the ionic radius is less than the atomic radius (because the electrons are more tightly bound); in anions the ionic radius is more than the atomic radius. Some elements, such as the transition metals, can have several different ionization states and their ionic radii vary according to the state involved.

ionization process of ion formation. It can be achieved in two ways. The first way is by the loss or gain of electrons by atoms to form positive or negative ions:

$$Na - e^- \rightarrow Na^+$$
$$1/2Cl_2 + e^- \rightarrow Cl^-$$

In the second mechanism, ions are formed when a covalent bond breaks, as when hydrogen chloride gas is dissolved in water. One portion of the molecule retains both electrons, forming a negative ion, and the other portion becomes positively charged. This bond-fission process is sometimes called dissociation:

$$HCl_{(g)} + aq \rightleftharpoons H^+_{(aq)} + Cl^-_{(aq)}$$

ionosphere ionized layer of Earth's outer atmosphere (60–1,000 km/38–620 mi) that contains sufficient free electrons to modify the way in which radio waves are propagated, for instance by reflecting them back to Earth. The ionosphere is thought to be produced by absorption of the Sun's ultraviolet radiation. The British Antarctic Survey estimates that the ionosphere is decreasing at a rate of 1 km/0.6 mi every five years, based on an analysis of data from 1960 to 1998. Global warming is the probable cause.

iron (Germanic *eis* 'strong') hard, malleable and ductile, silver-grey, metallic element, symbol Fe (from Latin *ferrum*), atomic number 26, relative atomic mass 55.847. It is the fourth most abundant element (the second most abundant metal, after aluminium) in the

Earth's crust. Iron occurs in concentrated deposits as the ores hematite (Fe_2O_3), spathic ore ($FeCO_3$), and magnetite (Fe_3O_4). It sometimes occurs as a free metal, occasionally as fragments of iron or iron–nickel meteorites.

irrational number number that cannot be expressed as an exact fraction. Irrational numbers include some square roots (for example, $\sqrt{2}$, $\sqrt{3}$, and $\sqrt{5}$ are irrational) and numbers such as π (for circles, which is approximately equal to the decimal 3.14159) and e (the base of natural logarithms, approximately 2.71828). If an irrational number is expressed as a decimal it would go on for ever without repeating. If an irrational number is divided or multiplied by another irrational number, it becomes a rational number.

irrigation artificial water supply for dry agricultural areas by means of dams and channels. Drawbacks are that it tends to concentrate salts at the surface, ultimately causing soil infertility, and that rich river silt is retained at dams, to the impoverishment of the land and fisheries below them.

isomer chemical compound having the same molecular composition and mass as another, but with different physical or chemical properties owing to the different structural arrangement of its constituent atoms. For example, the organic compounds butane ($CH_3(CH_2)_2CH_3$) and methyl propane ($CH_3CH(CH_3)CH_3$) are isomers, each possessing four carbon atoms and ten hydrogen atoms but differing in the way that these are arranged with respect to each other.

isoperimetry branch of geometry involving the study and measurement of figures with equal perimeters.

isostasy the condition of gravitational equilibrium of all parts of the Earth's crust. The crust is in isostatic equilibrium if, below a certain depth, the weight and thus pressure of rocks above is the same everywhere. The idea is that the lithosphere floats on the asthenosphere as a piece of wood floats on water. A thick piece of wood floats lower than a thin piece, and a denser piece of wood floats lower than a less dense piece. There are two theories of the mechanism of isostasy, the Airy hypothesis, and the Pratt hypothesis, both of which have validity. In the **Airy hypothesis** crustal blocks have the same density but different thicknesses: like ice cubes floating in water, higher mountains have deeper roots. In the **Pratt hypothesis**, crustal blocks have different densities allowing the depth of crustal material to be

the same. In practice, both mechanisms are at work.

isotherm line on a map, linking all places having the same temperature at a given time.

isotope one of two or more atoms that have the same atomic number (same number of protons), but which contain a different number of neutrons, thus differing in their atomic mass (see relative atomic mass). They may be stable or radioactive (see radio-isotope), naturally occurring, or synthesized. For example, hydrogen has the isotopes ^2H (deuterium) and ^3H (tritium). The term was coined by English chemist Frederick Soddy, a pioneer researcher in atomic disintegration.

isotropic having equal and uniform properties at all points and in all directions. In astronomy the term describes microwave background radiation.

Josephson effect a superconducting (see superconductivity) ring interrupted by a thin layer of insulating material gives rise to an alternating current (AC) in the barrier when a steady external voltage is applied to it. This AC effect has a direct-current analogue, occurring when a steady magnetic field is applied to the insulating material. The effect is named after Welsh physicist Brian Josephson.

joule SI unit (symbol J) of work and energy, replacing the calorie (one calorie equals 4.2 joules).

Joule–Kelvin effect or *Joule–Thomson effect*, fall in temperature of a gas as it expands adiabatically (without loss or gain of heat to the system) through a narrow jet. It can be felt when, for example, compressed air escapes through the valve of an inflated bicycle tyre. It is the basic principle of some refrigerators.

Joule's electrical law heat H in joules produced by the passing of a current of I amperes through a resistance of R ohms for a time t seconds is given by:

$$H = I^2Rt$$

Joule's thermal law internal energy of a gas at constant temperature is independent of its volume, provided the gas is ideal.

Jurassic period period of geological time 208–146 million years ago, the middle period of the Mesozoic era. Climates worldwide were equable, creating forests of conifers and ferns, dinosaurs were abundant and birds evolved.

Kelvin scale temperature scale used by scientists. It begins at absolute zero ($-273.15°C$) and increases in kelvins, the same degree intervals as the Celsius scale; that is, $0°C$ is the same as 273.15 K and $100°C$

485

is 373.15 K. It is named after the Irish physicist William Thomson, 1st Baron Kelvin.

Kennelly–Heaviside layer former term for the E layer of the ionosphere.

Kepler's laws three laws of planetary motion formulated in 1609 and 1619 by the German mathematician and astronomer Johannes Kepler: (1) the orbit of each planet is an ellipse with the Sun at one of the foci; (2) the radius vector of each planet sweeps out equal areas in equal times; (3) the squares of the periods of the planets are proportional to the cubes of their mean distances from the Sun.

kerosene thin oil obtained from the distillation of petroleum; a highly refined form is used in jet aircraft fuel. Kerosene is a mixture of hydrocarbons of the paraffin series.

Kerr cell device making use of the Kerr effect. The cell consists of a transparent container of a special liquid in which there are two electrodes, placed between two polarizing materials in the container. Only if the planes of polarization of all the various layers in the cell are aligned will light pass, and so the Kerr cell can be used as a shutter device.

Kerr effect elliptical polarization of light as a result of the beam of light being reflected from a pole of an electromagnet. A similar effect also exists for liquids, if a potential difference is applied to the liquid itself, the angle of polarization depending on the size of the potential difference.

kinematic relativity theory proposed by English astrophysicist Edward Milne as an alternative to Einstein's general theory of relativity, and based generally on kinematics (the science of pure motion, without reference to matter or force), from which Milne successfully derived new systems of dynamics and electrodynamics.

kinematics branch of mechanics that relates accelerations to the velocities and changes in distance they produce, without considering the forces that generate the accelerations involved. See also dynamics.

kinetic energy the energy of a body resulting from motion. It is contrasted with potential energy.

kinetic theory theory describing the physical properties of matter in terms of the behaviour – principally movement – of its component atoms or molecules. The temperature of a substance is dependent on the velocity of movement of its constituent particles, increased temperature being accompanied by increased movement. A gas consists of rapidly moving atoms or molecules and, according to kinetic theory,

it is their continual impact on the walls of the containing vessel that accounts for the pressure of the gas. The slowing of molecular motion as temperature falls, according to kinetic theory, accounts for the physical properties of liquids and solids, culminating in the concept of no molecular motion at absolute zero (0 K/$-273.15°$C).

knot closed curve that loops over or through itself; representations of such structures are commonly presented as congruency problems.

Kuiper band one of a number of bands in the spectra of Uranus and Neptune at wavelengths of 7,500 Å (7.5×10^{-7}m), indicating the presence of methane. They are named after Dutch-born US astronomer Gerard Kuiper.

Lamarckism theory of evolution, now discredited, advocated during the early 19th century by French naturalist Jean Baptiste Lamarck. Lamarckism is the theory that acquired characteristics, such as the increased body mass of an athlete, were inherited. It differs from the Darwinian theory of evolution.

lanthanide any of a series of 15 metallic elements (also known as rare earths) with atomic numbers 57 (lanthanum) to 71 (lutetium). One of its members, promethium, is radioactive. All occur in nature. Lanthanides are grouped because of their chemical similarities (most are trivalent, but some can be divalent or tetravalent), their properties differing only slightly with atomic number.

large-number hypothesis theory in cosmology that tries to understand the basic reason for an apparent coincidence of size in ratios of fundamental quantities in atomic and cosmological theory. If the apparent radius of the universe is divided by the radius of the electron, the resulting large number (about 10^{43}) is remarkably similar to the ratio of the strengths of the electrostatic and gravitational force between the electron and the proton. The reason for this is not clear, but if the relationship is to hold true for all time, it can be shown that on a Big Bang model of the universe the gravitational constant g must decrease with time; that is, the strength of gravity must decrease.

large numbers, laws of theorems in probability theory that predict that the observed frequencies of events for a large number of repeated trials are more and more likely to approach their theoretical probability as the number of repetitions increases.

laser acronym for *light amplification by stimulated emission of radiation*, device for producing a narrow beam of light, capable of travelling over vast distances without dispersion, and of being focused to give enormous power densities (10^8 watts per cm^2 for high-energy lasers). The laser operates on a principle similar to that of the maser (a high-frequency microwave amplifier or oscillator). The uses of lasers include communications (a laser beam can carry much more information than can radio waves), cutting, drilling, welding, satellite tracking, medical and biological research, and surgery.

latent heat heat absorbed or released by a substance as it changes state (for example, from solid to liquid) at constant temperature and pressure.

lathe tool used to produce objects with cylindrical symmetry, such as bars, screws, and barrels.

lattice network of straight lines.

lattice in chemistry, the arrangement of positions in, for example, a crystal in which the atoms, ions, or molecules remain virtually stationary.

law of universal attraction Isaac Newton's formulation of the law of gravity.

leaching process by which substances are washed through or out of the soil. Fertilizers leached out of the soil drain into rivers, lakes, and ponds and cause water pollution. In tropical areas, leaching of the soil after the destruction of forests removes scarce nutrients and can lead to a dramatic loss of soil fertility. The leaching of soluble minerals in soils can lead to the formation of distinct soil horizons as different minerals are deposited at successively lower levels.

least squares, method of method of deriving as exact an average value as possible from a set of approximate or inaccurate values by introducing the errors as unknown variables and requiring the sum of their squares to be minimized. The method was devised by German mathematician Karl Gauss as a precise way of best fitting a straight line through a set of plotted data points that are not collinear.

Legendre functions functions that satisfy the second-order differential equation:

$$(1 - x^2)\, d^2y/dx^2 - 2x\, dy/dx + n(n + 1)y = 0$$

lemniscate curve represented by the equation:

$$(x^2 + y^2)^2 = a^2(x^2 - y^2)$$

where a is constant and x and y are variables.

Lenz's law law stating that the direction of an electromagnetically induced current (generated by moving a magnet near a wire or a wire in a magnetic field) will be such as to oppose the motion producing it. It is named after the German physicist Heinrich Friedrich Lenz, who announced it in 1833.

leucocyte another name for a white blood cell.

lever simple machine consisting of a rigid rod pivoted at a fixed point called the fulcrum, used for shifting or raising a heavy load or applying force. Levers are classified into orders according to where the effort is applied, and the load-moving force developed, in relation to the position of the fulcrum.

Leyden jar early form of condenser consisting of a glass jar with an interior and exterior coating of metal foil, used to store static electricity.

libration slight, apparent wobble in the rotation of the Moon due to its variable speed of rotation and the tilt of its axis.

Lie groups or *Lie rings*, collections of mathematical objects in groups or rings that have further (topological) structure under which the collective operations are continuous, for example, vectors in the plane (where open discs define a topological structure).

ligand group that bonds symmetrically to a central atom or ion of a metal; the result is called a **coordination complex**. An example of a neutral ligand is ammonia; the nitrosyl ion NO^+ is a charged ligand. An example of a coordination complex is hexaminocobalt chloride, $[Co(NH_3)_6]Cl_3$, in which the central cobalt ion (Co^{3+}) is surrounded by covalent bonds with six ammonia molecules and ionic bonds with three chloride ions.

light electromagnetic waves in the visible range, having a wavelength from about 400 nanometres in the extreme violet to about 770 nanometres in the extreme red. Light is considered to exhibit particle and wave properties, and the fundamental particle, or quantum, of light is called the photon. The speed of light (and of all electromagnetic radiation) in a vacuum is approximately 300,000 km/186,000 mi per second, and is a universal constant denoted by c.

light reaction part of the photosynthesis process in green plants that requires sunlight (as opposed to the dark reaction, which does not). During the light reaction light energy is used to generate ATP (by the phosphorylation of ADP), which is necessary for the dark reaction.

light, speed of (symbol *c*) fundamental constant of nature, the speed of light is the limiting velocity that any body can travel at. It is equal to 2.997925×10^8 m/ 186,180 mi per second, and is the same for all observers, no matter how fast they move themselves.

light, theories of the nature of before the advent of modern physics, theories of the basic nature of light fell into two camps: those who viewed light as made up of a stream of particles (corpuscular theory), and those who viewed it as a wave motion (wave theory). Each attempted to explain all phenomena in optics, such as reflection, refraction, and so on, on the basis of these two viewpoints. It is now known that light exists in quanta known as photons that exhibit both corpuscular and wavelike behaviour in certain circumstances.

light year distance travelled by a beam of light in a vacuum in one year, approximately 9.4605×10^{12} km/5.9128×10^{12} mi.

limit in an infinite sequence, the final value towards which the sequence is tending. For example, the limit of the sequence 1/2, 3/4, 7/8, 15/16 . . . is 1, although no member of the sequence will ever exactly equal 1 no matter how many terms are added together. The limit of the ratios of a Fibonacci sequence is $(\sqrt{5} + 1)/2$. This number is also the golden section.

linear function or ***linear transformation***, in its simplest context, a transformation such as $y = mx$ that may be depicted as a line through the origin. More generally, a transformation T defined on a vector space with the property that:

$$T(\alpha x + \beta y) = \alpha T(x) + \beta T(y)$$

for x, y vectors and α, β scalars.

linear motor device that uses induction to produce forward motion along a track.

line spectrum spectrum made up of discrete lines of intensity at certain wavelengths, characterizing an atom in a particular state.

linkage association between two or more genes that tend to be inherited together because they are on the same chromosome. The closer together they are on the chromosome, the less likely they are to be separated by crossing over (one of the processes of recombination) and they are then described as being 'tightly linked'.

lipid any of a large number of esters of fatty acids, commonly formed by the reaction of a fatty acid with glycerol. They are soluble in alcohol but not in water. Lipids are the chief constituents of plant and animal waxes, fats, and oils.

liquid state of matter between a solid and a gas. A liquid forms a level surface and assumes the shape of its container. Its atoms do not occupy fixed positions as in a crystalline solid, nor do they have freedom of movement as in a gas. Unlike a gas, a liquid is difficult to compress since pressure applied at one point is equally transmitted throughout (Pascal's principle). Hydraulics makes use of this property.

Lissajous figures path followed by a point subjected to two or more simultaneous simple wave motions; for example, at right angles to one another. They were developed by French physicist Jules Lissajous.

lithium (Greek *lithos* 'stone') soft, ductile, silver-white, metallic element, symbol Li, atomic number 3, relative atomic mass 6.941. It is one of the alkali metals, has a very low density (far less than most woods), and floats on water (specific gravity 0.57); it is the lightest of all metals. Lithium is used to harden alloys, and in batteries; its compounds are used in medicine to treat manic depression.

ln symbol for natural logarithm.

Local Group cluster of about 30 galaxies that includes our own, the Milky Way. Like other groups of galaxies, the Local Group is held together by the gravitational attraction among its members, and does not expand with the expanding universe. Its two largest galaxies are the Milky Way and the Andromeda galaxy; most of the others are small and faint.

locus (Latin 'place') traditionally the path traced out by a moving point, but now defined as the set of all points on a curve satisfying given conditions. The locus of points a fixed distance from a fixed point is a circle. The locus of a point equidistant from two fixed points is a straight line that perpendicularly bisects the line joining them. The locus of points a fixed distance from a line is two parallel lines running either side.

log abbreviation for logarithm.

logarithm or ***log***, exponent or index of a number to a specified base – usually 10. For example, the logarithm to the base 10 of 1,000 is 3 because $10^3 = 1,000$; the logarithm of 2 is 0.3010 because $2 = 10^{0.3010}$. The whole-number part of a logarithm is called the **characteristic**; the fractional part is called the **mantissa**.

Before the advent of cheap electronic calculators, multiplication and division could be simplified by being replaced with the addition and subtraction of logarithms.

logic branch of philosophy that studies valid reasoning and argument. It is also the way in which one thing may be said to follow from, or be a consequence of, another (deductive logic). Logic is generally divided into the traditional formal logic of Aristotle and the symbolic logic derived from Friedrich Frege and Bertrand Russell.

logistic curve curve that represents logarithmic functions, from which logarithms of ordinary numbers can be read off.

longitudinal wave wave in which the displacement of the medium's particles is in line with or parallel to the direction of travel of the wave motion.

Lorentz transformation means of relating measurements of times and lengths made in one frame of reference to those made in another frame of reference moving at some velocity relative to the first. These formulae, which can be derived directly from Einstein's special theory of relativity, predict that at velocities approaching that of light, lengths appear to contract (the Lorentz contraction) and time intervals appear to increase (time dilation), as measured in a stationary frame of reference.

luminescence emission of light from a body when its atoms are excited by means other than raising its temperature. Short-lived luminescence is called fluorescence; longer-lived luminescence is called phosphorescence.

luminosity or *brightness*, amount of light emitted by a star, measured in magnitudes. The apparent brightness of an object decreases in proportion to the square of its distance from the observer. The luminosity of a star or other body can be expressed in relation to that of the Sun.

luteinization development of an ovum in a ruptured Graafian follicle within the ovary, initiated by oestrogen which in turn is activated by luteinizing hormone.

Lyman series series of lines in the spectrum of hydrogen that lie in the ultraviolet region of the spectrum. It is named after US physicist Theodore Lyman.

lysogeny presence of nonvirulent or temperate bacteriophages in a bacterium, that do not lyse it (damage the outer cell membrane). The phage does not replicate after entering the bacterial cell, although its DNA combines with that of the bacterium and is reproduced with it every time the bacterium multiplies. The basic characteristics of the host bacterium remain unchanged.

lysosome membrane-enclosed structure, or organelle, inside a cell, principally found in animal cells. Lysosomes contain enzymes that can break down proteins and other biological substances. They play a part in digestion, and in the white blood cells known as phagocytes the lysosome enzymes attack ingested bacteria.

machine device that allows a small force (the effort) to overcome a larger one (the load). There are three basic machines: the inclined plane (ramp), the lever, and the wheel and axle. All other machines are combinations of these three basic types. Simple machines derived from the inclined plane include the wedge, the gear, and the screw; the spanner is derived from the lever, the pulley from the wheel.

macromolecule very large molecule, generally a polymer.

Magellanic Clouds the two galaxies nearest to our own galaxy. They are irregularly shaped, and appear as detached parts of the Milky Way, in the southern constellations Dorado, Tucana, and Mensa.

magic numbers certain numbers of neutrons or protons (2, 8, 20, 28, 50, 82, 126) in the nuclei of elements of outstanding stability, such as lead and helium. Such stability is the result of neutrons and protons being arranged in completed 'layers' or 'shells'.

magma molten rock (either beneath or on a planetary surface) from which igneous rocks are formed. Lava is magma that has extruded on to the surface.

magnetic detector device used in early radio systems, in which high-frequency currents are detected through their demagnetizing effect on a magnetized iron core surrounded by the wire-carrying currents.

magnetic energy property of a magnet described by the multiplication of the flux density by the field strength on the demagnetization curve of a permanent magnet.

magnetic field region around a permanent magnet, or around a conductor carrying an electric current, in which a force acts on a moving charge or on a magnet placed in the field. The field can be represented by lines of force parallel at each point to the direction of a small compass needle placed on them at that point. A magnetic field's magnitude is given by the magnetic flux density, expressed in teslas.

magnetic field strength property measured to define the strength of a magnetic field. The SI unit is the ampere per metre.

magnetic flux measurement of the strength of the magnetic field around electric currents and magnets. Its SI unit is the weber; one weber per square metre is equal to one tesla.

magnetic flux density magnetic flux passing through one square metre of area of a magnetic field in a direction at right angles to the magnetic force. The SI unit is the tesla.

magnetic moment product of the strength and length of a magnet.

magnetic monopole prediction of one quantum theory (proposed by Dirac) that involves the existence of individual magnetic poles, analogous to the individual charges found in electrostatics (that is, electrons). Such an entity has not yet been definitely observed to exist, but is expected to be rare in any case.

magnetic permeability magnetic flux density in a body, divided by the external magnetic field strength producing it. It can be used as a way of classifying materials into different types of magnetism, such as ferromagnetic.

magnetism phenomena associated with magnetic fields. Magnetic fields are produced by moving charged particles: in electromagnets, electrons flow through a coil of wire connected to a battery; in permanent magnets, spinning electrons within the atoms generate the field.

magnetohydrodynamics study of the behaviour of plasmas under the influence of magnetic fields.

magnetosphere volume of space, surrounding a planet, in which the planet's magnetic field has a significant influence. The Earth's magnetosphere extends 64,000 km/40,000 mi towards the Sun, but many times this distance on the side away from the Sun. That of Jupiter is much larger, and, if it were visible, from the Earth it would appear to have roughly the same extent as the full Moon.

magnitude measure of the brightness of a star or other celestial object. The larger the number denoting the magnitude, the fainter the object. Zero or first magnitude indicates some of the brightest stars. Still brighter are those of negative magnitude, such as Sirius, whose magnitude is −1.46. **Apparent magnitude** is the brightness of an object as seen from Earth; **absolute magnitude** is the brightness at a standard distance of 10 parsecs (32.616 light years).

main sequence part of the Hertzsprung–Russell diagram that contains most of the stars, including the Sun. It runs diagonally from the top left of the diagram to the lower right. The most massive (and hence brightest) stars are at the top left, with the least massive (coolest) stars at the bottom right.

Malus's law law giving the intensity of light after having been polarized through a certain angle.

manifold in two-dimensional space, a regular surface that locally looks like a flat plane slightly distorted. It can be represented by differentiable functions. There are analogues in spaces of more dimensions.

mapping another name for transformation.

Markov chain ordered sequence of discrete states (random variables) $x_1, x_2, \ldots, x_i, \ldots, x_n$ such that the probability of x_i depends only on n and/or the state x_{i-1} which has preceded it. If independent of n, the chain is said to be homogeneous.

mascon contraction of mass concentration, one of a number of apparent regions on the surface of the Moon where gravity is somehow stronger. The effect is presumed to be due to localized areas of denser rock strata.

maser acronym for ***microwave amplification by stimulated emission of radiation***, high-frequency microwave amplifier or oscillator in which the signal to be amplified is used to stimulate excited atoms into emitting energy at the same frequency. Atoms or molecules are raised to a higher energy level and then allowed to lose this energy by radiation emitted at a precise frequency. The principle has been extended to other parts of the electromagnetic spectrum as, for example, in the laser.

mass quantity of matter in a body as measured by its inertia. Mass determines the acceleration produced in a body by a given force acting on it, the acceleration being inversely proportional to the mass of the body. The mass also determines the force exerted on a body by gravity on Earth, although this attraction varies slightly from place to place. In the SI system, the base unit of mass is the kilogram.

mass number or ***nucleon number***, sum (symbol A) of the numbers of protons and neutrons in the nucleus of an atom. It is used along with the atomic number (the number of protons) in nuclear notation: in symbols that represent nuclear isotopes, such as $^{14}_{6}\text{C}$, the lower number is the atomic number, and the upper number is the mass number.

mass spectrograph instrument for determining the masses of individual atoms by means of positive-ray analysis, which involves deflecting streams of positive ions using electric and magnetic fields. Ions with different masses are deflected by different amounts and can be detected (for example, photographically) to produce a mass spectrum. Isotopes were first discovered in this way.

mass spectrometer apparatus for analysing chemical composition. Positive ions (charged particles) of a substance are separated by an electromagnetic system, designed to focus particles of equal mass to a point where they can be detected. This permits accurate measurement of the relative concentrations of the various ionic masses present, particularly isotopes.

mathematical logic view of mathematics and logic that relates the two disciplines, and to do so uses mathematical or similar notation in the expression of axiomatic statements. Boolean algebra was the original form, and led to German philosopher Gottlob Frege's symbolic logic.

mathematical structure collection of objects that display (a) one or more relationships, and (b) one or more operations of which the properties may be summarized as a list of axioms, for example, a group, vector, space or ring.

matrix square ($n \times n$) or rectangular ($m \times n$) array of elements (numbers or algebraic variables) used to facilitate the study of problems in which the relation between the elements is important. They are a means of condensing information about mathematical systems and can be used for, among other things, solving simultaneous linear equations (see transformation).

matrix mechanics mathematical description of subatomic phenomena that views certain characteristics of the particles involved as being matrices, and hence obeying the rules of matrix mathematics that differ in significant ways from the rules obeyed by ordinary arithmetic.

matter, continuous creation of phenomenon invoked in certain cosmological theories, especially the steady-state model of the universe. In that particular theory, matter is considered to be constantly created, either evenly throughout space, or in localized regions of creation, so as to make up for the diluting effect the Hubble expansion of the universe has on the average density of matter. By continually creating matter, it is then possible for the universe to maintain a steady-state appearance.

Maxwell–Boltzmann distribution statistical equation describing the distribution of velocities among the molecules of a gas. It is named after Scottish physicist James Maxwell and Austrian physicist Ludwig Boltzmann, who derived the equation, independently of each other, in the 1860s.

Maxwell's equations set of four vector equations showing the interdependence of electricity and magnetism. The concepts of charge conservation are built into them, as are the experimental results of Faraday, Gauss, and Ampère.

mean measure of the average of a number of terms or quantities. The simple **arithmetic mean** is the average value of the quantities, that is, the sum of the quantities divided by their number. The **weighted mean** takes into account the frequency of the terms that are summed; it is calculated by multiplying each term by the number of times it occurs, summing the results and dividing this total by the total number of occurrences. The **geometric mean** of n quantities is the nth root of their product. In statistics, it is a measure of central tendency of a set of data.

measure theory extension of the notion of length, area, or volume (as appropriate) to general sets of points on the line, plane, or in space. Used commonly in analysis (especially integration theory), functional analysis, probability theory, and game theory (in the assessment of the size of coalitions), its definitive form was derived by French mathematician Henri Lebesgue.

mechanical advantage *MA*, ratio by which the load moved by a machine is greater than the effort applied to that machine. In equation terms: MA = load/effort.

mechanical curve curve that cannot be precisely expressed as an equation (unlike a geometric curve).

mechanical equivalent of heat constant factor relating the calorie (the c.g.s. unit of heat) to the joule (the unit of mechanical energy), equal to 4.1868 joules per calorie. It is redundant in the SI system of units, which measures heat and all forms of energy in joules.

mechanics branch of physics dealing with the motions of bodies and the forces causing these motions, and also with the forces acting on bodies in equilibrium. It is usually divided into dynamics and statics.

medium environment in which micro-organisms can be cultured. Common mediums include agar, broth, and gelatine, often with added salts and trace elements.

meiosis process of cell division in which the number of chromosomes in the cell is halved. It only occurs in eukaryotic cells, and is part of a life cycle that involves sexual reproduction because it allows the genes of two parents to be combined without the total number of chromosomes increasing.

melanin brown pigment that gives colour to the eyes, skin, hair, feathers, and scales of many vertebrates. In humans, melanin helps protect the skin against ultraviolet radiation from sunlight.

melanocyte cell that produces melanin.

membrane potential potential difference that exists across a membrane or cell wall, such as that across the wall of a nerve cell (neuron).

Mendelism theory of inheritance originally outlined by Austrian biologist Gregor Mendel. He suggested that, in sexually reproducing species, all characteristics are inherited through indivisible 'factors' (now identified with genes) contributed by each parent to its offspring.

meridian half a great circle drawn on the Earth's surface passing through both poles and thus through all places with the same longitude. Terrestrial longitudes are usually measured from the Greenwich Meridian.

mesoderm central layer of embryonic cells between the ectoderm and the endoderm.

meson group of unstable subatomic particles made up of a quark and an antiquark. It is found in cosmic radiation, and is emitted by nuclei under bombardment by very high-energy particles.

Mesozoic era of geological time 245–65 million years ago, consisting of the Triassic, Jurassic, and Cretaceous periods. At the beginning of the era, the continents were joined together as Pangaea; dinosaurs and other giant reptiles dominated the sea and air. By the end of the Mesozoic era, the continents had begun to assume their present positions, flowering plants were dominant, and many of the large reptiles and marine fauna were becoming extinct.

messenger RNA *mRNA*, single-stranded nucleic acid (made up of nucleotides) found in ribosomes, mitochondria, and nucleoli of cells that carries coded information for building chains of amino acids into polypeptides.

metabolism chemical processes of living organisms enabling them to grow and to function. It involves a constant alternation of building up complex molecules (**anabolism**) and breaking them down (**catabolism**). For example, green plants build up complex organic substances from water, carbon dioxide, and mineral salts (photosynthesis); by digestion animals partially break down complex organic substances, ingested as food, and subsequently resynthesize them for use in their own bodies. Within cells, complex molecules are broken down by the process of respiration. The waste products of metabolism are removed by excretion.

metallurgy science and technology of producing metals, which includes extraction, alloying, and hardening. **Extractive** or **process metallurgy** is concerned with the extraction of metals from their ores and refining and adapting them for use. **Physical metallurgy** is concerned with their properties and application. **Metallography** establishes the microscopic structures that contribute to hardness, ductility, and strength.

metamorphic rock rock that has been changed from its original form, texture, and/or mineral assemblage by pressure or heat. For example, limestone can be metamorphosed by heat into marble, and shale by pressure into slate. The term was coined in 1833 by Scottish geologist Charles Lyell.

meteor flash of light in the sky, popularly known as a **shooting** or **falling star**, caused by a particle of dust, a **meteoroid**, entering the atmosphere at speeds up to 70 kps/45 mps and burning up by friction at a height of around 100 km/60 mi. On any clear night, several **sporadic meteors** can be seen each hour.

meteorite piece of rock or metal from space that reaches the surface of the Earth, Moon, or other body. Most meteorites are thought to be fragments from asteroids, although some may be pieces from the heads of comets. Most are stony, although some are made of iron and a few have a mixed rock–iron composition.

meteorology scientific observation and study of the atmosphere, so that weather can be accurately forecast. Data from meteorological stations and weather satellites are collated by computer at central agencies, and forecast and weather maps based on current readings are issued at regular intervals. Modern analysis, employing some of the most powerful computers, can give useful forecasts for up to six days ahead.

methane, CH_4, the simplest hydrocarbon of the paraffin series. Colourless, odourless, and lighter than air, it burns with a bluish flame and explodes when mixed with air or oxygen. It is the chief constituent of natural gas and also occurs in the explosive firedamp of coal mines.

methyl bromide pesticide gas used to fumigate soil. It is a major ozone depleter. Industry produces 50,000 tonnes of methyl bromide annually (1995). The European Union (EU) promised a 25% reduction in manufacture by 1998, and the USA intends to ban use by 2001. The EU proposed a total ban on usage by 2001 at a meeting in July 1998.

micrometer instrument for measuring minute lengths or angles with great accuracy; different types of micrometer are used in astronomical and engineering work.

micro-organism or *microbe*, living organism invisible to the naked eye but visible under a microscope. Micro-organisms include viruses and single-celled organisms such as bacteria, protozoa, yeasts, and some algae. The term has no taxonomic significance in biology. The study of micro-organisms is known as **microbiology**.

microscope instrument for forming magnified images with high resolution for detail. Optical and electron microscopes are the ones chiefly in use; other types include acoustic, scanning tunnelling, and atomic force microscopes.

microsome minute particle occurring in the cytoplasm of a cell composed of vesicles with attached ribosomes, which are thought to derive from the endoplasmic reticulum. Microsomes are also thought to give rise to mitochondria.

microtome device for cutting extremely thin slices of tissue for microscopic examination. The tissue is embedded in wax or a synthetic resin (or is frozen) for ease of handling.

microwave radiation radiation in the electromagnetic spectrum between infrared and radio waves. This range has wavelengths of between about 20 cm/8 in and about 1 mm/0.04 in. Radiation of this type was detected as background radiation.

mimicry imitation of one species (or group of species) by another. The most common form is **Batesian mimicry** (named after English naturalist H W Bates), where the mimic resembles a model that is poisonous or unpleasant to eat, and has aposematic, or warning, coloration; the mimic thus benefits from the fact that predators have learned to avoid the model. Hoverflies that resemble bees or wasps are an example. Appearance is usually the basis for mimicry, but calls, songs, scents, and other signals can also be mimicked.

mineral naturally formed inorganic substance with a particular chemical composition and a regularly repeating internal structure. Either in their perfect crystalline form or otherwise, minerals are the constituents of rocks. In more general usage, a mineral is any substance economically valuable for mining (including coal and oil, despite their organic origins).

minor planet name sometimes given to the larger members of the asteroid belt.

mitochondria singular *mitochondrion*, membrane-enclosed organelles within eukaryotic cells, containing enzymes responsible for energy production during aerobic respiration. Mitochondria absorb O_2 and glucose and produce energy in the form of ATP by breaking down the glucose to CO_2 and H_2O. These rod-like or spherical bodies are thought to be derived from free-living bacteria that, at a very early stage in the history of life, invaded larger cells and took up a symbiotic way of life inside. Each still contains its own small loop of DNA called mitochondrial DNA, and new mitochondria arise by division of existing ones. Mitochondria each have 37 genes.

mitosis process of cell division by which identical daughter cells are produced. During mitosis the DNA is duplicated and the chromosome number doubled, so new cells contain the same amount of DNA as the original cell.

The genetic material of eukaryotic cells is carried on a number of chromosomes. To control movements of chromosomes during cell division so that both new cells get the correct number, a system of protein tubules, known as the spindle, organizes the chromosomes into position in the middle of the cell before they replicate. The spindle then controls the movement of chromosomes as the cell goes through the stages of division: **interphase**, **prophase**, **metaphase**, **anaphase**, and **telophase**. See also meiosis.

M number designation used in the Messier catalogue.

modulo two numbers are said to be equivalent modulo a fixed number if their difference is divisible by the fixed number.

molarity concentration of a solution expressed as the number of moles of solute per cubic decimetre of solution.

mole SI unit (symbol mol) of the amount of a substance. It is defined as the amount of a substance that contains as many elementary entities (atoms, molecules, and so on) as there are atoms in 12 g of the isotope carbon-12.

molecular biology study of the molecular basis of life, including the biochemistry of molecules such as DNA, RNA, and proteins, and the molecular structure and function of the various parts of living cells.

molecular formula formula indicating the actual number of atoms of each element present in a single molecule of a chemical compound. This is determined by two pieces of information: the empirical formula and the relative molecular mass, which is determined experimentally.

molecular weight see relative molecular mass.

molecule smallest particle of an element or compound that can exist independently. Hydrogen atoms, at room temperature, do not exist independently. They are bonded in pairs to form hydrogen molecules. A molecule of a compound consists of two or more different atoms bonded together. Molecules vary in size and complexity from the hydrogen molecule (H_2) to the large macromolecules of proteins. They may be held together by ionic bonds, in which the atoms gain or lose electrons to form ions, or by covalent bonds, where electrons from each atom are shared in a new molecular orbital. Each compound is represented by a chemical symbol, indicating the elements into which it can be broken down and the number of each type of atom present. The symbolic representation of a molecule is known as its formula. For example, one molecule of the compound water, having two atoms of hydrogen and one atom of oxygen, is shown as H_2O.

moment generalization to a higher power of the mean. The rth moment of a random variable is the expected value of the rth power of the variable less its mean.

moment of inertia sum of all the point masses of a rotating object multiplied by the squares of their respective distances from the axis of rotation.

momentum product of the mass of a body and its velocity. If the mass of a body is m kilograms and its velocity is v m s^{-1}, then its momentum is given by:

$$momentum = mv.$$

Its unit is the kilogram metre per second (kg m s^{-1}) or the newton second. The momentum of a body does not change unless a resultant or unbalanced force acts on that body. See Newton's laws of motion.

monochromatic literally, one colour, this is the property of some sources of radiation to emit waves of one frequency (or wavelength). Lasers and masers are examples.

monoclonal describing genetically identical cells produced from one clone.

monocotyledon angiosperm (flowering plant) having an embryo with a single cotyledon, or seed leaf (as opposed to dicotyledons, which have two). Monocotyledons usually have narrow leaves with parallel veins and smooth edges, and hollow or soft stems. Their flower

parts are arranged in threes. Most are small plants such as orchids, grasses, and lilies, but some are trees such as palms.

monotonic of a sequence, that its terms are increasing or decreasing along the sequence.

Mordell's equation equation named after US-born British mathematician Louis Mordell that is represented by $y^2 = x^3 + k$.

morphogenesis development of forms and structures in an organism.

morphology study of the physical structure and form of organisms, in particular their soft tissues.

Moseley's law law stating that the frequency of the characteristic X-ray line spectra for elements is directly proportional to the square of the atomic number, Z.

Mössbauer effect recoil-free emission of gamma rays from atomic nuclei under certain conditions. The effect was discovered in 1958 by German physicist Rudolf Mössbauer, and used in 1960 to provide the first laboratory test of Einstein's general theory of relativity.

motor nerve any nerve that transmits impulses from the central nervous system to muscles or organs. Motor nerves cause voluntary and involuntary muscle contractions, and stimulate glands to secrete hormones.

mural arc astronomical apparatus, used from the 16th to the 19th century, comprising a carefully oriented wall on which a calibrated device was fixed, by which the altitudes of celestial objects could be measured.

mutant gene, organism, or population that has undergone a change in character because of mutation.

mutation change in the genes produced by a change in the DNA that makes up the hereditary material of all living organisms. Mutations, the raw material of evolution, result from mistakes during replication (copying) of DNA molecules. Due to the redundancy built into genetic code many mutations have no effect upon DNA functions. Only a few improve the organism's performance and are therefore favoured by natural selection. Mutation rates are increased by certain chemicals and by radiation.

myelin fatty substance that forms a sheath around the nerve fibres of vertebrates.

myofibril one of many minute fibrils that together make up a fibre of smooth or striped muscle, running along the length of the muscle.

myosin protein made up of a chain of polypeptides that forms filaments in smooth (or striped) muscle fibrils. During muscle contraction it combines with actin (another muscle protein, contained in thinner filaments) to form actomyosin; the actin filaments are pulled into the myosin filaments, which shortens the myofibrils.

n variable used to denote an integer number.

n! factorial *n*, that is, $1 \times 2 \times 3 \times \ldots n$.

Napier's bones mechanical calculation device created by Scottish mathematician John Napier and consisting of a set of small bone rods with which to multiply or divide.

national grid network of cables, carried overhead on pylons or buried under the ground, that connects consumers of electrical power to power stations, and interconnects the power stations. It ensures that power can be made available to all consumers at any time, allowing demand to be shared by several power stations, and particular power stations to be shut down for maintenance work from time to time.

natural logarithm exponent of a number expressed to base e, where e represents the irrational number 2.71828. . . . Natural logarithms are also called Napierian logarithms, after their inventor, the Scottish mathematician John Napier.

natural number one of the set of numbers used for counting. Natural numbers comprise all the positive integers, excluding zero.

natural selection process whereby gene frequencies in a population change through certain individuals producing more descendants than others because they are better able to survive and reproduce in their environment. The accumulated effect of natural selection is to produce adaptations such as the insulating coat of a polar bear or the spadelike forelimbs of a mole. The process is slow, relying firstly on random variation in the genes of an organism being produced by mutation and secondly on the genetic recombination of sexual reproduction. It was recognized by Charles Darwin and English naturalist Alfred Russel Wallace as the main process driving evolution.

***n*-dimensional** having an unstated but finite number of dimensions. A typical example is the set of all runs of *n*-numbers (that is, $x_1, \ldots, x_n$), which is the basis of a coordinate geometry generalizing three-dimensional coordinates (x_1, x_2, x_3).

nebula cloud of gas and dust in space. Nebulae are the birthplaces of stars, but some nebulae are produced by gas thrown off from dying stars (see planetary nebula; supernova). Nebulae are classified depending on whether they emit, reflect, or absorb light.

nebula variable star or ***T Tauri variable***, variable star of spectral classification F, G, or K (a giant above the main sequence on the Hertzsprung–Russell diagram) that loses an appreciable proportion of its mass in its (irregular) more luminous periods, and is thus surrounded by volumes of gas and dust.

neighbourhood subsets of points in a topological space defining a 'locality' round a specific point, which includes an open set to which the specific point belongs. Axioms describing neighbourhoods were first formulated by German mathematician Felix Hausdorff.

neural fold one of two longitudinal (ectodermal) ridges along the dorsal surface of a vertebrate embryo. The ridges fuse to form the neural tube.

neural tube embryological structure from which the brain and spinal cord are derived. It forms when two edges of a groove in a plate of primitive nerve tissue fuse. Incomplete fusion results in neural tube defects such as spina bifida.

neurofibril one of the many fibrils in the cytoplasm of a neuron, which extend into its axon and dendrites.

neuron another name for a nerve cell.

neurosecretory cell nerve cell that secretes a chemical substance, such as a hormone.

neurotransmitter chemical that diffuses across a synapse, and thus transmits impulses between nerve cells, or between nerve cells and effector organs (for example, muscles). Common neurotransmitters are noradrenaline (which also acts as a hormone) and acetylcholine, the latter being most frequent at junctions between nerve and muscle. Nearly 50 different neurotransmitters have been identified.

neutrino any of three uncharged elementary particles (and their antiparticles) of the lepton class, having a mass that is very small. The most familiar type, the antiparticle of the electron neutrino, is emitted in the beta decay of a nucleus. The other two are the muon and tau neutrinos.

neutron one of the three main subatomic particles, the others being the proton and the electron. The neutron is a composite particle, being made up of three quarks, and therefore belongs to the baryon group of the hadrons. Neutrons have about the same mass as protons but no electric charge, and occur in the nuclei of all atoms except hydrogen. They contribute to the mass of atoms but do not affect their chemistry.

neutron star very small, 'superdense' star composed mostly of neutrons. They are thought to form when massive stars explode as supernovae, during which the protons and electrons of the star's atoms merge, owing to intense gravitational collapse, to make neutrons. A neutron star has the mass of 2–3 suns, compressed into a globe only 20 km/ 12 mi in diameter.

newton SI unit (symbol N) of force. One newton is the force needed to accelerate an object with mass of one kilogram by one metre per second per second. The weight of a medium size (100 g/3 oz) apple is one newton.

Newton's law of cooling law stating that the rate of loss of heat from a body is directly proportional to the instantaneous temperature difference between the body and the surroundings. This leads to an exponential law of temperature decline.

Newton's laws of motion three laws that form the basis of Newtonian mechanics. (1) Unless acted upon by an unbalanced force, a body at rest stays at rest, and a moving body continues moving at the same speed in the same straight line. (2) An unbalanced force applied to a body gives it an acceleration proportional to the force (and in the direction of the force) and inversely proportional to the mass of the body. (3) When a body A exerts a force on a body B, B exerts an equal and opposite force on A; that is, to every action there is an equal and opposite reaction.

Newton's rings interference effect occurring in observations made of thin films separated by an air gap, reflections between various surfaces being allowed.

niche 'place' occupied by a species in its habitat, including all chemical, physical, and biological components, such as what it eats, the time of day at which the species feeds, temperature, moisture, the parts of the habitat that it uses (for example, trees or open grassland), the way it reproduces, and how it behaves.

Nicol prism device which uses the optical properties of calcite to produce plane-polarized light by passing rays of light through it. It is named after Scottish physicist William Nicol who invented it in 1828.

nitrogen fixation process by which nitrogen in the atmosphere is converted into nitrogenous compounds by the action of micro-organisms, such as cyanobacteria and bacteria, in conjunction with certain legumes. Several chemical processes duplicate nitrogen fixation to produce fertilizers. See nitrogen cycle.

node in two- and three-dimensional geometry, where a curve intersects itself.

noise unwanted sound. Permanent, incurable loss of hearing can be caused by prolonged exposure to high noise levels (above 85 decibels). Over 55 decibels on a daily outdoor basis is regarded as an unacceptable level. In scientific and engineering terms, a noise is any random, unpredictable signal.

noncommutative in which the principle of commutation (see commutative operation) does not apply.

non-Euclidean geometry study of figures and shapes in three-or-more-dimensional (or curved) space, in which Euclid's postulates may not apply fully or at all. There are now many forms of non-Euclidean geometry, probably the best known being those propounded by Bernhard Riemann; the first proponents of such systems, however, were Karl Gauss, Nikolai Lobachevsky, and János Bolyai.

normal distribution distribution widely used to model variation in a set of data which is symmetrical about its mean value. It can be expressed in the form:

$$f(x) = \frac{1}{\sqrt{2\pi}} \exp\left\{\frac{-(x-\mu)^2}{2\sigma^2}\right\}$$

where $f(x)$ is the relative frequency of data value x, σ is the standard deviation, μ is the mean, exp is the exponential function, and π is a mathematical constant. The curve resulting when $f(x)$ is plotted against x is called the **normal distribution curve**.

North Polar sequence or *circumpolar stars*, comprises those stars which never set, from the viewpoint of an observer on Earth.

nova plural *novae*, faint star that suddenly erupts in brightness by 10,000 times or more, remains bright for a few days, and then fades away and is not seen again for very many years, if at all. Novae are believed to occur in close binary star systems, where gas from one star flows to a companion white dwarf. The gas ignites and is thrown off in an explosion at speeds of 1,500 kps/930 mps or more. Unlike a supernova, the star is not completely disrupted by the outburst. After a few weeks or months it subsides to its previous state; it may erupt many more times.

nuclear fusion process whereby two atomic nuclei are fused, with the release of a large amount of energy. Very high temperatures and pressures are required for the process. Under these conditions the atoms involved are stripped of all their electrons so that the remaining particles, which together make up a **plasma**, can come close together at very high speeds and overcome the mutual repulsion of the positive charges on the atomic nuclei. At very close range the strong nuclear force will come into play, fusing the particles to form a larger nucleus. As fusion is accompanied by the release of large amounts of energy, the process might one day be harnessed to form the basis of commercial energy production. Methods of achieving controlled fusion are therefore the subject of research around the world.

nuclear isomer atoms of an element of a given mass that differ in their rates of radioactive decay are known as nuclear isomers.

nucleic acid complex organic acid made up of a long chain of nucleotides, present in the nucleus and sometimes the cytoplasm of the living cell. The two types, known as DNA (deoxyribonucleic acid) and RNA (ribonucleic acid), form the basis of heredity. The nucleotides are made up of a sugar (deoxyribose or ribose), a phosphate group, and one of four purine or pyrimidine bases. The order of the bases along the nucleic acid strand contains the genetic code.

nucleophilic describing an atom, molecule, or ion that seeks a positive centre (for example, the atomic nucleus) during a chemical reaction. Nucleophiles react at centres of low electron density because they have electron pairs available for bonding. Common nucleophiles include the hydroxide ion (OH^-), ammonia (NH_3), water (H_2O), and halide anions. See also electrophilic.

nucleosynthesis cosmic production of all the species of chemical elements by large-scale nuclear reactions, such as those in progress in the Sun or other stars. One element is changed into another by reactions that change the number of protons or neutrons involved.

nucleotide organic compound consisting of a purine (adenine or guanine) or a pyrimidine (thymine, uracil, or cytosine) base linked to a sugar (deoxyribose or ribose) and a phosphate group. DNA and RNA are made up of long chains of nucleotides.

nucleus in biology, the central, membrane-enclosed part of a eukaryotic cell, containing threads of DNA. During cell division these coil up to form chromosomes. The nucleus controls the function of the cell by determining which proteins are produced within it (see DNA for details of this process). Because proteins are the chief structural molecules of living matter and, as enzymes, regulate all aspects of metabolism, it may be seen that the genetic code within the nucleus is effectively responsible for building and controlling the whole organism.

nucleus in physics, the positively charged central part of an atom, which constitutes almost all its mass. Except for hydrogen nuclei, which have only protons, nuclei are composed of both protons and neutrons. Surrounding the nuclei are electrons, of equal and opposite charge to that of the protons, thus giving the atom a neutral charge.

null hypothesis assumes that events occur on a purely random (chance) basis.

null set conventional set that has no members.

null vector another name for a zero vector.

number theory abstract study of the structure of number systems and the properties of positive integers (whole numbers). For example, the theories of factors and prime numbers fall within this area, as does the work of mathematicians Giuseppe Peano, Pierre de Fermat, and Karl Gauss.

nutation slight 'nodding' of the Earth in space, caused by the varying gravitational pulls of the Sun and Moon. Nutation changes the angle of the Earth's axial tilt (average 23.5°) by about 9 seconds of arc to either side of its mean position, a complete cycle taking just over 18.5 years.

objective system of lenses in a telescope or microscope nearest to the object being observed.

occultation temporary obscuring of a star by a body in the Solar System. Occultations are used to provide information about changes in an orbit, and the structure of objects in space, such as radio sources.

ohm SI unit (symbol Ω) of electrical resistance (the property of a conductor that restricts the flow of electrons through it).

Ohm's law law that states that, for many materials over a wide range of conditions, the current flowing in a conductor maintained at constant temperature is directly proportional to the potential difference (voltage) between its ends. The law was discovered by German physicist Georg Ohm in 1827.

Olbers's paradox question put forward in 1826 by German doctor, mathematician, and astronomer Heinrich Olbers, who asked: If the universe is infinite in extent and filled with stars, why is the sky dark at night? The answer is that the stars do not live infinitely long, so there is not enough starlight to fill the universe. A wrong answer, frequently given, is that the expansion of the universe weakens the starlight.

olefin common name for alkene.

oncogene gene carried by a virus that induces a cell to divide abnormally, giving rise to a cancer. Oncogenes arise from mutations in genes (proto-oncogenes) found in all normal

497

cells. They are usually also found in viruses that are capable of transforming normal cells to tumour cells. Such viruses are able to insert their oncogenes into the host cell's DNA, causing it to divide uncontrollably. More than one oncogene may be necessary to transform a cell in this way.

ontogeny process of development of a living organism, including the part of development that takes place after hatching or birth. The idea that 'ontogeny recapitulates phylogeny' (the development of an organism goes through the same stages as its evolutionary history), proposed by the German scientist Ernst Heinrich Haeckel, is now discredited.

open-hearth process or *Siemens–Martin process*, method of producing steel in which cast iron and steel scrap or iron ore are heated together in measured amounts with producer gas on a hearth in a furnace.

opposition moment at which a body in the Solar System lies opposite the Sun in the sky as seen from the Earth and crosses the meridian at about midnight.

optical activity ability of certain crystals, liquids, and solutions to rotate the plane of polarized light as it passes through them. The phenomenon is related to the three-dimensional arrangement of the atoms making up the molecules concerned. Only substances that lack any form of structural symmetry exhibit optical activity.

optical fibre thin thread of glass so constructed that it permits light to be transmitted down its length, even round corners, by total internal reflection.

optical isomer or *enantiomorph*, one of a pair of compounds whose chemical composition is similar but whose molecular structures are mirror images of each other. The presence of an asymmetric (usually carbon) atom makes each isomer optically active (that is, its crystals or solutions rotate the plane of polarized light); the direction of rotation is different (left or right, denoted by *d* or *l*) for each isomer.

optical isomerism type of stereoisomerism in which the isomers differ in their optical activity because of the different spatial arrangements of their atoms. Enantio-morphs (isomers with asymmetrical structures, each isomer being a mirror image of the other) are optical isomers.

optic vessel one of two bulges on each side of the anterior expansion of the neural tube in a vertebrate embryo, from which arise the essential nervous structures of the eyes.

orbit path of one body in space around another, such as the orbit of Earth around the Sun, or

the Moon around Earth. When the two bodies are similar in mass, as in a binary star, both bodies move around their common centre of mass. The movement of objects in orbit follows Johannes Kepler's laws, which apply to artificial satellites as well as to natural bodies.

orbital, atomic region around the nucleus of an atom (or, in a molecule, around several nuclei) in which an electron is likely to be found. According to quantum theory, the position of an electron is uncertain; it may be found at any point. However, it is more likely to be found in some places than in others, and this pattern of probabilities makes up the orbital.

Ordovician period period of geological time 510–439 million years ago, the second period of the Palaeozoic era. Animal life was confined to the sea: reef-building algae and the first jawless fish are characteristic.

ore body of rock, a vein within it, or a deposit of sediment, worth mining for the economically valuable mineral it contains. The term is usually applied to sources of metals. Occasionally metals are found uncombined (native metals), but more often they occur as compounds such as carbonates, sulphides, or oxides. The ores often contain unwanted impurities that must be removed when the metal is extracted.

organelle discrete and specialized structure in a living cell; organelles include mitochondria, chloroplasts, lysosomes, ribosomes, and the nucleus.

organizer part of the embryo that causes changes to occur in another part, through induction, thus 'organizing' development and differentiation.

organometallic compound any of a group of substances in which one or more organic radicals are chemically bonded to a metallic atom – excluding the ionic salts of metals and organic acids. A typical organometallic compound is tetraethyl lead, the 'antiknock' substance commonly added to petrol.

organophosphate insecticide insecticidal compounds that cause the irreversible inhibition of the cholinesterase enzymes that break down acetylcholine. As this mechanism of action is very toxic to humans, the compounds should be used with great care. Malathion and permethrin may be used to control lice in humans and have many applications in veterinary medicine and agriculture.

orogenesis in its original, literal sense, orogenesis means 'mountain building', but today it more specifically refers to the tectonics of mountain building (as opposed to mountain building by erosion).

orthogonal having a right angle, right-angled or perpendicular.

osmosis movement of water through a selectively permeable membrane separating solutions of different concentrations. Water passes by diffusion from a **weak solution** (high water concentration) to a **strong solution** (low water concentration) until the two concentrations are equal. The selectively permeable membrane allows the diffusion of water but not of the solute (for example, sugar molecules). Many cell membranes behave in this way, and osmosis is a vital mechanism in the transport of fluids in living organisms – for example, in the transport of water from soil (weak solution) into the roots of plants (stronger solution of cell sap).

osmotic pressure pressure that must be applied to a solution so that it no longer takes up pure solvent (usually water) across a membrane that is permeable to solvent but not to solute (the dissolved substance). See also osmosis.

oxidation number Roman numeral often seen in a chemical name, indicating the valency of the element immediately before the number. Examples are lead (II) nitrate, manganese (IV) oxide, and potassium manganate (VII).

oxyfuel fuel enriched with oxygen to decrease carbon monoxide (CO) emissions. Oxygen is added in the form of chemicals such as methyl tertiary butyl ether (MTBE) and ethanol.

oxygenation combination of (gaseous or dissolved) oxygen with a substance, such as with the blood in the lungs during respiration.

ozone layer thin layer of the gas ozone in the upper atmosphere that shields the Earth from harmful ultraviolet rays. A continent-sized hole has formed over Antarctica as a result of damage to the ozone layer. This has been caused in part by chlorofluorocarbons (CFCs), but many reactions destroy ozone in the stratosphere: nitric oxide, chlorine, and bromine atoms are implicated.

pair production creation of an electron and antielectron (positron) pair by the interaction between a high-energy particle or photon and the electrostatic field around a nucleus. It can also be used to describe the creation out of the vacuum state of particle–antiparticle pairs as allowed by Heisenberg's uncertainty principle.

Palaeozoic era era of geological time 570–245 million years ago. It comprises the Cambrian, Ordovician, Silurian, Devonian, Carboniferous, and Permian periods. The Cambrian, Ordovician, and Silurian constitute the Lower or Early Palaeozoic; the Devonian, Carboniferous, and Permian make up the Upper or Late Palaeozoic. The era includes the evolution of hard-shelled multicellular life forms in the sea; the invasion of land by plants and animals; and the evolution of fish, amphibians, and early reptiles.

panchromatic film photographic film that is reasonably sensitive to all frequencies within the visible light spectrum.

Pangaea or *Pangea*, (Greek 'all-land') single land mass, made up of all the present continents, believed to have existed between 300 and 200 million years ago; the rest of the Earth was covered by the Panthalassa ocean. Pangaea split into two land masses – Laurasia in the north and Gondwanaland in the south – which subsequently broke up into several continents. These then moved slowly to their present positions (see plate tectonics).

papilla small growth from the surface of a tissue, such as the papillae on the surface of the tongue.

parabola curve formed by cutting a right circular cone with a plane parallel to the sloping side of the cone. A parabola is one of the family of curves known as conic sections. The graph of $y = x^2$ is a parabola.

paraffin see **alkane**.

parahydrogen molecules of hydrogen in which the spins of the two atoms are opposed, in contrast to orthohydrogen in which spins are parallel.

parallax change in the apparent position of an object against its background when viewed from two different positions. In astronomy, nearby stars show a shift owing to parallax when viewed from different positions on the Earth's orbit around the Sun. A star's parallax is used to deduce its distance from the Earth.

parallelogram of forces method of calculating the resultant (combined effect) of two different forces acting together on an object. Because a force has both magnitude and direction it is a vector quantity and can be represented by a straight line. A second force acting at the same point in a different direction can be represented by another line drawn at an angle to the first. By completing the parallelogram (of which the two lines are sides) a diagonal may be drawn from the original angle to the opposite corner to represent the resultant force vector.

paramagnetism property of most elements and some compounds (but excluding the ferromagnetic substances – iron, cobalt, nickel, and their alloys) whereby they are weakly magnetized by relatively strong magnetic fields. See also ferromagnetism.

parity concept in quantum theory related to the mirror-symmetry of the functions describing mathematically the behaviour of the particles.

parsec unit (symbol pc, from *par*allax *sec*ond) used for distances to stars and galaxies. One parsec is the distance at which one astronomical unit subtends one arcsecond, and is equal to 3.2616 light years, 2.063×10^5 astronomical units, and 3.0857×10^{13} km.

parthenogenesis development of an ovum (egg) without any genetic contribution from a male. Parthenogenesis is the normal means of reproduction in a few plants (for example, dandelions) and animals (for example, certain fish). Some sexually reproducing species, such as aphids, show parthenogenesis at some stage in their life cycle to accelerate reproduction to take advantage of good conditions.

partial differentiation form of differential calculus in which instead of a function $y = f(x)$ – which has one variable – a function of two or more variables is considered, for example, $z = f(x, y)$. Such functions represent a surface in three-dimensional space.

partitioning of an integer in number theory, breaking the integer down into its constituent parts in as many ways as possible; for example, the number 6 can be partitioned in three ways: $5 + 1$, $4 + 2$, and $3 + 3$. Each of these contributory numbers is known as a summand.

pascal SI unit (symbol Pa) of pressure, equal to one newton per square metre. It replaces bars and millibars (10^5 Pa equals one bar). It is named after the French mathematician Blaise Pascal.

Pascal's principle law of hydrostatics stating that the application of pressure to a fluid results in that pressure being equally transmitted throughout the fluid, in all directions.

pasteurization treatment of food to reduce the number of micro-organisms it contains and so protect consumers from disease. Harmful bacteria are killed and the development of others is delayed.

pathogen (Greek 'disease producing') any micro-organism that causes disease. Most pathogens are parasites, and the diseases they cause are incidental to their search for food or shelter inside the host. Nonparasitic organisms, such as soil bacteria or those living in the human gut and feeding on waste foodstuffs, can also become pathogenic to a person whose immune system or liver is damaged. The larger parasites that can cause disease, such as nematode worms, are not usually described as pathogens.

Pauli exclusion principle principle of atomic structure. See exclusion principle.

Peano axioms axioms that formally introduce the properties of the positive whole numbers (originally devised, despite being named after Italian mathematician Guiseppe Peano, by German mathematician Julius Dedekind).

penumbra region of partial shade between the totally dark part (umbra) of a shadow and the fully illuminated region outside. It occurs when a source of light is only partially obscured by a shadow-casting object. The darkness of a penumbra varies gradually from total darkness at one edge to full brightness at the other. In astronomy, a penumbra is a region of the Earth from which only a partial eclipse of the Sun or Moon can be seen.

peptide molecule comprising two or more amino acid molecules (not necessarily different) joined by **peptide bonds**, whereby the acid group of one acid is linked to the amino group of the other (–CO.NH). The number of amino acid molecules in the peptide is indicated by referring to it as a di-, tri-, or polypeptide (two, three, or many amino acids).

perfect number number that is equal to the sum of all its factors (except itself); for example, 6 is a perfect number, being equal to $1 + 2 + 3$.

perfusion *in vitro*-induced passage of blood or a nutrient fluid through the blood vessels of an organism to keep it supplied with oxygen and nutrients.

perihelion point at which an object, travelling in an elliptical orbit around the Sun, is at its closest to the Sun. The point at which it is furthest from the Sun is the aphelion.

periodic law generalization that there is a recurring pattern in the properties of elements when they are arranged in order of increasing atomic number. The law is most apparent when the elements are arranged in the periodic table, in which the elements in each vertical column (group) show similar properties.

periodic table of the elements table in which the elements are arranged in order of their atomic number. The table summarizes the major properties of the elements and enables predictions to be made about their behaviour.

period–luminosity curve graph depicting the variation in luminosity of a Cepheid variable star with time. In general, the longer the period, the greater the luminosity. By measuring the period it is possible thus to derive an absolute magnitude; comparison of this with the star's observed (apparent) magnitude gives an indication of the distance.

peristalsis wavelike contractions, produced by the contraction of smooth muscle, that pass along tubular organs, such as the intestines. The same term describes the wavelike motion of earthworms and other invertebrates, in which part of the body contracts as another part elongates.

permanent magnet magnet that retains its magnetism permanently, and not just when subject to some external energy source (as is the case with an electromagnet).

permeability degree to which the presence of a substance alters the magnetic field around it. Most substances have a small constant permeability. When the permeability is less than 1, the material is a diamagnetic material; when it is greater than 1, it is a paramagnetic material. Ferrimagnetic materials have very large permeabilities. See also magnetism.

Permian period of geological time 290–245 million years ago, the last period of the Palaeozoic era. Its end was marked by a dramatic change in marine life – the greatest mass extinction in geological history – including the extinction of many corals and trilobites. Deserts were widespread, terrestrial amphibians and mammal-like reptiles flourished, and cone-bearing plants (gymnosperms) came to prominence. In the oceans, 49% of families and 72% of genera vanished in the late Permian. On land, 78% of reptile families and 67% of amphibian families disappeared.

permutation group group consisting of all the transformations (operations) permutating a fixed number of objects among themselves. Such groups were studied first by French mathematician Evariste Galois in connection with permutations of roots (solutions) of a polynominal equation.

perpetual motion idea that a machine can be designed and constructed in such a way that, once started, it will do work indefinitely without requiring any further input of energy (motive power). Such a device would contradict at least one of the two laws of thermodynamics that state that (1) energy can neither be created nor destroyed (the law of conservation of energy) and (2) heat cannot by itself flow from a cooler to a hotter object. As a result, all practical (real) machines require a continuous supply of energy, and no heat engine is able to convert all the heat into useful work.

persistence of vision brief retention of the sensation of light by the brain, once the initial stimulus has been removed. Essential phenomenon for the success of televisual and cinematographic images.

perturbation apparent irregularity in an orbit, or occasionally in a star's proper motion, caused by the gravitational effects of a nearby celestial body.

pH scale from 0 to 14 for measuring acidity or alkalinity. A pH of 7.0 indicates neutrality, below 7 is acid, while above 7 is alkaline. Strong acids, such as those used in car batteries, have a pH of about 1; strong alkalis such as sodium hydroxide are pH 13.

phagocyte type of white blood cell, or leucocyte, that can engulf a bacterium or other invading micro-organism. Phagocytes are found in blood, lymph, and other body tissues, where they also ingest foreign matter and dead tissue. A macrophage differs in size and life span.

phagocytosis process by which phagocytes surround foreign particles (by an amoeboid movement), engulf and digest them. See also endocytosis.

Phanerozoic eon (Greek *phanero* 'visible') eon in Earth history, consisting of the most recent 570 million years. It comprises the Palaeozoic, Mesozoic, and Cenozoic eras. The vast majority of fossils come from this eon, owing to the evolution of hard shells and internal skeletons. The name means 'interval of well-displayed life'.

phase in astronomy, the apparent shape of the Moon or a planet when all or part of its illuminated hemisphere is facing the Earth.

The Moon undergoes a full cycle of phases from new (when between the Earth and the Sun) through first quarter (when at 90° eastern elongation from the Sun), full (when opposite the Sun), and last quarter (when at 90° western elongation from the Sun).

phase in physics, a stage in an oscillatory motion, such as a wave motion: two waves are in phase when their peaks and their troughs coincide. Otherwise, there is a **phase difference**, which has consequences in interference phenomena and alternating current electricity.

phenol member of a group of aromatic chemical compounds with weakly acidic properties, which are characterized by a hydroxyl (OH) group attached directly to an

501

aromatic ring. The simplest of the phenols, derived from benzene, is also known as phenol and has the formula C_6H_5OH. It is sometimes called **carbolic acid** and can be extracted from coal tar.

phenotype visible traits, those actually displayed by an organism. The phenotype is not a direct reflection of the genotype because some alleles are masked by the presence of other, dominant alleles (see dominance). The phenotype is further modified by the effects of the environment (for example, poor nutrition stunts growth).

philosophy of mathematics has three main schools: the logical (in which mathematics is simply a branch of logic), the formalist (involving the study of the structure of objects and the property of symbols), and the intuitional (grounded on the basic premise of the possibility of constructing an infinite series of numbers).

phloem tissue found in vascular plants whose main function is to conduct sugars and other food materials from the leaves, where they are produced, to all other parts of the plant.

phosphor any substance that is phosphorescent, that is, gives out visible light when it is illuminated by a beam of electrons or ultraviolet light. A television screen is coated on the inside with phosphors that glow when beams of electrons strike them. Fluorescent lamp tubes are also phosphor-coated. Phosphors are also used in Day-Glo paints, and as optical brighteners in detergents.

phosphorescence type of luminescence in which a substance exposed to radiation emits light, this emission continuing after the radiation has been removed.

phosphorylation chemical addition of a phosphate group ($-PO_3^{2-}$) to a (organic) molecule. One of the most important biochemical phosphorylations is the addition of phosphate to ADP to form the energy-rich ATP. Phosphorylation involving light (as in photosynthesis) is termed photophosphorylation.

photochemical reaction any chemical reaction in which light is produced or light initiates the reaction. Light can initiate reactions by exciting atoms or molecules and making them more reactive: the light energy becomes converted to chemical energy. Many photochemical reactions set up a chain reaction and produce free radicals.

photoelectric effect transfer of energy from light rays falling onto a substance to the electrons within the substance. If the frequency, and hence the energy, of the radiation is high enough, it is possible to 'boil' electrons, then known as photo-electrons, out of the substance. Being charge-carriers, these electrons can constitute a photoelectric current.

photoelectric filtering means of measuring the astronomical colour index of a star, involving colour filters on photoelectric cells to define the colour index between two set wavelengths. The filters correspond to the UBV photometry system.

photoelectricity phenomenon whereby certain materials, such as selenium, can produce electrical output if exposed to light.

photolysis chemical reaction that is driven by light or ultraviolet radiation. For example, the light reaction of photosynthesis (the process by which green plants manufacture carbohydrates from carbon dioxide and water) is a photolytic reaction.

photometry measurement of the magnitudes of celestial bodies, originally carried out by expertise of eye alone, but now generally making use of photographic or photoelectric apparatus.

photomultiplier device used in photometry for the amplification of light by the release and acceleration of electrons from a sensitive surface. The result is a measurable electric current that is proportional to the intensity of received radiation.

photon elementary particle or 'package' (quantum) of energy in which light and other forms of electromagnetic radiation are emitted. The photon has both particle and wave properties; it has no charge, is considered massless but possesses momentum and energy. It is one of the gauge bosons, and is the carrier of the electromagnetic force, one of the fundamental forces of nature.

photosphere visible surface of the Sun, which emits light and heat. About 300 km/200 mi deep, it consists of incandescent gas at a temperature of 5,800 K (5,530°C/9,990°F).

photosynthesis process by which green plants trap light energy from the Sun. This energy is used to drive a series of chemical reactions which lead to the formation of carbohydrates. The carbohydrates occur in the form of simple sugar, or glucose, which provides the basic food for both plants and animals. For photosynthesis to occur, the plant must possess chlorophyll and must have a supply of carbon dioxide and water. Photosynthesis takes place inside chloroplasts which are found mainly in the leaf cells of plants.

phylogeny historical sequence of changes that occurs in a given species during the course of its evolution. It was once erroneously associated with ontogeny (the process of development of a living organism).

phylum plural *phyla*, major grouping in biological classification. Mammals, birds, reptiles, amphibians, fishes, and tunicates belong to the phylum Chordata; the phylum Mollusca consists of snails, slugs, mussels, clams, squid, and octopuses; the phylum Porifera contains sponges; and the phylum Echinodermata includes starfish, sea urchins, and sea cucumbers. In classifying plants (where the term 'division' often takes the place of 'phylum'), there are between four and nine phyla depending on the criteria used; all flowering plants belong to a single phylum, Angiospermata, and all conifers to another, Gymnospermata. Related phyla are grouped together in a kingdom; phyla are subdivided into classes.

piezoelectric effect property of some crystals (for example, quartz) to develop an electromotive force or voltage across opposite faces when subjected to tension or compression, and, conversely, to expand or contract in size when subjected to an electromotive force. Piezoelectric crystal oscillators are used as frequency standards (for example, replacing balance wheels in watches), and for producing ultrasound.

pigeon-hole principle another name for Schubfachprinzip.

pinocytosis ingestion of the contents of a vesicle by a cell (see endocytosis).

pion or *pi meson*, subatomic particle with a neutral form (mass 135 MeV) and a charged form (mass 139 MeV). The charged pion decays into muons and neutrinos and the neutral form decays into gamma-ray photons. They belong to the hadron class of elementary particles.

pioneer species those species that are the first to colonize and thrive in new areas. Coal pits, recently cleared woodland, and new roadsides are areas where pioneer species will quickly appear. As the habitat matures other species take over, a process known as **succession**.

pitch distance between the adjacent threads of a screw or bolt. When a screw is turned through one full turn it moves a distance equal to the pitch of its thread. A screw thread is a simple type of machine, acting like a rolled-up inclined plane, or ramp (as may be illustrated by rolling a long paper triangle around a pencil). A screw has a mechanical advantage greater than one.

placenta organ that attaches the developing embryo or fetus to the uterus in placental mammals (mammals other than marsupials, platypuses, and echidnas). Composed of maternal and embryonic tissue, it links the blood supply of the embryo to the blood supply of the mother, allowing the exchange of oxygen, nutrients, and waste products. The two blood systems are not in direct contact, but are separated by thin membranes, with materials diffusing across from one system to the other. The placenta also produces hormones that maintain and regulate pregnancy. It is shed as part of the afterbirth.

Planck's quantum constant constant (symbol h) with a value of about 6.63×10^{-34} joule seconds, which is the fundamental unit of spin angular momentum.

Planck's radiation law law stating that the energy of electromagnetic radiation of a certain frequency is given by the product of the frequency and Planck's constant, h.

plane figure two-dimensional figure. All polygons are plane figures.

plane-polarized light light in which the electric and magnetic vibrations of the waves are restricted to a single plane, the plane of the magnetic vibration being at right angles to that of the electric one.

planetary nebula shell of gas thrown off by a star at the end of its life. Planetary nebulae have nothing to do with planets. They were named by William Herschel, who thought their rounded shape resembled the disc of a planet. After a star such as the Sun has expanded to become a red giant, its outer layers are ejected into space to form a planetary nebula, leaving the core as a white dwarf at the centre.

plankton small, often microscopic, forms of plant and animal life that live in the upper layers of fresh and salt water, and are an important source of food for larger animals. Marine plankton is concentrated in areas where rising currents bring mineral salts to the surface.

plasma in biology, the liquid component of the blood. It is a straw-coloured fluid, largely composed of water (around 90%), in which a number of substances are dissolved. These include a variety of proteins (around 7%) such as fibrinogen (important in blood clotting), inorganic mineral salts such as sodium and calcium, waste products such as urea, traces of hormones, and antibodies to defend against infection.

plasma in physics, an ionized gas produced at extremely high temperatures, as in the Sun and other stars. It contains positive and negative charges in equal numbers. It is a

503

good electrical conductor. In thermonuclear reactions the plasma produced is confined through the use of magnetic fields.

plasmolysis shrinkage of plant cell contents away from the cell wall as a result of osmosis in which water diffuses out of the cell.

plastic any of the stable synthetic materials that are fluid at some stage in their manufacture, when they can be shaped, and that later set to rigid or semi-rigid solids. Plastics today are chiefly derived from petroleum. Most are polymers, made up of long chains of identical molecules.

plate tectonics theory formulated in the 1960s to explain the phenomena of continental drift and seafloor spreading, and the formation of the major physical features of the Earth's surface. The Earth's outermost layer, the lithosphere, is regarded as a jigsaw puzzle of rigid major and minor plates that move relative to each other, probably under the influence of convection currents in the mantle beneath. At the margins of the plates, where they collide or move apart or slide past one another, major landforms such as mountains, rift valleys, volcanoes, ocean trenches, and **ocean ridges** are created. The rate of plate movement is at most 15 cm/6 in per year.

Platonic solid another name for a regular polyhedron, one of five possible three-dimensional figures with all its faces the same size and shape.

pluteus advanced larval stage in echinoderms characterized by bilateral symmetry.

PM10 abbreviation for particulate matter less than 10 micrometres across, clusters of small particles, such as carbon particles, in the air that come mostly from vehicle exhausts.

point sets sets comprising some or all of the points of the space under study.

poisoning reduction in effectiveness of a catalyst as a result of its being contaminated by a reactant or a product of the reaction it catalyses. Although in theory a catalyst is unaffected by the reaction it catalyses, in practice particles ('poisons') accumulate on the surface of the catalyst and reduce its effectiveness.

Poisson's ratio ratio of the lateral strain to the longitudinal strain in a wire held under tension.

polar body in meiosis, any of three (haploid) egg nuclei that develop from a secondary oocyte (the fourth nucleus is the ovum); the three polar bodies degenerate.

polarimetry any technique for measuring the degree of polarization of radiation from stars, galaxies, and other objects.

polariscope instrument that enables the rotation of the plane of polarization of polarized light to be determined.

polarized light light in which the electromagnetic vibrations take place in one particular plane. In ordinary (unpolarized) light, the electric fields vibrate in all planes perpendicular to the direction of propagation. After reflection from a polished surface or transmission through certain materials (such as Polaroid), the electric fields are confined to one direction, and the light is said to be **linearly polarized**. In **circularly polarized** and **elliptically polarized** light, the electric fields are confined to one direction, but the direction rotates as the light propagates. Polarized light is used to test the strength of sugar solutions and to measure stresses in transparent materials.

polybasic describing an acid that has more than two atoms of replaceable acidic hydrogen in each of its molecules.

polyhedron solid figure with four or more plane faces. The more faces there are on a polyhedron, the more closely it approximates to a sphere. Knowledge of the properties of polyhedra is needed in crystallography and stereochemistry to determine the shapes of crystals and molecules.

polymer compound made up of a large long-chain or branching matrix composed of many repeated simple units (**monomers**) linked together by polymerization. There are many polymers, both natural (cellulose, chitin, lignin) and synthetic (polyethylene and nylon, types of plastic). Synthetic polymers belong to two groups: thermo-softening and thermosetting (see plastic).

polymerization chemical union of two or more (usually small) molecules of the same kind to form a new compound. **Addition polymerization** produces simple multiples of the same compound. **Condensation polymerization** joins molecules together with the elimination of water or another small molecule.

polysaccharide long-chain carbohydrate made up of hundreds or thousands of linked simple sugars (monosaccharides) such as glucose and closely related molecules.

population group of animals of one species, living in a certain area and able to interbreed; the members of a given species in a community of living things.

population genetics branch of genetics that studies the way in which the frequencies of different alleles (alternative forms of a gene) in populations of organisms change, as a result of natural selection and other processes.

population I stars younger stars, generally formed towards the edge of a galaxy, of the dusty material in the spiral arms, including the heavy elements. The brightest of this population are hot, white stars.

population II stars older stars, generally formed towards the centre of a galaxy, containing few heavier elements. The brightest of this population are red giants.

porosity percentage of empty space existing within a material.

positron antiparticle of the electron; an elementary particle having the same mass as an electron but exhibiting a positive charge. The positron was discovered in 1932 by US physicist Carl Anderson at the California Institute of Technology, USA, its existence having been predicted by the British physicist Paul Dirac in 1928.

potential difference *PD*, difference in the electrical potential of two points, being equal to the electrical energy converted by a unit electric charge moving from one point to the other. The SI unit of potential difference is the volt (V). The potential difference between two points in a circuit is commonly referred to as voltage. See also Ohm's law.

potential energy, energy possessed by an object by virtue of its relative position or state (for example, as in a compressed spring or a muscle). It is contrasted with kinetic energy, the form of energy possessed by moving bodies. An object that has been raised up is described as having gravitational potential energy.

power that which is represented by an exponent or index, denoted by a superior small numeral. A number or symbol raised to the power of 2 – that is, multiplied by itself – is said to be squared (for example, 3^2, x^2), and when raised to the power of 3, it is said to be cubed (for example, 2^3, y^3). Any number to the power zero equals 1.

power series infinite series of the general form:

$$a_0 + a_1x^1 + a_2x^2 + a_3x^3 + \ldots + a_nx^n + \ldots$$

This is called a power series in x. Examples of power series are the exponential series:

$$e^x = 1 + x + x^2/2! + x^3/3! + \ldots$$

the logarithmic series, and the series for the trigonometric functions (sine, cosine, and tangent).

Poynting–Robertson effect interaction of dust particles in interplanetary space with solar radiation causing a loss in orbital velocity of the dust around the Sun. This causes the dust particles to spiral into the Sun, if the effect is unopposed. However, under certain circumstances, radiation pressure is large enough to oppose the effect.

Precambrian period time from the formation of Earth (4.6 billion years ago) up to 570 million years ago. Its boundary with the succeeding Cambrian period marks the time when animals first developed hard outer parts (exoskeletons) and so left abundant fossil remains. It comprises about 85% of geological time and is divided into two eons: the Archaean and the Proterozoic.

precession slow wobble of the Earth on its axis, like that of a spinning top. The gravitational pulls of the Sun and Moon on the Earth's equatorial bulge cause the Earth's axis to trace out a circle on the sky every 25,800 years. The position of the celestial poles (see celestial sphere) is constantly changing owing to precession, as are the positions of the equinoxes (the points at which the celestial equator intersects the Sun's path around the sky).

precession of the equinoxes apparent movement per year of the two points in the sky representing the equinoxes; 50.26 seconds of arc per year, also called the constant of precession.

Precession is caused mainly by the gravitational effect of the Moon on the Earth's equatorial 'bulge'.

predicate calculus part of the theory of devising models involving deduction and the use of variables and negatives in systems of sentences; the term occurs commonly in mathematical logic.

pressure in a fluid, the force that would act normally (at right angles) per unit surface area of a body immersed in the fluid. The SI unit of pressure is the pascal (Pa), equal to a pressure of one newton per square metre. In the atmosphere, the pressure declines with height from about 100 kPa at sea level to zero where the atmosphere fades into space. Pressure is commonly measured with a barometer, manometer, or Bourdon gauge. Other common units of pressure are the bar and the torr.

primary factors most reduced (lowest) form of numbers, which, when multiplied together, produce the principal numbers or expression.

prime number number that can be divided only by 1 and itself, that is, having no other factors. There are an infinite number of primes, the first ten of which are 2, 3, 5, 7, 11, 13, 17, 19, 23, and 29 (the number 1 is excluded from the set of prime numbers).

505

The number 2 is the only even prime because all other even numbers have 2 as a factor. Numbers other than primes can be expressed as a product of their prime factors.

probability likelihood, or chance, that an event will occur, often expressed as odds, or in mathematics, numerically as a fraction or decimal. In general, the probability that n particular events will happen out of a total of m possible events is n/m. A certainty has a probability of 1; an impossibility has a probability of 0.

Probability = number of successful events/total possible number of events

product result of multiplying one quantity by another.

projective geometry form of two- and three-dimensional geometry concerned with the geometrical properties that remain constant (invariant) under projection, that is, extended on a single plane, or projected from one plane onto another. Perspective – two-dimensional representation of three-dimensional reality – uses the basic theory of projective geometry.

proof set of arguments used to deduce a mathematical theorem from a set of axioms.

proper motion gradual change in the position of a star that results from its motion in orbit around our Galaxy, the Milky Way. Proper motions are slight and undetectable to the naked eye, but can be accurately measured on telescopic photographs taken many years apart. Barnard's Star is the star with the largest proper motion, 10.3 arc seconds per year.

prophage DNA of a nonvirulent bacteriophage that has become linked with the bacterial host's DNA and which is replicated with it. See lysogeny.

proprioceptor one of the sensory nerve endings that are located in muscles, tendons, and joints. They relay information on the position of the body and the state of muscle contraction.

protein complex, biologically important substance composed of amino acids joined by peptide bonds. Proteins are essential to all living organisms. As enzymes they regulate all aspects of metabolism. Structural proteins such as **keratin** and **collagen** make up the skin, claws, bones, tendons, and ligaments; **muscle** proteins produce movement; **haemoglobin** transports oxygen; and **membrane** proteins regulate the movement of substances into and out of cells. For humans, protein is an essential part of the diet, and is found in greatest quantity in soy beans and other grain legumes, meat, eggs, and cheese.

protein synthesis manufacture, within the cytoplasm of the cell, of the proteins an organism needs. The building blocks of proteins are amino acids, of which there are 20 types. The pattern in which the amino acids are linked decides what kind of protein is produced. In turn it is the genetic code, contained within DNA, that determines the precise order in which the amino acids are linked up during protein manufacture.

Interestingly, DNA is found only in the nucleus, yet protein synthesis occurs only in the cytoplasm. The information necessary for making the proteins is carried from the nucleus to the cytoplasm by another nucleic acid, RNA.

Proterozoic Eon eon of geological time, 3.5 billion to 570 million years ago, the second division of the Precambrian. It is defined as the time of simple life, since many rocks dating from this eon show traces of biological activity, and some contain the fossils of bacteria and algae.

proton (Greek 'first') a positively charged subatomic particle, a constituent of the nucleus of all atoms. It belongs to the baryon group of hadrons and is composed of two up quarks and one down quark. A proton is extremely long-lived, with a lifespan of at least 10^{32} years. It carries a unit positive charge equal to the negative charge of an electron. Its mass is almost 1,836 times that of an electron, or 1.673×10^{-24} g. The number of protons in the atom of an element is equal to the atomic number of that element.

proton–proton cycle process of nuclear fusion by which relatively cooler stars produce and radiate energy; hotter stars commonly achieve the same result by means of the carbon–nitrogen cycle.

proto-planet early stage in the formation of a planet according to the theory by which planetary systems evolve through the condensation of gas clouds surrounding a young star. The theory is not, however, generally accepted.

protoplasm contents of a living cell. Strictly speaking it includes all the discrete structures (organelles) in a cell, but it is often used simply to mean the jellylike material in which these float. The contents of a cell outside the nucleus are called cytoplasm.

protostar early formation of a star that has recently condensed out of an interstellar cloud and which is not yet hot enough for hydrogen burning to start. Protostars derive their energy from gravitational contraction.

protozoa group of single-celled organisms without rigid cell walls. Some, such as amoeba, ingest other cells, but most are saprotrophs or parasites. The group is polyphyletic (containing organisms which have different evolutionary origins).

pulmonary pertaining to the lungs.

pulsar celestial source that emits pulses of energy at regular intervals, ranging from a few seconds to a few thousandths of a second. Pulsars are thought to be rapidly rotating neutron stars, which flash at radio and other wavelengths as they spin. They were discovered in 1967 by Jocelyn Bell-Burnell and Antony Hewish at the Mullard Radio Astronomy Observatory, Cambridge, England. By 1998 1,000 pulsars had been discovered since the initial identification in 1967.

pulsating nova or *recurrent nova*, variable star, probably not a true nova, in which the change between more and less luminous stages is extreme.

pulsating universe or *oscillating universe*, theory that the universe constantly undergoes a Big Bang, expands, gradually slows and stops, contracts, and gradually accelerates once more to a Big Bang. Alternative theories include an ever expanding universe and the steady-state universe.

pyramid of numbers diagram that shows quantities of plants and animals at different levels of a food chain. This may be measured in terms of numbers (how many animals) or biomass (total mass of living matter), though in terms of showing transfer of food, biomass is a more useful measure. There is always far less biomass at the top of the chain than at the bottom, because only about 10% of the food an animal eats is turned into flesh – the rest is lost through metabolism and excretion. The amount of food flowing through the chain therefore drops with each step up the chain, hence the characteristic 'pyramid' shape.

pyrometer any instrument used for measuring high temperatures by means of the thermal radiation emitted by a hot object. In a **radiation pyrometer** the emitted radiation is detected by a sensor such as a thermocouple. In an **optical pyrometer** the colour of an electrically heated filament is matched visually to that of the emitted radiation. Pyrometers are especially useful for measuring the temperature of distant, moving, or inaccessible objects.

pyrophoric finely powdered metals, or mixtures of metals and their oxides, which have a tendency to burst into flame, or oxidize when exposed to air.

Pythagoras' theorem theorem stating that in a right-angled triangle, the square of the hypotenuse (the longest side) is equal to the sum of the squares of the other two sides. If the hypotenuse is h units long and the lengths of the other sides are a and b, then $h^2 = a^2 + b^2$.

quadrant type of early sextant, with which the observer's latitude could be calculated.

quadrat square structure used to study the distribution of plants in a particular place, for instance a field, rocky shore, or mountain-side. The size varies, but is usually 0.5 or 1 metre square, small enough to be carried easily. The quadrat is placed on the ground and the abundance of species estimated. By making such measurements a reliable understanding of species distribution is obtained.

quadratic equation polynomial equation of the second degree, that is, an equation containing as its highest power the square of a variable, such as x^2. The general formula of such equations is:

$$ax^2 + bx + c = 0$$

in which a, b, and c are real numbers, and only the coefficient a cannot equal 0.

In coordinate geometry, a quadratic function represents a parabola.

quadratic extension extension of a field to a larger field by adjoining the root of a quadratic equation; for example, the set of all $p + q\sqrt{2}$ with both p and q as rational numbers is a quadratic extension of the field of rationals.

quadrilateral plane (two-dimensional) figure with four straight sides. The sum of all interior angles is 360°. The following are all quadrilaterals, each with distinguishing properties: **square** with four equal angles and sides, and four axes of symmetry; **rectangle** with four equal angles, opposite sides equal, and two axes of symmetry; **rhombus** with four equal sides, and two axes of symmetry; **parallelogram** with two pairs of parallel sides, and rotational symmetry; **kite** with two pairs of adjacent equal sides, and one axis of symmetry; and **trapezium** with one pair of parallel sides.

qualitative analysis procedure for determining the identity of the component(s) of a single substance or mixture. A series of simple reactions and tests can be carried out on a compound to determine the elements present.

quantitative analysis a procedure for determining the precise amount of a known

507

component present in a single substance or mixture. A known amount of the substance is subjected to particular procedures.

Gravimetric analysis determines the mass of each constituent present; **volumetric analysis** determines the concentration of a solution by titration against a solution of known concentration.

quantum chromodynamics *QCD*, theory describing the interactions of quarks, the elementary particles that make up all hadrons (subatomic particles such as protons and neutrons). In quantum chromodynamics, quarks are considered to interact by exchanging particles called gluons, which carry the strong nuclear force, and whose role is to 'glue' quarks together.

quantum electrodynamics *QED*, theory describing the interaction of charged sub-atomic particles within electric and magnetic fields. It combines quantum theory and relativity, and considers charged particles to interact by the exchange of photons. QED is remarkable for the accuracy of its predictions; for example, it has been used to calculate the value of some physical quantities to an accuracy of ten decimal places, a feat equivalent to calculating the distance between New York and Los Angeles to within the thickness of a hair. The theory was developed by US physicists Richard Feynman and Julian Schwinger and by Japanese physicist Sin-Itiro Tomonaga in 1948.

quantum mechanics branch of physics dealing with the interaction of matter and radiation, the structure of the atom, the motion of atomic particles, and with related phenomena (see elementary particle and quantum theory).

quantum number one of a set of four numbers that uniquely characterize an electron and its state in an atom. The **principal quantum number** n defines the electron's main energy level. The **orbital quantum number** l relates to its angular momentum. The **magnetic quantum number** m describes the energies of electrons in a magnetic field. The **spin quantum number** m_s gives the spin direction of the electron.

quantum physics branch of physics that takes into account the quantum nature of matter, energy, and radiation.

quantum theory general mathematical theory based on Max Planck's discovery that radiant energy is quantized, that is, emitted in discrete quanta ('packets') of energy. The original theory has been extended to inter-pret a wide range of physical phenomena; for example, quantum mechanics and wave mechanics are now extensively used to give quantitative accounts of the behaviour of small particles, such as electrons.

quark elementary particle that is the funda-mental constituent of all hadrons (subatomic particles that experience the strong nuclear force and divided into baryons, such as neutrons and protons, and mesons). Quarks have electric charges that are fractions of the electronic charge ($+2/3$ or $-1/3$ of the electronic charge). There are six types, or 'flavours': up, down, top, bottom, strange, and charmed, each of which has three varieties, or 'colours': red, green, and blue (visual colour is not meant, although the analogy is useful in many ways). To each quark there is an antiparticle, called an antiquark. See quantum chromodynamics (QCD).

quartic to the power of four; it is occasionally replaced by the word 'biquadratic'.

quasar contraction of quasi-stellar object; or *QSO*, one of the most distant extragalactic objects known, discovered in 1963. Quasars appear starlike, but each emits more energy than 100 giant galaxies. They are thought to be at the centre of galaxies, their brilliance emanating from the stars and gas falling towards an immense black hole at their nucleus. Most quasars are found in elliptical galaxies.

quaternary ammonium ion an ion in which the hydrogen atoms of the normal ammonium ion $(NH_4)^+$ have been replaced by organic alkyl or aryl radicals; it therefore has the formula $(NR_4)^+$.

Quaternary period period of geological time from 1.64 million years ago through to the present. It is divided into the Pleistocene and Holocene epochs.

quaternion one of an extended system of complex numbers, representable in the generalized form:

$$a + bi + cj + dk$$

where a, b, c, and d are real numbers, and in which i, j, and k are additional objects that multiply according to specific rules such that:

$$i^2 = j^2 = k^2 = -1,$$

$$ij = -ji = k,$$

$$jk = -kj = i \text{ and}$$

$$ki = -ik = j.$$

quintic to the power of five.

quotient result after dividing one quantity by another.

raceme type of inflorescence.

racemic mixture mixture of equal quantities of two enantiomorphs (isomers with mirror-image molecular structures). Because the optical activity of each component exactly cancels that of the other, the racemic mixture as a whole is optically inactive.

radar acronym for *radio direction and ranging*, device for locating objects in space, direction finding, and navigation by means of transmitted and reflected high-frequency radio waves.

radial pulsation periodic expansion and contraction of a star that may be merely an optical effect of recession.

radial velocity velocity of an object, such as a star or galaxy, along the line of sight, moving towards or away from an observer. The amount of Doppler shift (apparent change in wavelength) of the light reveals the object's velocity. If the object is approaching, the Doppler effect causes a blue shift in its light. That is, the wavelengths of light coming from the object appear to be shorter, tending toward the blue end of the spectrum. If the object is receding, there is a red shift, meaning the wavelengths appear to be longer, toward the red end of the spectrum.

radiant point in the sky from which a meteor shower appears to emanate.

radiant heat energy that is radiated by all warm or hot bodies. It belongs to the infrared part of the electromagnetic spectrum and causes heating when absorbed. Radiant heat is invisible and should not be confused with the red glow associated with very hot objects, which belongs to the visible part of the spectrum.

radiation emission of radiant energy as particles or waves – for example, heat, light, alpha particles, and beta particles (see electromagnetic waves and radioactivity). See also atomic radiation.

radiative equilibrium in a star, represents an even process by which energy (heat) is transferred from the core to the outer surface without affecting the overall stability of the star.

radical in mathematics, operation principle in the extraction of roots, that is, square roots, cube roots, and so on.

radical in chemistry, a group of atoms forming part of a molecule, which acts as a unit and takes part in chemical reactions without disintegration, yet often cannot exist alone for any length of time, for example, the methyl radical $-CH_3$, or the carboxyl radical $-COOH$.

radio transmission of electromagnetic radiation of frequencies from about 10^4 hertz to 10^{11} hertz. Transmission of communication signals frequently involves the use of either amplitude- or frequency-modulated carrier waves.

radioactive decay process of disintegration undergone by the nuclei of radioactive elements, such as radium and various isotopes of uranium and the transuranic elements. This changes the element's atomic number, thus transmuting one element into another, and is accompanied by the emission of radiation. Alpha and beta decay are the most common forms.

radioactivity spontaneous alteration of the nuclei of radioactive atoms, accompanied by the emission of radiation. It is the property exhibited by the radioactive isotopes of stable elements and all isotopes of radioactive elements, and can be either natural or induced. See radioactive decay.

radioimmunoassay *RIA*, technique for measuring small quantities of circulating hormones. The assay depends upon the ability of a hormone to inhibit the binding of the same hormone (which has been labelled with a radioactive isotope) to a specific antibody by competition for the binding sites.

radio interferometer type of radio telescope that relies on the use of two or more aerials at a distance from each other to provide a combination of signals from one source which can be analysed by computer. Such an analysis results in a resolution that is considerably better than that of a parabolic dish aerial by itself because of the greater effective diameter.

radioisotope or *radioactive isotope*, naturally occurring or synthetic radioactive form of an element. Most radioisotopes are made by bombarding a stable element with neutrons in the core of a nuclear reactor (see fission). The radiations given off by radioisotopes are easy to detect (hence their use as tracers), can in some instances penetrate substantial thicknesses of materials, and have profound effects (such as genetic mutation) on living matter.

radiometer device for the detection (and also measurement) of radiant electromagnetic radiation.

radiomicrometer an extremely sensitive detector of infrared radiation.

radio scintillation scintillation in received radio emission; the equivalent of 'twinkling' in visible light from the stars.

radiotherapy treatment of disease by radiation from X-ray machines or radioactive sources.

Radiation, which reduces the activity of dividing cells, is of special value for its effect on malignant tissues, certain nonmalignant tumours, and some diseases of the skin.

radio wave electromagnetic wave possessing a long wavelength (ranging from about 10^{-3} to 10^4 m) and a low frequency (from about 10^5 to 10^{11} Hz). Included in the radio wave part of the spectrum are microwaves, used for both communications and for cooking; ultra high- and very high-frequency waves, used for television and FM (frequency modulation) radio communications; and short, medium, and long waves, used for AM (amplitude modulation) radio communications. Radio waves that are used for communications have all been modulated to carry information. Certain astronomical objects emit radio waves, which may be detected and studied using radio telescopes.

radius straight line from the centre of a circle to its circumference, or from the centre to the surface of a sphere.

radius vector imaginary line connecting the centre of an orbiting body with the centre of the body (or point) that it is orbiting.

Raman effect change in the wavelength of light scattered by molecules.

rare-earth element alternative name for lanthanide.

rational number any number that can be expressed as an exact fraction (with a denominator not equal to 0), that is, as $a \div b$ where a and b are integers; or an exact decimal. For example, $2/1$, $1/4$, $15/4$, $-3/5$ are all rational numbers, whereas π (which represents the constant $3.141592\ldots$) is not. Numbers such as π are called irrational numbers.

Rayleigh–Jeans law formula giving the intensity of black-body radiation at long wavelengths for a radiator at a certain temperature. It is thus an approximation to Planck's full formula for the black-body intensity based on quantum concepts.

reaction coming together of two or more atoms, ions, or molecules with the result that a chemical change takes place, that is, a change that occurs when two or more substances interact with each other, resulting in the production of different substances with different chemical compositions. The nature of the reaction is portrayed by a chemical equation.

real number any of the rational numbers (which include the integers) or irrational numbers. Real numbers exclude imaginary numbers, found in complex numbers of the general form $a + b$i where $i = \sqrt{-1}$, although these do include a real component a.

recursive very general description of a function with natural numbers, as arguments, corresponding to the intuitive notion of computability. The basic recursive functions are $x + y$, xy, $f(x, y, z, \ldots) = x$, $g(x, y, z, \ldots) = y$, and a restricted form of the taking of a minimum.

red giant any large bright star with a cool surface. It is thought to represent a late stage in the evolution of a star like the Sun, as it runs out of hydrogen fuel at its centre and begins to burn heavier elements, such as helium, carbon, and silicon. Because of more complex nuclear reactions that then occur in the red giant's interior, it eventually becomes gravitationally unstable and begins to collapse and heat up. The result is either explosion of the star as a supernova, leaving behind a neutron star, or loss of mass by more gradual means to produce a white dwarf.

red shift lengthening of the wavelengths of light from an object as a result of the object's motion away from an observer. It is an example of the Doppler effect. The red shift in light from galaxies is evidence that the universe is expanding.

reduction gain of electrons, loss of oxygen, or gain of hydrogen by an atom, ion, or molecule during a chemical reaction.

reflecting telescope telescope in which light is collected and brought to a focus by a concave mirror. Cassegrain and Newtonian telescopes are examples.

reflection throwing back or deflection of waves, such as light or sound waves, when they hit a surface. The **law of reflection** states that the angle of incidence (the angle between the ray and a perpendicular line drawn to the surface) is equal to the angle of reflection (the angle between the reflected ray and a perpendicular to the surface).

reflection, laws of incident ray of light, the reflected ray of light, and the normal to the reflecting surface all lie in the same plane. Secondly, the angle between the incident ray and the normal is the same as that between the reflected ray and the normal to the surface.

reflex in animals, a very rapid involuntary response to a particular stimulus. It is controlled by the nervous system. A reflex involves only a few nerve cells, unlike the slower but more complex responses produced by the many processing nerve cells of the brain.

reflex action involuntary response to a stimulus.

reflex arc route of nervous impulses from the point of stimulation, along sensory nerve fibres to the central nervous system, and back along motor nerve fibres to the effector organ or muscle.

refracting telescope or **refractor**, telescope that uses lenses to magnify and focus an image onto an eyepiece.

refraction bending of a wave when it passes from one medium into another. It is the effect of the different speeds of wave propagation in two substances that have different densities. The amount of refraction depends on the densities of the media, the angle at which the wave strikes the surface of the second medium, and the amount of bending and change of velocity corresponding to the wave's frequency (dispersion). Refraction occurs with all types of progressive waves – electromagnetic waves, sound waves, and water waves – and differs from reflection, which involves no change in velocity.

refraction, laws of incident ray, the refracted ray, and the normal to the surface between the two media at the point of incidence lie in the same plane. In addition, Snell's law states that the ratio of the sines of the angles of incidence and refraction is equal to a constant. See refractive index.

refractive index measure of the refraction of a ray of light as it passes from one transparent medium to another. If the angle of incidence is i and the angle of refraction is r, the ratio of the two refractive indices n_1 and n_2 is given by $n_1/n_2 = \sin i/\sin r$. It is also equal to the speed of light in the first medium divided by the speed of light in the second, and it varies with the wavelength of the light.

refractive material materials able to withstand very high temperature, such as brick and concrete.

refractometer device for determining the refractive index of a substance.

relative atomic mass mass of an atom relative to one-twelfth the mass of an atom of carbon-12. It depends primarily on the number of protons and neutrons in the atom, the electrons having negligible mass. If more than one isotope of the element is present, the relative atomic mass is calculated by taking an average that takes account of the relative proportions of each isotope, resulting in values that are not whole numbers. The term **atomic weight**, although commonly used, is strictly speaking incorrect.

relative molecular mass mass of a molecule, calculated relative to one-twelfth the mass of an atom of carbon-12. It is found by adding the relative atomic masses of the atoms that make up the molecule. The term **molecular weight** is often used, but strictly this is incorrect.

relativity theory of the relative rather than absolute character of mass, time, and space, and their interdependence, as developed by German-born US physicist Albert Einstein in two phases.   **special theory of relativity** (1905) Starting with the premises that (1) the laws of nature are the same for all observers in unaccelerated motion, and (2) the speed of light is independent of the motion of its source, Einstein arrived at some rather unexpected consequences. Intuitively familiar concepts, like mass, length, and time, had to be modified. For example, an object moving rapidly past the observer will appear to be both shorter and more massive than when it is at rest (that is, at rest relative to the observer), and a clock moving rapidly past the observer will appear to be running slower than when it is at rest.

These predictions of relativity theory seem to be foreign to everyday experience merely because the changes are negligible at speeds less than about 1,500 kps/930 mps, and they only become appreciable at speeds approaching the speed of light.   **general theory of relativity** (1915) The geometrical properties of space-time were to be conceived as modified locally by the presence of a body with mass. A planet's orbit around the Sun (as observed in three-dimensional space) arises from its natural trajectory in modified space-time. Einstein's general theory accounts for a peculiarity in the behaviour of the motion of the perihelion of the orbit of the planet Mercury that cannot be explained in Newton's theory. The new theory also said that light rays should bend when they pass by a massive object. The predicted bending of starlight was observed during the eclipse of the Sun in 1919. A third corroboration is found in the shift towards the red in the spectra of the Sun and, in particular, of stars of great density – white dwarfs such as the companion of Sirius.

replication production of copies of the genetic material DNA; it occurs during cell division (mitosis and meiosis). Most mutations are caused by mistakes during replication.

repressor substance substance produced by a DNA regulator gene. When the repressor is inactivated by an inducer, the DNA structural genes are freed for the synthesis of messenger RNA.

511

resistance that property of a conductor that restricts the flow of electricity through it, associated with the conversion of electrical energy to heat; also the magnitude of this property. Resistance depends on many factors, such as the nature of the material, its temperature, dimensions, and thermal properties; degree of impurity; the nature and state of illumination of the surface; and the frequency and magnitude of the current. The SI unit of resistance is the ohm. The statement that current is proportional to voltage (resistance is constant) at constant temperature is known as Ohm's law. It is approximately true for many materials that are accordingly described as 'ohmic'.

resolution of a telescope, the clarity of the final presentation to the observer (in image, radio picture, or X-ray read-out).

resolving power ability of an optical device to discern two closely spaced light sources as independent entities.

resonance concept that, in certain molecules, the electrons involved in linking the constituent atoms are not associated with a specific bond (or bonds) but oscillate between atoms. Thus such molecules are not represented by a single valence-bond structure but by two or more alternative structures; the molecule 'resonates' between these alternative structures – that is, its structure is a resonance hybrid of the alternatives. The best known example of resonance is benzene, which, according to Kekulé's original formulation, resonates between two forms in which the double and single bonds are transposed. In Robinson's later modification, the six carbon atoms are linked by single bonds and the extra electrons are distributed equally among the carbon atoms, this being represented diagrammatically by a circle within the hexagonal carbon ring.

respiration metabolic process in organisms in which food molecules are broken down to release energy. The cells of all living organisms need a continuous supply of energy, and in most plants and animals this is obtained by **aerobic** respiration. In this process, oxygen is used to break down the glucose molecules in food. This releases energy in the form of energy-carrying molecules (ATP), and produces carbon dioxide and water as by-products. Respiration sometimes occurs without oxygen, and this is called **anaerobic** respiration. In this case, the end products are energy and either lactose acid or ethanol (alcohol) and carbon dioxide; this process is termed fermentation.

resting potential potential difference across a nerve or muscle membrane in the absence of a stimulus.

reticulo-endothelial system group of cells that exist in continual contact with the blood and lymph, that is, in the bone marrow, spleen, liver, and lymph nodes. They ingest bacteria, other foreign particles and dead tissue, and aid tissue repair.

retrograde orbit or rotation of a planet or satellite if the sense of rotation is opposite to the general sense of rotation of the Solar System. On the celestial sphere, it refers to motion from east to west against the background of stars.

reversing layer lower chromosphere of the Sun, a comparatively cool region in which radiation at certain wavelengths is absorbed from the continuous spectrum emitted from the Sun's photosphere.

ribonucleic acid full name of RNA.

ribosome protein-making machinery of the cell. Ribosomes are located on the endoplasmic reticulum (ER) of eukaryotic cells, and are made of proteins and a special type of RNA, ribosomal RNA. They receive messenger RNA (copied from the DNA) and amino acids, and 'translate' the messenger RNA by using its chemically coded instructions to link amino acids in a specific order, to make a strand of a particular protein.

Riemann geometry system of non-Euclidean geometry devised by Bernhard Riemann, developed primarily as elliptic geometry, but then extended to hyperbolic geometry.

Riemann hypothesis statement that has not as yet been proved (or disproved), that the zeta function takes the value zero in the right-half plane of the Argand diagram only for complex numbers of the form $1/2 + ia$, where $i = \sqrt{-1}$ and a is real.

Riemann space non-Euclidean geometry, using n-dimensional coordinates $(x_1, \ldots, x_n)$ and calculating length according to the formula:

$$ds^2 = \sum g_{ij}\, dx^i\, dx^{ij}$$

where ds is the limiting incremental length along a curve, dx^i is 'a limiting increment in the i coordinate', and i, j run through the values $1, 2, 3, \ldots, n$.

Riemann surface multi-connected, many-sheeted surface that can be dissected by cross-cuts into a singly connected surface. Such a representation of a complex algebraic function is used to study the 'behaviour' of other complex functions as they are mapped conformally (transformed) onto it. A

Riemann surface has been described as topologically equivalent to a box with holes in it.

right ascension coordinate on the celestial sphere that corresponds to longitude on the surface of the Earth. It is measured in hours, minutes, and seconds eastwards from the point where the Sun's path, the ecliptic, once a year intersects the celestial equator; this point is called the **vernal equinox**.

ring mathematical structure that constitutes a restricted form of field in which division might be unavailable.

RNA abbreviation for *ribonucleic acid*, nucleic acid involved in the process of translating the genetic material DNA into proteins. It is usually single-stranded, unlike the double-stranded DNA, and consists of a large number of nucleotides strung together, each of which comprises the sugar ribose, a phosphate group, and one of four bases (uracil, cytosine, adenine, or guanine). RNA is copied from DNA by the formation of base pairs, with uracil taking the place of thymine.

root of an equation, a value that satisfies the equality. For example, $x = 0$ and $x = 5$ are roots of the equation $x^2 - 5x = 0$.

rose symmetrical curve represented by the equation:

$$(x^2 + y^2)^3 = 4a^2x^2y^2$$

where a is constant.

rotation transformation in which a figure is turned about a given point, known as the **centre of rotation**. A rotation of 180° is known as a half turn.

Rydberg constant constant that relates atomic spectra to the spectrum of hydrogen. Its value is 1.0977×10^7 per metre.

saccharide another name for a sugar molecule.

salt any member of a group of compounds containing a positive ion from a metal or ammonica and a negative ion from an acid. **Common salt** is sodium chloride.

saltpetre former name for potassium nitrate (KNO_3), the compound used in making gunpowder (from about 1500). It occurs naturally, being deposited during dry periods in places with warm climates, such as India.

saponification hydrolysis (splitting) of an ester by treatment with a strong alkali, resulting in the liberation of the alcohol from which the ester had been derived and a salt of the constituent fatty acid. The process is used in the manufacture of soap.

satellite any small body that orbits a larger one, either natural or artificial. Natural satellites that orbit planets are called moons. The first **artificial satellite**, *Sputnik 1*, was launched into orbit around the Earth by the USSR in 1957. Artificial satellites are used for scientific purposes, communications, weather forecasting, and military applications. The brightest artificial satellites can be seen by the naked eye.

scalar quantity quantity that has magnitude but no direction, as distinct from a vector quantity, which has a direction as well as a magnitude. Temperature, mass, and volume are scalar quantities.

Schrödinger's equation equation that considers the electron in terms of a wave of probability, and which enables the behaviour of the electron in atoms and electric potentials to be calculated, and also the spectra of atoms to be predicted. It is the basis of wave mechanics.

scintillation rapid oscillation in the detected intensity of radiation emitted by stellar radio sources, caused by disturbances in ionized gas at some point between the source and the Earth's surface (usually in the Earth's own upper atmosphere).

seafloor spreading growth of the ocean crust outwards (sideways) from ocean ridges. The concept of seafloor spreading has been combined with that of continental drift and incorporated into plate tectonics.

secant *sec*, function of a given angle in a right-angled triangle obtained by dividing the length of the hypotenuse (the longest side) by the length of the side adjacent to the angle. It is the reciprocal of the cosine.

second order of a differential equation, involving only first and second derivatives. The term is also used in algebraic contexts to mean 'of the second degree', that is, involving expressions raised to at most the power of two (squares).

secular gradual, taking aeons to accomplish.

secular acceleration of the Moon, of the Sun apparent acceleration of the Moon and Sun across the sky, caused by extremely gradual reduction in speed of the Earth's rotation (one 50-millionth of a second per day).

sedimentary rock rock formed by the accumulation and cementation of deposits that have been laid down by water, wind, ice, or gravity. Sedimentary rocks cover more than two-thirds of the Earth's surface and comprise three major categories: clastic, chemically precipitated, and organic (or biogenic). Clastic sediments are the largest group and are composed of fragments of pre-existing rocks; they include clays, sands, and gravels.

Chemical precipitates include some limestones and evaporated deposits such as

513

gypsum and halite (rock salt). Coal, oil shale, and limestone made of fossil material are examples of organic sedimentary rocks.

seismology study of earthquakes, the seismic waves they produce, the processes that cause them, and the effects they have. By examining the global pattern of waves produced by an earthquake, seismologists can deduce the nature of the materials through which they have passed. This leads to an understanding of the Earth's internal structure.

selenium (Greek *Selene* 'Moon') grey, nonmetallic element, symbol Se, atomic number 34, relative atomic mass 78.96. It belongs to the sulphur group and occurs in several allotropic forms that differ in their physical and chemical properties. It is an essential trace element in human nutrition.

Obtained from many sulphide ores and selenides, it is used as a red colouring for glass and enamel.

self-inductance or *self-induction*, creation of an electromotive force opposing the current. See inductance.

self-pollination pollination of a plant by itself, whether intra- or interfloral.

semiconductor material with electrical conductivity intermediate between metals and insulators and used in a wide range of electronic devices. Certain crystalline materials, most notably silicon and germanium, have a small number of free electrons that have escaped from the bonds between the atoms. The atoms from which they have escaped possess vacancies, called holes, which are similarly able to move from atom to atom and can be regarded as positive charges. Current can be carried by both electrons (negative carriers) and holes (positive carriers). Such materials are known as **intrinsic semiconductors**.

semipermeable membrane membrane that allows certain substances in solution, such as crystalloids, to pass through it but is impervious to others, such as colloids. Semipermeable membranes are used in dialysis.

sensory nerve afferent nerve of the peripheral nervous system, made up of sensory neurons, which carries impulses to the central nervous system.

sepal part of a flower, usually green, that surrounds and protects the flower in bud. The sepals are derived from modified leaves, and are collectively known as the calyx.

sequence list of mathematical objects indexed by the natural numbers, following one another in some defined relationship (but with no mathematical operation implied). Sequences are said to increase (to higher values) or decrease (to lower), and may be finite (if the list terminates), convergent (to a limit), or divergent. If the values increase or decrease along the sequence, the sequence is said to be monotonic or monotone.

sequencing in biochemistry, determining the sequence of chemical subunits within a large molecule. Techniques for sequencing amino acids in proteins were established in the 1950s, insulin being the first for which the sequence was completed. The Human Genome Project was set up to attempt to determine the sequence of the 3 billion base pairs within human DNA.

sere plant succession developing in a particular habitat. A **lithosere** is a succession starting on the surface of bare rock. A **hydrosere** is a succession in shallow freshwater, beginning with planktonic vegetation and the growth of pondweeds and other aquatic plants, and ending with the development of swamp. A **plagiosere** is the sequence of communities that follows the clearing of the existing vegetation.

series sum of a list that constitutes a sequence. Series may be represented in an abbreviated form using the summation sign – for example, the series $a_1 + a_2 + a_3 + \ldots + a_n$ to infinity may be represented as:

$$\sum_{k=1}^{\infty} a_k$$

The word 'series' is for historical reasons occasionally misused to mean 'sequence', as in the Fibonacci series. Strictly speaking, however, a series is the limit of the sequence of partial sums.

set or *class*, any collection of defined things (elements), provided the elements are distinct and that there is a rule to decide whether an element is a member of a set. It is usually denoted by a capital letter and indicated by curly brackets { }.

sextic to the power of six.

shooting star another name for a meteor.

sidereal period orbital period of a planet around the Sun, or a moon around a planet, with reference to a background star. The sidereal period of a planet is in effect a 'year' for that planet. A synodic period is a full circle as seen from Earth.

sidereal time time measured by the rotation of the Earth with respect to the stars. A sidereal day is the time taken by the Earth to turn once with respect to the stars, namely 23 h 56 min 4 s. It is divided into sidereal hours, minutes, and seconds, each of which is proportionally shorter than the corresponding SI unit.

sigma Σ, summation symbol.

sigma bond type of chemical bond in which an electron pair (regarded as being shared by the two atoms involved in the bond) occupies a molecular orbital situated between the two atoms; the orbital is located along a hypothetical line linking the atoms' nuclei.

silicon (Latin *silex* 'flint') brittle, nonmetallic element, symbol Si, atomic number 14, relative atomic mass 28.086. It is the second-most abundant element (after oxygen) in the Earth's crust and occurs in amorphous and crystalline forms. In nature it is found only in combination with other elements, chiefly with oxygen in silica (silicon dioxide, SiO_2) and the silicates. These form the mineral quartz, which makes up most sands, gravels, and beaches.

Silurian Period period of geological time 439–409 million years ago, the third period of the Palaeozoic era. Silurian sediments are mostly marine and consist of shales and limestone. Luxuriant reefs were built by coral-like organisms. The first land plants began to evolve during this period, and there were many ostracoderms (armoured jawless fishes). The first jawed fishes (called acanthodians) also appeared.

silver nitrate white, crystalline substance used in chemical analysis and inks.

sine function of an angle in a right-angled triangle which is defined as the ratio of the length of the side opposite the angle to the length of the hypotenuse (the longest side).

single phase electrical power transmission involving a single sinusoidally varying potential difference.

singularity point in space-time at which the known laws of physics break down. Singularity is predicted to exist at the centre of a black hole, where infinite gravitational forces compress the infalling mass of a collapsing star to infinite density. It is also thought, according to the Big Bang model of the origin of the universe, to be the point from which the expansion of the universe began.

SI units French *Système International d'Unités*, standard system of scientific units used by scientists worldwide. Originally proposed in 1960, it replaces the m.k.s., c.g.s., and f.p.s. systems. It is based on seven basic units: the metre (m) for length, kilogram (kg) for mass, second (s) for time, ampere (A) for electrical current, kelvin (K) for temperature, mole (mol) for amount of substance, and candela (cd) for luminosity.

smoke suspension of a solid in a gas; the solid is in the form of extremely small particles and the smoke may be a colloid.

Snel's law of refraction rule that when a ray of light passes from one medium to another, the sine of the angle of incidence divided by the sine of the angle of refraction is equal to the ratio of the indices of refraction in the two media. For a ray passing from medium 1 to medium 2: $n_2/n_1 = \sin i/\sin r$ where n_1 and n_2 are the refractive indices of the two media. The law was devised by the Dutch physicist Willebrord Snel.

sodium pump hypothetical mechanism that maintains the asymmetry of the ionic (concentration) balance across a nerve cell membrane, reflected in the cell's resting potential.

sodium theory theory which proposes that the excitation of a nerve results from momentary changes in the selective permeability of a nerve cell membrane, which admits sodium and chloride ions into the cell and allows potassium ions to diffuse out. The ion movements briefly reverse the polarization of the membrane, resulting in an action potential which constitutes the nerve impulse.

solar constant mean radiation received from the Sun at the top level of Earth's atmosphere: 8.15 joules per sq cm per minute.

solar energy energy produced by nuclear fusion and comprising almost entirely electromagnetic radiation (particularly in the form of light and heat); particles are also radiated forming the solar wind.

solar flare sudden and dramatic release of a huge burst of solar energy through a break in the Sun's chromosphere in the region of a sunspot. Effects on Earth include auroras, magnetic storms, and radio interference.

solar parallax parallax of the Sun, now measured as 8.794 in.

solar prominence mass of hot hydrogen rising from the Sun's chromosphere, best observed during a total eclipse. Eruptive prominences are violent in force and may reach heights of 2,000,000 km/1,243,000 mi; quiescent prominences are relatively pacific but may last for months.

solar rotation rotation of the Sun about its axis. It is differential, the equatorial rotation taking less time than the polar by up to 9.4 Earth-days.

Solar System the Sun (a star) and all the bodies orbiting it: the nine planets (Mercury, Venus, Earth, Mars, Jupiter, Saturn, Uranus, Neptune, and Pluto), their moons, the asteroids, and the comets. The Sun contains 99.86% of the mass of the Solar System.

solar wind stream of atomic particles, mostly protons and electrons, from the Sun's corona, flowing outwards at speeds of between 300 kps/200 mps and 1,000 kps/600 mps.

solder alloy that is heated and used to join two metals together.

solenoid device consisting of a series of wires wound around a cylinder, which produces a magnetic field within the cylinder when a current flows through the wire windings. The intensity of the magnetic field depends directly on the number of turns of the wire.

solid state of matter that holds its own shape (as opposed to a liquid, which takes up the shape of its container, or a gas, which totally fills its container). According to kinetic theory, the atoms or molecules in a solid are not free to move but merely vibrate about fixed positions, such as those in crystal lattices.

solid-state physics study of materials in the solid state, investigating the magnetic, thermal, and electrical properties, for example.

solstice either of the days on which the Sun is farthest north or south of the celestial equator each year. The **summer solstice**, when the Sun is farthest north, occurs around 21 June; the **winter solstice** around 22 December.

soluble group notion introduced to extend theorems concerning commutative groups to a wider class of groups that, in an intuitive sense, can be constructed out of commutative 'pieces'. Given a group G the commutator subgroup C is introduced (generated by elements of the form $xyx^{-1}y^{-1}$); G can then be 'collapsed' to an Abelian group by a process that reduces C to the identity element; C is then collapsed in an identical fashion, and the process continually repeated. If the process, through a finite number of steps, eventually leads to a last group that is already commutative, G is said to be soluble.

solution two or more substances mixed to form a single, homogenous phase. One of the substances is the **solvent** and the others (**solutes**) are said to be dissolved in it.

solvent substance, usually a liquid, that will dissolve another substance (see solution). Although the commonest solvent is water, in popular use the term refers to low-boiling-point organic liquids, which are harmful if used in a confined space. They can give rise to respiratory problems, liver damage, and neurological complaints.

somatic cell any cell in an organism, excluding the reproductive cells.

sound physiological sensation received by the ear, originating in a vibration that communicates itself as a pressure variation in the air and travels in every direction, spreading out as an expanding sphere. All sound waves in air travel with a speed dependent on the temperature; under ordinary conditions, this is about 330 m/1,070 ft per second. The pitch of the sound depends on the number of vibrations imposed on the air per second (frequency), but the speed is unaffected. The loudness of a sound is dependent primarily on the amplitude of the vibration of the air.

space collection of mathematical objects (referred to as points) with an associated structure resembling (or analogous to) the properties of the space of everyday experience.

space constant term characterizing the line in non-Euclidean geometry that relates the angle of parallelism of two lines – a concept formulated by Hungarian mathematician János Bolyai.

space-time combination of space and time used in the theory of relativity. When developing relativity, Albert Einstein showed that time was in many respects like an extra dimension (or direction) to space. Space and time can thus be considered as entwined into a single entity, rather than two separate things.

speciation formation and development of species.

species distinguishable group of organisms that resemble each other or consist of a few distinctive types (as in polymorphism), and that can all interbreed to produce fertile offspring. Species are the lowest level in the system of biological classification.

specific heat capacity quantity of heat required to raise unit mass (1 kg) of a substance by one kelvin (1 K). The unit of specific heat capacity in the SI system is the joule per kilogram per kelvin ($J\,kg^{-1}\,K^{-1}$).

spectral classification classification of stars according to their surface temperature and luminosity, as determined from their spectra. Stars are assigned a spectral type (or class) denoted by the letters O, B, A, F, G, K, and M, where O stars (about 40,000 K/ 39,700°C/71,500°F) are the hottest and M stars (about 3,000 K/2,700°C/4,900°F) are the coolest.

spectral line dark line visible in an absorption spectrum, or one of the bright lines that make up an emission spectrum. Spectral lines are caused by the transference of an electron in an atom from one energy level to another; strong lines are produced at levels at which such transference occurs easily, weak where it occurs with difficulty. Ionization of certain elements can affect such transferences and cause problems in spectral analysis.

spectroheliograph device with which spectra of the various regions of the Sun are obtained and photographed.

spectroscope instrument that produces a spectrum for study or analysis. An object that produces radiation, such as a heated substance, forms an emission spectrum. Elements have characteristic spectra and spectroscopy is used in chemical analysis to identify the elements in a substance or mixture. Molecules or their constituent atoms or components of atoms can be made to absorb various types of energy in a characteristic way and give rise to such analytical techniques as infrared, ultraviolet, X-ray, and nuclear magnetic resonance spectroscopy. See also mass spectroscope.

spectroscopy study of spectra (see spectrum) associated with atoms or molecules in solid, liquid, or gaseous phase. Spectroscopy can be used to identify unknown compounds and is an invaluable tool in science, medicine, and industry (for example, in checking the purity of drugs).

spectrum plural *spectra*, pattern of frequencies or wavelengths obtained when electromagnetic radiations are separated into their constituent parts. Visible light is part of the electromagnetic spectrum and most sources emit waves over a range of wavelengths that can be broken up or 'dispersed'; white light can be separated into red, orange, yellow, green, blue, indigo, and violet.

speed of light speed at which light and other electromagnetic waves travel through empty space. Its value is 299,792,458 m / 186,282 mi per second. The speed of light is the highest speed possible, according to the theory of relativity, and its value is independent of the motion of its source and of the observer. It is impossible to accelerate any material body to this speed because it would require an infinite amount of energy.

spherical aberration optical error occurring when a lens or curved mirror does not bring all the incident rays of light to a sharp focus. See also aberration.

spherical collapse initial stage in the collapse of a star, followed by gravitational collapse and finally singularity.

spherical geometry system of non-Euclidean geometry devised by German mathematician

Bernhard Riemann as an extension of elliptic geometry and comprising two-dimensional geometry as effected on the outer surface of a sphere.

spin intrinsic angular momentum of a subatomic particle, nucleus, atom, or molecule, which continues to exist even when the particle comes to rest. A particle in a specific energy state has a particular spin, just as it has a particular electric charge and mass. According to quantum theory, this is restricted to discrete and indivisible values, specified by a spin quantum number. Because of its spin, a charged particle acts as a small magnet and is affected by magnetic fields.

spin-orbit interaction interaction between electrons orbiting a nucleus in an atom that arises from the magnetic field produced by the nucleus interacting with the spinning electron. The effect is to split the individual spectral lines in a spectrum into a number of components.

spiral nebula spiral galaxy – not really a nebula at all (although many do appear nebulous).

spirillium bacterium with a spiral shape (see bacteria).

spore small reproductive or resting body, usually consisting of just one cell. Unlike a gamete, it does not need to fuse with another cell in order to develop into a new organism. Spores are produced by the lower plants, most fungi, some bacteria, and certain protozoa. They are generally light and easily dispersed by wind movements. Plant spores are haploid and are produced by the sporophyte, following meiosis.

sporozoite protozoon of the class Sporozoa, such as the malarial parasite *Plasmodium*.

squaring the circle ancient Greek problem in geometrical construction, to describe a square of exactly the same area as a given circle, using ruler and compass only. German mathematician Ferdinand von Lindemann established that π was a transcendental number; unable thus to be the root of an equation, it cannot be constructed by ruler and compass – and the problem is therefore not solvable.

stain coloured compound that will bind to other substances. Stains are used extensively in microbiology to colour micro-organisms and in histochemistry to detect the presence and whereabouts in plant and animal tissue of substances such as fats, cellulose, and proteins.

standard deviation measure (symbol σ or s) of the spread of data. The deviation (difference)

517

of each of the data items from the mean is found, and their values squared. The mean value of these squares is then calculated. The standard deviation is the square root of this mean.

standard gauge distance between the steel rails on the majority of the world's railways: 4 ft 8 in/1.4224 m.

standing crop total number of individuals of a given species alive in a particular area at any moment. It is sometimes measured as the weight (or biomass) of a given species in a sample section.

Stark effect splitting of spectral lines into a number of components by a strong electric field.

starlight energy (seen as light) produced by a star through nuclear fusion.

statics branch of mechanics concerned with the behaviour of bodies at rest and forces in equilibrium, and distinguished from dynamics.

Steady-Stake theory rival theory to that of the Big Bang, which claims that the universe has no origin, but is expanding because new matter is being created continuously thourghout the universe.

steam turbine turbine that is powered by a jet of high-pressure steam.

Stefan–Boltzmann law law that relates the energy, E, radiated away from a perfect emitter (a black body), to the temperature, T, of that body. It has the form $E = \sigma T^4$, where E is the energy radiated per unit area per second, T is the temperature, and σ is the **Stefan–Boltzmann constant**. Its value is $5.6697 \times 10^{-8}\,\mathrm{W\,m^{-2}\,K^{-4}}$. The law was derived by the Austrian physicists Josef Stefan and Ludwig Boltzmann.

stellar of a star, of the stars.

stereochemistry branch of chemistry that is concerned with a study of the shapes of molecules.

stereoisomerism type of isomerism in which two or more substances differ only in the way that the atoms of their molecules are oriented in space.

stereoscope device that can produce the effect of three-dimensional images using only two-dimensional images, using the human brain to carry out the merging necessary.

steroid any of a group of cyclic, unsaturated alcohols (lipids without fatty acid components), which, like sterols, have a complex molecular structure consisting of four carbon rings. Steroids include the sex hormones, such as testosterone, the corticosteroid hormones produced by the adrenal gland, bile acids, and cholesterol.

strain extent to which a body is distorted when a deforming force (stress) is applied to it. It is a ratio of the extension or compression of that body (its length, area, or volume) to its original dimensions (see Hooke's law). For example, linear strain is the ratio of the change in length of a body to its original length.

strangeness quantum number assigned to certain unstable elementary particles that decay much more slowly than was originally expected. Stable particles, such as protons, have a strangeness quantum number of zero. Others, such as the hyperons, have nonzero quantum numbers.

stratigraphy branch of geology that deals with sedimentary rock layers (strata) and their sequence of formation. The basic principle of superposition establishes that upper layers or deposits accumulated later in time than the lower ones.

streamline line in a fluid such that the tangent to it at every point gives the direction of flow, and its speed, at any instant.

streptococcus any one of a genus of round or oval Gram-positive bacteria that have a tendency to form pairs or chains. They are widely distributed in nature, living mainly as parasites in the bodies of animals and humans. Some are harmless, but others are implicated in a number of infections, including scarlet fever.

stress and strain measures of the deforming force applied to a body (stress) and of the resulting change in its shape (strain). For a perfectly elastic material, stress is proportional to strain (Hooke's law).

Strömgren sphere or *H II zone*, zone of ionized hydrogen gas that surrounds hot stars embedded in interstellar gas clouds.

strong nuclear force one of the four fundamental forces of nature, the other three being the gravitational force or gravity, the electromagnetic force, and the weak nuclear force. The strong nuclear force was first described by the Japanese physicist Hideki Yukawa in 1935. It is the strongest of all the forces, acts only over very small distances within the nucleus of the atom (10^{-13} cm), and is responsible for binding together quarks to form hadrons, and for binding together protons and neutrons in the atomic nucleus. The particle that is the carrier of the strong nuclear force is the gluon, of which there are eight kinds, each with zero mass and zero charge.

subduction zone region where two plates of the Earth's rigid lithosphere collide, and one plate descends below the other into the weaker asthenosphere. Subduction results in the formation of ocean trenches, most of which encircle the Pacific Ocean.

substitution reaction replacement of one atom or functional group in an organic molecule by another.

substrate compound or mixture of compounds acted on by an enzyme. The term also refers to a substance such as agar that provides the nutrients for the metabolism of micro-organisms. Since the enzyme systems of micro-organisms regulate their metabolism, the essential meaning is the same.

succession series of changes that occur in the structure and composition of the vegetation in a given area from the time it is first colonized by plants (**primary succession**), or after it has been disturbed by fire, flood, or clearing (**secondary succession**).

summation symbol sign (Σ) representing the sum taken over all instances that accompany the sign, indicated above and below it. Thus:

$$\sum_{i=1}^{n} a_i$$

means $a_1 + a_2 + a_3 + \ldots + a_n$. The symbol is the capital form of the Greek letter sigma.

sunspot dark patch on the surface of the Sun, actually an area of cooler gas, thought to be caused by strong magnetic fields that block the outward flow of heat to the Sun's surface. Sunspots consist of a dark central **umbra**, about 4,000 K (3,700°C/6,700°F), and a lighter surrounding **penumbra**, about 5,500 K (5,200°C/9,400°F). They last from several days to over a month, ranging in size from 2,000 km/1,250 mi to groups stretching for over 100,000 km/62,000 mi.

supercluster grouping of several clusters of galaxies to form a structure about 100–300 million light years across. Our own Galaxy and its neighbours lie on the edge of the local supercluster of which the Virgo cluster is the dominant member.

superconductivity increase in electrical conductivity at low temperatures. The resistance of some metals and metallic compounds decreases uniformly with decreasing temperature until at a critical temperature (the superconducting point), within a few degrees of absolute zero (0 K/ −273.15°C/−459.67°F), the resistance suddenly falls to zero. The phenomenon was discovered by Dutch scientist Heike Kamerlingh Onnes in 1911.

superfluid fluid that flows without viscosity or friction and has a very high thermal conductivity. Liquid helium at temperatures below 2 K (−271°C/−456°F) is a superfluid: it shows unexpected behaviour; for instance, it flows uphill in apparent defiance of gravity and, if placed in a container, will flow up the sides and escape.

superheterodyne or *supersonic heterodyne*, means of receiving radio transmissions involving the changing of the frequency of the carrier wave to an intermediate frequency above the limit of audible sound by a heterodyne process. In this, the received wave is combined with a slightly different frequency wave produced within the receiver. Once the intermediate frequency has been formed, the combined waves are amplified and the signal taken off by a demodulator.

supernova explosive death of a star, which temporarily attains a brightness of 100 million Suns or more, so that it can shine as brilliantly as a small galaxy for a few days or weeks. Very approximately, it is thought that a supernova explodes in a large galaxy about once every 100 years. Many supernovae – astronomers estimate some 50% – remain undetected because of obscuring by interstellar dust.

supersonic speed speed greater than that at which sound travels, measured in Mach numbers. In dry air at 0°C/32°F, sound travels at about 1,170 kph/727 mph, but decreases its speed with altitude until, at 12,000 m/39,000 ft, it is only 1,060 kph/ 658 mph.

supersynthesis radio interferometer system in which two synthesis aerials are used; one is static and utilizes the rotation of the Earth to provide a field of scan, the other is mobile.

surface area area of the outside surface of a solid.

surface tension property that causes the surface of a liquid to behave as if it were covered with a weak elastic skin; this is why a needle can float on water. It is caused by the exposed surface's tendency to contract to the smallest possible area because of cohesive forces between molecules at the surface.

sustainable capable of being continued indefinitely. For example, the sustainable yield of a forest is equivalent to the amount that grows back.

synapse junction between two nerve cells, or between a nerve cell and a muscle (a neuromuscular junction), across which a nerve impulse is transmitted. The two cells are separated by a narrow gap called the **synaptic cleft**. The gap is bridged by a

519

chemical neurotransmitter, released by the nerve impulse.

synchrotron particle accelerator in which particles move, at increasing speed, around a hollow ring. The particles are guided around the ring by electromagnets, and accelerated by electric fields at points around the ring. Synchrotrons come in a wide range of sizes, the smallest being about 1 m/3.3 ft across while the largest is 27 km/17 mi across. The Tevatron synchrotron at Fermilab is some 6 km/4 mi in circumference and accelerates protons and antiprotons to 1 TeV.

synchrotron radiation polarized form of radiation produced by high-speed electrons in a magnetic field; it is this radiation that is emitted by the Crab nebula (in the constellation Taurus).

synodic period time taken for a planet or moon to return to the same position in its orbit as seen from the Earth, that is, from one opposition to the next. It differs from the sidereal period because the Earth is moving in orbit around the Sun.

synthesis aerial radio interferometer system utilizing a number of small aerials to achieve the effect of an impossibly large single one.

systole contraction of the heart. It alternates with diastole, the resting phase of the heart beat.

tangent straight line that touches a curve and gives the gradient of the curve at the point of contact. At a maximum, minimum, or point of inflection, the tangent to a curve has zero gradient. Also, in trigonometry, a function of an acute angle in a right-angled triangle, defined as the ratio of the length of the side opposite the angle to the length of the side adjacent to it; a way of expressing the gradient of a line.

tautomerism form of isomerism in which two interconvertible isomers are in equilibrium. It is often specifically applied to an equilibrium between the keto ($-CH_2-C=O$) and enol ($-CH=C-OH$) forms of carbonyl compounds.

Taylor's theorem or ***Brook Taylor's theorem***, expands a function of x as an infinite power series in powers of x.

tectonics study of the movements of rocks on the Earth's surface. On a small scale tectonics involves the formation of folds and faults, but on a large scale plate tectonics deals with the movement of the Earth's surface as a whole.

tensor calculus position in ordinary space usually requires specification of three coordinates, singly indexed; for example, x_1, x_2, x_3. In describing mathematical objects more complicated than position, a generalized type of coordinate system may be used: for example, X_{ijk} where i, j, k can each take values 1, 2, or 3. Tensor calculus is a systematized use of such objects.

terminal velocity or ***terminal speed***, the maximum velocity that can be reached by a given object moving through a fluid (gas or liquid) under the action of an applied force. As the speed of the object increases so does the total magnitude of the forces resisting its motion. Terminal velocity is reached when the resistive forces exactly balance the applied force that has caused the object to accelerate; because there is now no resultant force, there can be no further acceleration.

terpene any of a class of organic compounds, originally derived from plant oils, that contain only carbon and hydrogen and are empirically regarded as derivatives of isoprene (C_5H_8). They are classified according to the number of isoprene units in the molecule – for example, monoterpenes contain two isoprene units and have the formula $C_{10}H_{16}$, sesquiterpenes contain three units ($C_{15}H_{24}$), and diterpenes contain four units ($C_{20}H_{32}$). Turpentine consists of a mixture of several monoterpenes. Rubber is a polyterpene with between 1,000 and 5,000 isoprene units.

Tertiary period period of geological time 65–1.64 million years ago, divided into five epochs: Palaeocene, Eocene, Oligocene, Miocene, and Pliocene. During the Tertiary period, mammals took over all the ecological niches left vacant by the extinction of the dinosaurs, and became the prevalent land animals. The continents took on their present positions, and climatic and vegetation zones as we know them became established.

tesla SI unit (symbol T) of magnetic flux density. One tesla represents a flux density of one weber per square metre, or 10^4 gauss. It is named after the Croatian-born US physicist Nikola Tesla.

tesselation covering of a plane surface by regular congruent quadrilaterals in a side-by-side pattern; the first quadrilateral is derived by joining the mid-points of the sides of a given (regular or irregular) quadrilateral.

tetrahedron plural ***tetrahedra***, solid figure (polyhedron) with four triangular faces, that is, a pyramid on a triangular base. A regular tetrahedron has equilateral triangles as its faces.

thermal column of warm air, which is of a lower density than its surroundings, and contains rising currents of air.

thermal conductivity ability of a substance to conduct heat. Good thermal conductors, like good electrical conductors, are generally materials with many free electrons (such as metals).

thermionic tube or ***thermionic valve***, evacuated metal or glass container enclosing a system of electrodes. The cathode emits electrons when heated, and these are attracted to a positively charged anode. Perforated grid electrodes within the tube can be used to control the electron current.

thermochemistry branch of chemistry that deals with the heat changes that accompany chemical reactions.

thermocouple device comprising two wires of different metals (such as copper and iron) joined, and with their other ends held at different temperatures. A small current is set up within the wire proportional to the size of the temperature difference between the two ends. This makes thermocouples useful in the determination of temperature.

thermodynamic equilibrium a system is said to be in thermodynamic equilibrium if no heat flows between its component parts.

thermodynamics branch of physics dealing with the transformation of heat into and from other forms of energy. It is the basis of the study of the efficient working of engines, such as the steam and internal-combustion engines. The three laws of thermodynamics are: (1) energy can be neither created nor destroyed, heat and mechanical work being mutually convertible; (2) it is impossible for an unaided self-acting machine to convey heat from one body to another at a higher temperature; and (3) it is impossible by any procedure, no matter how idealized, to reduce any system to the absolute zero of temperature (0 K/−273.15°C/−459.67°F) in a finite number of operations. Put into mathematical form, these laws have widespread applications in physics and chemistry.

thermopile instrument for measuring radiant heat, consisting of a number of thermocouples connected in series with alternate junctions exposed to the radiation. The current generated (measured by an ammeter) is proportional to the radiation falling on the device.

thermoplastic or ***thermosoftening plastic***, type of plastic that always softens on repeated heating. Thermoplastics include polyethylene (polyethene), polystyrene, nylon, and polyester.

theta functions four types of elliptic function devised by German mathematician Carl Jacobi. Each function is defined as a Fourier series, and written θ_1, θ_2, θ_3, and θ_4; any θ can be converted by translation of the argument into another θ multiplied by a simple factor. (The quotient of any two θ is then periodic twice.)

thixotropy property of a substance that enables it to form a jellylike colloid that reverts to a liquid on mechanical agitation. Nondrip paints are common thixotropic materials.

three-body problem mathematical problem in astronomy, to describe the gravitational effects of three interacting celestial bodies on each other, and the shape of their orbits round each other.

thrust propulsive pressure exerted by, for example, a jet or rocket engine.

tide rhythmic rise and fall of the sea level in the Earth's oceans and their inlets and estuaries due to the gravitational attraction of the Moon and, to a lesser extent, the Sun, affecting regions of the Earth unequally as it rotates.

titration technique to find the concentration of one compound in a solution by determining how much of it will react with a known amount of another compound in solution.

TNT abbreviation for ***trinitrotoluene***, $CH_3C_6H_2(NO_2)_3$, a powerful high explosive. It is a yellow solid, prepared in several isomeric forms from toluene by using sulphuric and nitric acids.

tolerance range in the physical dimensions of an object within which the true dimensions lie. Often expressed in the form of, for example, ± 3 mm for a length.

topography surface shape and composition of the landscape, comprising both natural and artificial features, and its study. Topographical features include the relief and contours of the land; the distribution of mountains, valleys, and human settlements; and the patterns of rivers, roads, and railways.

topology branch of geometry that deals with those properties of a figure that remain unchanged even when the figure is transformed (bent, stretched) – for example, when a square painted on a rubber sheet is deformed by distorting the sheet. Topology has scientific applications, as in the study of turbulence in flowing fluids.

torque turning effect of force on an object. A turbine produces a torque that turns an electricity generator in a power station. Torque is measured by multiplying the force by its perpendicular distance from the turning point.

521

torus ring-shaped chamber used to contain plasma in nuclear fusion reactors such as the Joint European Torus (JET) reactor.

total internal reflection complete reflection of a beam of light that occurs from the surface of an optically 'less dense' material. For example, a beam from an underwater light source can be reflected from the surface of the water, rather than escaping through the surface. Total internal reflection can only happen if a light beam hits a surface at an angle greater than the critical angle for that particular pair of materials.

toxin any poison produced by another living organism (usually a bacterium) that can damage the living body. In vertebrates, toxins are broken down by enzyme action, mainly in the liver.

tracer small quantity of a radioactive isotope (form of an element) used to follow the path of a chemical reaction or a physical or biological process. The location (and possibly concentration) of the tracer is usually detected by using a Geiger–Muller counter.

trans- prefix used in stereochemistry to indicate that two groups or substituents lie on opposite sides of the main axis or plane of a molecule (as opposed to **cis-**, which indicates that they are on the same side).

transcendental curve curve for which there is no representative algebraic equation; examples are logarithmic curves and trigonometric curves.

transcendental number real number that is not an algebraic number, and can therefore not be expressed as a root (solution) of an algebraic equation with integral coefficients; an example is Euler's number (e).

transfer RNA (tRNA) a relatively small molecule of ribonucleic acid, the function of which is to carry amino acids to ribosomes where protein synthesis occurs. Each amino acid is borne by a different tRNA molecule. tRNA is complementary to messenger RNA (mRNA).

transformation in mathematics, a mapping or function, especially one which causes a change of shape or position in a geometric figure. Reflection, rotation, enlargement, and translation are the main geometrical transformations.

transformation in genetics, the substitution of one section of DNA by another. It requires at least two crossovers (or breaks) in the DNA and is a source of genetic variation.

transformer device in which, by electromagnetic induction, an alternating current (AC) of one voltage is transformed to another voltage, without change of frequency. Transformers are widely used in electrical apparatus of all kinds, and in particular in power transmission where high voltages and low currents are utilized.

transistor solid-state electronic component, made of semiconductor material, with three or more electrodes, that can regulate a current passing through it. A transistor can act as an amplifier, oscillator, photocell, or switch, and (unlike earlier thermionic valves) usually operates on a very small amount of power.

transit passage of a smaller object across the visible disc of a larger one. Transits of the inferior planets occur when they pass directly between the Earth and the Sun, and are seen as tiny dark spots against the Sun's disc.

transition metal any of a group of metallic elements that have incomplete inner electron shells and exhibit variable valency – for example, cobalt, copper, iron, and molybdenum. They are excellent conductors of electricity, and generally form highly coloured compounds.

transposon or *jumping gene*, segment of DNA able to move within or between chromosomes. Transposons trigger changes in gene expression by shutting off genes or causing insertion mutations.

transuranic element or *transuranium element*, chemical element with an atomic number of 93 or more – that is, with a greater number of protons in the nucleus than has uranium. All transuranic elements are radioactive. Neptunium and plutonium are found in nature; the others are synthesized in nuclear reactions.

transverse wave wave in which the displacement of the medium's particles, or in electromagnetic waves, the direction of the electric and magnetic fields, is at right angles to the direction of travel of the wave motion.

Triassic period period of geological time 245–208 million years ago, the first period of the Mesozoic era. The present continents were fused together in the form of the world continent Pangaea. Triassic sediments contain remains of early dinosaurs and other animals now extinct. By late Triassic times, the first mammals had evolved. There was a mass extinction of 95% of plants at the end of the Triassic possibly caused by rising temperatures.

trigonometry branch of mathematics that concerns finding lengths and angles in triangles. In a right-angled triangle the sides and angles are related by three trigonometric ratios: sine, cosine, and tangent. Trigonometry is used frequently in navigation, surveying, and simple harmonic motion in physics.

triode thermionic valve consisting of an anode, a cathode, and a grid.

tripeptide sequence of three amino acids, often occurring in the biochemical synthesis or breakdown of proteins.

trophic level position occupied by a species (or group of species) in a food chain. The main levels are **primary producers** (photosynthetic plants), **primary consumers** (herbivores), **secondary consumers** (carnivores), and **decomposers** (bacteria and fungi).

trophoblast outer layer of the ovum that supplies nutrition to the embryo and attaches the ovum to the wall of the uterus.

tropics area between the tropics of Cancer and Capricorn, defined by the parallels of latitude approximately 23.5° north and south of the Equator. They are the limits of the area of Earth's surface in which the Sun can be directly overhead. The mean monthly temperature is over 20°C/68°F.

turbulence irregular fluid (gas or liquid) flow, in which vortices and unpredictable fluctuations and motions occur. Streamlining reduces the turbulence of flow around an object, such as an aircraft, and reduces drag. Turbulent flow of a fluid occurs when the Reynolds number is high.

Tyndall effect scattering of light to produce a visible beam.

UBV photometry measurement of the astronomical colour index of a star, utilizing the ultraviolet, blue, and yellow visual images over two pre-set wavelengths obtained by photoelectric filtering. Other standardized filter wavebands are also used.

ultrasonics branch of physics dealing with the theory and application of ultrasound: sound waves occurring at frequencies too high to be heard by the human ear (that is, above about 20 kHz).

ultrasound scanning or ***ultrasonography***, use of ultrasonic pressure waves to create a diagnostic image. It is a safe, noninvasive technique that often eliminates the need for exploratory surgery.

ultraviolet excess screening technique devised by English astronomer Martin Ryle and US astronomer Allan Sandage to measure the spectral red shift of suspected quasars. It was this process that resulted in the discovery of quasi-stellar objects.

ultraviolet radiation electromagnetic radiation invisible to the human eye, of wavelengths from about 400 to 4 nm (where the X-ray range begins). Physiologically,

ultraviolet radiation is extremely powerful, producing sunburn and causing the formation of vitamin D in the skin.

umbra central region of a shadow that is totally dark because no light reaches it, and from which no part of the light source can be seen (compare penumbra). In astronomy, it is a region of the Earth from which a complete eclipse of the Sun or Moon can be seen.

uncertainty principle or ***indeterminacy principle***, principle that it is impossible to know with unlimited accuracy both the position and momentum of a particle. The principle arises because in order to locate a particle exactly, an observer must bounce light (in the form of a photon) off the particle, which must alter its position in an unpredictable way.

unconditioned reflex behavioural reflex or response that is natural and not acquired by training or conditioning.

unconditioned stimulus natural stimulus unassociated with behavioural training; it evokes a natural or unconditioned reflex or response (see conditioned stimulus).

unified field theory theory that attempts to explain the four fundamental forces (strong nuclear, weak nuclear, electromagnetic, and gravity) in terms of a single unified force.

uniformitarianism principle that processes that can be seen to occur on the Earth's surface today are the same in kind and degree as those that have occurred throughout geological time.

unit theory or ***theory of units***, in a field or ring, involves an element that possesses an inverse. In a field, every nonzero element is a unit; in a ring, 1 and −1 represent a unit.

universal set with regard to a mathematical structure, a set of objects in the structure. More generally, it represents the 'universe of discourse' appropriate to the discourse – for example, the set of vowels and the set of consonants are disjoint sets within the universal set comprising the alphabet.

urea cycle biochemical process by which nitrogenous waste is converted into urea, which is easily excreted. When proteins and amino acids break down, ammonia, which is highly toxic, is formed. Most of the ammonia is converted into glutamate, and becomes usable for the synthesis of more amino acids and proteins. Any excess is converted into the water-soluble compound, urea, which can be excreted as urine.

vaccine any preparation of modified pathogens (viruses or bacteria) that is introduced into the body, usually either orally or by a hypodermic syringe, to induce the specific antibody reaction that produces immunity against a particular disease.

vacuole fluid-filled, membrane-bound cavity inside a cell. It may be a reservoir for fluids that the cell will secrete to the outside, or may be filled with excretory products or essential nutrients that the cell needs to store. Plant cells usually have a large central vacuole containing sap (sugar and salts in solution) which serves both as a store of food and as a key factor in maintaining turgor. In amoebae (single-celled animals), vacuoles are the sites of digestion of engulfed food particles.

vacuum region completely empty of matter; in physics, any enclosure in which the gas pressure is considerably less than atmospheric pressure (101,325 pascals).

valency measure of an element's ability to combine with other elements, expressed as the number of atoms of hydrogen (or any other standard univalent element) capable of uniting with (or replacing) its atoms. The number of electrons in the outermost shell of the atom dictates the combining ability of an element.

valve, thermionic see thermionic tube.

Van Allen radiation belts two zones of charged particles around the Earth's magnetosphere, discovered in 1958 by US physicist James Van Allen. The atomic particles come from the Earth's upper atmosphere and the solar wind, and are trapped by the Earth's magnetic field. The inner belt lies 1,000–5,000 km/620–3,100 mi above the Equator, and contains protons and electrons. The outer belt lies 15,000–25,000 km/9,300–15,500 mi above the Equator, but is lower around the magnetic poles. It contains mostly electrons from the solar wind.

van de Graaff generator electrostatic generator capable of producing a voltage of over a million volts. It consists of a continuous vertical conveyor belt that carries electrostatic charges (resulting from friction) up to a large hollow sphere supported on an insulated stand. The lower end of the belt is earthed, so that charge accumulates on the sphere. The size of the voltage built up in air depends on the radius of the sphere, but can be increased by enclosing the generator in an inert atmosphere, such as nitrogen.

van der Waals' law modified form of the gas laws that includes corrections for the non-ideal behaviour of real gases (the molecules of ideal gases occupy no space and exert no forces on each other). It is named after Dutch physicist J D van der Waals.

vapour one of the three states of matter (with solid and liquid). The molecules in a vapour move randomly and are far apart, the distance between them, and therefore the volume of the vapour, being limited only by the walls of any vessel in which they might be contained. A vapour differs from a gas only in that a vapour can be liquefied by increased pressure, whereas a gas cannot unless its temperature is lowered below its critical temperature; it then becomes a vapour and may be liquefied.

vapour pressure pressure of a vapour given off by (evaporated from) a liquid or solid, caused by atoms or molecules continuously escaping from its surface. In an enclosed space, a maximum value is reached when the number of particles leaving the surface is in equilibrium with those returning to it; this is known as the **saturated vapour pressure** or **equilibrium vapour pressure**.

variable changing quantity (one that can take various values), as opposed to a constant. For example, in the algebraic expression $y = 4x^3 + 2$, the variables are x and y, whereas 4 and 2 are constants.

variable star star whose brightness changes, either regularly or irregularly, over a period ranging from a few hours to months or years. The Cepheid variables regularly expand and contract in size every few days or weeks.

variance square of the standard deviation, the measure of spread of data. Population variance and sample variance are denoted by σ^2 and s^2, respectively. Variance provides a measure of the dispersion of a set of statistical results about the mean or average value.

variation difference between individuals of the same species, found in any sexually reproducing population. Variations may be almost unnoticeable in some cases, obvious in others, and can concern many aspects of the organism. Typically, variations in size, behaviour, biochemistry, or colouring may be found. The cause of the variation is genetic (that is, inherited), environmental, or more usually a combination of the two. The origins of variation can be traced to the recombination of the genetic material during the formation of the gametes, and, more rarely, to mutation.

variety set of solutions of a simultaneous system of equations with a fixed number of variables. In two dimensions, examples are a circle, an ellipse, and a parabola. In three dimensions, a variety is a surface, for example, the surface of a sphere.

vector space or ***linear space***, mathematical structure comprising two types of objects: vectors and scalars. Vectors can be added by themselves; scalars lengthen or shorten them (that is, scale the vector length up or down) and are commonly either real or complex numbers. In general, scalars of a vector space constitute a field. The prime example of a vector space is the collection of elementary vectors in the two- or three-dimensional space of everyday experience. (The operation of addition follows the parallelogram law: to add two vectors, complete the parallelogram defined by the two lines – the sum is given by the diagonal through the common origin.) Vector spaces provide a framework for the study of linear transformations – which can be represented by matrices in finite-dimensional spaces – and are important in mathematical modelling of complicated systems (engineering, biological, and so on) where general transformations are approximated by linear ones, with recourse to the apparatus of differential calculus.

Venn diagram diagram representing a set or sets and the logical relationships between them. The sets are drawn as circles. An area of overlap between two circles (sets) contains elements that are common to both sets, and thus represents a third set. Circles that do not overlap represent sets with no elements in common (disjoint sets). The method is named after the English logician John Venn.

vernal equinox see equinox.

vertebrate any animal with a backbone. The 41,000 species of vertebrates include mammals, birds, reptiles, amphibians, and fishes. They include most of the larger animals, but in terms of numbers of species are only a tiny proportion of the world's animals. The zoological taxonomic group Vertebrata is a subgroup of the phylum Chordata.

vesicle small sac containing liquid. It is the medical term for a blister or elevation of the outer layer of the skin (epidermis) containing serous fluid.

vestigial organ organ that remains in diminished form after it has ceased to have any significant function in the adult organism. In humans, the appendix is vestigial, having once had a digestive function in our ancestors.

virtual particle theory theory devised by English physicist Stephen Hawking to account for apparent thermal radiation from a black hole (from which not even light can escape). It supposes that space is full of

'virtual particles' in a particle–antiparticle relationship, being created out of 'nothing' and instantly destroying each other. At an event horizon, however, one particle may be gravitationally drawn into the singularity, and the other appear to radiate as heat.

virus infectious particle consisting of a core of nucleic acid (DNA or RNA) enclosed in a protein shell. Viruses are acellular and able to function and reproduce only if they can invade a living cell to use the cell's system to replicate themselves. In the process they may disrupt or alter the host cell's own DNA. The healthy human body reacts by producing an antiviral protein, interferon, which prevents the infection spreading to adjacent cells.

There are around 5,000 species of virus known to science (1998), though there may be as many as 0.5 million actually in existence.

viscosity resistance of a fluid to flow, caused by its internal friction, which makes it resist flowing past a solid surface or other layers of the fluid. It applies to the motion of an object moving through a fluid as well as the motion of a fluid passing by an object.

viscous force drag that occurs on an object placed in a viscous medium. For example, air causes a viscous force on an aircraft travelling through it. For two parallel layers of fluid close to each other, the viscous force is proportional to the difference in velocity between the two layers.

vision, persistence of see persistence of vision.

vitamin any of various chemically unrelated organic compounds that are necessary in small quantities for the normal functioning of the human body. Many act as coenzymes, small molecules that enable enzymes to function effectively. They are generally classified as **water-soluble** (B and C) or **fat-soluble** (A, D, E, and K). See separate entries for individual vitamins, also nicotinic acid, folic acid, and pantothenic acid.

volcano crack in the Earth's crust through which hot magma (molten rock) and gases well up. The magma is termed lava when it reaches the surface. A volcanic mountain, usually cone shaped with a crater on top, is formed around the opening, or vent, by the build-up of solidified lava and ashes (rock fragments). Most volcanoes arise on plate margins (see plate tectonics), where the movements of plates generate magma or allow it to rise from the mantle beneath.

volt SI unit of electromotive force or electric potential symbol V. A small battery has a potential of 1.5 volts, whilst a high-tension transmission line may carry up to 765,000 volts. The domestic electricity supply in the UK is 230 volts (lowered from 240 volts in 1995); it is 110 volts in the USA.

voltaic cell another name for an electric cell.

volumetric analysis procedure used for determining the concentration of a solution. A known volume of a solution of unknown concentration is reacted with a solution of known concentration (standard). The standard solution is delivered from a burette so the volume added is known. This technique is known as titration. Often an indicator is used to show when the correct proportions have reacted. This procedure is used for acid–base, redox, and certain other reactions involving solutions.

vulcanization technique for hardening rubber by heating and chemically combining it with sulphur. The process also makes the rubber stronger and more elastic.

watt SI unit (symbol W) of power (the rate of expenditure or consumption of energy) defined as one joule per second. A light bulb, for example, may use 40, 60, 100, or 150 watts of power; an electric heater will use several kilowatts (thousands of watts). The watt is named after the Scottish engineer James Watt.

wave, electromagnetic see electromagnetic wave.

wavefront line of points in a wave motion that are all of equal phase.

wavelength distance between successive crests of a wave. The wavelength of a light wave determines its colour; red light has a wavelength of about 700 nanometres, for example. The complete range of wavelengths of electromagnetic waves is called the electromagnetic spectrum.

wave, longitudinal see longitudinal wave.

wave mechanics branch of quantum theory that derives the various properties of atoms on the basis of every particle having an associated wave existing in a multi-dimensional space, representing probabilities of certain properties of the particles involved. Schrödinger's equation is the basis of wave mechanics, which has been shown to be equivalent to matrix mechanics.

wave, transverse see transverse wave.

wax solid fatty substance of animal, vegetable, or mineral origin. Waxes are composed variously of esters, fatty acids, free alcohols, and solid hydrocarbons.

weak nuclear force or *weak interaction*, one of the four fundamental forces of nature, the other three being the gravitational force or gravity, the electromagnetic force, and the strong nuclear force. It causes radioactive beta decay and other subatomic reactions. The particles that carry the weak force are called weakons (or intermediate vector bosons) and comprise the positively and negatively charged W particles and the neutral Z particle.

weber SI unit (symbol Wb) of magnetic flux (the magnetic field strength multiplied by the area through which the field passes). It is named after German chemist Wilhelm Weber. One weber equals 10^8 maxwells.

white blood cell or *leucocyte*, one of a number of different cells that play a part in the body's defences and give immunity against disease. Some (neutrophils and macrophages) engulf invading micro-organisms, others kill infected cells, while lymphocytes produce more specific immune responses. White blood cells are colourless, with clear or granulated cytoplasm, and are capable of independent amoeboid movement. They occur in the blood, lymph, and elsewhere in the body's tissues.

white dwarf small, hot star, the last stage in the life of a star such as the Sun. White dwarfs make up 10% of the stars in the Galaxy; most have a mass 60% of that of the Sun, but only 1% of the Sun's diameter, similar in size to the Earth. Most have surface temperatures of 8,000°C/14,400°F or more, hotter than the Sun. Yet, being so small, their overall luminosities may be less than 1% of that of the Sun. The Milky Way contains an estimated 50 billion white dwarfs.

Wien's law relationship between the intensity and frequency of black-body radiation at the high-frequency end of the spectrum. Wien's displacement law relates the wavelength of maximum intensity to the temperature of the black body.

work measure of the result of transferring energy from one system to another to cause an object to move. Work should not be confused with energy (the capacity to do work, which is also measured in joules) or with power (the rate of doing work, measured in joules per second).

working fluid fluid that is used in such a way that its internal energy is converted into external energy. One example is water in hydroelectricity generation.

X chromosome larger of the two sex chromosomes, the smaller being the Y chromosome. These two chromosomes are involved in sex determination. Females have two X chromosomes, males have an X and a Y. Genes carried on the X chromosome produce the phenomenon of sex linkage.

X-ray band of electromagnetic radiation in the wavelength range 10^{-11} to 10^{-9} m (between gamma rays and ultraviolet radiation; see electromagnetic waves). Applications of X-rays make use of their short wavelength (as in X-ray diffraction) or their penetrating power (as in medical X-rays of internal body tissues). X-rays are dangerous and can cause cancer.

X-ray astronomy detection of X-rays from intensely hot gas in the universe. Such X-rays are prevented from reaching the Earth's surface by the atmosphere, so detectors must be placed in rockets and satellites. The first celestial X-ray source, Scorpius X-1, was discovered by a rocket flight in 1962.

X-ray diffraction method of studying the atomic and molecular structure of crystalline substances by using X-rays. X-rays directed at such substances spread out as they pass through the crystals owing to diffraction (the slight spreading of waves around the edge of an opaque object) of the rays around the atoms. By using measurements of the position and intensity of the diffracted waves, it is possible to calculate the shape and size of the atoms in the crystal. The method has been used to study substances such as DNA that are found in living material.

xylem tissue found in vascular plants, whose main function is to conduct water and dissolved mineral nutrients from the roots to other parts of the plant. Xylem is composed of a number of different types of cell, and may include long, thin, usually dead cells known as tracheids, fibres (schlerenchyma), thin-walled parenchyma cells, and conducting vessels.

Y chromosome smaller of the two sex chromosomes. In male mammals it occurs paired with the other type of sex chromosome (X), which carries far more genes. The Y chromosome is the smallest of all the mammalian chromosomes and is considered to be largely inert (that is, without direct effect on the physical body), apart from containing the genes that control the development of the testes. There are only 20 genes discovered so far on the human Y chromosome, much fewer than on all other human chromosomes.

yeast one of various single-celled fungi (see fungus) that form masses of tiny round or oval cells by budding. When placed in a sugar solution the cells multiply and convert the sugar into alcohol and carbon dioxide.

ylem hypothetical primordial state of matter – neutrons and their decay products (protons and electrons) – that might have existed before the Big Bang. The term was taken from Aristotle and forms part of the alpha-beta-gamma theory.

Zeeman effect splitting of spectral lines of a substance placed in an intense magnetic field.

zenith uppermost point of the celestial horizon, immediately above the observer; the nadir is below, diametrically opposite. See celestial sphere.

zeta function function that may be represented as the value of the infinite series:

$$\xi(s) = 1 + 1/2^s + 1/3^s + 1/4^s + \ldots$$

where s is a complex number. The function was significant in Swiss mathematician Leonhard Euler's study of prime numbers. See also the Riemann hypothesis.

Z_{He} symbol for the total mass density (in a star) of elements heavier than helium; values of Z_{He} are small for population II stars, large for population I stars.

zygote ovum (egg) after fertilization but before it undergoes cleavage to begin embryonic development.

Index

Note: Page references in **bold** indicate illustration captions. Page references in *italic* indicate special essays.

531